Pipeline Planning and Construction Field Manual

Pipeline Planning and Construction Field Manual

E. Shashi Menon, Ph.D., P.E.
SYSTEK Technologies, Inc.

AMSTERDAM • BOSTON • HEIDELBERG • LONDON
NEW YORK • OXFORD • PARIS • SAN DIEGO
SAN FRANCISCO • SINGAPORE • SYDNEY • TOKYO

Gulf Professional Publishing is an imprint of Elsevier

Gulf Professional Publishing is an imprint of Elsevier
225 Wyman Street, Waltham, MA 02451, USA
The Boulevard, Langford Lane, Kidlington, Oxford, OX5 1GB, UK

Notices

Knowledge and best practice in this field are constantly changing. As new research and experience broaden our understanding, changes in research methods, professional practices, or medical treatment may become necessary.

Practitioners and researchers must always rely on their own experience and knowledge in evaluating and using any information, methods, compounds, or experiments described herein. In using such information or methods they should be mindful of their own safety and the safety of others, including parties for whom they have a professional responsibility.

To the fullest extent of the law, neither the Publisher nor the authors, contributors, or editors, assume any liability for any injury and/or damage to persons or property as a matter of products liability, negligence or otherwise, or from any use or operation of any methods, products, instructions, or ideas contained in the material herein.

Library of Congress Cataloging-in-Publication Data

Pipeline planning and construction field manual / [edited by] E. Shashi Menon, Ph.D., P.E.
p. cm.
Includes index.
ISBN 978-0-12-383867-4
1. Pipelines. I. Menon, E. Shashi.
TA660.P55P575 2011
621.8'672–dc22 2010052967

British Library Cataloguing-in-Publication Data

A catalogue record for this book is available from the British Library.

For information on all Gulf Professional Publishing publications visit our Web site at *www.elsevierdirect.com*

Typeset by: diacriTech, Chennai, India

Printed in the United States of America
11 12 13 14 15 16 10 9 8 7 6 5 4 3 2 1

Contents

3. Pipeline Regulatory and Environmental Permits

William E. Bauer

4. Right-of-Way

William E. Bauer

5. Alignment Sheets

Hal S. Ozanne

8. Pipeline Hydraulic Analysis

E. Shashi Menon, Ph.D., P.E.

9. Series and Parallel Piping and Power Required

E. Shashi Menon, Ph.D., P.E.

10. Valve Stations

Barry G. Bubar, P.E.

11. Pump Stations

E. Shashi Menon, Ph.D., P.E.

List of Contributors

E. Shashi Menon, Ph.D., P.E.

William E. Bauer

Barry G. Bubar, P.E.

Hal S. Ozanne

Glenn A. Wininger

Author Biography

E. Shashi Menon, Ph.D., P.E.

E. Shashi Menon is the vice president of SYSTEK Technologies, Inc. in Lake Havasu City, Arizona, USA. He has worked in the oil and gas and manufacturing industry for over 37 years. He held positions of design engineer, project engineer, engineering manager, and chief engineer with major oil and gas companies in the United States. He has authored four technical books for major publishers and coauthored over a dozen engineering software applications. He conducts training workshops in liquid and gas pipeline hydraulics at various locations in the United States and South America.

Barry G. Bubar, P.E.

Barry Bubar graduated from University of California with a BS degree in mechanical engineering. He has worked in the petroleum pipeline industry as a district engineer, project engineer and staff engineer and has over 35 years experience in oil, gas, and power companies. He has taught classes in pipeline hydraulics and pipeline welding and now works as a mechanical engineering consultant.

William E. Bauer

Bill Bauer has been associated with right-of-way acquisition projects for over 35 years. He has managed the acquisition of pipeline rights-of-way, regulatory permits, and associated actions throughout the continental United States, Alaska, Europe, and Russia. He is a graduate of Lamar University, Beaumont, Texas, with a BS degree in Math and has written and/or edited numerous books, articles, and videos relating to right-of-way. He is also a certified instructor for the International Right-of-Way Association. Bill has seen right-of-way acquisition move from a hand shake, a signature, and a nominal payment to a highly technical effort sometimes approaching 25% or more of the total cost of a pipeline project.

Hal S. Ozanne

Hal S. Ozanne, BSME, is the vice president of Denver Operations of ENGlobal Engineering, Inc. in Denver, Colorado, USA.

He has worked in the oil and gas industry for over 42 years. His experience has included managing a division office for a consulting engineering firm providing engineering services to the oil and gas industry, serving as project manager for various pipeline projects throughout the United States, and working for a pipeline operating company in various capacities.

Glenn A. Wininger

Glenn Wininger graduated from Oklahoma State University with a BS degree in civil engineering in 1984 and a BS degree in biology in 1990 from Ohio State University. He worked for numerous engineering firms with emphasis in cross-country pipeline projects and as a consultant for engineering firms related to local area gas distribution companies. He held positions within gas companies in engineering and construction management, as well as operations. He assisted companies in compiling data for the Federal Energy Regulatory Committee (FERC) applications, as well as providing support for Draft Environmental Impact Statement (DEIS) and Final Environmental Impact Statement (FEIS) response related to various requests. He held a Registered Professional Land Surveyor (RPLS) license from 1986 to 1990.

Preface

There are thousands of pipelines crisscrossing the globe, both onshore and offshore. Designing, constructing, and operating these pipelines and their appurtenant facilities require special skills along with experience. Design criteria and construction techniques differ from area to area and knowing where and how to access such criteria is essential for pipeline professionals.

This book was prepared in order to give engineers and technicians a working knowledge of the processes of planning, designing, and construction of a pipeline system. The idea for the book was conceived by Elsevier Senior Acquisitions Editor, Kenneth McCombs, in consultation with Shashi Menon, a professional engineer with over 37 years of experience in the US Oil and Gas industry. In addition, we assembled a team of experts with over 180 years combined experience throughout the United States and the world to collaborate on the book and produce a relevant and useful reference manual for pipeline planning and construction.

Chapter 1 covers the design basis that forms the foundation for the design of pipelines, pump stations, compressor stations, valves, and other facilities that comprise the pipeline system.

Chapter 2 introduces the various things that must be taken into consideration in selecting a pipeline route and how a route may be selected and changed as it is being developed.

Chapter 3 reviews pipeline regulatory and environmental permits. This includes numerous permits and approvals that must be obtained from state, federal, and local agencies.

Chapter 4 covers the right-of-way (ROW) aspects including the responsibility of ROW team to provide the project a continuous constructible strip of land for the construction of the pipeline and all related surface facilities, including a continuous pipeline right-of-way, all additional work spaces, surface sites for compressor stations, pump stations, meters, valves, and storage sites.

Chapter 5 describes how pipeline alignment sheets are prepared, the information that is included on them and their use.

Chapter 6 is an overview of pipeline materials. The chapter describes how materials for a pipeline are selected taking into consideration the pipeline service, operating conditions, and the appropriate regulations that must be followed.

Chapter 7 is a discussion of the strength capabilities of a pipeline that is subject to internal pressure and how the required pipe wall thickness is calculated.

Chapter 8 explains pipeline hydraulic analysis for both liquid and gas pipelines. The chapter reviews the different types of flow, Reynolds number, and pressure drop due to friction and determining pumping pressure requirements and location of pump stations and compressor stations.

Chapter 9 covers the calculation of the pressure required in series and parallel piping. In addition, the pumping power required and the number of pumps or compressor stations needed for a long transmission pipeline are discussed.

Chapter 10 reviews requirements of multiple valve stations along a pipeline necessary for isolating segments of pipelines for repair work and in case of a leak, damage, or rupture. In addition, valves installed at pipeline branch connections for delivery or receipt of product being shipped on the mainline are also discussed.

Chapter 11 explains the pump stations and pumping configurations in liquid pipelines along with the optimum locations of pump stations for hydraulic balance. Centrifugal pumps and positive displacement pumps and their performance characteristics are reviewed. The use of variable speed pumps to save pumping power under different operating conditions is also discussed.

Chapter 12 explains the approach to sizing compressor stations in gas pipelines. The optimum locations and pressures at which compressor stations operate are reviewed. Centrifugal and positive displacement compressors used in natural gas transportation are compared with reference to their performance characteristics and cost.

Chapter 13 discusses pipeline corrosion, how corrosion occurs, and the method employed to protect liquid and gas pipelines and associated facilities from corrosion damage.

Chapter 14 introduces the provisions for leak detection for a pipeline. Pipeline operators must take the necessary preparations to eliminate or greatly reduce the possibility of a leak from their system.

Chapter 15 discusses pipeline pigging and internal inspection. Pigging of a pipeline is essential for effective and efficient operation and maintenance. This results in increased pipeline efficiency and extends its useful life.

Chapter 16 discusses pipeline construction with reference to federal, state, district, and local regulations.

Chapter 17 discusses welding and nondestructive testing (NDT) of liquid and gas pipelines. Pipe welding procedures, double jointing, welder qualification, automatic welding, radiography, weld rejection criteria are reviewed.

Chapter 18 discusses hydrostatic testing to ensure integrity of pipeline in service. The federal regulations such as CFR Title 49, Part 195 for Hazardous Liquid Pipelines and CFR Title 49, Part 192 for Gas Pipelines are reviewed.

Chapter 19 describes the preparation and steps to commission or place a pipeline into operation.

Chapter 20 covers specification writing, data sheet production, requisition development, and bid analysis for pipeline materials and equipment.

Chapter 21 describes the information that is included in operations and maintenance manuals and the preparation of these manuals.

The authors would like to acknowledge the many suggestions and constructive comments received from their peers who reviewed portions of the manuscript. Special thanks to David W. Sinclair for his assistance in the review of Chapters 3 and 4 of this manual. Mr. Sinclair, a right-of-way executive for more than 30 years, has been a strong supporter of education and professionalism through the International Right of Way Association (IRWA). In addition, the authors would like to thank their families for being understanding during the many hours spent writing, revising, and proofreading the manuscript and subsequent page proofs.

We would like to take this opportunity to thank Kenneth McCombs, Senior Acquisitions Editor of Elsevier Publishing, for suggesting the subject matter and format for the book. We enjoyed working with him, as well as others, at Elsevier such as Jill Leonard (Editorial Project Manager) and Heather Tighe (Associate Project Manager).

Authors have exercised care and diligence to contact copyright holders for permission to use published reference materials. We have also worked hard to eliminate errors and omissions. Readers are encouraged to independently check calculations and verify results prior to using them in their projects. We welcome notifications of corrections and suggestions for improvement of this field manual in subsequent edition.

E. Shashi Menon
Barry G. Bubar
William E. Bauer
Hal S. Ozanne
Glenn A. Wininger

Chapter 1

Design Basis

E. Shashi Menon, Ph.D., P.E.

Chapter Outline

INTRODUCTION

In this chapter, we outline the design basis that forms the foundation for the design of pipelines, pump stations, compressor stations, valves, and other facilities that comprise the pipeline system. The Design Basis Manual or Memorandum (DBM) is a document that is initially developed following discussions between the pipeline owner company and the engineering firm that is responsible for the designing and (in many cases) construction management of the pipeline. This document is continuously revised and updated during the project life. All participants in the project must have access to the DBM so that a consistent documented basis for all aspects of the pipeline will be followed throughout the design and construction of the project.

First, we review the units of measurement used in the pipeline industry. The various units of measurement and calculations used in the United States of America, Canada, and other countries will be discussed and the conversion between the commonly used units explained. Next, we address the physical properties of fluids (liquids and gases) that are transported in the pipeline. Chapters 7–9 will further describe the details of the pipeline design basis by analyzing the major components such as pipes, valves, pumps, compressors, and ancillary equipment. An outline of the various components that constitute a DBM is also provided in Appendix 1.

1.1 UNITS OF MEASUREMENT

The units of measurement employed in the pipeline transportation industry consist mainly of the English or USCS system of units (US Customary System) and the metric or SI (Système International) system of units. USCS units are used exclusively in the United States of America, whereas SI units are used in the countries that use metric units, such as Europe, Asia, Australia, and South America. In Canada and some South American countries, a combination of USCS and SI units are used.

In USCS units, measurements are derived from the old foot-pound-second (FPS) and foot-slug-second (FSS) system that originated in England. The basic units are foot (ft) for length, slug (slug) for mass, and second (s) for measurement of time.

In SI units, the corresponding units for length, mass, and time are meter (m), kilogram (kg), and second (s), respectively. In both USCS and SI units, time has a common unit of second.

Units of measurement are generally divided into three classes as follows:

Base units
Supplementary units
Derived units

Base units are units that are dimensionally independent, such as units of length, mass, time, electric current, temperature, amount of substance, and luminous intensity.

Supplementary units include those used to measure plain angles and solid angles, such as radian and steradian.

Derived units are formed by combining base units, supplementary units, and other derived units. Examples are force, pressure, and energy.

1.1.1 Base Units

In USCS units, the base units are as follows:

Length – foot (ft)
Mass – slug (slug)
Time – second (s)
Electric current – ampere (A)
Temperature – degree Fahrenheit (°F)
Amount of substance – mole (mol)
Luminous intensity – candela (cd)

In SI units, the base units are as follows:

Length – meter (m)
Mass – kilogram (kg)
Time – second (s)

Electric current – ampere (A)
Temperature – Kelvin (K)
Amount of substance – mole (mol)
Luminous intensity – candela (cd)

1.1.2 Supplementary Units

In USCS and SI units, the supplementary units are as follows:

Plain angle – radian (rad)
Solid angle – steradian (sr)

Radian is defined as the plain angle between two radii of a circle with an arc length equal to the radius. Thus, it represents the angle of a sector of a circle with the arc length equal to its radius.

$$\text{One radian} = (180/\pi)\text{ degrees} = 57.3\text{ degrees (deg)}$$

Since a circle contains 360 degrees, this is equivalent to

$$(360/57.3) = 2\pi\text{ radians} = 6.28\text{ rad}$$

The steradian is the solid angle having its apex at the center of a sphere such that the area of the surface of the sphere that it cuts out is equal to that of a square with sides equal to the radius of this sphere.

1.1.3 Derived Units

Derived units are those that are formed by combining base units, supplementary units, and other derived units. For example, area and volume are derived units formed by combination of the base unit length. Similarly, velocity (or speed) is derived from the base unit of length and time. It is important to note that numerically velocity and speed are the same, but velocity is a vector quantity, whereas speed is a scalar quantity. A vector has both magnitude and direction, whereas a scalar has only magnitude.

In USCS units, the following derived units are used:

Area – square inches (in^2), square feet (ft^2)
Volume – cubic inches (in^3), cubic feet (ft^3), gallons (gal), and barrels (bbl)
Speed/velocity – feet per second (ft/s)
Acceleration – feet per second per second (ft/s^2)
Density – slug per cubic foot ($slug/ft^3$)
Specific weight – pound per cubic foot (lb/ft^3)
Specific volume – cubic feet per pound (ft^3/lb)
Dynamic viscosity – pound second per square foot ($lb \cdot s/ft^2$)
Kinematic viscosity – square feet per second (ft^2/s)
Force – pounds (lb)

Pressure – pounds per square inch (lb/in^2 or psi)
Energy/work – foot pound (ft·lb)
Quantity of heat – British Thermal Units (Btu)
Power – Horsepower (HP)
Specific heat – Btu per pound per °F (Btu/lb/°F)
Thermal conductivity – Btu per hour per foot per °F (Btu/h/ft/°F)

In SI units, the derived units are as follows:

Area – square meters (m^2)
Volume – cubic meters (m^3)
Speed/velocity – meter per second (m/s)
Acceleration – meter per second per second (m/s^2)
Density – kilogram per cubic meter (kg/m^3)
Specific volume – cubic meters per kilogram (m^3/kg)
Force – Newton (N)
Pressure – Newton per square meter (N/m^2) or Pascal (Pa)
Dynamic viscosity – Pascal second (Pa·s)
Kinematic viscosity – square meters per second (m^2/s)
Energy/work – Newton meter (N·m) or joule (J)
Quantity of heat – joule (J)
Power – joule per second (J/s) or watt (W)
Specific heat – joule per kilogram per Kelvin (J/kg/K)
Thermal conductivity – joule per second per meter per Kelvin (J/s/m/K) or (W/m/K)

Other derived units used in USCS and SI units and the conversion between various units are listed in Appendix 1.

1.2 PHYSICAL PROPERTIES OF LIQUIDS AND GASES

Since pipelines are used to transport liquids or gases (collectively referred to as fluids), we discuss some important physical properties of fluids that affect pipeline transportation. In liquid pipelines, these include specific gravity, viscosity, specific heat, bulk modulus, and vapor pressure. In compressible fluids, such as natural gas pipelines, the important properties are specific gravity, viscosity, molecular composition, heating value, specific heat, and the compressibility factor. These physical properties and how they are calculated including methods between various units will be illustrated using examples. The variation of these properties with the temperature and pressure of the fluid is important in both liquid and gas pipelines. In heavy crude oil pipelines, sometimes, the crude oil is heated to reduce viscosity and thus improve pumpability. This, in turn, reduces power requirements and hence cost of transportation. Therefore, the variation in viscosity and gravity with temperature become very important. Sometimes, a low-viscosity product (such as a diluent or light crude oil) is blended with a heavy crude oil to reduce the viscosity and enhance pumpability. We explain the methods commonly used to determine the blended properties of two or more liquids. Similarly for gases, knowing the molecular composition of individual gases, we explain the method of calculating the composition of the gas mixture and the corresponding gravity and viscosity.

This chapter forms the foundation for all calculations for designing and planning the pipelines used to transport liquids and gases. These include pressure drop due to friction in pipes, valves, and fittings, as well as pump and compressor power requirements, all of which will be addressed in Chapters 8 through 12. In Appendix 1, tables are included listing physical properties of commonly transported liquids and gases such as water, refined petroleum products, crude oils, and natural gas.

1.2.1 Liquid Properties

Mass, Weight, Volume, and Density

For both liquids and gases, mass, weight, volume, and density are discussed in this section and the related terms specific volume and specific weight are also explained.

Mass is defined as the quantity of matter in a substance and it does not vary with temperature or pressure. It is a scalar quantity and hence has magnitude but no direction, compared to a vector quantity that has both magnitude and direction. Mass is measured in slug (slug) in USCS units and kilograms (kg) in SI units. The term **weight** depends on the mass and acceleration due to gravity at a particular location and is a vector quantity. Weight is actually the force acting on a mass and hence is a derived unit. In USCS units, weight is stated in pounds (lb) and in SI units it is measured in Newton (N). The quantity of liquid contained in a storage tank may be referred to as 5000 lb weight. This is sometimes referred to incorrectly as 5000 lb mass of liquid. The correct term would be to say the mass of liquid contained in the tank is $5000/32.17 = 155.4$ slug. The factor 32.17 represents the acceleration due to gravity (32.17 ft/s^2). This is based on Newton's second law of motion, represented by the following relationship:

$$\text{Force} = \text{mass} \times \text{acceleration} \tag{1.1}$$

Since force has the units of lb, from Eq. (1.1) it is clear that slug has the units of $\text{lb} \cdot \text{s}^2/\text{ft}$.

Similarly, in SI units, if a storage tank contains 170 kg of crude oil, this is the mass of the crude oil. Its weight in Newton is $170 \times 9.81 = 1667.7\,\text{N}$.

The factor 9.81 is the acceleration due to gravity (9.81 m/s^2) in SI units. However, in common usage we tend to say (incorrectly) that the weight of crude oil in the tank is 170 kg.

Volume is defined as the space occupied by a given mass. In the case of a liquid in a tank, the liquid fills the tank up to a certain height. In comparison, a compressible fluid such as natural gas will fill an entire sphere or bullet used as a storage vessel. Thus, gas expands to fill its container. Consider a cylindrical storage tank for gasoline, if the inside diameter of the tank is 100 ft, the cross-sectional area is

$$A = (\pi/4) \times (100)^2 = 7854\,\text{ft}^2$$

If the liquid level in the tank is 20 ft, the volume of gasoline contained in the tank is given by

$$V = A \times \text{height} = 7854 \times 20 = 157{,}080\,\text{ft}^3$$

In USCS units, the volume of a liquid may be stated in cubic feet (ft^3), gallons (gal) or barrels (bbl). In the US petroleum industry, a barrel and a gallon are defined as follows:

$$1\,\text{bbl} = 42\,\text{US gal} \quad 1\,\text{US gal} = 231\,\text{in}^3$$

The imperial gallon, used in Canada and the United Kingdom, is 20% larger than the US gallon.

In SI units, liquid or gas volume is stated in cubic meters (m^3) or liters (L). These are related to each other and the US gallon as follows:

$$1\,\text{m}^3 = 1000\,\text{L} \quad 1\,\text{US gal} = 3.785\,\text{L}$$

Also the USCS and SI units for volume are related as follows:

$$1\,\text{m}^3 = 35.32\,\text{ft}^3 \quad 1\,\text{bbl} = 0.159\,\text{m}^3 = 158.97\,\text{L}$$

Most liquids are practically incompressible, and they take the shape of their container and have a free surface. This is true if the container is pressurized or is an atmospheric tank. The volume of a liquid varies with temperature and very slightly with pressure. Compared to gases, which are compressible, liquids are almost incompressible, and therefore, pressure has little effect on the volume of a liquid. In the gasoline storage tank example, if the temperature of gasoline increases from 60°F to 80°F, the liquid volume will increase slightly thereby raising the height of the liquid level in the tank. The amount of increase in volume per unit temperature rise depends on the coefficient of expansion of the liquid. Therefore, when measuring petroleum liquids, for the purpose of custody transfer, it is customary to correct volumes to a fixed temperature such as 60°F (15.6°C). Volume correction factors from American Petroleum Institute (API) publications are commonly used in the petroleum industry.

The volume of a gas is also sensitive to both temperature and pressure. Therefore, a standard temperature, such as 60°F (15.6°C), is used when referring to gas volumes in pipelines.

The volume flow rate in a liquid pipeline (or pipeline throughput) is stated in cubic feet per second (ft^3/s), gallons per minute (gal/min), barrels per hour (bbl/h), or barrels per day (bbl/day) in USCS units.

In SI units, liquid flow rate is stated in cubic meters per hour (m^3/h) or liters per second (L/s). Similarly, gas flow rate is measured in cubic meters per hour or million cubic meters per day (Mm^3/day). Also, in gas pipeline terminology, standard volumes are used based on a reference temperature of 60°F (15.6°C). This is discussed further in section 1.2.2.

In a liquid pipeline, it is customary to talk about the "line fill volume" of the pipeline. This is the volume of liquid contained between any two points along the length of the pipeline. The volume of liquid contained between two valves on a pipeline can be calculated by knowing the internal diameter of the pipe and the length of the pipe segment between the two valves.

For example, consider a pipeline of NPS 20 with an outside diameter of 20 in. and 0.500 in. wall thickness. The line fill volume in a 5000-foot-long section of the pipeline is

$$\text{Line fill volume} = (\pi/4) \times (20 - 2 \times 0.500)^2 \times 5000/144 = 9844.77\ \text{ft}^3$$

This is also equal to

$$(9844.77 \times 1728/231) = 73{,}644\ \text{gal or } 1753.43\ \text{bbl}$$

The above calculation is based on conversion factors of 1728 in^3/ft^3 and 42 gal/bbl.

Since the volume of a liquid varies with temperature, in a liquid transmission pipeline, the inlet flow rate measured at the inlet temperature and the outlet flow rate measured at the outlet temperature may be different in a long-distance pipeline, even if there are no intermediate flow injections or deliveries. This is because the inlet temperature of the liquid will be different from the outlet temperature due to heat loss or gain between the pipeline liquid and the surrounding soil in a buried pipeline. Significant variation in temperature may be observed when pumping crude oils or other products that are heated at the pipeline inlet. In refined petroleum products pipelines, such as gasoline and diesel, no heating occurs at the pipe inlet and therefore the temperature variations along the pipeline may be insignificant. Regardless, if the volume flow rate measured at the pipeline inlet is corrected to a standard temperature such as 60°F (or 15°C), the corresponding outlet volume flow rate can also be corrected to the same standard temperature. With this temperature correction, the flow rate throughout the pipeline from inlet to outlet will be the same, provided there are no intermediate injections or deliveries along the pipeline. Due to the principle of conservation of mass, the mass flow rate throughout the pipeline will be the same regardless of temperature variation from inlet to outlet, provided there are no intermediate injections or deliveries along the pipeline.

Density of a liquid is a measure of how densely packed its molecules are in a given volume. Mass density is calculated by dividing the mass by its volume. In USCS units, mass density is stated as $slug/ft^3$. In SI units, mass density is stated as kg/m^3. Similarly, the weight density is defined as the weight per unit volume. Weight density is more commonly called specific weight.

For example, the specific weight of water is 62.4 lb/ft^3 at 60°F. In comparison, a typical diesel fuel has a weight density of 53.6 lb/ft^3.

In SI units, the mass density of a sample of water may be stated as 1000 kg/m^3 at 15°C.

$$\text{Mass density} = \text{mass/volume} \tag{1.2}$$

$$\text{Weight density (specific weight)} = \text{weight/volume} \tag{1.3}$$

Although mass does not change with temperature, compared to volume, the mass density will vary with the temperature of the liquid. Density and volume are inversely related from Eq. (1.3). Therefore, as temperature increases, liquid volume increases while its density decreases. Similarly, as temperature decreases, liquid volume decreases and its density increases. Similar to volume, the density of a liquid varies very slightly with pressure. This is because most liquids are practically incompressible, compared to gases.

Specific Gravity and API Gravity of Liquids

The specific gravity of a liquid is a measure of how heavy a liquid is compared to water. Therefore, it is the ratio of the liquid density to that of water at the same temperature. Being a ratio of similar units, specific gravity is dimensionless and has no units. By definition, the specific gravity of water is 1.00, since the density of water compared to itself is the same.

The term relative density is also used synonymously with specific gravity. For example, at 60°F water has a density of 62.4 lb/ft^3 compared to a density of 46.2 lb/ft^3 for gasoline. The relative density of gasoline at this temperature is therefore 46.2/62.4 = 0.74. This is also called the specific gravity of gasoline at 60°F. As another example, at 15°C, a certain crude oil has a density of 890 kg/m^3. At this temperature, water weighs 1000 kg/m^3. Therefore, the specific gravity of crude oil is 890/1000 or 0.89.

Specific gravity, like density, varies with temperature. With increase in temperature, both density and specific gravity decrease. Similarly, as temperature decreases, density and specific gravity increase. Similar to volume and density, pressure has very little effect on liquid specific gravity, in the normal range of pressures encountered in liquid transmission pipelines.

In the petroleum industry, in addition to specific gravity, the term API gravity (°API) is also used. The API gravity is a scale of measurement, such that for water API = 10 at 60°F. Liquids lighter than water have API values higher than 10. Thus, a typical diesel has an API gravity of 35°API. The API scale is thus an inverse scale compared to specific gravity and is always stated at 60°F. The API value is determined in the laboratory comparing the density of a liquid with the density of water at 60°F. As another example, gasoline has an API gravity of 59.7°API, whereas a typical crude oil has 35°API.

The relationship between API gravity and specific gravity are stated in Eqs (1.4) and (1.5):

$$\text{Specific gravity Sg} = 141.5/(131.5 + \text{API}) \tag{1.4}$$

$$\text{API} = 141.5/\text{Sg} - 131.5 \tag{1.5}$$

Setting API gravity equals 10 for water in Eq. (1.4) gives a specific gravity of 1.00 for water, as expected. Note that, if specific gravity > 1.076, the API value is negative; therefore, to use the above equations, the specific gravity must be less than 1.076.

The specific gravity of a typical gasoline at 60°F is 0.736. Therefore, its API gravity can be calculated from Eq. (1.5) as follows:

$$\text{API gravity} = 141.5/0.736 - 131.5 = 60.76°\text{API}$$

Conversely, if a certain crude oil has an API gravity of 35, its specific gravity can be calculated from Eq. (1.4) as follows:

$$\text{Specific gravity} = 141.5/(131.5 + 35) = 0.8498$$

It is important to remember that API gravity is always measured at 60°F. Therefore in Eqs (1.4) and (1.5), the value of specific gravity used must be at 60°F. It is meaningless to say that the API gravity of a liquid is 35°API at 70°F, because by definition API is always measured in the laboratory at 60°F.

The API gravity of a liquid is measured in the laboratory in accordance with the method described in the ASTM D1298 standard, using a calibrated glass hydrometer. For further discussion on API gravity, refer to the API Manual of Petroleum Measurements.

Liquid Specific Gravity: Variation with Temperature

The specific gravity of a liquid varies with temperature. It increases with decrease in temperature and vice versa. For commonly encountered temperatures in liquid transmission pipelines, the specific gravity of a liquid varies approximately linearly with temperature. Therefore, a 10% increase in temperature results in a 10% decrease in specific gravity. The specific gravity versus temperature can therefore be expressed approximately as follows:

$$S_T = S_{60} - a(T - 60) \tag{1.6}$$

where

S_T – Specific gravity at temperature T
S_{60} – Specific gravity at 60°F
T – Temperature, °F
a – A constant that depends on the liquid

Suppose the specific gravity of a liquid at 70°F and 80°F are known. These two sets of temperature and specific gravity can be substituted in Eq. (1.6) resulting in two simultaneous equations in the unknowns S_{60} and a. By solving the two simultaneous equations, the values of S_{60} and a can be obtained. We can then determine the specific gravity of the liquid at any other temperature using Eq. (1.6).

Liquid Specific Gravity: Blended Products

Sometimes, two liquids are mixed together to form a homogeneous liquid mixture. If we know the specific gravity of each component, at a common temperature, the specific gravity of the blended mixture can be calculated at the same temperature.

Consider a crude oil of specific gravity 0.895 at 70°F blended with a lighter crude oil of specific gravity 0.815 at 70°F in equal volumes. What is the specific gravity of the blended mixture? Common sense suggests that since equal volumes are used, the resultant mixture should have a specific gravity of the average of the two liquids or

$$(0.895 + 0.815)/2 = 0.855$$

When two or more liquids are blended to form a homogenous mixture, the specific gravity of the mixture can be calculated using the weighted average method. For example, if a mixture is formed by blending 10% of liquid A (specific gravity = 0.85) and 90% of liquid B (specific gravity = 0.89), the specific gravity of the resulting blended liquid is

$$(0.1 \times 0.85) + (0.9 \times 0.89) = 0.886$$

When calculating the blended specific gravity of two or more products, the specific gravity values must be measured at the same temperature. A general equation to calculate the specific gravity of a blended mixture of two or more products is as follows:

$$S_b = [(Q_1 S_1) + (Q_2 S_2) + \cdots]/(Q_1 + Q_2 + \cdots) \tag{1.7}$$

where

S_b – Specific gravity of the blended liquid
Q_1, Q_2, etc. – Volume of each component
S_1, S_2, etc. – Specific gravity of each component

The above method of calculating the specific gravity of a mixture of two or more liquids cannot be used directly with API gravities. The API gravities must first be converted to the corresponding specific gravities at 60°F and the weighted average method applied. After calculating the blended specific gravity, the API gravity of the mixture can be determined using Eq. (1.5).

Example Problem 1.1

Three liquids A, B, and C are blended homogenously in the ratio of 15%, 20%, and 65%, respectively, by volume. Calculate the specific gravity of the blended liquid, if the individual liquids have the following specific gravities at 70°F:

Specific gravity of liquid A: 0.815
Specific gravity of liquid B: 0.850
Specific gravity of liquid C: 0.895

Solution

Using Eq. (1.7), we get the specific gravity of the blended liquid as

$$S_b = (15 \times 0.815 + 20 \times 0.850 + 65 \times 0.895)/100 = 0.874$$

Liquid Viscosity

Viscosity of a liquid is a measure of the sliding friction between successive layers of the liquid as it flows through a pipeline. The higher the viscosity the more difficult it is to flow. Lower viscosity fluids flow easily in pipes and cause less pressure drop due to friction. Liquids also have much higher viscosity compared to gases. For example, water has a viscosity of 1.0 centipoise (cP), whereas the viscosity of natural gas is approximately 0.0008 cP. Viscosity has an impact on the type of flow in pipelines. The Reynolds number (discussed in Chapter 8) is a dimensionless parameter that is used to classify the type of flow (laminar or turbulent) in pipelines, and it depends on the liquid viscosity, flow rate, and pipe diameter. In practice, two types of viscosities are used: dynamic (also called absolute) viscosity and kinematic viscosity.

Kinematic viscosity of a liquid is obtained by dividing the absolute viscosity by its density at the same temperature.

$$\nu = \mu/\rho \tag{1.8}$$

where, in USCS units,

ν – Kinematic viscosity, ft^2/s
μ – Dynamic viscosity, $lb \cdot s/ft^2$
ρ – Density, $slug/ft^3$

In SI units

ν – Kinematic viscosity, m^2/s
μ – Dynamic viscosity, $kg/m \cdot s$
ρ – Density, kg/m^3

In USCS units, the dynamic viscosity μ is measured in $lb \cdot s/ft^2$.

In SI units, μ is stated as kg/m·s, Pascal·s (or Poise), or centipoise (cP). The term Pascal (Pa) is the unit of pressure in SI units and is equal to N/m^2.

Kinematic viscosity ν is stated as ft^2/s in USCS units and m^2/s in SI units. Appendix 1 lists various conversion factors for viscosity units. Other commonly used units for kinematic viscosity include stokes (St) and centistokes (cSt).

If the viscosities are in cP and cSt, the relationship with the specific gravity of the liquid Sg is

$$\text{Viscosity, cSt} = (\text{Viscosity, cP})/\text{Sg} \tag{1.9}$$

Therefore, for a crude oil with a specific gravity of 0.89 and dynamic viscosity $\mu = 38$ cP, kinematic viscosity is calculated from Eq. (1.9) as $\nu = 38/0.89 = 42.7$ cSt.

In the petroleum industry, two additional kinematic viscosity units for liquids are used. These are Saybolt Seconds Universal (SSU) and Saybolt Seconds Furol (SSF). These are used in conjunction with heavy crude oils and fuel oils. Similar to specific gravity, the dynamic and kinematic viscosities of a liquid also vary with temperature. The viscosity of a liquid decreases as temperature increases and vice versa. However, unlike specific gravity, the viscosity versus temperature is not a linear relationship. In addition, the viscosity of a liquid also varies slightly with pressures. In the normally encountered range of pressures in liquid transmission pipelines, the variation of liquid viscosity with pressures is insignificant. However, at pressures in the range of 2000–5000 psi or more, liquid viscosity increases with pressure.

Water has a viscosity of 1.0 cP (dynamic viscosity) or 1.0 cSt (kinematic viscosity) at 60°F. In comparison, Alaskan North Slope (ANS) crude oil has a viscosity of 200 SSU (43.33 cSt) at 60°F. Values of viscosity in SSU and SSF may be converted to their equivalent kinematic viscosity in centistokes using the following equations.

Conversion from SSU to centistokes:

$$\text{Centistokes} = 0.226\,(\text{SSU}) - 195/(\text{SSU}) \quad \text{for } 32 \le \text{SSU} \le 100 \tag{1.10}$$

$$\text{Centistokes} = 0.220\,(\text{SSU}) - 135/(\text{SSU}) \quad \text{for SSU} > 100 \tag{1.11}$$

Conversion from SSF to centistokes:

$$\text{Centistokes} = 2.24\,(\text{SSF}) - 184/(\text{SSF}) \quad \text{for } 25 < \text{SSF} \le 40 \tag{1.12}$$

$$\text{Centistokes} = 2.16\,(\text{SSF}) - 60/(\text{SSF}) \quad \text{for SSF} > 40 \tag{1.13}$$

The viscosities of common liquids are shown in Appendix 1. Also, Appendix 1 includes conversion factors for converting viscosity from one set of units to another.

Liquid Viscosity: Variation with Temperature

The viscosity of a liquid decreases as the temperature increases and vice versa. However, the variation is not linear, but it is logarithmic in nature as follows:

$$\log_e(\nu) = A - B(T) \tag{1.14}$$

where

ν – Viscosity of liquid, cSt
T – Absolute temperature, °R or K

$$T = (t + 460)\ °\text{R for temperature } t \text{ in °F} \tag{1.15}$$

$$T = (t + 273)\ \text{K for temperature } t \text{ in °C} \tag{1.15a}$$

A and B are constants that depend on the specific liquid.

It can be seen from Eq. (1.14) that a graphic plot of $\log_e(\nu)$ against the temperature T will result in a straight line with a slope of $-B$. If we are given two sets of viscosity and temperature data for a liquid, A and B values can be determined by substituting the two sets of viscosity and temperature data in Eq. (1.14) and solving the resulting simultaneous equations. Once A and B are calculated, the viscosity of the liquid at any other temperature can be determined using Eq. (1.14).

Example Problem 1.2

The kinematic viscosities of a liquid at 60°F and 100°F are 43 cSt and 10 cSt, respectively. Using Eq. (1.13), determine the values of constants A and B and the viscosity of the liquid at 80°F.

$$\log_e(43) = A - B\,(60 + 460)$$

and

$$\log_e(10) = A - B\,(100 + 460)$$

Solution

Solving the above two equations for A and B results in

$$A = 22.7232 \quad B = 0.0365$$

Having found A and B, we can now calculate the viscosity of this liquid at any other temperature using Eq. (1.13). Therefore, the viscosity at 80°F is calculated as follows:

$$\log_e(\nu) = 22.7232 - 0.0365\,(80 + 460) = 3.0132$$

$$\text{Viscosity at 80°F} = 20.35\ \text{cSt}$$

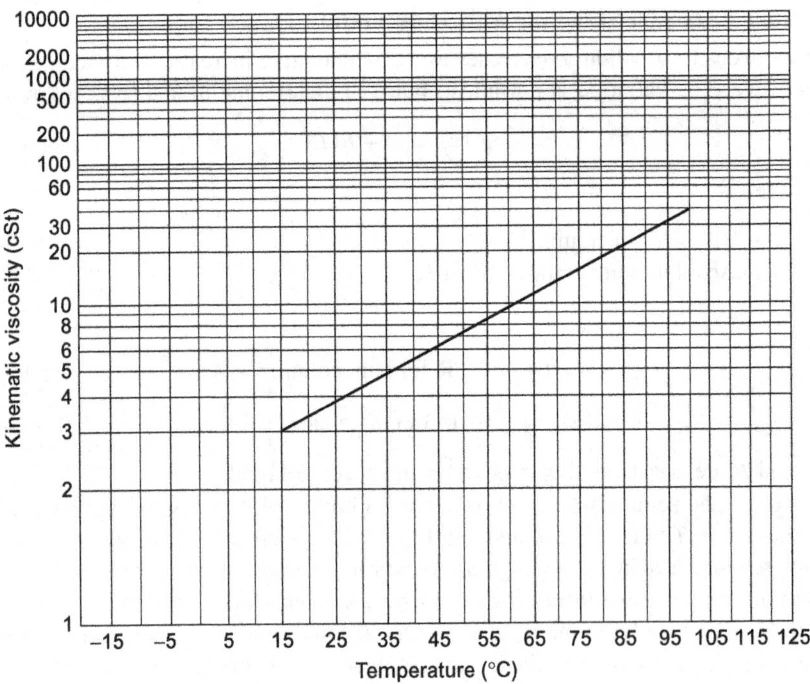

FIGURE 1.1 ASTM D341: Viscosity–temperature chart.

Several other methods are available to predict the viscosity variation of petroleum liquids with temperature. The most popular of these is known as the ASTM D341 method. In this method, a special graph paper with logarithmic scales is used to plot the viscosity of a liquid at two known temperatures. A line is then drawn connecting the two points on the graph. The viscosity at any intermediate temperature can then be interpolated. To some extent, viscosity outside the range may also be extrapolated from this chart. Figure 1.1 illustrates the method.

Appendix 1 provides the equations to manually calculate the viscosity versus temperature using the ASTM method, without using the special logarithmic graph paper.

Example Problem 1.3

A crude oil sample has the following viscosities at the two listed temperatures:

Temperature, °F	**60**	**180**
Viscosity, cSt	750	25

Using the ASTM method, calculate the viscosity of this liquid at 85°F.

Solution

1. At the first temperature 60°F

   C, *D*, and *Z* are calculated using Eqs (A 1.6) through (A 1.8) in Appendix 1.

$$C_1 = \exp[-1.14883 - 2.65868 \times 750] = 0$$
$$D_1 = \exp[-0.0038138 - 12.5645 \times 750] = 0$$
$$Z_1 = (750 + 0.7) = 750.7$$

Similarly, at the second temperature 180°F, the corresponding values of *C*, *D*, and *Z* are calculated to be

$$C_2 = \exp[-1.14883 - 2.65868 \times 25] = 0$$
$$D_2 = \exp[-0.0038138 - 12.5645 \times 25] = 0$$
$$Z_2 = (25 + 0.7) = 25.7$$

Substituting in Eq. (A 1.5) we get

$$\begin{aligned} \log_{10}\log_{10}(750.7) &= A - B\log(60+460) \\ 0.4587 &= A - 2.716(B) \end{aligned} \tag{1.16}$$

and for the second temperature

$$\begin{aligned} \log_{10}\log_{10}(25.7) &= A - B\log(180+460) \\ 0.1492 &= A - 2.8062(B) \end{aligned} \tag{1.17}$$

Solving the simultaneous Eqs (1.16) and (1.17) for *A* and *B*, we get

$$A = 9.778$$
$$B = 3.4313$$

2. Using Eq. (A 1.5), at 85°F, we calculate the value of *Z* as follows:

$$\log_{10}\log_{10}(Z) = A - B\log(85+460)$$
$$\log_{10}\log_{10}(Z) = 9.778 - 3.4313 \times 2.7364 = 0.3886$$
$$Z = 279.78$$

Therefore,

$$\text{Viscosity at } 85°\text{F} = 279.78 - 0.7 = 279.08 \text{ cSt}$$

Liquid Viscosity: Blended Products

Frequently, in the petroleum industry, two or more products are blended to form a homogenous mixture. A heavy crude oil may be mixed with a lighter product to form a mixture with an intermediate viscosity, which will then be easier to pump through a pipeline. Consider a crude oil with a viscosity 89 cSt at 60°F blended with a lighter crude oil with a viscosity 15 cSt at 60°F, in equal volumes. We need to determine the viscosity of the blended mixture. Due to the nonlinear nature of viscosity with mass and volume, we cannot average the viscosities as we did with specific gravities blending earlier. We must resort to a different approach.

If the viscosities are given in SSU, the following methods can be used to calculate the viscosity of blended liquids.

$$\sqrt{V_b}/(Q_1 + Q_2 + \cdots) = 1/[(Q_1/\sqrt{V_1}) + (Q_2/\sqrt{V_2}) + \cdots] \tag{1.18}$$

where

V_b – Viscosity of blend, SSU
Q_1, Q_2, etc. – Volumes of each component
V_1, V_2, etc. – Viscosity of each component, SSU

Since this method requires all component viscosities to be in SSU, we cannot use this approach to calculate the blended viscosity when any of the component viscosities are less than 32 SSU (1.0 cSt), the lower limit of the SSU scale. Also, the individual viscosities must all be at the same temperature. Therefore, the viscosity of a mixture consisting of product A (50 SSU at 60°F) and product B (100 SSU at 70°F) cannot be calculated using this method. All component liquid viscosities must first be calculated at some base temperature.

Another method of calculating the viscosity of blended products uses the Blending Index (BI) method. In this method, a BI parameter is calculated for each component liquid based on its viscosity. Next, the BI of the mixture is calculated from the individual BI using the weighted average of the composition of the mixture. Finally, the viscosity of the blended mixture is calculated using the BI of the mixture as described in the equations in Appendix 1.

Example Problem 1.4

Calculate the blended viscosity obtained by mixing 20% of liquid A with a viscosity of 10 cSt and 80% of liquid B with a viscosity of 30 cSt at 70°F.

Solution

First, convert the given viscosities to SSU using Eq. (1.10).

Since the SSU equivalent of kinematic viscosity in cSt is approximately five times the cSt value, the viscosity of 10 cSt of liquid A in SSU is calculated using Eq. (1.10) as

$$10 = 0.226(V_A) - (195/V_A)$$

Rearranging the equation, we get

$$0.226V_A^2 - 10V_A - 195 = 0$$

Solving the quadratic equation for V_A we get

$$V_A = 58.90\,\text{SSU}$$

Similarly, viscosity of liquid B in SSU is calculated using Eq. (1.11) as

$$V_B = 140.72\,\text{SSU}$$

From Eq. (1.18), the blended viscosity is

$$\sqrt{V_{\text{blnd}}} = \frac{20+80}{(20/\sqrt{58.9})+(80/\sqrt{140.72})} = 10.6953$$

Therefore, the viscosity of the blend using Eq. (1.10) is

$$V_{\text{blnd}} = 114.39\ \text{SSU}$$

$$\text{Viscosity of the blend} = 23.99\ \text{cSt, converting from SSU to cSt}$$

Liquid Viscosity: Graphic Blending Method

A graphical method using ASTM D341-77 is also available to calculate the blended viscosities of two liquids. This method involves using a logarithmic chart with viscosity scales on the left and right sides of the paper. The horizontal axis is for selecting the percentage of each product as shown in Fig. 1.2. This chart is also available in handbooks such as the *Crane Handbook* and the *Hydraulic Institute Engineering Data Book*. Note that the viscosities of both products must be plotted at the same temperature.

Specific Heat of Liquids

The specific heat of a liquid is the amount of heat or thermal energy required to increase its temperature by 1 degree. For water, at atmospheric pressure and 68°F, the specific heat is 1 Btu/lb/°F (4.182 kJ/kg/°C at 20°C). In comparison, a sample of crude oil has a Cp value of 0.45 Btu/lb/°F (1.88 kJ/kg/°C). Specific heat of a liquid at a certain temperature is referred to at constant pressure and denoted by the symbol Cp. This property is important in heavy crude or heated liquid pipeline systems, where heat transfer between the pipeline fluid and the surrounding medium (soil or ambient air) is taken into account.

Bulk Modulus of Liquids

The bulk modulus of a liquid is related to its compressibility. It is defined as the pressure required to cause a unit change of volume of a liquid. Since most liquids are practically incompressible, they require very large pressures to cause any significant volume change. For most liquids, the bulk modulus is approximately in the range of 250,000–300,000 psi. The fairly high number demonstrates the incompressibility of liquids.

The inverse of the bulk modulus (K) is called the compressibility. Water has a bulk modulus of approximately 300,000 psi (2.1 GPa) and therefore a compressibility of 3.3×10^{-6} $(\text{psi})^{-1}$. A typical diesel fuel has $K = 243{,}700$ psi at 85°F and a pressure of 285 psi. Generally, there are two K values specified: isothermal and adiabatic. The adiabatic K value is typically used in pipeline calculations involving line pack and surge analysis.

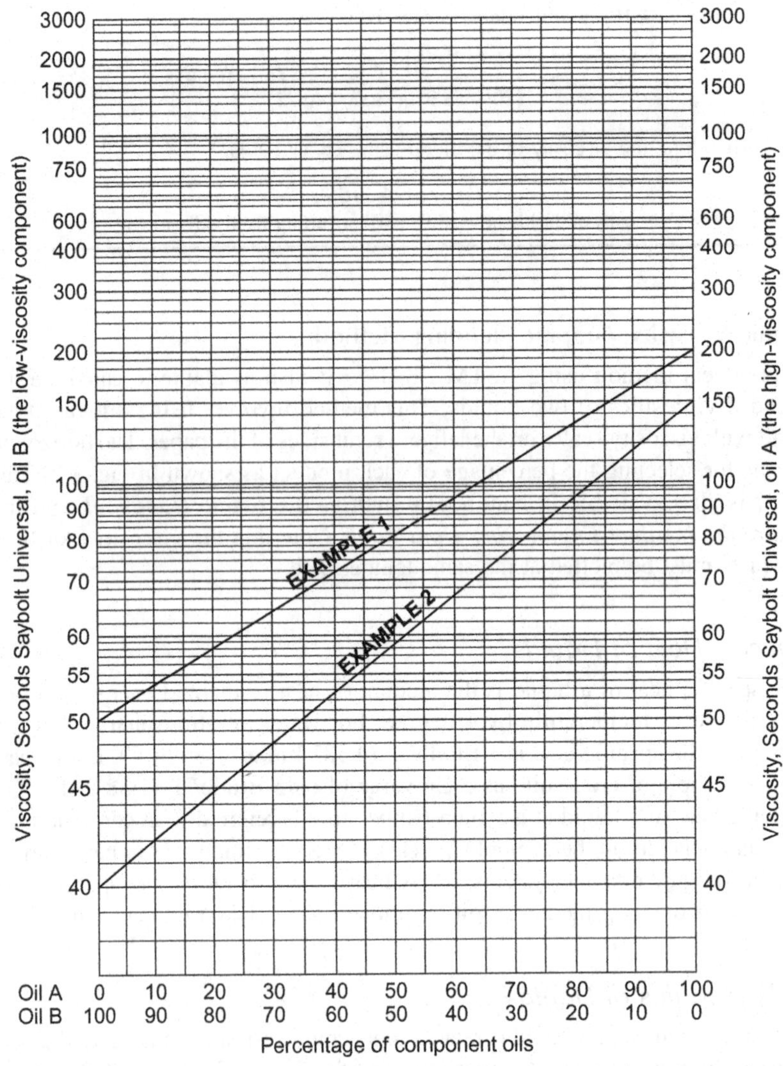

FIGURE 1.2 Viscosity blending chart.

The adiabatic bulk modulus can be calculated using the equation

$$K = A + B(P) - C(T)^{1/2} - D(\text{API}) - E(\text{API})^2 + F(T)(\text{API}) \tag{1.19}$$

where

$A = 1.286 \times 10^6$; $B = 13.55$; $C = 4.122 \times 10^4$; $D = 4.53 \times 10^3$; $E = 10.59$; $F = 3.228$
P – Pressure in psig
T – Temperature in °R
API – API gravity of liquid

Example Problem 1.5

The gravity of a petroleum product is 62°API. Calculate its adiabatic bulk modulus at a temperature of 60°F and a pressure of 1000 psig.

Solution

Using Eq. (1.19), the adiabatic bulk modulus is calculated as follows:

$$K = 1.286\times10^6 + (13.55\times1000) - 4.122\times10^4(60+460)^{1/2} - 4.53\times10^3(62)$$
$$- 10.59(62)^2 + 3.228\,(60+460)\,(62) = 142{,}092\,\text{psi}$$

Vapor Pressure of Liquids

The vapor pressure of a liquid is defined as the pressure at a given temperature at which the liquid and vapor exist in equilibrium. The normal boiling point of a liquid is thus defined as the temperature at which the vapor pressure equals the atmospheric pressure. In the laboratory, the vapor pressure is measured at a fixed temperature of 100°F and is then called the Reid vapor pressure. The vapor pressure of a liquid increases with temperature, as shown in Fig. 1.3. The actual vapor pressure of a liquid at any temperature can be found using Fig. 1.3 once its Reid vapor pressure is known.

The vapor pressure of a liquid is important when using centrifugal pumps to transport the liquid in a pipeline. To prevent cavitation damage in pumps, the liquid vapor pressure at the flowing temperature must be taken into account in the calculation of net positive suction head (NPSH) available at the pump suction. Centrifugal pumps are discussed in Chapter 9.

1.2.2 Gas Properties

The **volume** of a gas is the space occupied by a given mass of the gas at a certain temperature and pressure. Since gas is compressible, it will expand to fill the available space. Gas volume will vary with temperature and pressure. If gas is contained in a sphere or a bullet storage tank, on heating the gas volume cannot change since it fills the container. Since the mass remains the same, the increase in temperature is accompanied by an increase in pressure with volume remaining constant. This is referred to as Charles' law for gases.

Consider a spherical tank containing propane gas. If the inside diameter of the sphere is 20 ft, the volume of gas contained is the volume of the sphere and is equal to

$$V = (\pi/6)\times(20)^3 = 4189\,\text{ft}^3$$

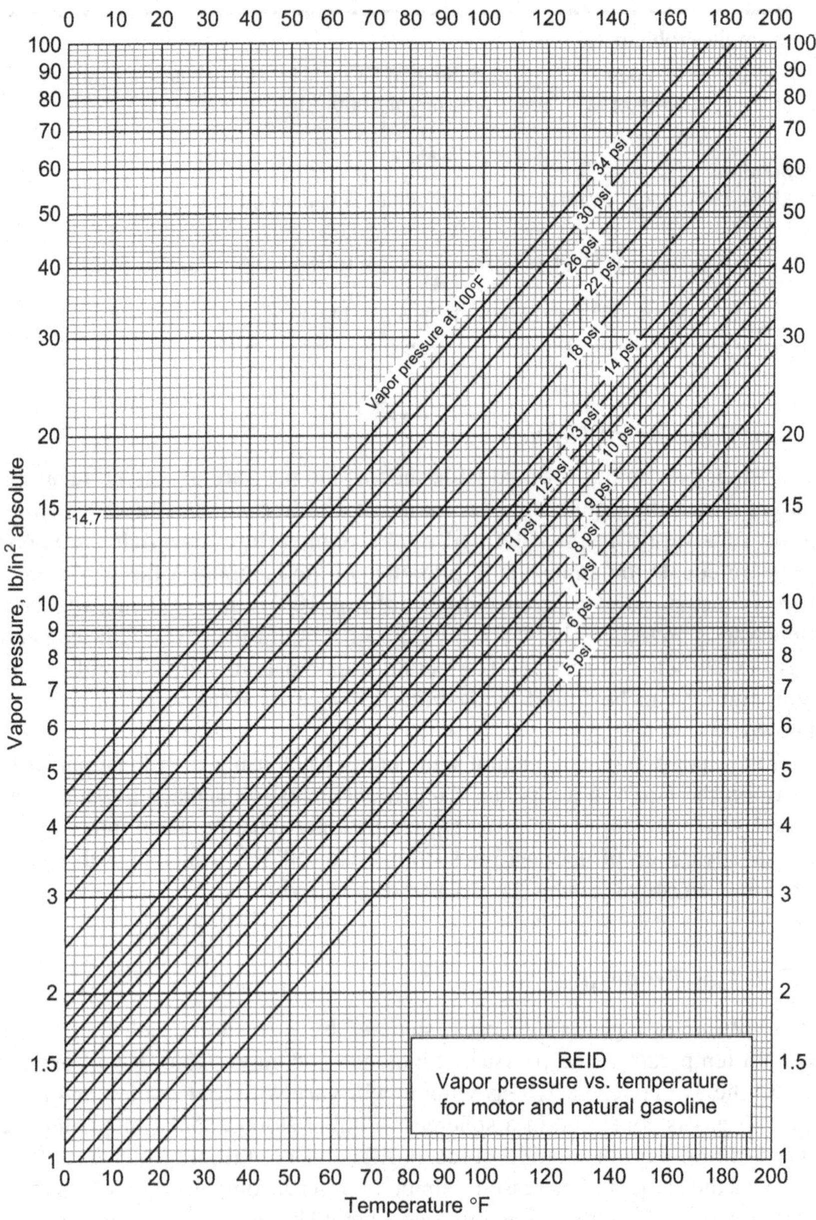

FIGURE 1.3 Vapor pressure of liquids.

Similarly, a bullet storage tank, 10 m long, 5 m in diameter with hemispherical ends containing natural gas has the following volume:

$$V = (\pi/6) \times (5)^3 + (\pi/4) \times (5)^2 \times 10 = 261.8\,\text{m}^3$$

Suppose a quantity of gas is contained in a volume of 500 ft^3 at a temperature of 80°F and a pressure of 400 psi. If the temperature is increased to 100°F, the volume remains constant, but the pressure will increase. Conversely, if the temperature is reduced to 60°F, the gas pressure will also reduce since its volume remains constant. Charles' law states that for constant volume the pressure of a fixed mass of gas will vary directly with the temperature. Thus, if temperature increases by 10%, the pressure will also rise by 10%. Similarly, if pressure is maintained constant, and the gas volume allowed to change, the volume of gas will increase or decrease in direct proportion with temperature. Charles' law, Boyle's law, and other gas laws will be discussed in detail later in this chapter.

In USCS units, the volume of gas is measured in cubic feet (ft^3). In SI units, the volume of gas is stated in cubic meters (m^3). In gas transmission pipelines, other units for volume include thousand ft^3 (Mft^3) and million ft^3 (MMft^3) in USCS units. In SI units, gas volumes may also be stated in million m^3 (Mm^3). Gas volumes are referred to at certain standard conditions of temperature and pressure, also called base conditions. In USCS units, the base conditions are at 60°F and 14.7 psia. The volume is then stated as standard volume and referred to as standard ft^3 (SCF) or million standard ft^3 (MMSCF). It must be noted that in USCS units, the practice has been to use M to represent a thousand and therefore MM refers to a million. In SI units, the base conditions are usually at 15°C and 101 kPa pressure. The standard volume in SI units is measured in this base condition of temperature and pressure. In SI units, the letter k (for kilo) is used for thousand and the letter M (for mega) is used for a million. Therefore, 100 MSCF in USCS units refers to 100 thousand standard cubic feet, whereas 20 Mm^3 means 20 million cubic meters in SI units. This distinction in the use of the letter M to denote a thousand in USCS units to denote a million in SI units must be carefully noted.

The volume flow rate in a gas transmission pipeline in USCS units is stated as cubic feet per minute (ft^3/min), cubic feet per hour (ft^3/h), and cubic feet per day (ft^3/day). Since gas volume is measured under standard conditions, the term standard cubic feet per day is used often. Thus, SCFD for standard cubic feet per day, MSCFD for thousand standard cubic feet per day, and MMSCFD for million standard cubic feet per day are generally used in USCS units. In SI units, gas flow rate is expressed in cubic meter per hour (m^3/h) or million cubic meters per day (Mm^3/day). Under standard conditions, the terms std m^3/h and std Mm^3/day are used.

The **density** of a gas represents the amount of gas that can be packed in a given volume. It is measured in terms of mass per unit volume. If 5 lb of a gas is contained in 100 ft^3 of volume, at some temperature and pressure, the gas density is 5/100 = 0.05 lb/ft^3. Strictly speaking in USCS units density must be expressed as slug/ft^3 since mass in USCS units is customarily referred to in slug. The weight density or specific weight of a gas is stated in lb/ft^3 in USCS units.

In USCS units, density is defined as follows:

$$\rho = \frac{m}{V} \tag{1.20}$$

where

ρ – Density of gas, slug/ft^3
m – Mass of gas, slug
V – Volume of gas, ft^3

In SI units, gas density is stated in kg/m^3.

Specific weight or weight density, represented by the symbol γ, is the weight of gas per unit volume. In USCS units, it is measured in lb/ft^3, compared to the mass density, which is measured in slug/ft^3. In SI units, the specific weight is expressed in Newton per cubic meter (N/m^3). The reciprocal of the specific weight is known as the specific volume, and it represents the volume occupied by a unit weight of gas. In USCS units, specific volume is measured in ft^3/lb. In SI units, specific volume is stated in m^3/N. For example, if the specific weight of a particular gas is 0.06 lb/ft^3 at some temperature and pressure, its specific volume is 1/0.06 or 16.67 ft^3/lb.

The specific gravity of a gas, also simply called gravity, is a measure of how heavy the gas is compared to air (air = 1.00) at a particular temperature. Usually, 60°F (15.6°C) is used as the standard temperature. Gas gravity may also be called relative density and is expressed as the ratio of the density of gas to the density of air at the standard temperature. Relative density, being a ratio, has no units since it is a dimensionless quantity as shown in Eq. (1.21):

$$G = \frac{\rho_g}{\rho_{air}} \tag{1.21}$$

where

G – Gas gravity, dimensionless
ρ_g – Density of gas
ρ_{air} – Density of air

Both densities in Eq. (1.21) must be in the same units and measured at the same temperature.

For example, a typical natural gas has a specific gravity of 0.60 (compared to air = 1.00) at 60°F. This means that the gas is 60% as heavy as air at the specified temperature.

If we are given the molecular weight of the gas, its gravity (air = 1.00) can be calculated from its molecular weight and the molecular weight of air, as follows:

$$G = \frac{M_g}{M_{air}} = \frac{M_g}{28.9625} \tag{1.22}$$

or

$$G = \frac{M_g}{29} \tag{1.23}$$

where the molecular weight of air is rounded off to 29 in Eq. (1.23) and where

G – Specific gravity of gas
M_g – Molecular weight of gas
M_{air} – Molecular weight of air = 28.9625

Typically, natural gas consists of a mixture of several gases such as methane, ethane, etc. Therefore, the molecular weight M_g used in Eq. (1.23) is called the apparent molecular weight of the gas mixture. It is calculated by taking into account the molecular weight and percentage composition of each component gas in the natural gas mixture.

Knowing the molecular weight and the percentage or mole fractions of the individual components of a natural gas mixture, we can calculate the molecular weight of the gas mixture using the weighted average method.

Consider a natural gas mixture that consists of 90% methane, 8% ethane, and 2% propane. The specific gravity of the natural gas mixture is calculated as follows:

$$G = \frac{(0.9 \times M1) + (0.08 \times M2) + (0.02 \times M3)}{29} \tag{1.24}$$

where $M1$, $M2$, and $M3$ are the molecular weights of methane, ethane, and propane, respectively, and 29 represents the molecular weight of air.

Appendix 1 lists the molecular weights and other properties of several hydrocarbon gases.

The density of a gas is very sensitive to changes in pressure and temperature. Therefore, the specific gravity also varies with temperature and pressure. At constant pressure, as temperature increases, the gas gravity decreases and vice versa.

Gas Viscosity

Even though the viscosity of a gas is a small number, compared to that of a liquid (0.0008 cP for gas versus 1.0 cP for water), it does have an influence on the Reynolds number and hence type of flow in a gas pipeline. The Reynolds number (discussed in Chapter 8) is a dimensionless parameter that depends on the gas viscosity, flow rate, pipe diameter, temperature, and pressure. The value of the Reynolds number determines whether the flow is laminar or turbulent, which in turn affects the pressure drop due to friction in a gas pipeline. The absolute viscosity of a gas, also called the dynamic viscosity, is expressed in $lb \cdot s/ft^2$ in USCS units and poise (P) or centipoise (cP) in SI units. The kinematic viscosity of a gas is simply the dynamic viscosity divided by the density. The two viscosities are related as follows:

$$\nu = \frac{\mu}{\rho} \tag{1.25}$$

where, in USCS units,

ν – Kinematic viscosity, ft^2/s
μ – Dynamic viscosity, $lb \cdot s/ft^2$
ρ – Density, $slug/ft^3$

In SI units,

ν – Kinematic viscosity, m^2/s
μ – Dynamic viscosity, $kg/m \cdot s$
ρ – Density, kg/m^3

The viscosities of common hydrocarbon gases are shown in Appendix 1. The appendix also includes conversion factors for converting viscosity from one set of units to another.

The **viscosity** of a gas depends on its temperature and pressure. Unlike liquids, the viscosity of a gas increases as its temperature increases and vice versa. Since viscosity represents resistance to flow in a pipeline, as the gas temperature increases, it requires more pressure to overcome the friction in a pipeline. Therefore, for the same inlet pressure, the quantity of gas flow through a pipeline will decrease as the gas temperature increases, due to the increase in viscosity. In contrast, at lower gas temperatures, the pipeline throughput is increased. This is opposite to liquid flow in a pipeline where the throughput increases with increase in liquid temperature due to lowering of its viscosity and vice versa. Unlike liquids, the viscosity of a gas is also substantially affected by pressure. Similar to temperature, increase in gas pressure results in increase in gas viscosity and vice versa.

Figure 1.4 shows the variation of gas viscosity with temperature.

Appendix 1 lists the viscosities of common gases.

Viscosity of Gas Mixtures

Typically, natural gas is a mixture of two or more pure gases such as methane and ethane. The viscosity of such a mixture can be determined from the

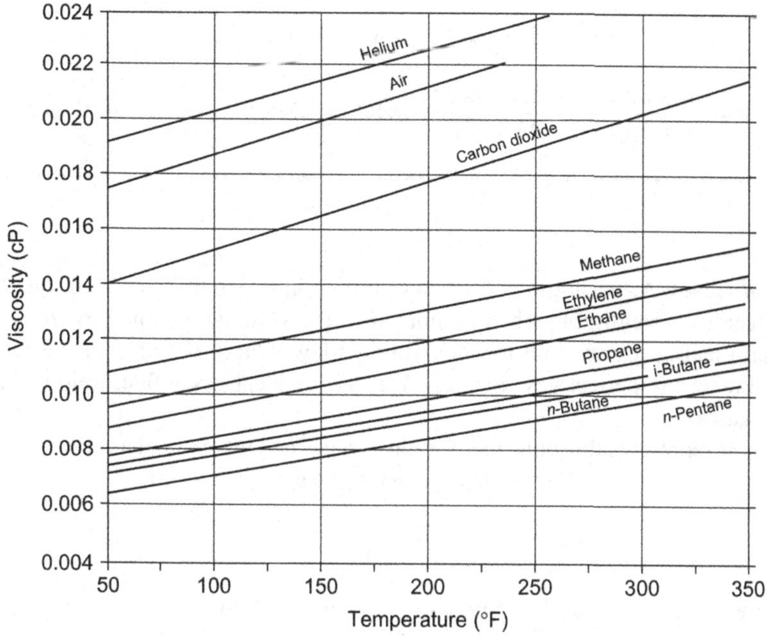

FIGURE 1.4 Variation of gas viscosity with temperature.

viscosities of the component gases and their respective percentage in the mixture using the following formula:

$$\mu = \frac{\Sigma(\mu_i y_i \sqrt{M_i})}{\Sigma(y_i \sqrt{M_i})} \tag{1.26}$$

where

μ – Dynamic viscosity of gas mixture
μ_i – Dynamic viscosity of gas component i
y_i – Mole fraction or percent of gas component i
M_i – Molecular weight of gas component i

Note that all viscosities must be measured at the same temperature and pressure.

Example Problem 1.6

Consider a homogeneous mixture consisting of 20% of gas A (molecular weight = 18) that has a viscosity 6×10^{-6} poise and 80% of gas B (molecular weight = 17) that has a viscosity 8×10^{-6} poise. Calculate the resultant viscosity of the gas mixture.

Solution

Using Eq. (1.26), the viscosity of the gas mixture is

$$\mu = \frac{(0.2\times 6\times\sqrt{18})+(0.8\times 8\times\sqrt{17})}{(0.2\times\sqrt{18})+(0.8\times\sqrt{17})}\times 10^{-6} = 7.59\times 10^{-6}\,\text{poise} = 0.000759\,\text{cP}$$

Ideal Gases An ideal gas is defined as a fluid in which the volume of the gas molecules is negligible when compared to the volume occupied by the gas. Such ideal gases are said to obey Boyle's law, Charles' law, and the ideal gas law or the perfect gas equation. We discuss ideal gases first, followed by real gases.

If M represents the molecular weight of a gas and the mass of a certain quantity of gas is m, the number of moles n is given by

$$n = \frac{m}{M} \tag{1.27}$$

where n is the number that represents the number of moles in the given mass. For example, the molecular weight of methane is 16.043. Therefore, 50 lb of methane will contain approximately 3 moles.

The ideal gas law, also called the perfect gas equation, states that the pressure, volume, and temperature of the gas are related to the number of moles by the following equation:

$$PV = nRT \tag{1.28}$$

where, in USCS units,

P – Absolute pressure, pounds per square inch absolute (psia)
V – Gas volume, ft^3
n – Number of lb moles as defined in Eq. (1.27)
R – Universal gas constant, psia $ft^3/lb\cdot mol\cdot °R$
T – Absolute temperature of gas, °R (°F + 460)

The universal gas constant R has a value of 10.73 psia $ft^3/lb\cdot mol\cdot °R$ in USCS units.

In SI units, the perfect gas equation is as follows:

$$PV = nRT \tag{1.29}$$

where

P – Absolute pressure, kPa
V – Gas volume, m^3
n – Number of kg moles as defined in Eq. (1.27)

R – Universal gas constant, kPa·m^3/kg·mol·K
T – Absolute temperature of gas, K (°C + 273)

The universal gas constant R has a value of 8.314 J/mol·K in SI units.

We can combine Eq. (1.27) with Eq. (1.28) and express the ideal gas equation as follows:

$$PV = \frac{mRT}{M} \tag{1.30}$$

The constant R is the same for all ideal gases and hence it is called the universal gas constant.

It has been found that the ideal gas equation is correct only at low pressures close to the atmospheric pressure (14.7 psia or 101 kPa). Since gas pipelines generally operate at pressures higher than atmospheric pressures, we must modify Eq. (1.30) to take into account the effect of compressibility. The latter is accounted for by using a term called the compressibility factor or gas deviation factor. We discuss real gases and the compressibility factor under the heading Real Gases.

It must be noted that in the ideal gas equation (Eq. [1.30]), the pressures and temperatures must be in absolute units. Absolute pressure is defined as the gauge pressure (as measured by a pressure gauge) plus the local atmospheric pressure at the specific location. Therefore,

$$P_{abs} = P_{gauge} + P_{atm} \tag{1.31}$$

Thus, if the gas pressure is 200 psig (measured by a pressure gauge) and the atmospheric pressure is 14.7 psia, the absolute pressure of the gas is

$$P_{abs} = 200 + 14.7 = 214.7\,\text{psia}$$

Absolute pressure is expressed as psia while the gauge pressure is referred to as psig. The adder to the gauge pressure, which is the local atmospheric pressure, is also called the base pressure. In SI units, 500 kPa gauge pressure is equal to 601 kPa absolute pressure if the base pressure is 101 kPa. Pressure in USCS units is stated in pounds per square inch (lb/in^2) or psi. In SI units, pressure is expressed in kilopascal (kPa), megapascal (MPa), or bar. Refer to Appendix 1 for unit conversion tables.

The absolute temperature of a gas is measured above a certain datum. In USCS units, the absolute scale of temperature is designated as degree Rankin (°R) and is obtained by adding the constant 460 to the gas temperature in °F. In SI units, the absolute temperature scale is referred to as Kelvin (K). Absolute temperature in K is equal to (°C + 273).

Therefore,

$$°\text{R} = °\text{F} + 460 \tag{1.32}$$

$$\text{K} = °\text{C} + 273 \tag{1.33}$$

Note that unlike temperatures in degree Rankin (°R), there is no degree symbol for absolute temperature in Kelvin (K).

Ideal gases also obey Boyle's law and Charles' law. Boyle's law relates the pressure and volume of a given quantity of gas when the temperature is kept constant. Constant temperature is called isothermal condition. According to Boyle's law, for a given quantity of gas under isothermal conditions, the pressure is inversely proportional to the volume. In other words, the volume of a gas will double, if its pressure is halved and vice versa. Since density and volume are inversely related, Boyle's law also means that the pressure is directly proportional to the density at a constant temperature. Thus, a given quantity of gas at a fixed temperature will double in density when the pressure is doubled. Similarly, a 10% reduction in pressure will cause the density to also decrease by the same amount. Boyle's law may be expressed as follows:

$$\frac{P_1}{P_2} = \frac{V_2}{V_1} \quad \text{or} \quad P_1V_1 = P_2V_2 \tag{1.34}$$

where P_1 and V_1 are the pressure and volume of the gas at condition 1 and P_2 and V_2 are the corresponding value at some other condition 2, where the temperature is the same.

Charles' law states that for constant pressure, the gas volume is directly proportional to its temperature. Similarly, if volume is kept constant, the pressure varies directly as the temperature, as indicated by the following equations:

$$\frac{V_1}{V_2} = \frac{T_1}{T_2} \text{ at constant pressure} \tag{1.35}$$

$$\frac{P_1}{P_2} = \frac{T_1}{T_2} \text{ at constant volume} \tag{1.36}$$

where T_1 and V_1 are the temperature and volume of the gas at condition 1 and T_2 and V_2 are the corresponding values at some other condition 2, where the pressure is the same. Similarly, at constant volume, T_1 and P_1 and T_2 and P_2 are the temperatures and pressures of the gas at conditions 1 and 2, respectively.

Therefore, according to Charles' law for an ideal gas at constant pressure, the volume will change in the same proportion as its temperature. Thus, a 20% increase in temperature will cause a 20% increase in volume as long as the pressure does not change. Similarly, if volume is kept constant, a 20% increase in temperature will result in the same percentage in increase in gas pressure. Constant pressure is also known as isobaric condition.

Example Problem 1.7 (USCS)

An ideal gas occupies a tank volume of 400 ft^3 at a pressure of 200 psig and a temperature of 100°F.

1. What is the gas volume at standard conditions of 14.73 psia and 60°F? Assume atmospheric pressure is 14.6 psia.
2. If the gas is cooled to 80°F, what is the gas pressure?

Solution

1. Initial conditions

$$P_1 = 200 + 14.6 = 214.6\,\text{psia}$$
$$V_1 = 400\,\text{ft}^3$$
$$T_1 = 100 + 460 = 560°\text{R}$$

Final conditions

$$P_2 = 14.73\,\text{psia}$$

V_2 is to be calculated.

$$T_2 = 60 + 460 = 520°\text{R}$$

Using the ideal gas equation (Eq. [1.30]), we can state that

$$\frac{214.6 \times 400}{560} = \frac{14.73 \times V_2}{520}$$

$$V_2 = 5411.3\,\text{ft}^3$$

2. When the gas is cooled to 80°F, the final conditions are to be determined.

$$T_2 = 80 + 460 = 540°\text{R}$$
$$V_2 = 400\,\text{ft}^3$$

P_2 is to be calculated.
The initial conditions are

$$P_1 = 200 + 14.6 = 214.6\,\text{psia}$$
$$V_1 = 400\,\text{ft}^3$$
$$T_1 = 100 + 460 = 560°\text{R}$$

It can be seen that the volume of gas is constant (tank volume) and the temperature reduces from 100°F to 80°F. Therefore, using the Charles' law equation (Eq. [1.36]), we can calculate the final pressure as follows:

$$\frac{214.6}{P_2} = \frac{560}{540}$$

Solving for P_2, we get

$$P_2 = 206.94\,\text{psia} = 206.94 - 14.6 = 192.34\,\text{psig}$$

Real Gases When dealing with real gases, we can use the ideal gas equation discussed in the preceding sections but we need to modify it to include the gas deviation factor, to account for the fact that the gas pressures in a pipeline are higher than the atmospheric pressure.

Before we calculate the gas deviation factor, two terms called critical temperature and critical pressure need to be defined. The critical temperature of a pure gas is defined as the temperature above which a gas cannot be compressed to form a liquid, regardless of the pressure. The critical pressure is defined as the minimum pressure that is required at the critical temperature to compress a gas into a liquid.

The modifying factor for real gases is called the gas deviation factor or compressibility factor Z. It can be defined as the ratio of the gas volume at a given temperature and pressure to the volume the gas would occupy if it were an ideal gas at the same temperature and pressure. Z is a dimensionless number less than 1.0 and it varies with temperature, pressure, and composition of the gas.

Using the compressibility factor Z, the ideal gas equation is modified for real gases as follows:

$$PV = ZnRT \tag{1.37}$$

where in USCS units

P – Absolute pressure of gas, psia
V – Volume of gas, ft^3
Z – Gas compressibility factor, dimensionless
T – Absolute temperature of gas, °R
n – Number of lb moles as defined in Eq. (1.27)
R – Universal gas constant, 10.73 psia ft^3/lb · mole · °R

We also need to define two other parameters called the reduced temperature and reduced pressure. The reduced temperature is the ratio of the temperature of the gas to its critical temperature. Similarly, the reduced pressure is the ratio of the gas pressure to its critical pressure as described in the following equations:

$$T_r = \frac{T}{T_c} \tag{1.38}$$

$$P_r = \frac{P}{P_c} \tag{1.39}$$

where

P – Absolute pressure of gas, psia
T – Absolute temperature of gas, °R
T_r – Reduced temperature, dimensionless
P_r – Reduced pressure, dimensionless
T_c – Critical temperature, °R
P_c – Critical pressure, psia

In SI units, these equations remain the same, except the temperatures are in K and the pressures are in kPa (absolute).

For example, the critical temperature and critical pressure of methane are 343°R and 666 psia, respectively. Therefore, the reduced temperature and pressure of the gas at 80°F and 1000 psia pressure are as follows:

$$T_r = \frac{80 + 460}{343} = 1.57$$

and

$$P_r = \frac{1000}{666} = 1.50$$

According to the theorem of corresponding states, two gases A and B may be at different temperatures and pressures. But, if their reduced temperature and reduced pressure are the same, then their gas deviation factors (Z) will be the same. Therefore, generalized plots showing the variation of Z with reduced temperature and reduced pressure may be used for most real gases for calculating the compressibility factor. Such a plot is shown in Fig. 1.5. This is called the

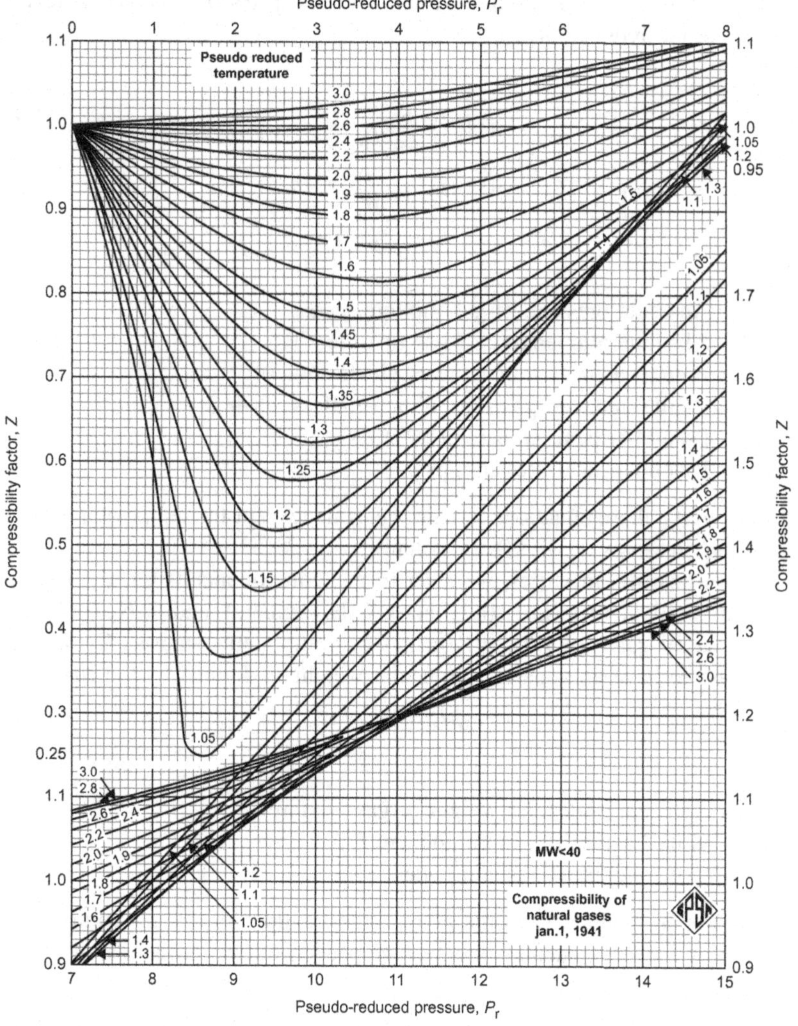

FIGURE 1.5 Compressibility factor chart for natural gases. *Source: Courtesy of the GPSA. Used with permission.*

Standing–Katz correlation for the compressibility factor. The calculation of the compressibility factor Z will be discussed in detail under the heading Compressibility Factor.

Natural Gas Mixtures When a gas consists of mixtures of different components, the critical temperature and critical pressure are called the pseudo-critical temperature and pseudo-critical pressure, respectively. If we know the composition of the gas mixture, we can calculate these pseudo-critical values of the mixture, using the critical pressure and temperature values of the pure components that constitute the gas mixture.

Similar to the pseudo-critical temperature and pseudo-critical pressure discussed above, for a gas mixture we can define the pseudo-reduced temperature and the pseudo-reduced pressure.

Thus,

$$T_{pr} = \frac{T}{T_{pc}} \tag{1.40}$$

$$P_{pr} = \frac{P}{P_{pc}} \tag{1.41}$$

where

P – Absolute pressure of gas mixture, psia
T – Absolute temperature of gas mixture, °R
T_{pr} – Pseudo-reduced temperature, dimensionless
P_{pr} – Pseudo-reduced pressure, dimensionless
T_{pc} – Pseudo-critical temperature, °R
P_{pc} – Pseudo-critical pressure, psia

In SI units, these equations remain the same, except the temperatures are in K and the pressures are in kPa (absolute).

In hydrocarbon mixtures, we refer to gas components as C_1, C_2, C_3, etc. These are equivalent to CH_4 (methane), C_2H_6 (ethane), and C_3H_8 (propane), etc. A natural gas mixture that consists of components such as C_1, C_2, C_3, etc. is said to have an apparent molecular weight as defined by the equation

$$M_a = \Sigma y_i M_i \tag{1.42}$$

where

M_a – Apparent molecular weight of gas mixture
y_i – Mole fraction of gas component i
M_i – Mole weight of gas component i

In a similar manner, from the given mole fractions of the gas components, we use Kay's rule to calculate the average pseudo-critical properties of the gas mixture as follows:

$$T_{pc} = \Sigma y_i T_c \tag{1.43}$$

$$P_{pc} = \Sigma y_i P_c \tag{1.44}$$

where T_c and P_c are the critical temperature and pressure, respectively, of the pure component (C_1, C_2, etc.) and y_i refers to the mole fraction of the component. T_{pc} and P_{pc} are the average pseudo-critical temperature and pseudo-critical pressure, respectively, of the gas mixture.

Example Problem 1.8

Calculate the apparent molecular weight of a natural gas mixture that has 80% methane, 12% ethane, 5% propane, and 3% normal butane as shown below.

Component	Percent	Molecular Weight
C1	80	16.01
C2	12	30.10
C3	5	44.10
n-C4	3	58.10
Total	100	

Solution

Using Eq. (1.42) we get

$$M_a = (0.80 \times 16.01) + (0.12 \times 30.1) + (0.05 \times 44.1) + (0.03 \times 58.1) = 20.37$$

Therefore, the apparent molecular weight of the gas mixture is 20.37.

Example Problem 1.9

Calculate the pseudo-critical temperature and the pseudo-critical pressure of a natural gas mixture consisting of 85% methane, 10% ethane, and 5% propane.

The critical properties of the C_1, C_2, and C_3 components are as follows:

Components	Critical Temperature, °R	Critical Pressure, psia
C1	343	666
C2	550	707
C3	666	617

Solution

Using the given data, from Eq. (1.43) and Eq. (1.44), we calculate the pseudo-critical properties as follows:

$$T_{pc} = (0.85 \times 343) + (0.10 \times 550) + (0.05 \times 666) = 379.85°\text{R}$$

and

$$P_{pc} = (0.85 \times 666) + (0.10 \times 707) + (0.05 \times 617) = 667.65\,\text{psia}$$

Therefore, the pseudo-critical temperature of the gas mixture = 379.85°R and the pseudo-critical pressure of the gas mixture = 667.65 psia.

Example Problem 1.10

If the temperature of the gas in the previous example is 80°F and the average gas pressure is 1000 psig, what is the pseudo-reduced temperature and pseudo-reduced pressure of this gas? Use 14.7 psia for base pressure.

Solutions

From Eqs (1.40) and (1.41), we get

$$\text{The pseudo-reduced temperature } T_{pr} = \frac{80+460}{379.85} = 1.42$$

$$\text{The pseudo-reduced pressure } P_{pr} = \frac{1000+14.7}{667.65} = 1.52$$

Pseudo-Critical Properties from Gas Gravity

If the percentages of the various components in the natural gas mixture are not available, we can calculate approximate values of the pseudo-critical properties of the gas mixture, if we know the gas gravity. The pseudo-critical properties are calculated approximately from the following equations:

$$T_{pc} = 170.491 + 307.344\,G \tag{1.45}$$

$$P_{pc} = 709.604 - 58.718\,G \tag{1.46}$$

where

G – Gas gravity (air = 1.00)
T_{pc} – Pseudo-critical temperature, °R
P_{pc} – Pseudo-critical pressure, psia

In SI units the pseudo-critical properties are calculated as follows:

$$T_{pc} = 94.38 + 170.747\,G \tag{1.47}$$

$$P_{pc} = 4894 - 404.95\,G \tag{1.48}$$

where

G – Gas gravity (air = 1.00)
T_{pc} – Pseudo-critical temperature, K
P_{pc} – Pseudo-critical pressure, kPa

Example Problem 1.11

Calculate the gravity of a natural gas mixture consisting of 85% methane, 10% ethane, and 5% propane. From the gas gravity, calculate the pseudo-critical temperature and pseudo-critical pressure for this natural gas mixture.

Solution

Using Kay's rule for multicomponent mixtures and Eq. (1.22) for gas gravity, we get

$$G = \frac{(0.85 \times 16.04) + (0.10 \times 30.07) + (0.05 \times 44.10)}{29} = 0.6499$$

Therefore, the gas gravity is 0.6499.

From Eqs (1.45) and (1.46), we calculate the pseudo-critical properties as follows:

$$T_{pc} = 170.491 + 307.344 \times (0.6499) = 370.23°R$$

$$P_{pc} = 709.604 - 58.718 \times (0.6499) = 671.44\,\text{psia}$$

Comparing the above values with the values calculated using the more accurate method in the previous example we find that the value of T_{pc} is off by 2.6% and P_{pc} is off by 0.6%. These differences are small enough for most calculations related to natural gas pipeline hydraulics.

Impact of Sour Gas and Nonhydrocarbon Components

The Standing–Katz chart used for determining the compressibility factor of a gas mixture is accurate only if the amount of nonhydrocarbon components are small. Since sour gases contain carbon dioxide and hydrogen sulfide, adjustments must be made to take into account these components in calculations of the pseudo-critical temperature and pseudo-critical pressure. Depending on the amounts of carbon dioxide and hydrogen sulfide present in the sour gas, we calculate an adjustment factor ε for the pseudo-critical temperature from the equation

$$\varepsilon = 120(A^{0.9} - A^{1.6}) + 15(B^{0.5} - B^{4.0}) \tag{1.49}$$

where

ε – Adjustment factor, °R
A – Sum of the mole fractions of CO_2 and H_2S
B – Mole fraction of H_2S

In SI units ε in K is calculated from

$$\varepsilon = 66.667\,(A^{0.9} - A^{1.6}) + 8.333\,(B^{0.5} - B^{4.0}) \tag{1.50}$$

The pseudo-critical temperature is modified to get the adjusted pseudo-critical temperature T'_{pc} from the following equation:

$$T'_{pc} = T_{pc} - \varepsilon \tag{1.51}$$

Similarly, the pseudo-critical pressure is adjusted as follows:

$$P'_{pc} = \frac{P_{pc} \times T'_{pc}}{T_{pc} + B(1-B)\varepsilon} \tag{1.52}$$

where

P'_{pc} – Adjusted pseudo-critical pressure, psia

In SI units,

$$P'_{pc} = \frac{P_{pc} \times T'_{pc}}{T_{pc} + B(1-B)\varepsilon} \tag{1.53}$$

P'_{pc} – Adjusted pseudo-critical pressure, kPa

Example Problem 1.12

The pseudo-critical temperature and the pseudo-critical pressure of a natural gas mixture were calculated as 370°R and 670 psia, respectively. If the CO_2 content is 10% and H_2S is 20%, calculate the adjustment factor ε and the adjusted values of the pseudo-critical temperature and pressure.

Solution

$$A = 0.10 + 0.20 = 0.30$$
$$B = 0.20$$

The adjustment factor ε from Eq. (1.49) is

$$\varepsilon = 120(0.30^{0.9} - 0.30^{1.6}) + 15(0.20^{0.5} - 0.20^{4}) = 29.8082°\text{R}$$

Therefore, the adjustment factor ε is 29.81°R.

The adjusted values of the pseudo-critical properties are found using Eqs (1.51) and (1.52) as follows:

$$T'_{pc} = 370.0 - 29.81 = 340.19°\text{R}$$

and

$$P'_{pc} = \frac{670 \times 340.19}{370 + 0.20(1-0.20) \times 29.8082} = 608.18\ \text{psia}$$

Therefore, the adjusted pseudo-critical temperature = 340.19°R.
The adjusted pseudo-critical pressure = 608.18 psia.

Compressibility Factor

As explained earlier, the compressibility factor (or gas deviation factor) is a measure of how close a real gas is to an ideal gas. The compressibility factor is a dimensionless number close to 1.00 and is a function of the gas gravity, gas temperature, gas pressure, and the critical properties of the gas. For example, a particular natural gas mixture may have a compressibility factor equal to 0.87 at 1000 psia and 80°F. Charts have been constructed that depict the variation of Z with the reduced temperature and reduced pressure. Another term called the supercompressibility factor, F_{pv}, which is related to the compressibility factor Z, is defined as follows:

$$F_{pv} = \frac{1}{\sqrt{Z}} \tag{1.54}$$

For example, if the compressibility factor $Z=0.85$, using Eq. (1.54) we calculate the supercompressibility factor, F_{pv}, as follows:

$$F_{pv} = \frac{1}{\sqrt{0.85}} = 1.0847$$

There are several approaches to calculating the compressibility factor for a particular gas temperature T and pressure P. One method uses the critical temperature and critical pressure of the gas mixture. First, the reduced temperature T_r and reduced pressure P_r are calculated from the given gas temperature and gas pressure and the critical temperature and critical pressure using Eqs (1.38) and (1.39).

Once we know the values of Tr and Pr, the compressibility factor can be found from charts similar to the Standing–Katz chart.

The following methods are available to calculate compressibility factor.

1. Standing–Katz Method
2. Dranchuk, Purvis, and Robinson Method
3. AGA Method
4. CNGA Method

Appendix 1 lists the various formulas pertaining to the above methods for the compressibility factor.

In a gas pipeline, the pressure and temperature of the gas vary along the length of the pipeline, due to the frictional pressure drop in the pipeline. This is discussed in Chapter 8. The compressibility factor Z also varies along the pipe length and must therefore be calculated for an average pressure at any location on the pipeline. If the gas pressure at two points along the pipeline are P_1 and P_2, the average pressure in the pipe segment is $\frac{P_1+P_2}{2}$. This is true only if the segment length is very short. Instead, the following formula is used for a more accurate value of the average pressure in a pipe segment:

$$P_{avg} = \frac{2}{3}\left(P_1 + P_2 - \frac{P_1 \times P_2}{P_1 + P_2}\right) \tag{1.55}$$

By mathematical manipulation, Eq. (1.55) may also be written as follows:

$$P_{avg} = \frac{2}{3}\left(\frac{{P_1}^3 - {P_2}^3}{{P_1}^2 - {P_2}^2}\right) \tag{1.56}$$

Example Problem 1.13

Using the Standing–Katz compressibility chart, calculate the compressibility factor for the gas in Example Problem 1.10 at 70°F and 1200 psig. Use the values of T_{pc} and P_{pc} calculated in Example Problem 1.10.

Solutions

From Example Problem 1.10

Pseudo-reduced temperature $T_{pr} = 1.42°R$
Pseudo-reduced pressure $P_{pr} = 1.52\,psia$

Using the Standing–Katz chart (Fig. 1.5), we read the value of Z as 0.825.

Example Problem 1.14

A natural gas mixture consists of the following components:

Component	Mole Fraction y
C_1	0.780
C_2	0.005
C_3	0.002
N_2	0.013
CO_2	0.016
H_2S	0.184

1. Calculate the apparent molecular weight of the gas, gas gravity, and the pseudo-critical temperature and pseudo-critical pressure.
2. Calculate the compressibility factor of the gas at 90°F temperature and 1200 psia pressure.

Solution

Using Appendix 1, we create the following table showing the molecular weight M, critical temperature T_c, and critical pressure P_c for each of the component gases. The molecular weight of the mixture and the pseudo-critical temperature and pseudo-critical pressure are then calculated using Eqs (1.43) and (1.44).

Component	y	M	yM	Tc	Pc	yTc	yPc
C1	0.780	16.04	12.5112	343.34	667.00	267.81	520.26
C2	0.005	30.07	0.1504	550.07	707.80	2.75	3.54
C3	0.002	44.10	0.0882	665.93	615.00	1.33	1.23
N2	0.013	28.01	0.3641	227.52	492.80	2.96	6.41
CO2	0.016	44.01	0.7042	547.73	1070.00	8.76	17.12
H2S	0.184	34.08	6.2707	672.40	1300.00	123.72	239.20
Total	1.000		20.0888			407.33	787.76

Therefore, the apparent molecular weight of the natural gas is

$$\mathrm{Mw} = \Sigma yM = 20.09$$

The gas gravity is calculated using Eq. (1.23).

$$G = \frac{20.09}{29} = 0.6928$$

Next, we calculate the pseudo-critical temperature and pressure values.

Pseudo-critical temperature $= \Sigma y T_c = 407.33°R$
Pseudo-critical pressure $= \Sigma y P_c = 787.76$ psia

Since this sour gas contains more than 5% nonhydrocarbons, we will adjust the pseudo-critical properties using Eqs (1.49) and (1.50). The temperature adjustment factor ε is calculated using Eq. (1.49) as follows:

$$A = (0.016 + 0.184) = 0.20$$

and

$$B = 0.184$$

Therefore, the adjustment factor is

$$\varepsilon = 120[(0.2)^{0.9} - (0.2)^{1.6}] + 15[(0.184)^{0.5} - (0.184)^{4.0}] = 25.47°R$$

The adjusted pseudo-critical temperature from Eq. (1.51) is

$$T'_{pc} = 407.33 - 25.47 = 381.86°R$$

The adjusted pseudo-critical pressure from Eq. (1.52) is

$$P'_{pc} = \frac{787.76 \times 381.86}{407.33 + 0.184 \times (1 - 0.184) \times 25.47} = 731.63 \text{ psia}$$

Next, we calculate the compressibility factor at 90°F and 1200 psia pressure using these values as follows:
From Eq. (1.40)

$$\text{Pseudo-reduced temperature} = \frac{90 + 460}{381.86} = 1.44$$

From Eq. (1.41)

$$\text{Pseudo-reduced pressure} = \frac{1200}{731.63} = 1.64$$

Finally, from the Standing–Katz chart, we get the compressibility factor for the reduced temperature and reduced pressure as $Z = 0.825$.

Specific Heat of Gases

The specific heat of a gas represents the amount of heat required to increase its temperature by 1 degree. There are two values of specific heat defined for gases: specific heat at constant pressure (Cp) and specific heat at constant volume (Cv). The ratio Cp/Cv is called the adiabatic or isentropic exponent of the gas and is denoted by γ and is an important parameter used to calculate the power required to compress a gas. Appendix 1 lists the specific heat of various common gases.

Heating Value

The heating value of a gas is defined as the thermal energy per unit volume of the gas. It is expressed in Btu/ft^3 in USCS units. In SI units it is stated as gigajoules/m^3 (GJ/m^3). Sometimes, it is stated on a mass basis: Btu/lb in USCS units, MJ/kg in SI units. For natural gas, it is approximately in the range of 900–1000 Btu/ft^3 (33.5–37.3 GJ/m^3). There are two heating values used in the industry. These are the lower heating value (LHV) and higher heating value (HHV). The LHV is calculated by subtracting the heat of vaporization of the water vapor from the HHV.

For a gas mixture, the term gross heating value is used. It is calculated based on the heating values of the component gases and their mole fractions using the following equation:

$$H_{\mathrm{m}} = \Sigma(y_i H_i) \tag{1.57}$$

where

H_{m} – Gross heating value of mixture, Btu/ft^3 or GJ/m^3
y_i – Mole fraction or percent of gas component i
H_i – Heating value of gas component, Btu/ft^3 or GJ/m^3

For example, a natural gas mixture consisting of 80% of gas A (heating value = 900 Btu/ft^3) and 20% of gas B (heating value = 1000 Btu/ft^3) will have a gross heating value of

$$H_{\mathrm{m}} = (0.8 \times 900) + (0.2 \times 1000) = 920\,\mathrm{Btu/ft^3}$$

SUMMARY

In this chapter, we first introduced the various units of measurement employed in the transmission pipeline industry. The USCS units used in the United States and the SI units used in the rest of the world were discussed and compared. Appropriate conversion factors were reviewed. Next, the important physical properties of liquids and gases that form the basis of transmission pipeline calculations in later chapters were introduced. The specific gravity and viscosity of liquids and gases were explained along with how to calculate these properties in liquid mixtures and gas mixtures at various temperatures. We discussed several gas properties that influence gas pipeline transportation. The ideal gas equation was introduced along with Boyle's law and Charles' law and an explanation followed of how they can be applied with modifications to real gases and real gas mixtures. The gas deviation factor or compressibility factor that modifies ideal gas behavior was introduced. Critical properties of hydrocarbon gases and mixtures and the reduced temperature and pressure that determine the state of a gas was explained. Variation of the compressibility factor with pressure and temperature was explored, and calculation methodologies using analytical and graphical approaches were covered. The influence of nonhydrocarbon components in a natural gas mixture was also discussed along with correction factors for CO_2 and H_2S in sour gas.

BIBLIOGRAPHY

A.E. Uhl, et al., Steady Flow in Gas Pipelines, AGA, AGA Report No. 10, New York, 1965.

Compressibility Factors, AGA, AGA Report No. 8, New York, 1992.

Flow of Fluids through Valves, Fittings and Pipes, Crane Company, New York, 1976.

Gas Processors Suppliers Association, Engineering Data Book, tenth ed., Tulsa, OK, 1994.

E. Shashi Menon, Liquid Pipeline Hydraulics, Marcel-Dekker, New York, 2004.

E. Shashi Menon, Gas Pipeline Hydraulics, CRC Press, Boca Raton, FL, 2005.

D.L. Katz, et al., Handbook of Natural Gas Engineering, McGraw-Hill, New York, 1959.

W.D. McCain Jr., The Properties of Petroleum Fluids, Petroleum Publishing Company, Tulsa, OK, 1973.

E. Shashi Menon, Piping Calculations Manual, McGraw-Hill, New York, 2005.

Pipeline Design for Hydrocarbon Gases and Liquids, American Society of Civil Engineers, New York, 1975.

Chapter 2

Route Selection

Hal S. Ozanne

Chapter Outline

INTRODUCTION

There are several steps to be taken in developing a pipeline project. Once it has been determined that a pipeline is planned to be constructed between two or more locations, then one of the next critical steps is to determine what the possible options are in routing the pipeline between the beginning and ending points of the pipeline.

Several criteria are generally used to select the route(s). These criteria include the following:

- Community and local agencies
- Technical and project necessities
- Constructibility of the pipeline along the route
- Right-of-way acquisition or landowner issues
- Environmental issues
- Archeological issues
- Threatened and endangered species issues
- Beginning and end points
- Mapping systems
- Field review

- Ability to parallel other pipelines or linear infrastructure
- Existence of established corridors
- Required connections along the route (injection or delivery points)
- Accessibility via existing roads
- Population density (especially for high-pressure gas systems)

The shortest distance between the beginning (source(s) of product) and the end (delivery point(s)) presides over the likelihood analysis of developing pipeline projects. A few alternative routes are necessary while considering certain relative factors, including deliverable pressures or amount of product, required diameter and grade or wall thickness of pipe, and number of required compression stations or pumping stations. Detailed and accurate environmental impact assessments and statements along with public impact evaluations provide the necessary approvals from the federal, state, and local governing authorities in the shortest period. Pipeline route selection is directed by, but not limited to, the following objectives:

- Establishing the shortest possible route to reduce material and construction costs while minimizing any intermediate compression and/or pumping facilities
- Minimizing environmental (wetlands, cultural resources, and other protected areas) damage and impacts; routing the pipeline to minimize clearing of trees as much as possible
- Keeping road, highway, rail, and waterway crossings to a minimum
- Avoiding inhabited areas when realistic
- Utilizing an existing right-of-way containing existing pipelines, overhead transmission or power lines, fiber optic lines, etc. when possible
- Avoiding sidehill construction and choosing terrain that circumvents obstacles, such as wells, houses, orchards, lakes, or ponds
- Provide access along the entire route for personnel and equipment to conduct normal operations and maintenance functions.

The route selection should be developed by taking the relevant legal, environmental, technical right-of-way, and project requirements into consideration. In addition, detailed consultation with local community concerns that are affected by the route selection should be pursued. Safety is more predominantly a factor of concern by the community during consultation and is usually the first priority. It will behoove the project to address safety up-front with a prearranged and well-thought-out process that will provide the community and officials with a sense of calm that will hopefully promote preapproval.

In many cases, the route selection affects certain local environmental conservatories that cannot be avoided without considerable rerouting. Where possible, these areas should be avoided to minimize impact to the conservatory areas. When impact to these areas cannot be avoided, certain construction techniques may need to be implemented to satisfy the local conservatory board and mitigate the impact to the conservatory areas. This can be costly, so it should be planned and included in estimated construction costs.

Generally, the routing of a pipeline will start with review of US Geological Survey (USGS) quadrangle maps for the area in consideration, which, depending on how long it has been since the maps were updated, will indicate roads, other pipelines, power lines, and topographical information.

Recently, online mapping systems, such as Google Earth and Bing, have been utilized as effective tools in the initial stages of routing pipelines. These tools use satellite imagery that is updated every few years. Typically, this information is more current and reliable than most other mapping sources. Right-of-way scars for pipelines and electric power lines are usually evident within the imagery because either the scars from construction are evident or the maintenance of the cleared right-of-way is clearly identified. Roads and water bodies are also easily identified.

2.1 COMMUNITY AND LOCAL AGENCIES

The local community and agencies in the area of the proposed pipeline need to be consulted while planning the project. They will be impacted by the project, and if they are involved from the beginning, they can have a positive influence on the approval of the project. Items to be considered include the following:

- Safety: It is considered during construction activities and operational safety.
- Impacts on people including proximity to dwellings and public centers: When impacts occur (as they normally do unless the pipeline is routed through an exceedingly rural area), site-specific plans outlining constrained construction techniques may be required.
- Land use (the zoned use of the land): When selecting a route, land use needs to be addressed so that construction methods can be identified and documented. In addition, the land use is important for the subsequent operation or maintenance of the pipeline.
- Planning: When selecting a route for a proposed pipeline, significant analysis of any potential development that may be in consideration should include, but not be limited to: residential subdivisions, business parks, recreational facilities, shopping centers, landfills, etc. A study of any potential development that requires avoidance will save considerable time and expense in finalizing the route.

2.2 POPULATION DENSITY

The Code of Federal Regulations 49 (CFR 49), Part 192, Transportation of Natural and Other Gas by Pipeline, and Part 195, Transportation of Hazardous Liquids by Pipeline, governs the engineering and construction of pipelines classified as gas transmission pipelines and hazardous liquid pipelines, respectively. Gas transmission pipelines are also regulated by the Federal Energy Regulatory Commission (FERC).

The Code of Federal Regulations, Title 49, Part 192, Transportation of Natural Gas and Other Gas by Pipeline: Minimum Federal Safety Standards state the following:

2.2.1 Subpart A: General

192.5 Class Locations: This section classifies locations for purposes of this part.

A class location unit is an onshore unit that extends 220 yards on either side of the centerline of any continuous 1-mile length of pipeline.

Class 1: 10 or fewer buildings intended for human occupancy

Class 2: More than 10 but fewer than 46 buildings intended for human occupancy

Class 3: 46 or more buildings intended for human occupancy; or an area where the pipeline lies within 100 yards of either a building or defined outside area that is occupied by 20 or more persons on at least 5 days a week for 10 weeks in any 12-month period

Class 4: Any class location where buildings with four or more stories above ground are prevalent.

Federal regulations, for natural gas pipelines, mandate classifications based on population density along the route. The different classifications result in an application of different factors of safety to the maximum allowable operating calculations, which may necessitate greater pipe wall thickness or yield strength, or both, in higher populated areas along the route.

In routing a pipeline, the population density is a significant consideration. Not only will the construction be more costly in heavier populated areas but also the pipe will be more costly as well, due to greater required wall thickness.

2.3 TECHNICAL AND PROJECT NECESSITIES

- Pipeline length: Although the shortest route from supply and demand (point A to point B) is predominantly important in an effort to minimize pipe and construction costs, pressure drops, and right-of-way acquisition costs, to name a few, the route may need to be modified to minimize impacts as described above. However, route selection should always stay as close to the shortest route as possible.
- Pipeline construction and optimization of pipeline design and operation: When selecting a route, consideration of the size and grade of pipe is extremely important, for example, a 4″ line can be constructed much more easily in many more places than a 30″ line. In addition, operational and maintenance knowledge is critical, so the pipeline location does not encumber these two issues later.
- Specific project requirements: A detailed proposed budget and schedule should be established prior to final route selection. From this, the selection of the final route with the deviations to the initial concept should be documented, and any additional costs or effect on schedule should be incorporated into the final budget.

2.4 CONSTRUCTIBILITY

This term refers to taking into consideration how a route will actually be constructed and what issues might play into modifying your initial route from a construction prospective. Things to be considered are access for construction equipment and personnel, stringing of pipe along the right-of-way, the proximity of other infrastructure adjacent to the proposed route, and the type of construction techniques that will be required.

Some other things to be considered are as follows:

- Soils: Whether a trench can be cut with standard equipment without blasting
- Terrain: Whether there are slopes along the route that would be difficult, and if it is possible for construction to take place using normal construction equipment and techniques
- Physical Features: Whether there are any obstructions (for example, wide water bodies) that may require a long directional drill in order to make the crossing

In other words, can the pipeline be constructed along the proposed route?

The pipeline route should allow for continuous uninterrupted construction and associated activities including hauling of water, construction inspection, and the establishment of environmental mitigation measures and maintenance. The right-of-way should be wide enough for the size of equipment required for construction to operate efficiently and safely. In addition, there should be sufficient space within the working area to allow for vehicles to traverse in both directions safely and without impacting adjacent areas outside the working area. As much as is practical, a disruption in the continuous flow of construction equipment or move-arounds should be avoided as this reduces productivity while increasing construction costs. Furthermore, the working side for the construction equipment should remain on one side of the right-of-way, as much as possible, to eliminate the need to readjust certain equipment and maximize productivity. Figure 2.1 shows an example of the typical width of right-of-way that is required for construction.

There are certain important things to adhere to during the layout of a pipeline route to ensure constructibility. Some of the key issues that should invariably apply when establishing a route, from a preliminary selection on paper (or digitally) to field engineering the final route for permitting and approvals, are represented as follows:

- Depending on circumstances related to the size and grade of pipe, always establish pipeline bends or engineered bends (segmented or nonsegmented) far enough from roads, railroads, streams, or other obstacles to allow for a constructible approach to the obstacles. This is required to provide ample room for safe and appropriate methods of construction for the obstacle. By not allowing for sufficient straight-level approach to a known crossing, the method of crossing for the obstacle, by either bore or open cut, is not attainable and may be out of code or compliance to an obligatory permit.

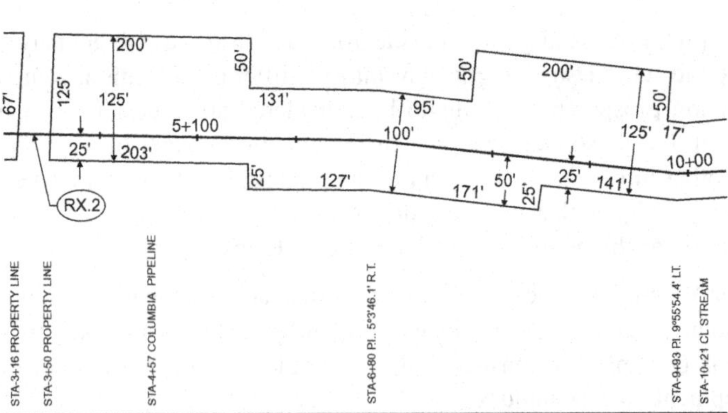

FIGURE 2.1 Typical construction right-of-way requirements.

- When establishing a route, from either the desktop or field engineering, ensure that any road or railroad crossing that is required to be bored is established so that the slope of the terrain is not too severe on either side of the crossing. Severe slopes prohibit the establishment of necessary bore pits without dedicating considerable work space for safely shoring the ditch or alternately installing costly shoring boxes, either of which translates into greater time requirements and construction costs.

Mobilization is a major construction activity. One factor in pipeline routing is the provision for effective mobilization by the contractor. Distance to market, the availability of power and water, and the number of unskilled laborers are typical requirements for starting effective construction activities.

Pipeline construction methods vary greatly with terrain conditions. For example, laying pipeline across a river requires horizontal direction drilling (HDD), whereas laying across a rocky area requires rock-trenching techniques. Therefore, location characteristics are a major cost component for pipeline construction. Inappropriate terrain selection can cause major time and cost overruns.

2.5 RIGHT-OF-WAY

When selecting a route, one of the most significant factors to consider is the easements (or right-of-way) that should be obtained to construct the pipeline. The ownership of the land in which the route will traverse has to be determined so that proper legal owners are contacted and negotiated with for easement acquisition. Typical owners may include federal, state, county, and other government agencies or private ownership. In some instances, there may be reasons why the right-of-way cannot be obtained across some of the land and the route may need to be changed.

Another thing to consider is if the project will have the right of eminent domain. This means that if a landowner does not want to grant an easement but the pipeline operator can show that the pipeline will be a benefit for the general public, the land can be condemned, and the landowner will be required to grant the pipeline easement for a fair market value. Not all pipelines have the right of eminent domain. This should be determined before the project starts because it may have a significant influence on the route selection.

In most cases, a permit will be required from a governmental agency. In those instances, the agency may require the pipeline operator to go through an extensive process before consideration will be given to granting an easement or a permit to construct the line on publicly owned land. This agency will be responsible to issue a right-of-way grant or easement to the pipeline prior to construction.

The width of the proposed right-of-way is also critical in the route selection process. Determining the width of the permanent right-of-way along with the width of the temporary construction right-of-way is important in selecting the pipeline route. These widths are heavily dependent on the size of the pipeline. The greater the diameter of the pipeline, the wider the right-of-way that will be required for construction and maintenance.

The permanent right-of-way width will also have to be wide enough so that the pipeline can be maintained after construction is complete and after the pipeline has been placed in operation. This includes the passage of vehicles on the right-of-way and the room to excavate the pipe if required.

Consideration must be given for accessing the right-of-way during construction. Arrangements will need to be made for the use of existing roads or to construct temporary roads for access to the pipeline right-of-way for use by the construction contractor during construction and for the delivery and stringing of the pipe.

When selecting routes, consideration must be given to determine if there is adequate room for all of the right-of-way requirements along the route.

2.6 ENVIRONMENTAL ISSUES

In most cases, pipeline projects will require a permit to be issued by a governmental agency before construction can proceed. This may be at the local, county, state, or federal level. The permitting process is usually very extensive and requires an environmental assessment to be conducted. This is used to determine what environmental impact the project may have in the area surrounding the project, both direct and indirect.

The environmental permitting process may require a considerable amount of time and cost to obtain. The process can require data research, and research along the proposed route, to determine if any items such as threatened and endangered plant or animal species, cultural resources, or wetlands may be impacted by the project, and if so, to what extent. If it is determined that one or more of the above may be impacted, it is possible that a permit may not be granted or that mitigation measures may be required to eliminate or reduce the impact.

It is best to take these items into consideration when selecting a route or route options. The more front-end work that is done in this area of selecting the route to avoid these issues, the better the chance of success in obtaining the required project permits.

The social economic impact in the area surrounding a pipeline project is a major concern for regulatory agencies when they are considering the permit application.

Other considerations include the following:

- Impacts on wildlife, wetlands, and waterways: A study of any potential impacts on these areas and subsequent minimization of impacted areas prior to route finalization will often lead to satisfying the concerns many federal, state, and local environmental agencies have.
- Impacts on archaeology and cultural heritage: During route selection and prior to field work, a detailed study of any impacts should be implemented, so avoidance can be arranged.
- Visual impacts: The route selection should consider any visible impact regarding aesthetics.

2.7 ROUTE BEGINNING AND ENDING POINTS

The reason to consider constructing a pipeline is to move a commodity (crude oil, natural gas, refined products, etc.) from one or more points to one or more delivery locations. The operation of a pipeline is normally a more economical means to transport product of a significant daily volume than by other means such as truck or rail. In some instances, there is a requirement or an opportunity either to receive additional product into the pipeline or to deliver product out of the pipeline along the route. These factors will determine the general route of the pipeline.

2.8 CONNECTIONS

The route will be selected to start and finish at the end points and pass as close as possible to the proposed connections, if any, along the route. As the route is being developed, these factors will be taken into consideration.

2.9 MAPPING SYSTEM

The end points of the proposed pipeline will be identified and pinpointed on a map. The proposed route will then be selected and drawn on the map taking into consideration the factors previously described. A versatile tool for laying out preliminary route options is an online mapping system like Google Earth, which is free to use. Routes can be drawn, saved, and printed on maps created with satellite photography. USGS quadrangle maps are also often used because they show topography, physical features, and other pipelines.

Alternate routes can be drawn, compared, and distributed to project team members to get their input. During this process, some options may be added or eliminated.

2.10 FIELD REVIEW

On completion of a general routing of a proposed pipeline through desktop studies utilizing relational information including, but not limited to, quadrangle maps, aerial photography, satellite imagery, Google Earth, etc., a field review of the proposed route is imperative. A comprehensive review of the entire route will identify areas that require minor deviations to the proposed route while confirming the applicability of major changes to the route indicated through the desktop study.

Some indicators that trigger deviations during the field review are mentioned below:

- New construction: Often, this cannot be identified during the desktop study due to out-of-date aerial imagery. Many of these areas require slight modification to the route, whereas others may dictate more substantial deviations. Simply identifying these areas during the field review is not sufficient; a detailed plan to minimize the impact to these areas, along with a route modification, is of utmost importance.
- Road and railroad crossings: During the field review, each road and railroad crossing should be studied to determine if the crossing, as defined by the desktop study, is constructible. The field review should take into account the terrain on both sides of a crossing. Should the terrain have steep banks on either or both sides prohibiting the safe installation of a bore pit, the alignment should be adjusted to one side or the other of the proposed route such that accommodation for a constructible crossing is achieved. The review should identify any existing utilities that may impede the crossing. In most cases, a desktop study cannot determine where utilities cross a road. There may be storm drains, sewer lines, water lines, gas distribution lines, underground electric lines, etc., which are located in the same place as the proposed route. In this case, as with the terrain, the alignment should be adjusted to one side or the other to eliminate impact to these utilities.
- Rural construction versus urban and/or suburban construction: During a desktop study, the route is identified more predominantly by following as straight a line as possible. Many times this alignment is identified as passing through urban and suburban areas that dictate a very different approach to construction methods. These urban and suburban areas usually take a longer

time during the construction stage due to minimizing work space, minimizing ditching operations, and requiring the contractor often to build the pipeline in the ditch with one joint of pipe at a time (sometimes called "stovepiping"). All these issues increase the cost of construction, often by twice that of standard construction. In addition, the cost for mitigation after construction with regards to damage to roads, environmental, and aesthetics can be significant as most communities will require that the landscape be returned to as good or better than before the installation of the pipeline. By reviewing the possibility of routing around urban and/or suburban areas through more rural areas, the cost for construction and mitigation could be much less even though the pipeline footage may increase.

- Deciduous forested areas versus planted timber areas: During the desktop study, the route cannot clearly identify, in many cases, where deciduous forested areas are versus planted timber areas. During the field review, it is ascertained that the route does indeed go through a planted timber area and the route could be shifted slightly into deciduous forested areas; the review should note this and recommend the appropriate shift. The cost of the timber in the planted areas more than likely will be significantly higher than that of a deciduous forested area.
- Side slopes: Where the route from the desktop study navigates along steep side slopes and the initiation of wider working space for the "two-tone" method of construction should be identified during the field review. When possible and applicable, the field review should suggest a modification to minimize the side slope areas.
- Stream and river crossings: During the desktop study, the alignment of the proposed pipeline where it crosses a stream or river is not always clear. The field review should always include, when applicable, a thorough assessment of the crossings to determine the size of the banks, crossing angle, and a precursory determination of method of crossing. Many times, the route can be shifted slightly to either side of the proposed route to allow for a more suitable crossing.

At the conclusion of the field review, the project team should get together to review the results of the field review, the initial agency meetings, and the other information that has been gathered regarding the route options. The goal of this meeting should be that the team agrees on the primary route and perhaps one option that will be used to further develop the project.

2.11 PARALLEL OTHER LINES

The Federal Energy Regulatory Commission (FERC) along with other federal, state, and local agencies many times encourage and sometimes mandate that a pipeline route follows or parallels an existing pipeline corridor. This method of construction does have advantages where the newly proposed pipeline can reduce many construction costs when it comes to the clearing of trees. When applicable, the proposed route centerline should be situated outside of the existing pipeline right-of-way limits and along the old working side of the existing pipeline. When this can be accomplished, the amount of trees to be cleared for the construction of the new pipeline can be minimized.

Paralleling other existing pipelines has some advantages because it minimizes clearing, provides a general understanding of preexisting environmental conditions, and identifies routing issues and urban or suburban routing parameters. There are many disadvantages as well.

Many times paralleling an existing right-of-way is not advisable due to the increase of construction costs along the existing route. In these instances, it is necessary to deviate away from the existing route to minimize impact to dwellings and commercial developments.

2.12 INTEGRITY

Certain cathodic protection measures need to be addressed when paralleling existing pipelines. In all new pipelines where cathodic protection is applied while the existing parallel pipeline is not protected, current can be inducted into the existing line in some places from the protected line while discharging that current into the earth to the newly protected line in other places. This can cause "hot spots" that may result in corrosion and the possibility of a leak.

Many times existing pipelines can be damaged during the construction of the new pipeline. Instances where the new route must be moved to the opposite side of the existing pipeline increase any likelihood of damage to the existing line when excavating for the new line. In circumstances where the new route requires the use of blasting for the excavation of the ditch, the development of explicit blasting plans will ensure the integrity of the existing pipeline. Road bore pits and bell holes that encroach much closer than that of the standard ditch pose potential impacts to the existing pipeline as well.

2.13 ESTABLISHED CORRIDORS

As with paralleling existing pipeline rights-of-way, following established corridors also allows for ease of routing, provides for easier identification of environmental issues and conditions from the established corridor, and allows for fewer burdens for construction in urban and suburban conditions. Although the possibility of damage to the structure(s) within the established corridor is diminished in relationship to a new pipeline corridor, there are still many factors that need to be considered and addressed during construction of the new pipeline.

In the case of following a corridor with electric overhead transmission lines, the pipe is subject to induced voltages and currents from the overhead electric lines. The induced voltages are hazardous to construction personnel working around the pipeline. During construction, equipment should be equipped with grounding cables. Pipeline sections that are welded and ready for lowering into the ditch should be grounded at each end with grounding rods driven into the ground at a sufficient depth and connected to the pipe with copper cable. Parked vehicles that are left for any appreciable time can collect static electricity; therefore, measures should be taken to park vehicles more than 200 ft perpendicular from the base of the overhead electric towers. When operating cranes, side booms, backhoes, etc. during pipeline construction, extreme care must be observed when such equipment is under the power lines. Certain guidelines for these scenarios should be established prior to construction.

Established corridors may include pipelines, electric transmission lines, or roads. Permitting agencies prefer the use of established corridors because the

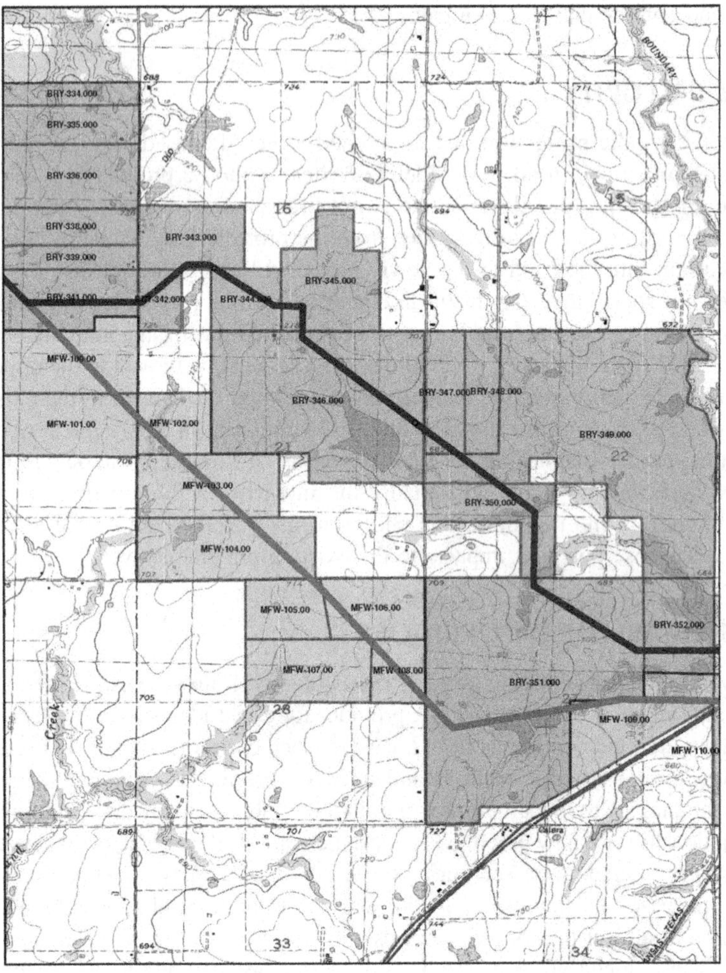

FIGURE 2.2 Route map.

construction of the new line will have less impact on the environment, landowners, and other infrastructure if it is routed in an established corridor.

Obviously, this is not always possible, but an investigation should be made into the location of any existing routes and corridors in the area of the proposed new pipeline. An attempt should be made to parallel existing corridors in areas where it is practical. This may actually add some length to a proposed route, but it may mean the difference in whether the line will be permitted.

These routes can be overlaid on the map, so they can be referenced as the new route is being selected by both the desktop review and the field review.

Examples of route options and their details are shown in Figs 2.2, 2.3, and 2.4.

FIGURE 2.3 Route options.

FIGURE 2.4 Route details.

BIBLIOGRAPHY

ASME B31.4-2006, Pipeline Transportation Systems for Liquid Hydrocarbons and Other Liquids, American Society of Mechanical Engineers, New York, NY, 2006.

ASME B31.8-2007, Gas Transportation and Distribution Piping Systems, American Society of Mechanical Engineers, New York, NY, 2007.

CFR 49, Part 195, Transportation of Hazardous Liquids by Pipeline: Minimum Federal Safety Standards, U.S. Government Printing Office, Washington, DC.

CFR 49, Part 192, Transportation of Natural or Other Gas by Pipeline: Minimum Federal Safety Standards, U.S. Government Printing Office, Washington, DC.

Chapter 3

Pipeline Regulatory and Environmental Permits

William E. Bauer

Chapter Outline

INTRODUCTION

New construction, transportation rates, operation, and environmental impacts of pipelines are highly regulated. Statutory authorities govern all aspects of pipeline systems and their related facilities. When first constructed, there are numerous permits and approvals that must be obtained from state, federal, and local agencies.

Companies constructing and operating such pipelines generally have personnel and/or consultants who specialize in these permitting processes and support the project manager on such new pipeline projects. While the ultimate responsibility for obtaining such permits often rests with the project manager, the company specialists and/or consultants perform the actual permitting tasks.

A complete examination of the permits required for any particular project and the processes of obtaining them is beyond the scope of this manual; however, we have provided, below, a brief discussion of the major permits that may be required for pipelines carrying various commodities. The permits are broken into federal, state, and local jurisdictions.

It should be noted that those permits, generally acquired from local agencies, are most often the direct responsibility of the project manager, through his or her right-of-way group. Only a thorough field examination can identify all of the local permits required for a proposed pipeline.

3.1 REGULATION OF INTERSTATE PIPELINES

Pipeline regulations are generally divided into two categories, rates and safety. The Federal Energy Regulatory Commission (FERC) regulates transportation rates for natural gas and oil pipelines, whereas the Department of Transportation's Pipeline and Hazardous Materials Safety Administration (DOT/PHMSA), through the Office of Pipeline Safety, regulates the safety of natural gas, oil, and hazardous materials pipelines. FERC's authority over oil pipelines is almost specifically limited to rates; however, FERC exerts greater authority over natural gas pipelines. The Natural Gas Act of 1938 granted FERC's predecessor agency, the Federal Power Commission, authority over certificates for construction, operation, and ancillary facilities for natural gas pipelines.

3.1.1 FERC-Regulated Natural Gas Pipelines

FERC regulations began with the passing of The Natural Gas Act of 1938. This act gave the Federal Power Commission (FPC), forerunner of FERC, jurisdiction over the regulation of interstate natural gas sales. Therefore, to build a pipeline, companies were required to first receive approval from the FPC.

In 1954, the Supreme Court in its *Phillips Petroleum Co. v. Wisconsin* decision ruled that natural gas producers who sold gas into the interstate market fell under the classification of "natural gas companies" and were subject to regulatory oversight by the FPC. This gave FERC authority over the price of all gas flowing into the interstate market.

The Natural Gas Policy Act of 1978, recognizing that price controls at the wellhead had eventually hurt consumers, allowed market forces to establish the wellhead price of natural gas and attempted to equalize the supply with demand. This was the beginning of decontrol.

FERC Order No. 436 changed how interstate pipelines were regulated. Generally known as the Open Access Order, Order No. 436 allowed pipeline customers the choice of purchasing their natural gas and their transportation separately.

FERC Order No. 636, known as the Final Restructuring Rule, states that pipelines must separate their transportation and sales services so that all pipeline customers have a choice in selecting their gas sales and transportation independently.

As a result of The Natural Gas Act of 1938 and subsequent rulings and orders, FERC has been given the authority to grant certificates for the construction and operation of natural gas pipelines. In so doing, a "Certificate of Public Convenience and Necessity" pursuant to Section 7 of the Natural Gas Act is issued. Prior to the certificate, however, a proposed project must undergo an extensive pre-filing and filing process that includes the approval of the route, review of new lines, environmental assessments, and coordination with various other federal and state agencies.

3.1.2 FERC-Regulated Oil Pipelines

FERC-regulated oil pipelines have a much simpler process in obtaining a construction and operation permit. FERC regulates oil transportation rates but does not regulate the oversight of oil pipeline construction, abandonment of service, or safety.

The Hepburn Act of 1906 classified interstate oil pipelines as common carriers. The Interstate Commerce Commission (ICC) was responsible for regulating such pipelines; however, in 1977, the Department of Energy (DOE) Organization Act of 1977 transferred the regulation of oil pipelines from the ICC to DOE and subsequently to FERC.

The Energy Policy Act of 1992 mandated that FERC provide a "simplified and generally applicable" ratemaking methodology for oil pipelines and authorized the agency to streamline oil pipeline proceedings. The safety of interstate oil pipelines remains with the Department of Transportation.

3.1.3 Safety Regulations of Oil, Gas, and Hazardous Materials Pipelines

The Department of Transportation is the major regulator of the construction and operation of both oil and natural gas pipelines primarily as a result of two statutes: the Hazardous Liquid Pipeline Safety Act of 1979 and the Natural Gas Pipeline Safety Act of 1978. The Pipeline and Hazardous Materials Safety Administration (PHMSA), through the Office of Pipeline Safety (OPS), is responsible for establishing and enforcing proper design, construction, and maintenance of both oil and natural gas pipelines. The operating regulations for hazardous liquid pipelines are set forth at 49 CFR Part 195.

3.2 REGULATION OF INTRASTATE PIPELINES

Pipelines located totally within a state are regulated by the state. With some exceptions, states have adopted similar ratemaking procedures and safety standards as those of the federal government. Energy producing states have generally been more active in regulating pipelines than states that produce little or no energy. The agencies governing pipeline transportation vary from state to state and should be investigated to determine the appropriate jurisdictional authority. Making this determination is most commonly the responsibility of the Regulatory Department within a pipeline company. However, agencies typically charged with oversight duties of pipelines are commonly know as the Public Utilities Commission (PUC). In Texas, the agency with oversight of pipelines is known as the Railroad Commission of Texas.

3.3 ENVIRONMENTAL PERMITS FOR INTERSTATE PIPELINES

It is important that the project manager of a proposed project consult with his or her company environmental specialist or external environmental consultant experienced in linear facilities to identify and process the necessary environmental permits, clearances, or approvals. These specialists/consultants can provide accurate estimates as to the costs and time required to process required permits. Permits required for most interstate pipelines typically include but are not limited to compliance with the provisions of the following federal and state acts:

- National Environmental Policy Act of 1969: NEPA requires the lead federal agency exercising jurisdiction over the project to consider the environmental impacts of the proposed project. The agency may prepare or issue an Environmental Impact Statement (EIS), an Environmental Assessment (EA), or a "finding of no significant impact" depending upon the projects' impact on the environment.
- Federal Water Pollution Control Act ("Clear Water Act" or CWA): Section 401 of the CWA requires that the pipeline company obtain a certification from any state in which any discharge into navigable waters of the United States is made. In the event the state fails to act within 1 year, the requirement is considered waived.
- Section 404 of the CWA authorizes the United States Army Corps of Engineers (USACE) to issue permits for the discharge of dredged or fill material in waters of the United States including wetland area, streams, rivers, lakes, coastal waters, or other water bodies or aquatic areas.
- Additional permits such as a National Pollutant Discharge Elimination System (NPDES) Permit may be required for discharge of test water during construction.
- Coastal Zone Management Act (CZMA): The CZMA manages the nation's coastal resources, including the Great Lakes, and is administered by the National Oceanic and Atmospheric Administration (NOAA) under the Department of Commerce. An applicant must certify that the action is in compliance with the enforceable policies of the state's federally approved coastal zone management program.

- Endangered Species Act (ESA): Section 7 of the ESA requires federal agencies to ensure that the proposed project does not harm threatened or endangered species or critical habitats of such species. Section 9 of the ESA makes it unlawful to harm such endangered species during construction or operation of the project.
- Clean Air Act (CAA): The CAA exercises jurisdiction over the construction and operation of pipelines. While pipelines in operation do not generally cause air pollution, construction of the pipeline and operation of compressor or pump stations generally affect air quality. Specific requirements under CAA programs include the New Source Review (NSR) and Prevention of Significant Deterioration (PSD) program and the permitting program for major stationary sources under Title V of the CAA.
- National Historic Preservation Act (NHPA): The NHPA requires federal agencies to consider the effects of a construction project upon historic artifacts and structures. This is accomplished by federal agencies consulting with state historic preservation officers (SHPOs). The federal Advisory Council on Historic Preservation (ACHP) may participate in this process.
- The Pipeline Safety Improvement Act (PSIA): The PSIA applies to natural gas pipelines and requires each company to prepare and implement an "integrity management program." This program addresses primarily "high consequence areas" (HCA) and requires a baseline integrity assessment. The program generally applies to pipelines in place.
- The Pipeline Inspection, Protection, Enforcement, and Safety Act of 2006: This act further addresses the "integrity management program" and focuses on better use of the state "one-call" systems. Like the PSIA above, this program generally applies to pipelines in place.

3.4 ENVIRONMENTAL PERMITS FOR INTRASTATE PIPELINES

States vary in their implementation of environmental regulations for pipeline construction and operation. Some states have less stringent regulations than the federal government while other states such as California have more stringent regulations. Likewise, the agencies administering the environmental regulations vary from state to state. In addition, many states bordering oceans or the Great Lakes have separate agencies controlling areas along the water. In many instances, these state agencies share jurisdictional authority with federal agencies. Often, one of these agencies takes the lead during the permit process and will condition their issuance of a permit upon when the other agency has issued one. As with the federal permitting process, a qualified environmental permitting expert is necessary to navigate and shepherd this process to acquire all required state permits in a reasonable time period. Typical state agencies requiring permits or approvals for pipelines are State Historic Preservation Offices (SHPO), Department of Environmental Protection (DEP), or Department of Environmental Quality (DEQ).

3.5 LOCAL PERMITS

Local permits are generally applied for and acquired during the initial phase of a construction program. The project manager through his or her right-of-way group is usually responsible for obtaining these permits. Local permits include but are not limited to the following:

- Federal highways: Local offices of the state departments of transportation (state DOTs) generally issue crossing permits for federal highways, with some exceptions. Certain highways such as the Blue Ridge Parkway are administered and maintained directly by the federal government and as such require a permit from the appropriate regulating agency. Once a road crossing application is reviewed and all conditions are met, a crossing permit similar to the one typically granted by the state is then issued.
- State highways: As with federal highways, local offices of state DOTs generally issue permits for state highway crossings. Recently, pipeline companies have generally been trying to eliminate or minimize the utilization of casings when crossing roads. The trend has been for state DOTs to consider relaxing the pipeline casing requirements by allowing the pipeline company to upgrade the wall thickness and grade of the pipeline and/or requiring alternate methods for crossing such as utilizing a horizontal directional drill (HDD) versus a conventional bore. It is noted that some of the alternate methods required by the state may be more problematical than casing the pipeline. Once a road crossing application is reviewed and all conditions are met, a road crossing permit is then issued by the state DOT.
- County roads: Counties (parishes in Louisiana) throughout the United States generally issue permits for crossing their roads. Requirements typically range from boring to open cutting of the roads depending upon the county and type of road. County road departments or county engineers normally issue these types of permits. Once all requirements are met and any administrative procedure is complete, the appropriate authority within the county will then issue a permit.
- City streets: City road crossing permits are similar to county permits in that requirements can range from boring to open cutting depending upon the type of road and city. Often but not always, the agency issuing the street permit will issue other crossing permits such as water and sewer lines. Once all requirements are met and any administrative procedure is complete, a street encroachment or crossing permit is generally issued.
- Utility districts: Numerous utilities including electricity, canal, irrigation, levee, drainage, and other facilities may exist along a pipeline route. Each of these entities will generally require a crossing agreement or permit and must be contacted early in the project to determine crossing requirements.
- Railroads: Railroads typically require a permit or license agreement before crossing their tracks and right-of-way with a pipeline. Many pipeline companies with the right of condemnation now seek an easement agreement for

the crossing in lieu of a permit or license. It is noted that some railroads have recently cancelled permits and licenses citing the age of the permit, sale of the rail asset, or simply wanting to "rework the deal" as their reason to cancel. A permit or license agreement is simply an agreement that gives permission to a company to use property for a particular purpose and does not convey a recordable right in the property. Consequently, it typically contains language that allows the party granting the permit or license to cancel it for any reason, while an easement is a recordable right that typically gives the pipeline company a perpetual right, thereby preventing a gap in pipeline surface rights in the event the railroad company wants to renegotiate the permit or license or the tracks are abandoned.

- Use permits: Many city and counties, especially in the more urban areas, have planning departments that control most types of construction. New construction projects must undergo a filing and hearing process before a planning commission or zoning board. However, most planning commissions and zoning boards have provisions for special uses that preclude the need for a public hearing. Pipelines are many times included in these special uses. If a pipeline does not qualify for any exceptions, a Conditional Use Permit or Special Use Permit is generally required. In addition to these permits, some planning departments issue temporary use permits before construction.
- Other easement crossings: A pipeline of any distance will cross fiber optic cables, other pipelines, major power lines, and other linear facilities. These entities are often referred to as "foreign utilities" and may require a crossing agreement. The easements for these foreign utilities detail the rights they possess across a parcel of land and the extent to which they can demand certain requirements of a pipeline wishing to cross or colocate in their easement. These easements should be researched as a part of the title work phase of the project to determine the extent of their rights.
- Other considerations: Numerous governmental and quasi-governmental agencies have districts in states and counties throughout the United States that may or may not issue permits for pipeline construction. Some states have created clearinghouses for pipeline permits, often run by the commerce agency or department of that state. In practicality, these agencies have proven to be inefficient and, in some cases, incorrect in assisting to identify necessary permits. It is advised that these agencies be used with caution.

3.5.1 Identifying Permits and Determining Requirements along a Proposed Linear Facility

Omitting a required permit along a pipeline route will generally result in substantial construction delays. It is the responsibility of the project manager, through the right-of-way group to investigate, determine requirements, and apply for and obtain all local permits. The types of agencies issuing permits vary from one part of the United States to the other. For example, many farming areas have irrigation districts, others areas have flood control or levee districts, and yet other parts of the country have drainage districts. Below is a simple process for identifying permits and determining their requirements along a pipeline route.

- Roadways: Identify all road crossings and determine their jurisdictional control. Of importance to the engineering and construction groups is the required depth and casing requirements. Roads are typically crossed via the use of a conventional bore; however, if the road is to be open cut, the burial depth along with compaction requirements will also need to be considered. Additionally, the bond requirements for a bored versus open cut crossing may be different and will need to be identified and reported. A sample crossing drawing or plat obtained during the first meeting with the permit agency is important in identifying drawing or plat requirements for the group responsible for drafting permit drawings or plats.
- Railroads: Identify the owners of all railroads crossed and determine the crossing requirements. As previously stated, most railroad companies issue licenses that often contain stringent construction and operational restrictions. The method most commonly utilized to cross railroads is a conventional bore. Requirements for crossing railroads can typically be found in an application for crossing or sample permit drawing. Copies of these documents should be obtained from the railroad and provided to the pipeline company engineering, construction, and survey groups to determine impact to the project and drawing requirements for permit application. Once all requirements are met, the permit application is completed and submitted to the railroad for review.
- Canals: Identify the owners of all of the canals along the proposed right-of-way. Canal companies are sensitive to digging under their canals because of breaks and consequently often have strict crossing regulations. In some instances, canal companies may demand overhead crossings as opposed to boring. Obtaining requirements and sample drawings are important to the engineering and drafting groups to assure that the pipeline is properly designed and that crossing drawings contain the information required to expeditiously obtain the required permit.
- Ditches: Nearly every ditch along a pipeline route is controlled by a district or agency. Especially in areas of heavy precipitation and/or irrigation, numerous ditches may be crossed. Drainage districts and agencies often oppose an open cut crossing to minimize any detrimental impacts to the structural integrity of the drainage ditch. This may result in either a conventional bore or HDD being utilized to cross the ditch. Once again, crossing requirements and sample drawings are helpful to the engineering and drafting groups to prepare supporting documents for the permit application.

- Overhead power lines: Electric power companies are extremely sensitive to work within their easements and working near their power lines. These companies often have restrictions relating to tractor booms and heavy equipment working with their right-of-ways. It is important to determine the type of agreement necessary for the crossing and obtain any crossing requirements and restrictions that could impact the construction and operation of the pipeline.
- Underground pipelines: Generally speaking, any pipeline crossing another pipeline will be required to locate the proposed pipeline under the existing pipeline a prescribed distance. This may vary from company to company. Crossing requirements are important and must be communicated to the engineering group so that design information can be included on the construction drawings.
- Underground cables: Telecommunication companies and other underground cable entities normally have crossing requirements. Any new facility is generally required to lay under the existing facility. All crossing requirements must be identified and communicated to the engineering, construction, and survey groups so that engineering design and requirements for construction and/or permit drawings can be captured.

Chapter 4

Right-of-Way

William E. Bauer

Chapter Outline

INTRODUCTION

It is the responsibility of a right-of-way group to provide the project a continuous constructible strip of land for the construction of a pipeline and all related surface facilities. This includes, but is not limited to, a continuous pipeline right-of-way; all additional work spaces; and all surface sites for compressor stations, pump stations, meters, valves, storage sites, and other necessary sites.

4.1 RIGHT-OF-WAY DELIVERABLES AND REQUIREMENTS

The right-of-way group is responsible for providing certain deliverables and information to the proposed project. In turn, other groups or departments within the project must provide certain deliverables and information to the right-of-way group.

4.1.1 Right-of-Way Deliverables

A list of items that a right-of-way group must deliver to the project is as follows:

- Route selection: There are numerous land obstacles and structures that can substantially affect the project budget from a right-of-way perspective. An example of these obstructions could be a permitted subdivision approved by a planning commission but with no construction activities. It may appear to be as good a route as any for a pipeline route; however, negotiating for each individual lot would have a substantial impact on the project budget. It is the responsibility of the right-of-way group to identify unforeseen route "snags" and consult with the route selection party.
- Field survey support: It is the responsibility of the right-of-way group to contact all property owners and agencies along a proposed route and obtain access for survey studies.
- Field environmental studies: Various environmental consultants to the project require access to properties for specific studies. It is the responsibility of the right-of-way group to obtain this access from owners and/or agencies.
- Permit requirements: Generally, all agencies having facilities along a proposed pipeline will have special requirements that affect design of the pipeline and the budget. The right-of-way group must contact these agencies and provide the engineering and construction groups with all special construction and operational requirements.
- Right-of-way: As mentioned in the introduction, it is the responsibility of the right-of-way group to provide the construction operation a continuous and constructible strip of land along with sites for all surface facilities such as valves, compressors, stations, and other facilities. The right-of-way group must also provide access to the right-of-way including remote areas.
- Condemnation support (if applicable): Provided condemnation is available to the proposed project, it is the responsibility of the right-of-way group to support the condemnation attorneys and provide them with expert witnesses, appraisals, documentation, and any other services helpful to the attorneys in reaching a successful conclusion.
- Construction restrictions: Most agencies and many private owners impose restrictions upon the construction operation. These may include things such as the manner in which a fence is to be replaced, the treatment of the soil during construction, and replanting of grass after construction. The right-of-way group will provide this information to the construction group in the form of a construction restriction or line list.
- Construction support: Just prior to construction, it is the responsibility of the right-of-way group to notify landowners and agencies of the impending access by the construction contractor's personnel; once begun, the group has to provide a right-of-way agent to observe the construction operation.

- Liaison support: The right-of-way group or a member thereof acts as liaison, the only contact between other project personnel and the property owners and agencies. This is necessary to have a single voice to nonproject personnel and entities.
- Operational restrictions: At the completion of the project, it is the responsibility of the right-of-way group to provide to the pipeline company all the requirements and restrictions affecting the operation of the pipeline.
- Documentation: At the end of the project, the right-of-way group provides all recorded easements, permits, databases, and other documentation to the pipeline company.

4.1.2 Right-of-Way Requirements

The right-of-way group cannot perform its duties without the assistance of other groups and departments within the project. Below is a list of the items and support that other groups must provide to the right-of-way group:

- A detailed pipeline route: The engineering and survey groups must provide the right-of-way group a specific and detailed route with which to begin title research and gain access to properties.
- Plats, maps, and drawings: Agencies and many private owners will require crossing and property plats prior to making applications or beginning negotiations. The drafting group must provide the required plats and drawings on a timely basis.
- Chain of command: Many owners ask engineering or environmental questions that cannot be answered by the right-of-way agent. Also, many owners request specific provisions or demand monies that are beyond the limits set by the company. Right-of-way agents must have quick and concise answers to provide to these owners. A chain of command that can provide definitive responses in a timely manner must be established.
- Approved forms: The pipeline company must provide all easements and right-of-way support forms or approval forms submitted by the right-of-way group at or prior to the beginning of the project.
- System for making payments: The pipeline company must provide right-of-way funds and a system for making payments to property owners and agencies.
- Legal assistance: Many agencies and owners submit agreements or changes to easements that are of a legal nature. The right-of-way group must have access to legal assistants that are qualified and authorized to make these decisions.

4.2 PROJECT PLANNING

Early project planning is the key to a successful right-of-way acquisition effort. Developing methods and systems during the acquisition of right-of-way will result in delays and additional costs. Establishing negotiating guidelines and methods of payment as soon as possible is imperative. The following items must be addressed at the onset of a pipeline project:

- Policy and procedures manual: A detailed policy and procedures manual that sets out all steps in the right-of-way effort should be prepared specific to the project. There are some schools of thought in the right-of-way field that oppose a manual of this type. These individuals feel that a policy or procedure not followed may lead to a lawsuit. The writer strongly believes that a manual should be prepared and followed accordingly.
- Chains of command and contacts: As stated in Section 4.1.1, a definitive project chain of title should be developed to respond to right-of-way agents' questions and requests in a timely manner.
- Right-of-way forms: As stated in Section 4.1.1, right-of-way forms should be developed early in the project. Easements and other property owner forms should be fair and balanced.
- Draft system or other method of payment: A right-of-way draft is a form that looks like a check but cannot be cashed. It must be sent through a bank to the pipeline company's bank for collection. There are many companies who believe that this system is too risky and issue checks instead, after an easement has been executed by a property owner and delivered to the company; however, the opposite is true. Both systems have their problems; however, the draft system is more expedient because a right-of-way agent has authority to issue a draft immediately upon execution of an easement. Otherwise, the agent would be required to request a check from the field office once negotiations are concluded. This gives landowners an opportunity to change their minds. With the draft system, the executed easement can be checked for correctness, and then the bank will be notified to pay the draft.
- Negotiating policies: Different companies have different philosophies regarding negotiations. Many companies have hard and fast rules where an offer to a property owner is made three times, and the file is turned over to the attorneys for condemnation when that option is available. Other companies tend to negotiate and increase their offers to avoid condemnation at all costs. A middle ground approach seems to be the most effective method. This is true of negotiations even when condemnation is not allowed.

4.3 RIGHT-OF-WAY BUDGETING

The cost of right-of-way is often between 10% and 25% of the total cost of a project depending upon the location of the pipeline. Right-of-way costs are normally higher in urban areas than in rural areas. Budgeting for right-of-way costs is extremely difficult. One condemnation award (if condemnation is available to the project) made by the court can deal a fatal blow to a budget if the owner's attorney prevails. Also, if condemnation is not available, one holdout landowner can also break a budget. These issues can be mitigated somewhat by researching prior condemnations and negotiations in the area and budgeting accordingly. (See Appendix 3 for a sample right-of-way budget.) Below is a list of major items of a right-of-way budget:

- Management: Management costs include the costs of anyone in the pipeline company who is not a part of the project but who, from time to time, will charge his or her personal time to the right-of-way budget. This may include the company's right-of-way manager and his or her staff.
- Personnel: All persons in the project right-of-way group including the right-of-way manager, agents, administrative assistants, and any other person on the team make up the pipeline project personnel.
- Land titles: Some companies prefer to have land title companies prepare the title ownership studies and issue a "Limited Title Certificate." These title companies generally charge a fee per tract. If this is the case, the budget should reflect these charges. Otherwise, the cost of the right-of-way personnel should be budgeted for such studies including the costs they generate through the use of abstract plants and making copies from the public records.
- Personnel expenses: Personnel expenses include all travel and related expenses by any of the right-of-way personnel and include motel, meals, and automobile expenses.
- Consultants: The right-of-way group utilizes appraisers, agriculture consultants, and other specialists to determine fair offers. These experts are, also, utilized in condemnations. These costs should be reflected on the budget.
- Right-of-way: Cost of right-of-way includes the cost for easements and agreements, construction damages (excluding right-of-way damages caused by the contractor), condemnation awards, permit fees, and surface site costs.
- Condemnation: Attorney fees for consulting and condemnation are the responsibility of the right-of-way group and as such, should be addressed in the budget.
- Office and miscellaneous: Office and miscellaneous costs include office space allocated to the right-of-way group, copiers, faxes, papers, utilities, printing, and any other costs set out in the categories above.

4.4 RIGHT-OF-WAY DATABASE AND RECORDS

4.4.1 Right-of-Way Database

A database is essential for all phases of right-of-way acquisition. There are numerous off-the-shelf right-of-way databases that provide excellent information to the management and right-of-way personnel. Many companies and consulting engineering companies have pipeline databases; however, many of these do not provide the type of data required by the right-of-way group during the acquisition program.

A recent development in right-of-way databases provides for access by other groups through the use of a website with some given access with a password for read only and others being able to add, change, or delete the data. The ability to view the database is useful, for example, during the environmental field studies. The environmental contractors can access certain parts of the database to determine which tract owners have authorized access without having to contact agents or staff. At a minimum, a right-of-way database should provide the following information:

- Tract numbers: Each specific tract of land should be given a tract number. Numbers generally run with the flow of the product. The tract numbering method is usually determined by the pipeline company to stay consistent with their other pipeline systems but is accomplished by the project right-of-way group in conjunction with the drafting group.
- Names/addresses/telephone numbers of agencies and owners: Each tract number should list the name, address, and telephone number of the owner(s) or agency.
- Names/addresses/telephone numbers of tenants: Names, addresses, and telephone numbers of all tenants should be provided.
- Parcel tax I.D. number: This is the tax number that can be obtained from the County Tax Assessor's office. It is used to get the tax certificate showing taxes paid current or delinquent.
- Property descriptions: A full or abbreviated property description should be provided depending upon the type of description and functionality of the database.
- Access status: This column will reflect whether permission has been obtained from the property owner or agency to access the proposed pipeline right-of-way. Some property owners will allow a civil survey but will not allow environmental access. This can be noted on the access status.
- Negotiating status: During negotiations, the management must know the status of each tract to predict problems or delays. Each agent must provide a complete state of each tract on a daily basis for inclusion into the database; however, giving agents field access to the database is discouraged. It is too easy for a property owner to gain this access, and negotiations and payments are private information. Each agent should prepare a daily status report and should send it through e-mail to the documents agent or title clerk for entry into the database.
- Acquisition data: Following successful negotiations, all acquisition data should be provided to the field office by means of a status report e-mailed to the field office personnel for inclusion in the database. This information should include total payment, cost per foot or rod or other unit, width of right-of-way, length of right-of-way, all additional work spaces, and other relevant information. Of special importance would be any clauses or provisions negotiated, which are not a part of the easement form approved by the company attorney.

- Construction restrictions: It is important that all agreements, provisions, and requirements made over and above the standard easement affecting construction be entered into the construction-restriction list. This provides the contractor with instructions he will need during construction.
- Operational restrictions: This is described in Section 4.8.2.
- Comments sections: Several comments sections should be provided for right-of-way agents, supervisors, or other parties to make comments or explain actions or agreements.

4.4.2 Land Title Research

If the option of having a "Limited Title Certificate" issued by a title company is not utilized, then land right-of-way agents perform the work. Land title research is a procedure by which the accurate names of the property owners and their lien holders are identified. This is necessary to assure that negotiations are conducted with the correct owners or agencies for each tract. There are two schools of thought on research. It is the policy of many companies to conduct a complete search and secure an easement from the owner and an approval of the easement from any of the owner's lien holders. In other words, if the lien holder is agreeable to subordinating its lien then it will have no recourse in the event of a loan default. Other companies are more lax on title research with the thought that any title problems can be resolved after construction. The complete title research method is advised. Appendix 3 provides a sample title research checklist for use by right-of-way agents when performing land title research. Appendix 3 also provides a title report setting out current owners and all lien holders. At a minimum, land title research for pipeline right-of-way should include the following:

- Current property owner or owners: All property owners, their interest, and their type of ownership (i.e., tenants in common, joint tenants, etc.) should be provided.
- Address of owner or owners: Address of all owners should be provided, if available.
- Current taxpayer: A taxpayer is not necessarily the owner of record. For example, a brother may be assessed and paying taxes for his five other brothers. Only the assessed brother would show up on the tax rolls.
- Status of property taxes: Ad valorem property taxes must be paid current on a tract of land, or the easement taken by the pipeline company would be subordinate to the tax lien. A tax certificate should be obtained for each tract showing taxes current or period of time and monetary amount in arrears.
- All liens and judgments against the owner or property: A mortgage or deed of trust securing a note is superior to an easement unless the mortgage or deed of trust is subordinated to the easement as mentioned earlier. Mortgage companies are generally agreeable to subordinating an easement, provided

the owner is current on his or her payments. If the borrower is not current, the mortgage company will usually subordinate their lien if the acquisition money is paid toward the lien.

- Copy of vesting document: The most current document providing ownership to the current owner is considered the vesting document. A copy of this document should be provided to the documents agent or title clerk in the right-of-way group.
- Chain of title: A chain of title is a chronological list of all the documents affecting the property for a specific period of time. Many states have 20–30 year statutes of limitation regarding land matters; therefore, a 25–30 year chain of title will generally identify all problems that could jeopardize a pipeline easement.

4.4.3 Right-of-Way Documents

Proper right-of-way documentation insures against costly claims and lawsuits during and after the completion of the project. Care should be taken to develop standard easements, damage releases, and other right-of-way documents that protect the company but are not so onerous that property owners will not accept them. Below is a list of primary documents that are typically used in a right-of-way acquisition program.

- Survey permit: Many companies require that property owners execute survey permits allowing for entry into their property. When time is critical, the companies will allow a verbal contact through phone. A survey permit is usually a one-page document signed by the property owner giving the survey group permission to enter the property. (See Appendix 3 for a sample survey permit.)
- Easement or grant: A right-of-way easement or grant is the most critical document of an acquisition program. If prepared without certain provisions, it will not protect the company. If the provisions are too stringent, property owners will not accept the company easement and will be more likely to have their own attorney prepare an easement. A company easement must be fair and balanced to provide the best document to present to a property owner while still protecting the pipeline company. (See Appendix 3 for a sample right-of-way easement.)
- Surface easements or leases: Various surface documents should be prepared for each pipeline project (i.e., meter and valve easements, compressor or pump station easements and/or deeds, storage site leases, construction yard leases, cathodic protection, and anode bed easements if located off the permanent right-of-way, etc.). (See Appendx 3 for sample surface easements. Also, see Appendix 3 for a sample deed and for a sample storage site lease.)
- Tenants' documents: Tenants' consent forms and other tenant-related documents should be prepared.

- Right-of-way draft: Provided the decision is made to utilize right-of-way drafts, they, along with draft report documents and other draft-related documents, should be printed in advance.
- Damage release: There are a number of damage releases that are useful in a right-of-way program, including advance damage releases, construction damage releases, tenant's damage releases, and others. (See Appendix 3 for sample damage releases.)

4.5 FIELD SUPPORT

As first recognized when discussing liaison support in Section 4.1.1, it is imperative that a single group or department interfaces between project personnel and property owners and agency representatives. Various project personnel talking to a property owner or agency can, and almost always will, result in claims of promises made by one party or another. The right-of-way group must be the only member of the project that makes contact with property owners and agency personnel. If the need arises for another project personnel member to speak to a property owner, a member of the right-of-way group must accompany that person and document the meeting.

In addition to conducting regular negotiations and being the only contact between the project personnel and owners and agencies, the right-of-way group supports the project by securing access permissions and information from owners and agencies for project studies. The owner and agency contacts made in regular support of the project are as follows:

- Field survey: It is the responsibility of the right-of-way group to contact and obtain access for the survey group to conduct the civil survey. More recently, with improved technology, surveyors need very little actual property access; however, the right-of-way group must stand ready to provide any access that could be required.
- Environmental and cultural resource studies: Studies performed by these two groups do require actual access to properties. The right-of-way group is responsible for contacting owners and agencies to gain this access.
- Engineering design: Agencies having facilities along the pipeline right-of-way, such as canals and ditches, often impose construction requirements that can affect the design and cost of a project. It is the responsibility of the right-of-way group to investigate and report to the engineering group all special requirements demanded by these agencies.
- Construction: Construction contacts are discussed in Section 4.7.

4.6 RIGHT-OF-WAY NEGOTIATIONS AND CONDEMNATION

4.6.1 Negotiations

Right-of-way negotiations are the key to a successful acquisition program. Agents must be qualified and provided with the proper tools. Right-of-way agents who are thoroughly trained in company policies and procedures will provide the project with a cost-effective acquisition and good property owner relations. A good acquisition program consists of the following:

- Agent training: Right-of-way agents cannot negotiate successfully unless they understand the project, its goals, and the philosophy of the pipeline company. Also, they may be the only member of the project that the property owner ever sees, and they need to represent the company well. It is imperative that training and follow-up training sessions be held regularly to discuss the project and answer agent's questions.
- Good tools: Right-of-way agents require good tools to negotiate successfully with property owners. This includes items such as accurate drawings, fair and equitable property appraisals, guidelines for company offers, and related information. Of great importance is the understanding of the property along the pipeline right-of-way and its uses and treatments.
- Timely support: If a right-of-way agent is in a property owner's home at 11:00 P.M. and needs clarification on a certain point or construction procedures, he or she must be able to contact someone so that the negotiations can be concluded at that time. The right-of-way group must provide names of managers and supervisors whom the agent can contact at all times. (See the chain of command discussion in Section 4.1.2.)
- Latitude: A strict negotiating policy will not produce the desired acquisition results. Right-of-way agents must be given a certain amount of latitude with which to negotiate. Such latitude should include a range of right-of-way payments and certain terms. Damage payments, however, should not be negotiable.

4.6.2 Condemnation Through the Power of Eminent Domain (Provided the Project Qualifies)

Condemnation is the right given to a company by a regulatory agency to take property in the event negotiations are not successfully concluded. Condemnation is a last resort in the negotiation process and often times detrimental to owner relations in that area. Condemnation should be avoided whenever possible; however, it is made available to projects that are intended to serve the general public.

At the point when condemnation appears unavoidable, the owner's file is turned over to the legal group, and the right-of-way group provides support to the condemnation process. Support is provided in the following manner:

- Appraisals and expert witnesses: While the condemnation process is placed in the hands of the company attorneys, it is the responsibility of the right-of-way group to provide appraisal and expert witness support to the process. These witnesses have credentials that prove that they have expertise in their field. They may include experts in agriculture, drainage, pipeline construction, and any others that the company attorney deems necessary or helpful in bringing the condemnation to a positive conclusion.

- Agent testimony: The company must show evidence of fair and earnest negotiations prior to condemnation. The right-of-way agent is responsible for swearing in the courtroom that he or she negotiated in earnest and made a fair offer for the easement.
- Documentation: In addition to the agent's testimony, the agent is responsible for providing documentation through contact reports and/or offer letters showing proof of earnest negotiations.

4.7 CONSTRUCTION SUPPORT

The right-of-way group is critical to the construction operation by providing the following services:

- Construction restriction list: Numerous owners and agencies will demand and receive special construction requirements or restrictions along the pipeline route. The pipeline contractor is made aware of these requirements and restrictions by means of a construction restriction list or line list. This document lists all of the tracts along the route along with restrictions, requests, and requirements. This document must be issued prior to the contractor bidding process so that all additional cost items can be identified.
- Owner or agency notifications: Just prior to construction, right-of-way agents are responsible for notifying all property owners, tenants, and agencies of the impending construction. This gives the parties time to prepare for construction, such as moving cows to another field, cutting the hay along the right-of-way, or performing some type of operations to protect property and/or animals.
- Restriction verification: Right-of-way agents should be present to observe the construction operation. With the help of the construction restriction list, the agent must observe and record all violations and/or omissions to the restriction list. Although agents do not have authority to stop construction or make demands upon the contractor, they can report such problems to the proper authority, generally the inspection group who subsequently reports to the right-of-way manager.
- Construction damages: Following the clean-up of the right-of-way after construction, the right-of-way agent is responsible for paying owners, tenants, and agencies for all damages caused as a result of the construction. In many instances, companies choose to negotiate these damages prior to construction. There is a distinct line between damages that are the responsibility of the company and damages that are the responsibility of the contractor. In general, the company's damage responsibility lies totally within the right-of-way, whereas the contractor is responsible for all damages outside the right-of-way.

- Clean-up verification: Prior to the contractor leaving the area, it is customary for right-of-way agents to contact owners, tenants, and agencies and accompany them to view the clean-up. If the owner is satisfied, he or she will execute a "clean-up slip" stating that the clean-up is satisfactory to them. If the owner notes a problem, the contractor can return to the tract and make the necessary repairs.

4.8 PROJECT COMPLETION AND PIPELINE OPERATIONS

4.8.1 Project Completion

At the completion of a construction project, the right-of-way group must insure that all damages have been paid, and all documentation is complete. After the project budget is closed, any further right-of-way costs, even if the costs were incurred as a result of construction, would have to be borne by the operational budget. Therefore, it is important that all pending matters are resolved prior to the closing of the project budget. The right-of-way group must address the following issues at the completion of construction:

- Construction damages: Following construction, and if not paid prior to construction, the right-of-way group must pay all construction damages that are the responsibility of the pipeline company.
- Documentation: The right-of-way group must insure that all easements, grants, and other recordable documents are properly recorded. In addition, the group must assure that files are in order and there are no gaps in the right-of-way or other problems that may jeopardize the pipeline later.
- Pending condemnations and lawsuits: During the condemnation process, through court order, construction is allowed to proceed even though the condemnation case has not been resolved. The right-of-way group must review all pending condemnation and lawsuits and assure that the project budget has sufficient funds to cover such actions.

4.8.2 Pipeline Operations

The new pipeline construction project is only the first phase of the life of a pipeline. The pipeline will be operated for a number of years and may be sold and merged into another company, or portions may be relocated or abandoned. All of these operations require right-of-way input. If the right-of-way group has provided the pipeline company with proper operational documentation during the life of the project, sales, mergers, or other actions will present no problem. The right-of-way group should provide an operational database with the following information:

- Operational restrictions: A right-of-way database must identify all restrictions and requirements relevant to the operations of the pipeline system agreed to during negotiations. Some of those restrictions and requirements may include maintaining pipeline markers, keeping the right-of-way free of weeds, and maintaining erosion control.

- Permit or easement term documents: Many agency and some private easements expire after a certain period of time. All expiring documents, either permits or easements, must be noted on an operational database with a notification method of at least 30 days prior to expiration to pipeline company personnel to start the renewal process.
- Rentals or other annual payments: Many easements and permits require an annual or longer payment rental period. The due date and amount must be noted on this database for future payments with means of notification in a timely manner.
- Permanent dimensions: The permanent right-of-way width and other dimensions should be provided for each tract.
- Assignability: In the event a pipeline is to be sold, it is important to note at least four items – this item and the three below. The database must state whether each easement, permit, license, or agreement in the operational restriction list is assignable and with permission or not. In other words, can the document be assigned from the pipeline company to another entity and, if assignable, is permission required from the current property owner.
- Commodity restrictions: Many easements and permits limit the commodity to a certain gas or liquid. All commodity restrictions must be listed in the database.
- Abandonment: Many times, an easement or permit requires extensive costly restoration to the property in the event of abandonment. The acquisition document may also contain a provision that gives a period of time of non-utility that renders the pipeline abandoned.
- Change size: Some easements and permits allow the company to change the size of the pipeline, for example, replace the existing pipeline with a larger one.
- Multiple pipelines: As with changing the size of the pipeline, it should be noted whether the acquisition document grants multiple line rights or allows for additional pipelines to be constructed at later dates for additional payment.

Chapter 5

Alignment Sheets

Hal S. Ozanne

Chapter Outline

INTRODUCTION

Alignment sheets or drawings graphically show the location of a pipeline system and related facilities. The drawings are normally georeferenced to a coordinate system with a drawing base of aerial images, USGS quadrangle maps, or other base map system to show where the pipeline is or will be located relative to the surrounding area. The data on the drawings includes the precise location of the line, the pipe material, the different land ownership through which the pipeline crosses, pipeline crossings, and valve locations.

5.1 USES

Pipeline alignment sheets or route drawings are used for various purposes including acquisition of permits; application for a FERC Certificate for Construction for a natural gas transmission pipeline; inclusion of a construction bid package for obtaining quotations for construction; construction of the pipeline; and as a record of the system for operation after it is placed into service. The drawings can also be displayed at public meetings, during the permitting process, to show the attendees where the proposed pipeline and related facilities will be located.

When maintenance is required on the pipeline, landowner information can be obtained from the drawings so that the affected landowner(s) can be contacted before entering their land for the maintenance. The location of intermediate block valves and cathodic protection test stations are included. Technicians can be dispatched to these locations for required operations and maintenance purposes.

All information for the pipeline shown on the plan and profile views of the alignment drawings are based on a geographic coordinate system so that the information is tied into the same real-world coordinate system. The location of any and all components for the pipeline can be identified in the field by using the coordinate location of each component of the system indicated on the drawings (Fig. 5.1).

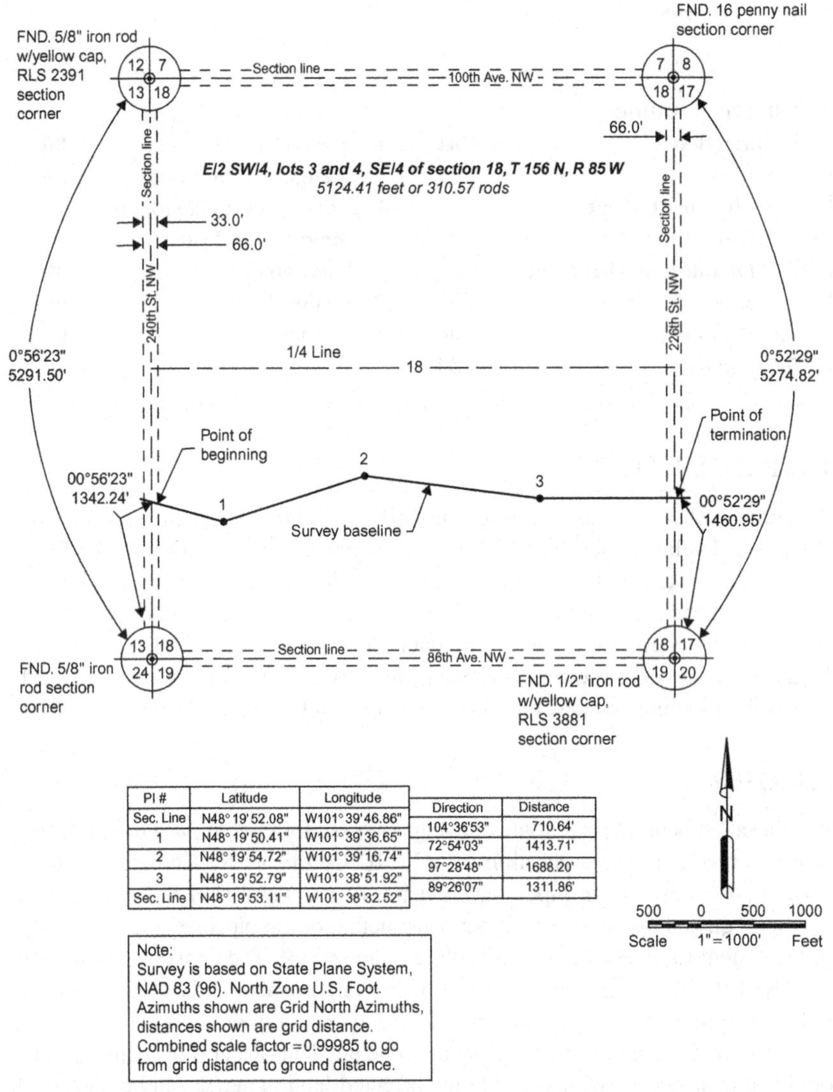

PI #	Latitude	Longitude
Sec. Line	N48° 19'52.08"	W101° 39'46.86"
1	N48° 19'50.41"	W101° 39'36.65"
2	N48° 19'54.72"	W101° 39'16.74"
3	N48° 19'52.79"	W101° 38'51.92"
Sec. Line	N48° 19'53.11"	W101° 38'32.52"

Direction	Distance
104°36'53"	710.64'
72°54'03"	1413.71'
97°28'48"	1688.20'
89°26'07"	1311.86'

FIGURE 5.1 Coordinate system example.

5.2 ALIGNMENT SHEET DEVELOPMENT

After the pipeline route has been selected, preparation of alignment sheets can begin. Following are some parameters that should be determined before the drawings are started.

- Drawing scale: A common scale is 1 inch = 500 feet. This will show good detail of the route surrounding the pipeline in most areas. In congested areas, a larger scale is sometimes used, such as 1 inch = 200 feet, which will show the greater detail of the pipeline route and obstructions in such areas.
- Background for the drawings: Examples are aerial imagery, road maps, or USGS quadrangle maps. Aerial photography is typically used since it is readily available.
- Drawing layout: It is the location on the drawing where the various information will be placed on the drawing.
- Information to be included: The types and amount of information and data are included on the drawings.
- Whether or not the ground and pipeline profile will be included.

The alignment drawings are normally prepared using either CAD or GIS computer software. There are alignment sheet software programs that can be used with either CAD or GIS software to simplify and automate some of the drawing processes in the preparation of the drawings.

The final basis for the centerline of the route of the pipeline is normally a civil survey. After the preferred route has been selected, as described in Chapter 2 a project pipeline engineer will accompany the survey crew to the field to show the route, where proposed points of intersection (PIs), valves, and other appurtenances are planned, and other information the crew will need to complete the survey. In some cases, the engineer will stay with the crew the entire time they are in the field. Minor route changes may be made during this activity.

As the survey work is being completed in the field, the electronic survey data is captured and transferred to the mapping group, usually on a daily basis, for use in preparing the drawings. The survey team or company is provided with survey specifications that are used by the crew(s) to obtain all the data required in the correct format so that drawing development will proceed smoothly.

In the early stages of route development and alignment sheet preparation, the route centerline may be established by a "paper survey." The centerline is prepared by drawing a route on USGS quadrangle maps or on aerial photography. This data can be loaded into a Global Positioning System (GPS) or other survey equipment for a field crew to use in locating the proposed route in the field. As described earlier, by using a coordinate system, the data can be loaded into a GPS so that personnel going to the project location can go directly to points along the route. The drawings prepared using a paper survey can be used for some of the early permitting and regulatory agency meetings and initial meetings with affected landowners.

5.3 QUANTITY OF ALIGNMENT SHEETS

The length of the pipeline and the drawing scale will determine the quantity of alignment sheets required for the project. Using a scale of 1 inch = 500 feet, approximately 24 inches of route can be included on each sheet, which equates to 12,000 feet or approximately 2.25 miles of the pipeline. When there is a significant change in direction of the line, or PI, of an angle of more than a few degrees, then less than 24 inches of route will fit on a sheet.

When the project requires more than one alignment drawing to cover the entire route, an index map is usually prepared. This map is developed at a small enough scale so the entire route can be shown on one drawing. Outlines of the individual alignment drawings, with the drawing numbers, are overlaid over the index map route. A person can then determine which map to review for a particular area.

5.4 STATIONING

A stationing system is used for the pipeline and is based on a surveyed or calculated distance in feet from the beginning point of the pipeline or the beginning point of a replacement or relocation of the line to any point on the pipeline. All points along the line are given a stationing such as a PI at station 1 + 22.5 or a block valve at station 1007 + 56.7, which are distances, in feet, from the established beginning point of the pipeline. This is a reference system that is used to communicate locations of specific points or locations along the pipeline. The stationing of all points is shown on the alignment drawings.

When changes are made to the route before construction or if a section of the pipe is replaced or relocated after the line is in operation and the new length of the line will change, then a stationing equation is developed to account for the change in length of the pipeline and where the change was made. An example of a stationing equation is 100 + 35.5 = 101 + 15.0. This indicates that a change in pipe length of 79.5 feet was added to the pipeline and the new station where the reroute is completed is 101 + 15.0 and the point where the line connects back into the existing line is at the old station 100 + 35.5.

5.5 SURVEY

The basis for the survey and the information shown on the alignment drawings is a geographic coordinate system. The system to be used needs to be compatible with the system that the pipeline operators use with their GIS system for permanent records, ongoing operation, and maintenance purposes. The system most commonly used is NAD 83 or North American Datum. This is a system in which all route information and components of the pipeline are tied to the same geographic system. Each point has a corresponding latitude, longitude, and elevation. USGS quadrangle maps and Google Earth, which are commonly used for routing pipelines, are based on the NAD 83 coordinate system.

There are normally at least three passes of the pipeline route by the civil survey team, two of which are used to gather data for use in the preparation of alignment sheets and other project exhibits and documents.

The first pass is to set and flag the proposed centerline of the pipeline route so that field teams, such as the environmental crew, have a basis to conduct their work. Survey stakes are used for this purpose. Different-colored flagging is tied around the stake to signify different information such as centerline of the route, centerline right-of-way, right-of-way width boundaries, and PI locations. The surveyor will also write information on the stake, such as the pipeline station, centerline, and property boundaries. The crew will also flag the proposed parallel boundaries of the route so that the environmental and other field crews have the route corridor area established for their effort. All these data are loaded into the GPS and survey equipment, which provides electronic data to be outputted.

The environmental and field crews will typically use their own GPS equipment to tie in their field-collected data into the agreed geographic coordinate system. The data can then be directly incorporated into the alignment drawings.

The centerline flags can also be reviewed with affected landowners by the right-of-way agents or permitting agencies to show them the proposed route and right-of-way width across the affected property.

The survey crew will make a second pass along the route to obtain other required data. That data include the following list for use in preparation of the alignment sheets and other project documents and details that require survey accuracy:

- Route or pipeline centerline
- Location of PIs (points of intersection) or changes in direction of the line (angles in degrees)
- Ground profile elevation
- Any significant structure in the route corridor, such as trees, power or communication structures, and fences
- Locations of anything the proposed pipeline will cross or tie into, such as other pipelines, power lines, communication lines, fences, roads, waterways, and railroads
- Property boundary lines and corners of the properties that are impacted by the pipeline
- A detailed profile and right-of-way widths for crossings of roads, railroads and waterways, and other crossings that require a detailed design and/or permit exhibit
- Environmental crews will locate and flag environmental features such as wetlands and their extents, cultural resources, and endangered species within the impacted route corridor. The civil survey crew will tie those locations into the coordinate system and capture the electronic survey data if the environmental crew has not used GPS equipment to locate their features themselves. That data will be forwarded to the mapping group for inclusion on the alignment sheets.

5.6 DRAWING ISSUANCE

Before the drawings are initially sent to the client and other project team members for review and comments, they are normally stamped with "Issued for Review" with the date they are issued. This will keep the drawings from being confused with the drawings issued for other purposes or on other dates.

Drawings that are issued for permit purposes will be stamped with "Issued for Permit Purposes" with the date.

The alignment drawings are typically included in the bid package that is provided to contractors for obtaining construction bids for the pipeline. These drawings along with construction details and specifications include the information that the contractors will need for the construction bid preparation. At this stage, the drawings are stamped with "Issued for Bid Purposes Only."

Prior to the drawings being issued for construction, any changes that have been made since the previous formal issue will be incorporated. The drawings will then be stamped with "Issued for Construction" with the corresponding date. This is to prevent the contractor from using an earlier version of the drawings for construction. The construction manager and inspectors should verify that the contractor is issued the revised drawings.

5.7 CHANGES TO THE ROUTE

During the bidding process, prior to being issued for construction, the alignment drawings will be revised as necessary to incorporate any changes to the route, right-of-way width, or any other changes that will impact the construction of the pipeline. The revised drawings may need to be resubmitted to the permitting agencies. The revised drawings will be issued for construction. The civil survey crew will use these drawings to stake the location of the centerline of the pipeline and other features such as right-of-way width, temporary construction areas, property lines, locations for construction bores, changes in pipe wall thickness, and other items that the construction contractor will need for construction purposes. These stakes will be used by the construction contractor to install the pipeline in the correct location and to limit the work area to where temporary construction right-of-way has been acquired from the landowners.

Alignment sheets will normally include the following information:

- Route centerline based on the civil survey including PIs or changes in direction with the associated degrees of the change in direction using a geographic coordinate system
- Permanent right-of-way boundaries
- Temporary construction right-of-way easement boundaries
- Extra workspace required for crossings such as roads, railroads, and waterways

- Pipe staging or storage yards
- Contractor temporary yards or land for equipment and material storage
- Pipeline stationing, in feet, of property lines, state and county boundaries, roads, railroads, waterways, PIs, section lines, and other features
- Mileposts of the pipeline in 0.1-mile increments
- All other infrastructures to be crossed including roads, communication lines, electric power lines, fences, waterways and railroads including the width of the corresponding right-of-way whether they are overhead or underground
- Property lines of all properties to be crossed with surveyed ties to section corners or other established survey monuments
- Access roads used during construction indicating the road route from a public road to the pipeline right-of-way
- Location of intermediate block valves and scraper launcher and receiver installations
- Location of any lateral connections to the pipeline
- References to detailed construction drawings
- Location of any intermediate pump or compressor stations
- A material band showing the pipe diameter, pipe grade of steel, pipe wall thickness, coating to be used, lengths of pipe and coating and any changes to the material on each drawing
- Location of cathodic protection test stations
- Location of pipeline location markers
- Location of aerial markers or signs
- Ownership band showing the pipeline stationing where the pipeline enters and exits each property, the landowner's names and the distances across the properties
- Class locations based on population density
- Environmental band showing the location of environmental features along the route, including the location of wetlands, cultural resources, and nesting areas; any mitigation requirements that the contractor is required to abide by or a reference to where that information can be found
- Ground elevation profiles and depth of pipe below the ground surface
- Location, entrance and exit points, profile for horizontal directional drills (HDDs) crossings for waterways and other long crossings
- Location of entrance and exit points for bored crossings under road, railroads, etc.

Figures 5.2 and 5.3 are examples of a typical alignment drawing, first with an aerial imagery background and second with a USGS quadrangle background.

When submitting an application for a Federal Energy Regulatory Commission (FERC) Certificate of Construction permit, FERC requires that an aerial imagery background be used and that the imagery be no older than 1 year at the time of filing the application.

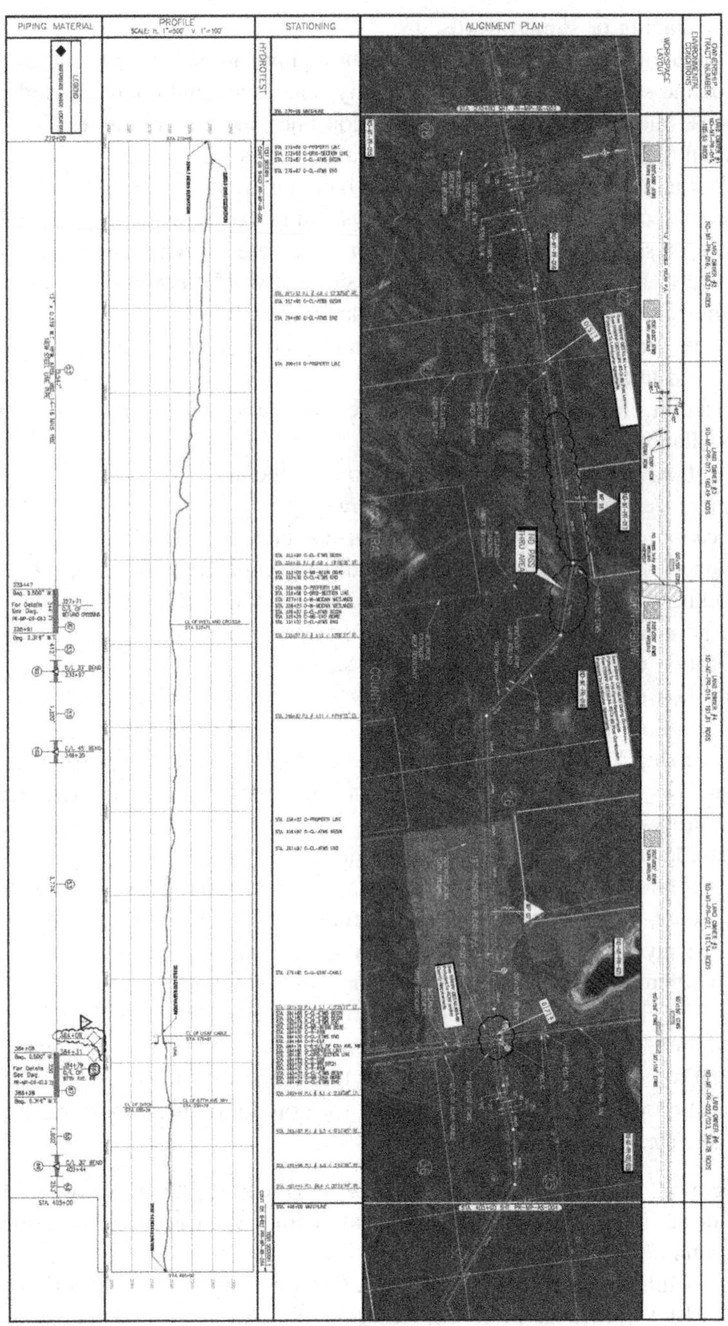

FIGURE 5.2 Alignment drawing with aerial photography background.

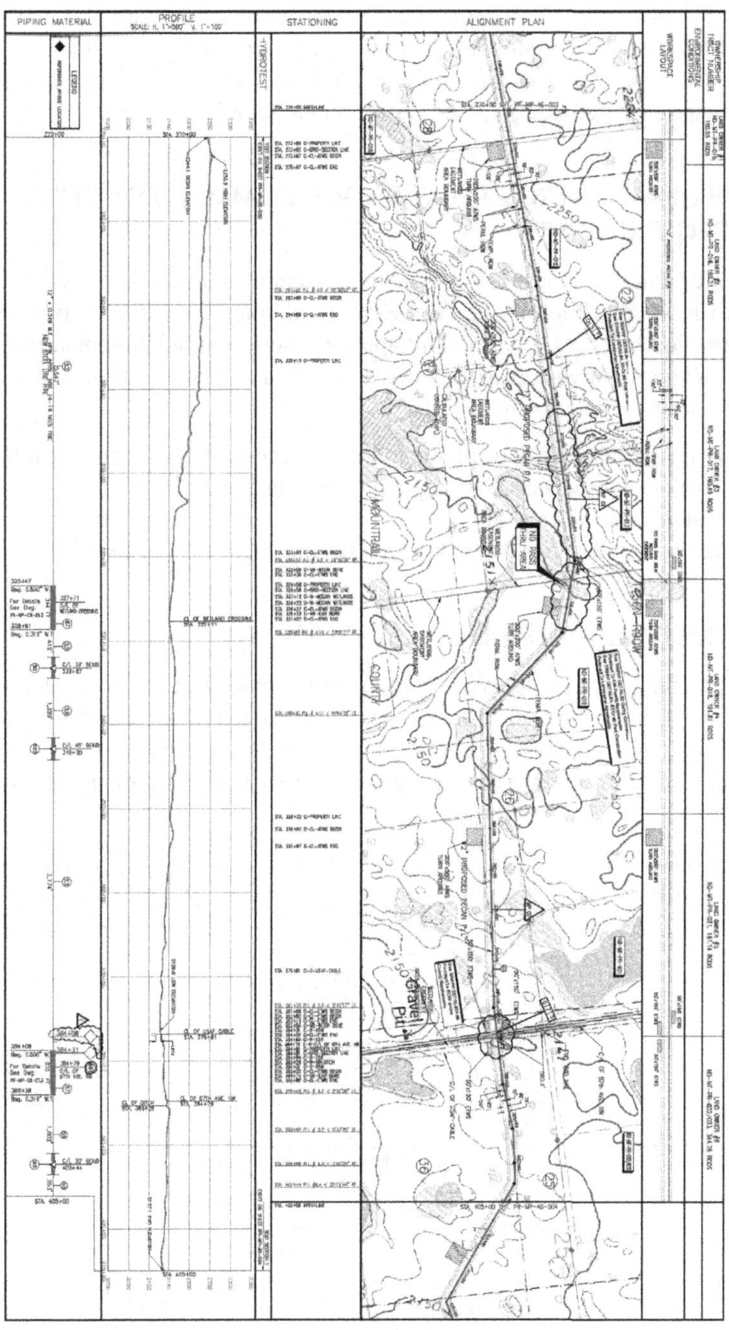

FIGURE 5.3 Alignment drawing with USGS quadrangle map background.

FERC also requires a set of USGS 7.5 minute topographical maps be submitted with the permit application for the proposed pipeline. The pipeline route, with mileposts, and all planned facilities, such as valves, scraper launchers, and compressor stations, are to be shown on the set of maps.

5.8 FEDERAL ENERGY REGULATORY COMMISSION (FERC) REQUIREMENTS

FERC regulates gas transmission pipelines. Alignment drawings are required as part of Resource Report 1 in the FERC 7C application for a Certificate of Construction for a gas transmission pipeline. There are minimum requirements for the alignment drawings in the application submittal to avoid rejection. Those requirements include the following:

- Provide aerial images or photographs or alignment sheets based on these sources with mileposts showing the project facilities
- No more than 1 year old
- Scale no smaller than 1:6000, which is equivalent to 1 inch = 500 feet

5.9 EXISTING SYSTEMS

When changes are made to an existing pipeline system, such as replacing a section of pipe, repairing a leak, lowering a section of the existing pipe, or installing a new appurtenance to the line, the affected alignment drawing or drawings need to be revised. Field notes are sent to the mapping department

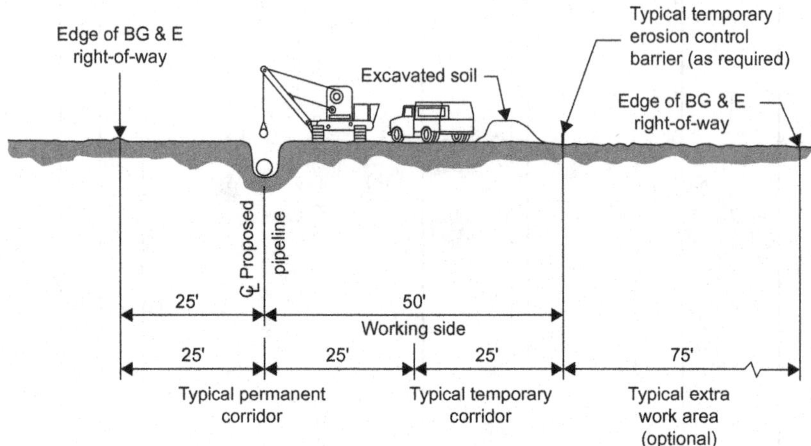

FIGURE 5.4 Right-of-way width example.

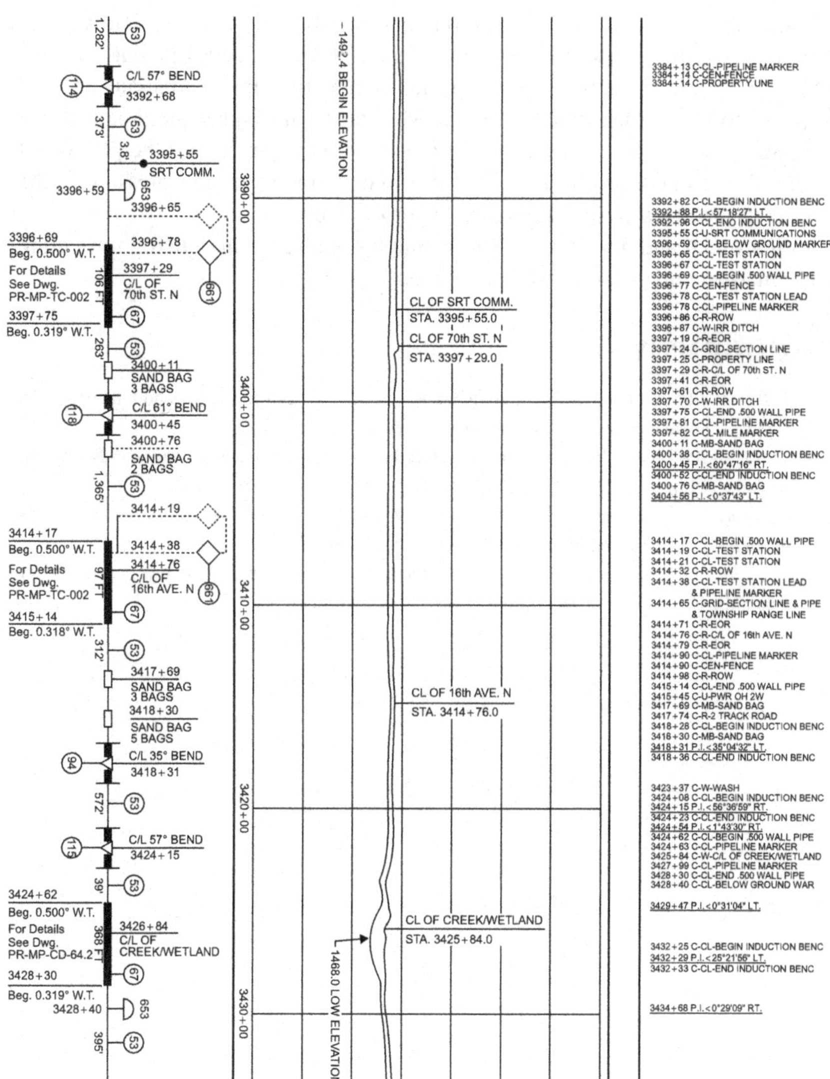

FIGURE 5.5 Alignment sheet stationing example.

showing the changes and location where the change was made. The alignment drawings are updated to reflect the change so that the pipeline operations department will have current information available.

The aerial photography for pipelines is updated periodically especially in developing areas or congested areas. The existing data is overlaid on the new photography. The updated drawings can be reviewed to determine that new development is approaching the pipeline. This is very important for gas transmission lines. The class locations are based on the number and type of buildings within a prescribed distance of the pipeline. If development is occurring, the class location may change. If this occurs, the pipeline operator may be required to replace the pipe in the impacted area with higher yield strength or wall thickness to add additional factor of safety to the pipeline. If the pipe is not replaced, the operator will be required to reduce the operating pressure in the impacted area to achieve the higher factor of safety.

Figures 5.4 and 5.5 show various examples of alignment drawing formatting and enlargements of portions of alignment drawings.

BIBLIOGRAPHY

ASME B31.4-2006, Pipeline Transportation Systems for Liquid Hydrocarbons and Other Liquids, American Society of Mechanical Engineers, New York, NY, 2006.

ASME B31.8-2007, Gas Transportation and Distribution Piping Systems, American Society of Mechanical Engineers, New York, NY, 2007.

CFR 49, Part 195, Transportation of Hazardous Liquids by Pipeline: Minimum Federal Safety Standards, U.S. Government Printing Office, Washington, DC.

CFR 49, Part 192, Transportation of Natural or Other Gas by Pipeline: Minimum Federal Safety Standards, U.S. Government Printing Office, Washington, DC.

Chapter 6

Overview of Pipeline Materials

Hal S. Ozanne

Chapter Outline

INTRODUCTION

An important decision to be made when planning the installation of a new pipeline or modifications to an existing pipeline is to determine the materials to be used in the design and construction of the line. Factors to be taken into consideration when making this determination include the following: product and daily volume to be transported, operating pressure, location where the line is to be installed, and types of construction to be employed. American Petroleum Institute (API) codes and recommended practices are used in the pipeline industry in the selection of materials for pipelines and components. These codes are incorporated by reference in regulations regulating the industry.

6.1 CRITERIA

Steel pipe is normally used for pipelines operating at a pressure of 100 psig or more. Steel pipe withstands high pressures, is durable, and has a long operating life cycle. Fiberglass, PVC, or high-density polyethylene (HDPE) pipe is used in certain instances for low-pressure gas gathering pipelines. Components of the pipeline include the following: pipes, valves, fittings, and equipment such as metering, pumps, and compressors.

In this book, we are concerned only with transporting hydrocarbons such as natural gas, refined petroleum products, crude oil, and liquefied petroleum gas in steel pipelines. Therefore, we will not deal with materials such as PVC pipe.

6.2 PRODUCT TO BE TRANSPORTED

The commodity to be transported plays a significant part in determining the type of material that will be used for the pipe. Steel pipe is normally used because of its strength and durability, and it can be operated at high pressures. Some products to be transported are corrosive and may cause internal corrosion of the pipe over time. In this case, pipe material with special metallurgy is required. An analysis is required to determine the metallurgical components of the pipe that will be needed. In some cases, a corrosion inhibitor may be injected into the pipe to prevent internal corrosion of the pipe.

6.3 OPERATING PRESSURE

Hydraulic calculations are completed for the pipeline to determine the pipe diameter, pipe wall thickness, and operating pressure based on the length of the pipeline and volume of product to be transported. There are several grades of steel available. The greater the numerical grade of steel, the greater the specified minimum yield strength of the pipe, which means that the maximum allowable operating pressure is higher.

An economic evaluation should be made to determine the most economic grade of steel compared with the pipe diameter and wall thickness. Current prices of pipe will need to be obtained for various scenarios in order to make the evaluation.

It is usually more economic to use a higher grade of steel and/or wall thickness and operate at a higher pressure than to use a larger pipe diameter and a lower operating pressure. Construction of larger pipe diameters with thinner wall thickness is usually more expensive than that of smaller pipe diameters with greater wall thickness.

In addition to the internal pressure, the pipe must withstand any external loads that may be present. This may include vehicular and railroad traffic over the pipe and crossings and loads imposed on the pipe during construction, such as when the pipe is pulled through a bore or horizontal directional bore hole.

6.4 OPERATING TEMPERATURE

In those cases where the pipe is installed above ground and the ambient temperature is going to be below 32°F, or where the pipe is going to be buried and the ground temperature will be below 32°F, then attention must be paid to the pipe materials. If the product to be transported is below 32°F, such as in the case of a CO_2 pipeline, or above 100°F, attention must be paid to the pipe and system components. An analysis will be required to determine the metallurgy of the steel that will be required to withstand the effects of the temperature.

6.5 HANDLING AND WELDING

In selecting the size, wall thickness, and grade of steel for a pipeline, consideration must be given to the handling of the pipe. The pipe will be transported from the coating yard either by truck or rail or both.

The pipe will normally be stored in a staging yard near the project site until it is delivered to the right-of-way. When the construction contractor is ready to move the pipe to the construction right-of-way, the pipe will be loaded onto trucks, transported to, and strung or laid along, the right-of-way so that the joints can be welded together.

As described earlier, the pipe diameter, grade of steel, and the wall thickness are determined for the volume and operating pressure of the product to be transported. Consideration must also be taken in handling the pipe. In some instances, a wall thickness that is adequate for the proposed operating pressure may not be sufficient for handling. When the pipe is picked up by slings for loading or unloading, buckling of the pipe can occur if the wall thickness is too thin. The ends of the pipe may also become elongated or "egged," preventing the pipe from being lined up with other joints for welding the joints together. In these cases, a pipe of greater wall thickness will be required, which will add cost to the material costs for the project.

The wall thickness and the steel composition also have an impact on the amount of welding and the welding procedures to be used. The subject of welding is discussed in detail in Chapter 17 of this book, but the intent here is to mention that greater wall thickness of pipe will require more welding passes, which does increase the cost and time required for welding each joint of pipe.

The grade of steel selected does have a significant impact on the welding process to be employed. The higher grades of steel do require more stringent welding procedures, more costly welding rods, and greater skill of the welders. Regulations require that welders pass tests for the type of steel that they will be welding. The success rate for welders passing tests is lower for the higher grades of steel.

All of these items should be taken into consideration when selecting the wall thickness and grade of steel for the pipe.

6.6 VOLUME OR THROUGHPUT

When the initial work begins for the design of a pipeline, criteria will be established for the volume to be transported in the pipeline. Consideration should also be given for any planned increases in the volume in the foreseeable future. If the volume is forecast to increase within, say, ten years from the installation of the line, it may be decided to increase the diameter of the pipe for the planned future greater volume. This is generally economically less expensive than constructing a second pipeline parallel to the first line within the same time period.

The pump station or compressor station throughput capacity can be increased as the system volume increases in planned increments by adding additional pumping or compression equipment to existing stations and/or adding new intermediate stations along the line.

6.7 CODES AND REGULATIONS

Codes and regulations specify the materials to be used for the pipe material depending on the pipeline operating conditions. It is recommended that the latest edition of the relative codes and regulation be consulted during project planning.

6.7.1 Gas Pipelines

The Code of Federal Regulations, Title 49, Part 192, Transportation of Natural and Other Gas by Pipeline: Minimum Federal Safety Standards require the following material considerations:

Subpart B – Materials

Minimum Requirements – This subpart prescribes minimum requirements for selection and qualification of pipe and components for use in pipelines.

Materials must be able to maintain the structural integrity of the pipeline under temperature and other environmental conditions that may be anticipated.

Materials must be chemically compatible with any gas that they transport and with any other material in the pipeline to which they are in contact.

Steel Pipe

New steel pipe is qualified for use under this subpart if:

It is manufactured in accordance with a listed specification.

It meets the requirements of Appendix B to Part 192 for steel pipe of unknown or unlisted specifications.

CFR Title 49 Part 192.103 General

Pipe must be designed with sufficient wall thickness to withstand anticipated external pressures and loads that will be imposed on the pipe after installation.

CFR Title 49 Part 192.105 Design Formula for Steel Pipe

The design pressure for steel pipe is determined in accordance with the following formula:

$$P = (2St/D) \times F \times E \times T$$

P = Design pressure
S = Yield strength
D = Nominal outside diameter
t = Wall thickness
F = Design factor (safety)
E = Longitudinal joint factor
T = Temperature derating factor

The design factor to be used in the calculation of the design pressure in the above formula is dependent on the Class location of where the pipe is to be or is installed. The design factors are as follows:

Class Location	Design Factor
1	0.72
2	0.60
3	0.50
4	0.40

ASME B31.8, Gas Transmission and Distribution Piping Systems, Chapter I, Materials and Equipment requires the following:

810 Materials and Equipment

810.1 – It is intended that all materials and equipment that will become a permanent part of any piping system constructed under this Code shall be suitable and safe for the conditions in which they are used. All such materials and equipment shall be qualified for the conditions of their use by compliance with certain specifications, standards, and special requirements of this Code, or otherwise as provided herein.

811 Qualification of Materials and Equipment

811.1 – Materials and equipment fall into the following six categories pertaining to methods of qualification for use under this Code:

1. Items that conform to standards or specifications referenced in this Code
2. Items that are important from a safety standpoint, of a type for which standards or specifications are referenced in this Code but specifically do not conform to a referenced standard (e.g., pipe manufactured to a specification not referenced in the Code).
3. Items of a type for which standards or specifications are referenced in this Code, but that do not conform to the standards and are relatively unimportant from a safety standpoint because of their small size or because of the conditions under which they are to be used.
4. Items of a type for which no standard or specification is referenced in this Code (e.g., gas compressor)
5. Proprietary items
6. Unidentified or used pipe

811.2 – Prescribed procedures for qualifying each of these six categories are given in the following paragraphs.

811.21 Items that conform to standards or specifications referenced in this Code may be used for appropriate applications, as prescribed and limited by this Code without further qualification.

811.22 Important items of a type for which standards or specifications are referenced in this Code, such as pipe, valves, and flanges, but that do not conform to standards or specifications referenced in this Code shall be qualified as described in para. 811.221, 811.222, or 811.24.

811.221 A material conforming to a written specification that does not vary substantially from a referenced standard or specification and that meets the minimum requirements of this Code with respect to quality of materials and workmanship may be used. This paragraph shall not be construed to permit deviations that would tend to affect weldability or ductility adversely. If the deviations tend to reduce strength, full allowance for the reduction shall be provided for in the design.

811.222 When petitioning the Section Committee for approval, the following requirements shall be met. If possible, the material shall be identified with a comparable material, and it should be stated that the material will comply with that specification, except as noted. Complete information as to chemical composition and physical properties shall be supplied to the Section Committee, and their approval shall be obtained before this material is used.

811.23 Relatively unimportant items that do not conform to a standard or specification [para. 811.1(3)] may be used, provided that

1. They are tested or investigated and found suitable for the proposed service.
2. They are used at unit stresses not greater than 50% of those allowed by this Code for comparable materials.
3. Their use is not specifically prohibited by the Code.

811.24 Items of a type for which no standards or specifications are referenced in this Code [para. 811.1(4)] and proprietary items [para. 811.1(5)] may be qualified by the user provided

1. The user conducts an investigation and tests (if needed) that demonstrate that the item of material or equipment is suitable and safe for the proposed service (e.g., clad or duplex stainless steel pipe); or
2. The manufacturer affirms the safety of the item recommended for that service (e.g., gas compressors and pressure relief devices).

811.25 Unidentified or used pipe may be used and is subject to the requirements of para. 817.

812 Materials for Use in Cold Climates

Some of the materials conforming to specifications referenced for use under this Code may not have properties suitable for the lower portion of the temperature band covered by this Code. Engineers are cautioned to pay attention to the low-temperature impact properties of the materials used for facilities to be exposed to unusually low ground temperatures or low atmospheric temperatures.

813 Marking

All valves, fittings, flanges, bolting, pipe, and tubing shall be marked in accordance with the marking sections of the standards and specifications to which the items were manufactured or in accordance with the requirements of MSS SP-2 (Manufacturer's Standard Society Standard Marking System).

814 Material Specifications

For a listing of all referenced material specifications, see Appendix A of the Code.

815 Equipment Specifications

Except for the piping components and structural materials listed in Appendices A and C of the Code, it is not intended to include complete specifications for equipment in the Code. Certain details of design and fabrication, however, necessarily refer to equipment, such as pipe hangers, vibration dampeners, electrical facilities, engines, compressors, etc. Partial specifications for such equipment items are given herein, particularly if they affect the safety of the piping system in which they are to be installed. In other cases where the Code gives no specifications for the particular equipment item, the intent is that the safety provisions of the Code shall govern, insofar as they are applicable. In any case, the safety of equipment installed in a piping system shall be equivalent to that of other parts of the same system.

816 Transportation of Line Pipe

Provisions should be made to protect the pipe, bevels, corrosioncoating, and weight coating (if applicable) from damage during any transportation (highway, rail, and/or water) of line pipe.

Any line pipe transported by railroad, inland waterway, or by marine transportation shall be loaded and transported in accordance with API RP5L1 or API RP5LW. In the case where it is not possible to establish that pipe was loaded and transported in accordance with the above referenced recommended practice, the pipe shall be hydrostatically tested for at least 2 hr to at least 1.25 times the maximum allowable operating pressure if installed in a Class 1 location, or to at least 1.5 times the maximum allowable operating pressure if installed in a Class 2, 3, or 4 location.

817 Conditions for the Reuse of Pipe

Refer to the Code for the conditions for the reuse of pipe.

6.7.2 Hazardous Liquid Pipelines

The Code of Federal Regulations, Title 49, Part 195 – Transportation of Hazardous Liquids by Pipeline: Minimum Federal Safety Standards requires the following:

The design pressure calculation is the same as for gas pipelines except the design factor (F) of 0.72 is always used except for some very specific situations. For those situations, the regulations are reviewed.

External loads such as the pipe crossing under railroads and highways must be considered. Refer to ASME/ANSI B31.4 for the procedures that must be followed in the design for external loads.

ASME B31.4, Pipeline Transportation Systems for Liquid Hydrocarbons and Other Liquids, Chapter III, Materials requires the following:

423 Materials – General Requirements

423.1 Acceptable Materials and Specifications

1. The materials used shall conform to the specifications listed in Table 423.1 or shall meet the requirements of this Code for materials not listed. Specific editions of standards incorporated in this Code by reference, and the names and addresses of the sponsoring organizations, are shown in Appendix A, since it is not practical to refer to a specific edition of each standard in Table 423.1 and throughout the Code text. Appendix A will be revised at intervals, as needed, and issued in Addenda to the Code. Materials and components conforming to a specification or standard listed earlier in Table 423.1 or to a superseded edition of a listed specification or standard, may be used.
2. Except as otherwise provided for in this Code, materials which do not conform to a listed specification or standard in Table 423.1 may be used provided they conform to a published specification covering chemical, physical, and mechanical properties; method and process of manufacture; heat treatment; and quality control, and otherwise meet the requirements of this Code. Allowable stresses shall be determined in accordance with the applicable allowable stress basis of this Code or a more conservative basis.

423.2 Limitations on Materials

423.2.1 General

1. The designer shall give consideration to the significance of temperature on the performance of the material.
2. Selection of material to resist deterioration in service is not within the scope of this Code. It is the designer's responsibility to select materials suitable for the fluid service under the intended operating conditions. An example of a source of information on materials' performance in corrosive environments is the Corrosion Data Survey published by the National Association of Corrosion Engineers.

423.2.3 Steel Steels for pipe are shown in Table 423.1 (except as noted in para. 423.2.5). Steel pipe designed to be operated above 20% SMYS shall be impact tested in accordance with the procedures of supplementary requirement SR5 of API 5L or ASTM A 333. The test temperature shall be one below 32°F (0°C) or the lowest expected metal temperature during service, with regard to past recorded temperature data and possible effects of lower air and ground temperatures. The average of the Charpy energy values from each heat shall meet or exceed the following:

1. For all grades with a SMYS equal to or greater than 42,000 psi (289 MPa), the required minimum average (set of three specimens) absorbed energy for each heat based on full-sized (10 mm × 10 mm) specimens shall be 20 lb-ft (27 J) for transverse specimens or 30 lb-ft (41 J) for longitudinal samples.
2. For all grades with SMYS less than 42,000 psi (289 MPa), the required minimum average (set of three specimens) absorbed energy for each heat based on full-sized (10 mm × 10 mm) specimens shall be 13 lb-ft (18 J).

423.2.4 Cast, Malleable, and Wrought Iron

1. Cast, malleable, and wrought iron shall not be used for pressure-containing parts except as provided in paras. 407.1(a) and (b), and para. 423.2.4(2).
2. Cast, malleable, and wrought iron are acceptable in pressure vessels and other equipment noted in para. 400.1.2(b) and in proprietary items [see para. 400.1.2(g)], except that pressure-containing parts shall be limited to pressures not exceeding 250 psi (17 bar).

423.2.5 Materials for Liquid Anhydrous Ammonia Pipeline Systems Only steel conforming to specifications listed in Appendix A shall be used for pressure-containing piping components and equipment in liquid anhydrous ammonia pipeline systems. However, internal parts of such piping components and equipment may be made of other materials suitable for the service. The longitudinal or spiral weld of electric resistance welded and electric induction welded pipe shall be normalized.

Cold-formed fittings shall be normalized after fabrication.

Except for the quantities permitted in steels by individual specifications for steels listed in Appendix A, the use of copper, zinc, or alloys of these metals is prohibited for all pressure piping components subject to a liquid anhydrous ammonia environment.

423.2.6 Materials for Carbon Dioxide Piping Systems Blow down and bypass piping in carbon dioxide pipelines shall be of a material suitable for the low temperatures expected.

6.8 COATING

After the pipe material has been selected, one of the next steps is to determine the type of pipe coating to be used. The type of coating to be used is normally based on soil conditions and moisture content of the soil that the pipe will be installed in. In most of the cases, fusion bond epoxy (FBE) coating will provide the protection for steel pipe that will last for many years and prevent the pipe from corroding. The coating thickness is normally 12–14 mils. This thickness will withstand most of the handling of the pipe during loading, shipment, and unloading. In locations where the soil, which the pipeline traverses through, has a high moisture content, FBE coating may not be the most effective, and other types of coating should be investigated.

Once the ditch has been cut in the areas where there is high concentration of rock, the ditch is padded with sand or other fine soil before the pipe is lowered into the ditch to protect the pipe and coating from being damaged by the rock. After the pipe has been lowered into the ditch, additional fine material or sand is placed, to a depth of at least 12 inches, over the pipe before native fill is placed into the ditch. This is referred to as padding the ditch.

In some instances, rock shield is placed over the pipe before native backfill is placed on top of the pipe to protect it.

In locations where the pipe is going to be pushed or pulled through a bore hole or a horizontal directional bore, the pipe has an additional coating of ARO, or abrasive-resistant coating placed over the fusion-bonded coating. The ARO is a thin layer of concrete coating that is very resistant to mechanical damage to the pipe. This does add cost to the project, so an evaluation should be made as to where and how much ARO should be used.

6.9 JOINT COATING

Coating that is applied at a coating plant is normally applied to within about six inches at each end of the joint of pipe. This allows for welding the joints of pipe together without damaging the coating. After the coated pipe has been transported and strung along the pipeline right-of-way, the joints are welded together. This leaves about 12 inches of uncoated pipe at each joint.

The joints are coated in the field before the pipe is "jeeped" to check for holidays, holes, or thin spots in the pipe coating. There are several methods for coating the joints.

One type of joint coating is shrink sleeves. The material is wrapped around the joint and then heated with a propane torch. The sleeve with shrink will be bonded to the pipe and the pipeline coating to provide a waterproof seal.

Another type of joint coating is tape. The joint is cleaned, coated with primer, and the tape is wrapped around the pipe with overlaps.

6.10 FITTINGS

All pipe components that are to be buried below ground must be coated. This includes elbows, bends, and valves. The coating for elbow and bends may be tape or field-applied FBE. Valves are normally coated with mastic.

BIBLIOGRAPHY

ASME B31.4-2006, Pipeline Transportation Systems for Liquid Hydrocarbons and Other Liquids, American Society of Mechanical Engineers, New York, NY, 2006.

ASME B31.8-2007, Gas Transportation and Distribution Piping Systems, American Society of Mechanical Engineers, New York, NY, 2007.

CFR 49, Part 195, Transportation of Hazardous Liquids by Pipeline: Minimum Federal Safety Standards, U.S. Government Printing Office, Washington, DC.

CFR 49, Part 192, Transportation of Natural or Other Gas by Pipeline: Minimum Federal Safety Standards, U.S. Government Printing Office, Washington, DC.

Chapter 7

Pipe Strength and Wall Thickness

E. Shashi Menon, Ph.D., P.E.

Chapter Outline

INTRODUCTION

In this chapter, we discuss the strength capabilities of a pipeline that is subject to internal pressure and how the required pipe wall thickness is calculated. In order to transport a liquid or a gas through a pipeline, pumping pressure is required at the origin of the pipeline and at intermediate pump and compressor stations along the length of the pipeline. At various locations along the pipeline, the origin pressure is gradually reduced due to frictional pressure drop. Further, the pressure will be increased at low elevation points and decreased at high elevation points. These internal pressures subject the pipe material to circumferential, axial, and radial stresses.

Therefore, we must select proper pipe material with adequate pipe wall thickness to withstand the internal pressure during the normal course of operation of the pipeline. Also, in a buried pipeline, the pipe is subject to external loads. External pressure can result from the weight of the soil above the pipe in a buried pipeline and also by the loads transmitted from vehicular traffic in areas where the pipeline is located below roads, highways, and railroads. These cause additional pipe stresses that have to be considered in selecting the required pipe wall thickness. We will discuss different materials used to construct pipelines, the design standards and codes that apply, and the method of calculating the internal pressure that a given pipe can withstand based on the strength of the pipe material, its diameter, and its wall thickness.

For a specific internal pressure, the minimum wall thickness required will be calculated depending on the pipe diameter and yield strength of the pipe material based on design and construction codes. Once the maximum allowable continuous operating pressure is established, the required hydrostatic test pressure can be defined to ensure that the pipeline can be safely operated. The hydrostatic testing of pipelines is discussed in detail in Chapter 18.

7.1 ALLOWABLE OPERATING PRESSURE

To transport a fluid (liquid or gas) through a transmission pipeline, the fluid must be under sufficient pressure to compensate for the pressure loss due to friction along the pipeline and the pressure required to overcome any elevation changes along the pipeline. Longer pipelines operating at high flow rates cause higher frictional pressure drop and, consequently, require higher internal pressure to transport the fluid from the beginning of the pipeline to its terminus. In many instances, long pipelines require multiple pump stations along the pipeline to provide the incremental pressures required to transport the fluid.

In gravity flow systems, liquid may flow from a storage tank located at a higher elevation through a pipeline down to a terminus at lower elevation without additional pump pressure. Even if no external pumping pressure is required for a gravity flow system, the pipeline will still be subject to internal pressure due to the static elevation difference between the two ends of the pipeline.

The allowable internal operating pressure in a pipeline is defined as the maximum safe continuous pressure at which the pipeline can be operated without causing pipe rupture. This is generally referred to as the maximum allowable operating pressure (MAOP). At this internal pressure, the pipe material is subjected to stresses that are less than the yield strength of the pipe material. As mentioned earlier, the pipe subjected to internal pressure results in stresses in the pipe material in three different directions as follows:

1. Circumferential (or hoop) stress
2. Longitudinal (or axial) stress
3. Radial stress

Among these, radial stress is significant only in thick-walled pipes. Because most transmission pipelines are considered to be thin walled, the radial stress component is neglected. Thus, the two important stresses in a transmission pipeline are hoop stress S_h and axial stress S_a, as shown in Fig. 7.1.

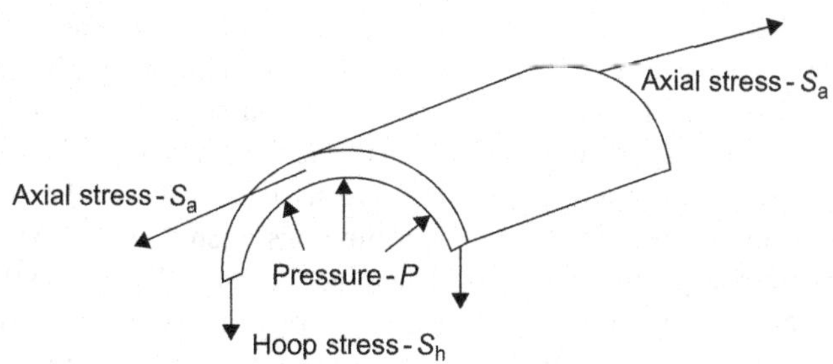

FIGURE 7.1 Stresses in a pipeline.

It is shown that the axial stress S_a equals one half of the hoop stress S_h. Therefore, the higher hoop stress is the controlling stress that determines the amount of internal pressure the pipeline can withstand. In liquid transmission pipelines, the hoop stress may be allowed to reach 72% of the yield strength of the pipe material.

If pipe material has 70,000 psi yield strength, the maximum hoop stress that the liquid pipeline can be subject to due to internal pressure is

$$0.72 \times 70{,}000 = 50{,}400\,\text{psi}$$

In order to ensure that the pipeline can be operated safely at a particular MAOP, the pipeline must be tested at a higher internal pressure prior to operation. Generally, this is done using water, and the process is called hydrostatic testing. The pipeline is divided into test sections and is filled with water. Each test section is subjected to the required hydrostatic test pressure and maintained for a specified period of 8 h, and the pipeline is inspected for leaks. Hydrostatic testing is discussed in more detail in Chapter 18.

The hydrostatic test pressure is the pressure (higher than MAOP) that the pipeline is tested to. Generally, the pipeline is tested for a specified period of time, such as 4 h (for above ground piping) or 8 h (buried pipeline), as required by the pipeline design code or by city, county, state, or federal government regulations. In the United States, the Department of Transportation (DOT) Code of Federal Regulations (CFR), Part 192 applies to gas pipelines. For liquid transmission pipelines, the corresponding code is Part 195. Generally, the hydrostatic test pressure must be a minimum of 125% of the MAOP. Thus, if the MAOP is 1440 psig, the pipeline will be hydrostatically tested at a minimum pressure of $1.25 \times 1440 = 1800$ psig. Because the MAOP is based on the hoop stress equal to 72% of the yield strength of the pipeline, S_h will reach a value of $1.25 \times 72\% = 90\%$ of pipe yield strength during hydrotesting. To summarize, under normal operating conditions, in liquid pipelines, the MAOP results in hoop stress S_h equal to 72% SMYS (the term SMYS represents the specified minimum yield strength of the pipe material), whereas during hydrotesting, it results in S_h equal to 90% SMYS.

In the United States, the transmission pipelines are constructed using pipe materials conforming to API 5LX standards; the commonly used pipe materials are designated as API 5LX-42, 46, 52, 60, 65, 70, and 80. API 5LX-42 pipe has an SMYS equal to 42,000 psi, whereas API 5LX-80 has an SMYS equal to 80,000 psi. The lowest grade of pipe material used is 5L Grade B, which has an SMYS of 35,000 psi. Seamless, steel pipe designated as ASTM A106 is also used for liquid pipeline systems. This pipe has an SMYS of 35,000 psi.

Calculation of internal design pressure in a pipeline is based on Barlow's equation for thin-walled cylindrical pipes and is discussed in the following section.

7.2 BARLOW'S EQUATION FOR INTERNAL PRESSURE

When a thin-walled cylindrical pipe is subject to internal pressure, the hoop stress in the pipe material is calculated using Barlow's equation as follows:

$$S_h = \frac{PD}{2t} \tag{7.1}$$

where, in USCS units,

S_h – Hoop (or circumferential) stress, psi
P – Internal pressure, psi
D – Pipe outside diameter, in.
t – Pipe wall thickness, in.

In SI units, the same Eq. (7.1) is used with S_h and P in kPa and D and t in mm.

In addition to the hoop stress, the axial stress that occurs in the longitudinal direction designated as S_a is calculated as follows:

$$S_a = \frac{PD}{4t} \tag{7.2}$$

In SI units, the same Eq. (7.2) is used with S_a and P in kPa and D and t in mm.

It can be seen from Eqs (7.1) and (7.2) that the hoop stress S_h is equal to twice the value of the axial stress S_a. Therefore, to determine the minimum wall thickness required for a pipe of diameter D subject to an internal pressure P, we use Eq. (7.1) based on the hoop stress. In practice, due to code requirements, a slightly different formula is used that considers some factors such as seam joint factor, design factor, and temperature derating factor. These are discussed later in this chapter.

For example, suppose the internal design pressure required is 1200 psig. If the outside diameter of the pipe is 20 in., and the allowable hoop stress is 50,400 psi (corresponding to 72% of API 5LX-70 pipe), the wall thickness required will be calculated from Eq. (7.1) as follows:

$$50{,}400 = 1200 \times 20/(2t)$$

Solving for the wall thickness,

$$t = 1200 \times 20/(2 \times 50{,}400) = 0.2381 \text{ in.}$$

Thus, a minimum wall thickness of $t = 0.2381$ in. is required to withstand the internal pressure of 1200 psi without exceeding a hoop stress of 50,400 psi in the 20-in. outside diameter pipeline. In this calculation, we used the hoop stress value of 50,400 psi. For liquid pipelines (and gas pipelines in Class 1 locations), 72% SMYS is the maximum allowable hoop stress. If the pipe material were API 5LX-80, the allowable hoop stress is

$$S_h = 0.72 \times 80{,}000 = 57{,}600 \text{ psi}$$

In this case, the minimum wall thickness required with X-80 pipe is, using Eq. (7.1),

$$t = 1200 \times 20/(2 \times 57{,}600) = 0.2083 \text{ in.}$$

Therefore, by changing the pipe material from X-70 to X-80, the required wall thickness for the 20-in. outside diameter pipe to withstand an internal pressure of 1200 psi is reduced from 0.2381 to 0.2083 in. This represents a percentage reduction of

$$(0.2381 - 0.2083)/0.2381 \times 100 = 12.52\%$$

If the pipe length is 100 mi, this could mean a considerable reduction in total pipe requirement and hence reduction in cost as well. However, the higher strength X-80 pipe is also more expensive than X-70 pipe and may require a slightly different welding procedure. Hence, this will require a more detailed analysis of cost to justify going with the X-80 pipe.

7.3 DERIVATION OF BARLOW'S EQUATION

The derivation of Barlow's Eq. (7.1) is as follows:

Consider one half of a length of pipe L as shown in Fig. 7.2. The internal pressure P causes a bursting force on one half of the pipe equal to pressure multiplied by the projected area as follows:

$$\text{Bursting force} = P \times D \times L$$

This bursting force is exactly balanced by the hoop stress S_h acting along both edges of the half-pipe section. Therefore,

$$(S_h \times t \times L) \times 2 = P \times D \times L$$

Solving for S_h, we get Eq. (7.1) discussed earlier.

$$S_h = \frac{PD}{2t}$$

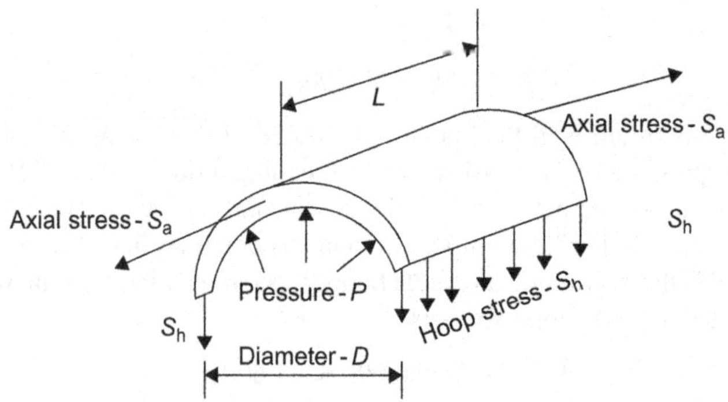

FIGURE 7.2 Derivation of Barlow's equation.

Eq. (7.2) for axial stress S_a is derived as follows:

The axial stress S_a acts on an area of cross section of pipe represented by πDt. This is balanced by the internal pressure P acting on the internal cross-sectional area of pipe equal to $\pi D^2/4$. Equating these two, we get

$$S_a \times \pi Dt = P \times \pi D^2/4$$

Solving for S_a, we get Eq. (7.2) discussed earlier

$$S_a = \frac{PD}{4t}$$

7.4 MODIFIED BARLOW'S EQUATION

In the design of transmission pipeline, the internal design pressure is calculated using Barlow's equation with some modifications mentioned below. We introduce three factors: seam joint factor (E), design factor (F), and temperature deration factor (T). In USCS units, the internal design pressure in a pipe is calculated using the following equation:

$$P = \frac{2tSEFT}{D} \tag{7.3}$$

where

P – Internal pipe design pressure, psig
D – Nominal pipe outside diameter, in.
t – Nominal pipe wall thickness, in.
S – Specified minimum yield strength of pipe material, psig
E – Seam joint factor, 1.0 for seamless and submerged arc welded (SAW) pipes (See Appendix 4.)
F – Design factor, usually 0.72 for liquid pipelines, except that a design factor of 0.60 is used for pipe, including risers, on a platform located offshore or on a platform in inland navigable waters, and 0.54 is used for pipe that

has been subjected to cold expansion to meet the SMYS and subsequently heated, other than by welding or stress relieving as a part of the welding, to a temperature higher than 900°F (482°C) for any period of time or more than 600°F (316°C) for more than 1 h.

T – Temperature deration factor equals 1.00 for temperatures less than 250°F (121°C). See Table 7.1 for details.

For gas transmission pipelines, the design factor F ranges from 0.72 for cross-country gas pipelines to as low as 0.4 in Class 4 locations. Class locations for gas pipelines depend on the population density in the vicinity of the pipeline, and the corresponding values of F are listed in Table 7.2.

The above form of Barlow's equation may be found in Parts 192 and 195 of Code of Federal Regulations, Title 49 and ASME standards B31.4 and B31.8 for liquid and gas pipelines, respectively.

In SI units, Barlow's equation can be written as follows:

$$P = \frac{2tSEFT}{D} \tag{7.4}$$

TABLE 7.1 Temperature Deration Factor

Temperature		Deration Factor, T
°F	°C	
250 or less	121 or less	1.000
300	149	0.967
350	177	0.033
400	204	0.900
450	232	0.867

TABLE 7.2 Class Locations for Gas Pipelines

Class Location	Design Factor, F
1	0.72
2	0.60
3	0.50
4	0.40

where

P – Pipe internal design pressure, kPa
D – Nominal pipe outside diameter, mm
t – Nominal pipe wall thickness, mm
S – Specified minimum yield strength of pipe material, kPa
E – Seam joint factor, 1.0 for seamless and submerged arc welded (SAW) pipes
F – Design factor, usually 0.72 for liquid pipelines, except that a design factor of 0.60 is used for pipe, including risers, on a platform located offshore or on a platform in inland navigable waters, and 0.54 is used for pipe that has been subjected to cold expansion to meet the SMYS and subsequently heated, other than by welding or stress relieving as a part of the welding, to a temperature higher than 900°F (482°C) for any period of time or more than 600°F (316°C) for more than 1 h.
T – Temperature deration factor equals 1.00 for temperatures less than 250°F (121°C).

7.5 GAS PIPELINES: CLASS LOCATIONS

The following definitions of Class 1 through Class 4 are taken from DOT 49 CFR, Part 192. The class location unit (CLU) is defined as an area that extends 220 yards (201 m) on either side of the centerline of a 1-mi (1.6-km) section of pipe as indicated in Fig. 7.3.

7.5.1 Class 1

Offshore gas pipelines are in Class 1 locations. For onshore pipelines, any class location unit that has 10 or less buildings intended for human occupancy is termed Class 1.

7.5.2 Class 2

This is any class location unit that has more than 10 but less than 46 buildings intended for human occupancy.

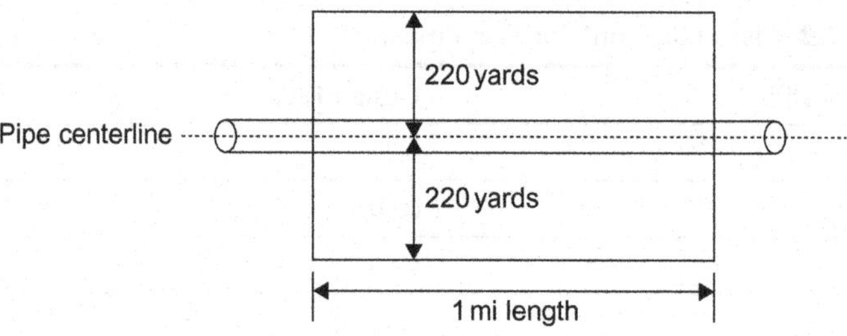

FIGURE 7.3 Class location unit.

7.5.3 Class 3

This is any class location unit that has 46 or more buildings intended for human occupancy or an area where the pipeline is within 100 yards of a building or a playground, recreation area, outdoor theater, or other places of public assembly that is occupied by 20 or more people on at least 5 days a week for 10 weeks in any 12-month period. The days and weeks need not be consecutive.

7.5.4 Class 4

This is any class location unit where buildings with four or more stories exist above the ground.

The temperature deration factor T for gas pipelines is equal to 1.00 up to gas temperature 250°F as indicated in Table 7.1.

Example Problem 7.1

A gas pipeline is constructed of API 5LX-70 steel, NPS 16, 0.250 in. wall thickness. Calculate the MAOP of this pipeline for Class 1 through Class 4 locations. Use a temperature deration factor of 1.00.

Solution

Using Eq. (7.3), the MAOP is given by

$$P = \frac{2 \times 0.250 \times 70000 \times 1.0 \times 0.72 \times 1.0}{16} = 1575.0\,\text{psig for Class 1}$$

Similarly, for

$$\text{Class 2, MAOP} = 1575.0 \times \frac{0.6}{0.72} = 1312.5\,\text{psig}$$

$$\text{Class 3, MAOP} = 1575.0 \times \frac{0.5}{0.72} = 1093.75\,\text{psig}$$

$$\text{Class 4, MAOP} = 1575.0 \times \frac{0.4}{0.72} = 875.0\,\text{psig}$$

It is obvious from Barlow's Eq. (7.3) that for a given pipe diameter, pipe material, and seam joint factor, the allowable internal pressure P is directly proportional to the pipe wall thickness. For example, 16-in. diameter pipe with a wall thickness of 0.250 in. made of 5LX-52 pipe has an allowable internal design pressure of 1170 psi calculated as follows:

$$P = (2 \times 0.250 \times 52000 \times 1.0 \times 0.72)/16 = 1170\,\text{psig}$$

Therefore, if the wall thickness is increased to 0.375 in., the allowable internal design pressure increases to

$$(0.375/0.250) \times 1170 = 1755\,\text{psig}$$

However, if the pipe material is changed to 5LX-80, keeping the wall thickness at 0.250 in., the new internal pressure is

$$(80{,}000/52{,}000)\ 1170 = 1800\,\text{psig}$$

Note that we used Barlow's equation to calculate the allowable internal pressure based on the pipe material being stressed to 72% of SMYS. In some situations, more stringent city or government regulations may require that the pipe be operated at a lower pressure. Thus, instead of using a 72% factor in Eq. (7.3), we may be required to use a more conservative factor (lower number) in place of $F = 0.72$.

As an example, in certain areas of the City of Los Angeles, liquid pipelines are only allowed to operate at a 66% factor instead of the 72% factor. Therefore, in the earlier example, the 16 in./0.250 in./X52 pipeline can only be operated at MAOP = 1170(66/72) = 1073 psig.

As mentioned before, in order to operate a pipeline at 1170 psig, it must be hydrostatically tested at 25% higher pressure. As 1170 psig internal pressure is based on the pipe material being stressed to 72% of SMYS, the hydrostatic test pressure will cause the hoop stress to reach 1.25(72) = 90% of SMYS.

Generally, the hydrostatic test pressure is specified as a range of pressures, such as 90–98% SMYS. This is called the hydrotest pressure envelope. Therefore, in the present example, the range of hydrostatic test pressure is

$$1.25 \times 1170 = 1463\ \text{psig} - \text{lower limit (90\% SMYS)}$$
$$(98/90) \times 1463 = 1593\ \text{psig} - \text{higher limit (98\% SMYS)}$$

To summarize, in this example, a pipeline with MAOP of 1170 psig needs to be hydrotested at a pressure range of 1463–1593 psig. According to the design code, the test pressure will be held for a minimum of 4 h for above ground pipelines and 8 h for buried pipelines. Hydrotesting is discussed in more detail in Chapter 18.

In calculating the allowable internal pressure in older pipelines, consideration must be given to wall thickness reduction because of the corrosion over the life of the pipeline. A pipeline that was installed 25 years ago with an initial wall thickness of 0.250 in. may have reduced in wall thickness to 0.200 in. or less due to corrosion. Therefore, the allowable internal pressure will have to be reduced in the ratio of the wall thickness compared with the original design pressure.

7.6 THICK-WALLED PIPES

Consider a thick-walled pipe with an outside diameter D_O and inside diameter D_i and internal pressure of P. The largest stress in the pipe wall will be found to occur in the circumferential direction near the inner surface of the pipe. According to standard text books on thick cylinders, this stress can be calculated from the following equation:

$$S_{max} = \frac{P(D_o^2 + D_i^2)}{(D_o^2 - D_i^2)} \tag{7.5}$$

The pipe wall thickness is

$$t = \frac{D_o - D_i}{2} \tag{7.6}$$

Rewriting Eq. (7.5) in terms of outside diameter and wall thickness, we get

$$S_{max} = P\left[\frac{D_o^2 + (D_o - 2t)^2}{D_o^2 - (D_o - 2t)^2}\right]$$

Simplifying further,

$$S_{max} = \frac{PD_o}{2t}\left[\frac{1 - \left(\frac{t}{D_o}\right) + 2\left(\frac{t}{D_o}\right)^2}{1 + \left(\frac{t}{D_o}\right)}\right] \tag{7.7}$$

In the limiting case, a thin-walled pipe is one in which the wall thickness is very small compared with the diameter D_O. In this case, (t/D) is small compared with 1 and, therefore, can be neglected in Eq. (7.7). Therefore, the approximation for thin-walled pipes from Eq. (7.7) becomes

$$S_{max} = \frac{PD_o}{2t},$$

which is the same as Barlow's Eq. (7.1) for hoop stress.

Example Problem 7.2

A gas pipeline is subjected to an internal pressure of 1400 psig. It is constructed of steel pipe with 24-in. outside diameter and 0.75 in. wall thickness. Calculate the maximum hoop stress in the pipeline considering both the thin-walled pipe approach and the thick-walled pipe equation. What is the error in assuming that the pipe is thin walled?

Solution

$$\text{Pipe inside diameter} = 24 - 2 \times 0.75 = 22.5\text{ in.}$$

From Eq. (7.1) for thin-walled pipe, Barlow's equation gives the maximum hoop stress as

$$S_h = \frac{1400 \times 24}{2 \times 0.75} = 22{,}400\text{ psig}$$

Considering the thick-walled pipe formula Eq. (7.5),

$$S_{max} = \frac{1400\,(24^2 + 22.5^2)}{(24^2 - 22.5^2)} = 21{,}723\text{ psig}$$

Therefore, by assuming thin-walled pipe, the hoop stress is overestimated by approximately $\frac{22{,}400 - 21{,}723}{21{,}723} = 0.0312$ or 3.12%, which is quite small.

Hence, this pipe can be assumed to be a thin-walled pipe for the purpose of calculating the hoop stress due to internal pressure.

7.7 MAINLINE VALVES

Valve stations are discussed in detail in Chapter 10. Mainline valves are installed in pipelines so that portions of the pipeline may be isolated for hydrostatic testing and maintenance. Valves are also necessary to separate sections of pipe and minimize liquid or gas loss that may occur as a result of pipe rupture from construction damage. Design codes specify the spacing of these valves based on class location, which in turn depends on the population density around the pipeline. The maximum spacing between mainline valves in gas transmission piping is listed below. These data are taken from ASME B31.8 code.

Class Location	Valve Spacing
1	20 mi
2	15 mi
3	10 mi
4	5 mi

It can be seen from the preceding that the valve spacing is shorter as the pipeline traverses high-population areas. This is necessary as a safety feature to protect the inhabitants in the vicinity of the pipeline by restricting the amount of gas that may escape because of the rupture of the pipeline.

7.8 BLOWDOWN CALCULATIONS

Blowdown valves are installed around the mainline valve in a gas transmission piping system to evacuate gas from sections of pipeline in the event of an emergency or for maintenance. The objective of the blowdown assembly is to remove gas from the pipeline once the pipe section is isolated by closing the mainline block valves in a reasonable period of time. The pipe size required to blowdown a section of pipe will depend on the gas gravity, pipe diameter, length of pipe section, the pressure in the pipeline, and the blowdown time. The American Gas Association (AGA) recommends the following equation to estimate the blowdown time:

$$T = \frac{0.0588 P_1^{\frac{1}{3}} G^{\frac{1}{2}} D^2 L F_c}{d^2} \text{ (USCS units)} \tag{7.8}$$

where

T – Blowdown time, min
P_1 – Initial pressure, psia
G – Gas gravity (air = 1.00)
D – Pipe inside diameter, in.
L – Length of pipe section, mi

d – Inside diameter of blowdown pipe, in.
F_c – Choke factor (as follows)

The choke factor list is as follows:

Ideal nozzle = 1.0
Through gate = 1.6
Regular gate = 1.8
Regular lube plug = 2.0
Venture lube plug = 3.2

In SI units,

$$T = \frac{0.0886 P_1^{\frac{1}{3}} G^{\frac{1}{2}} D^2 L F_c}{d^2} \text{ (SI units)} \tag{7.9}$$

where

P_1 – Initial pressure, kg/cm^2
D – Pipe inside diameter, mm
L – Length of pipe section, km
d – Pipe inside diameter of blowdown, mm

Other symbols are as defined before.

EXAMPLE PROBLEM 7.3

Calculate the blowdown time required for a NPS 8, 0.250 in. wall thickness blowdown assembly on an NPS 20 pipe, 0.500 in. wall thickness considering a 5-mi pipe section starting at a pressure of 1000 psia. The gas gravity is 0.6, and the choke factor is 1.8.

Solution

$$\text{Pipe inside diameter} = 20 - 2 \times 0.500 = 19.0 \text{ in.}$$
$$\text{Blowdown pipe inside diameter} = 8.625 - 2 \times 0.250 = 8.125 \text{ in.}$$

Using Eq. (7.8), we get

$$T = \frac{0.0588 \times (1000)^{\frac{1}{3}} (0.6)^{\frac{1}{2}} (19)^2 \times 5 \times 1.8}{8.125^2} = 22.4 \text{ min}$$

7.9 DETERMINING PIPE TONNAGE

Frequently, in pipeline design, we are interested in knowing the amount of pipe used so that we can determine the total cost of pipe. A convenient formula for calculating the weight per unit length of pipe, used by pipe vendors, is given in Eq. (7.10).

In USCS units, pipe weight in lb/ft is calculated for a given diameter and wall thickness as follows:

$$w = 10.68 \times t \times (D - t) \quad \text{(USCS units)} \tag{7.10}$$

where

w – Pipe weight, lb/ft
D – Pipe outside diameter, in.
t – Pipe wall thickness, in.

The constant 10.68 in Eq. (7.10) includes the density of steel and, therefore, the equation is only applicable to steel pipe. For other pipe material, we can multiply by the ratio of densities to obtain the pipe weight for non-steel pipe.

In SI units, the pipe weight in kg/m is found from

$$w = 0.0246 \times t \times (D - t) \quad \text{(SI units)} \tag{7.11}$$

where

w – Pipe weight, kg/m
D – Pipe outside diameter, mm
t – Pipe wall thickness, mm

Example Problem 7.4

Calculate the total amount of pipe in a 10-mi pipeline, NPS 20, 0.500 in. wall thickness. If pipe costs $1150 per ton, determine the total pipeline cost.

Solution

Using Eq. (7.10), the weight per foot of pipe is

$$w = 10.68 \times 0.500 \times (20 - 0.500) = 101.46 \text{ lb/ft}$$

Therefore, the total pipe tonnage in 10 mi of pipe is

$$\text{Tonnage} = 101.46 \times 5280 \times 10/2000 = 2679 \text{ tons}$$
$$\text{Total pipeline cost} = 2679 \times 1150 = \$3{,}080{,}850$$

Example Problem 7.5

A 50-km pipeline consists of 20 km of DN 500, 12 mm wall thickness pipe connected to a 30 km length of DN 400, 10 mm wall thickness pipe. What is the total metric tons of pipe?

Solution

Using Eq. (7.11), the weight per meter of DN 500 pipe is

$$w = 0.0246 \times 12 \times (500 - 12) = 144.06 \text{ kg/m}$$

The weight per meter of DN 400 pipe is

$$w = 0.0246 \times 10 \times (400 - 10) = 95.94\,\text{kg/m}$$

Therefore, the total pipe weight for 20 km of DN 500 pipe and 30 km of DN 400 pipe is

$$\text{Weight} = (20 \times 144.06) + (30 \times 95.94) = 5759.4$$
$$\text{Total metric tons} = 5759.4$$

Example Problem 7.6

Calculate the MAOP for NPS 20 pipeline, 0.375 in. wall thickness constructed of API 5LX-60 steel. What minimum wall thickness is required for an internal working pressure of 1440 psi? Use Class 2 construction with design factor $F = 0.60$ and for an operating temperature less than 250°F.

Solution

Using Eq. (7.3), the internal design pressure is

$$P = \frac{2 \times 0.375 \times 60{,}000 \times 0.60 \times 1.0 \times 1.0}{20} = 975\,\text{psig}$$

For an internal working pressure of 1,440 psi, the wall thickness required is

$$1440 = \frac{2 \times t \times 60{,}000 \times 0.6 \times 1.0}{20}$$

Solving for t, we get

$$\text{Wall thickness } t = 0.400\,\text{in.}$$

Example Problem 7.7

A natural gas pipeline, 600 km long, is constructed of DN 800 pipe and has a required operating pressure of 9 MPa. Compare the cost of using X-60 or X-70 steel pipe. The material cost of the two grades of pipe are as follows:

Pipe Grade	Material Cost, $/ton
X-60	800
X-70	900

Use Class 1 design factor and a temperature deration factor of 1.00.

Solution

We will first determine the wall thickness of pipe required to withstand the operating pressure of 9 MPa.

$$\text{X-60 pipe has SMYS} = 60{,}000\,\text{psi} = 60{,}000/145.0 = 414\,\text{MPa}$$
$$\text{X-70 pipe has SMYS} = (70/60) \times 414 = 483\,\text{MPa}$$

Using Eq. (7.3), the pipe wall thickness required for X-60 pipe is

$$t = \frac{9 \times 800}{2 \times 414 \times 1.0 \times 0.72 \times 1.0} = 12.08\,\text{mm (use 13 mm wall thickness)}$$

Similarly, the wall thickness required for X-70 pipe is

$$t = \frac{9 \times 800}{2 \times 483 \times 1.0 \times 0.72 \times 1.0} = 10.35\,\text{mm (use 11 mm wall thickness)}$$

Pipe weight in kg/m will be calculated using Eq. (7.11)
For X-60 pipe,

$$\text{weight per meter} = 0.0246 \times 13 \times (800 - 13) = 251.68\,\text{kg/m}$$

Therefore, total cost of the 600-km pipeline at $800 per ton of X-60 pipe is

$$\text{Total cost} = 600 \times 251.68 \times 800 = \$120.81\,\text{million}$$

Similarly, pipe weight in kg/m for X-70 pipe is

$$\text{weight per meter} = 0.0246 \times 11 \times (800 - 11) = 213.50\,\text{kg/m}$$

Therefore, total cost of the 600-km pipeline at $900 per ton of X-70 pipe is

$$\text{Total cost} = 600 \times 213.50 \times 900 = \$115.29\,\text{million}$$

Therefore, the X-70 pipe will cost less than the X-60 pipe. The difference in cost is $120.81 − $115.29 = $5.52 million

SUMMARY

In this chapter, we discussed how to calculate the allowable internal pressure in a pipeline, depending on pipe diameter, wall thickness, and the yield strength of the pipe material. We showed that for a pipe under internal pressure, the hoop stress in the pipe material will be a controlling factor.

The influence of the population density in the vicinity of the gas pipeline on the required pipe wall thickness was illustrated using the design factor based on class locations. The importance of design factor in selecting pipe wall thickness was illustrated using examples. Based on Barlow's equation, the internal design pressure calculation as recommended by ASME standards B31.4 and B31.8 and the 49 CFR, Parts 192 and 195 was illustrated.

We explored the range of pressures required for hydrostatic testing of pipeline sections to ensure safe operation of the pipeline. The need for isolating portions of the pipeline by properly spaced mainline valves and the method of calculating the time required for evacuating gas from the pipeline sections were also discussed. Finally, a simple method of calculating the total pipe tonnage was explained.

BIBLIOGRAPHY

G.M. Jones, R.L. Sanks, G. Tchobanoglous, B. Bossermann, Pumping Station Design, Revised third ed., Elsevier, Inc., 2008.

E. Shashi Menon, Liquid Pipeline Hydraulics, Marcel-Dekker, New York, 2004.

E. Shashi Menon, Gas Pipeline Hydraulics, CRC Press, Boca Raton, FL, 2005.

M. Mohitpour, H. Golshan, A. Murray, Pipeline Design and Construction, second ed., ASME Press, New York, 2003.

E. Shashi Menon, Piping Calculations Manual, McGraw-Hill, New York, 2005.

Chapter 8

Pipeline Hydraulic Analysis

E. Shashi Menon, Ph.D., P.E.

Chapter Outline

INTRODUCTION

In this chapter, we discuss the important category of pipeline hydraulic analysis for both liquid and gas pipelines. Building on the liquid and gas properties and the pipe strength analysis in earlier chapters, this chapter will review the different types of flows, Reynolds number, and pressure drop resulting from friction. Flow of liquids in pipelines will be discussed first, followed by gas flow in pipelines. Next, we will discuss how the pipe elevation profile affects the pumping pressure requirements and how to calculate the number and location of pumping stations and compressor stations.

8.1 VELOCITY OF FLOW IN LIQUID PIPELINES

The rate at which a fluid flows through a pipeline may be measured by the velocity of flow. In steady state flow, the average velocity V at any point in a pipeline is constant and it depends on the volume flow rate Q and the inside diameter D of the pipe. In USCS units, the velocity can be calculated as follows:

$$V = 0.4085(Q)/D^2 \text{ (USCS)} \tag{8.1}$$

where

V – Average velocity, ft/s
Q – Volume flow rate, gal/min
D – Pipe inside diameter, in.

If the flow rate Q is in bbl/h, the velocity is calculated from

$$V = 0.2859(Q)/D^2 \text{ (USCS)} \tag{8.2}$$

In SI units, the average velocity of flow can be calculated as follows:

$$V = 353.6777(Q)/D^2 \text{ (SI)} \tag{8.3}$$

where

V – Average velocity, m/s
Q – Volume flow rate, m^3/h
D – Inside diameter, mm

For example, liquid flowing through a 20-in. pipeline, 0.500 in. wall thickness at the rate of 100,000 bbl/day, has an average velocity of

$$V = 0.2859(100{,}000/24)/(20 - 2 \times 0.5)^2 = 3.3 \text{ ft/s}$$

Similarly, a DN 400 pipe, 8 mm wall thickness flowing gasoline at 500 m^3/h has an average velocity calculated using Eq. (8.3) as follows:

$$V = 353.6777(500)/(400 - 2 \times 8)^2 = 1.2 \text{ m/s}$$

This represents the average velocity at a particular cross section of pipe. The velocity at the centerline will be higher than this, depending on whether the flow is turbulent or laminar. This brings up an important question. How high can the flow velocity be? For a given pipe size (diameter and wall thickness), if the flow rate increases, the velocity also increases. However, as we shall see in subsequent sections of this chapter, there are practical limits to the velocity and hence the flow rate possible in a given pipe diameter.

Table 8.1 shows the average velocity in pipes of various diameters at different flow rates.

TABLE 8.1 Average Liquid Velocity in Pipes

Flow Rate (gal/min)	In USCS Units (ft/s)											
	Inside Diameter (in.)											
	4	**6**	**8**	**10**	**12**	**13.5**	**15.5**	**17.5**	**19**	**23**	**29**	**35**
100	2.55	1.13	0.64	0.41	0.28	0.22	0.17	0.13	0.11	0.08	0.05	0.03
200	5.11	2.27	1.28	0.82	0.57	0.45	0.34	0.27	0.23	0.15	0.10	0.07
500	12.77	5.67	3.19	2.04	1.42	1.12	0.85	0.67	0.57	0.39	0.24	0.17
1000	25.53	11.35	6.38	4.09	2.84	2.24	1.70	1.33	1.13	0.77	0.49	0.33
2000	51.06	22.69	12.77	8.17	5.67	4.48	3.40	2.67	2.26	1.54	0.97	0.67
3000	76.59	34.04	19.15	12.26	8.51	6.72	5.10	4.00	3.39	2.32	1.46	1.00
4000	102.13	45.39	25.53	16.34	11.35	8.97	6.80	5.34	4.53	3.09	1.94	1.33
5000	127.66	56.74	31.91	20.43	14.18	11.21	8.50	6.67	5.66	3.86	2.43	1.67
6000	153.19	68.08	38.30	24.51	17.02	13.45	10.20	8.00	6.79	4.63	2.91	2.00
Flow Rate (m^3/h)	**In SI Units (m/s)**											
	Inside Diameter (mm)											
	100	**150**	**200**	**250**	**300**	**338**	**394**	**445**	**483**	**584**	**737**	**889**
10	0.35	0.16	0.09	0.06	0.04	0.03	0.02	0.02	0.02	0.01	0.01	0.00
20	0.71	0.31	0.18	0.11	0.08	0.06	0.05	0.04	0.03	0.02	0.01	0.01

(Continued)

TABLE 8.1 Average Liquid Velocity in Pipes—cont'd

Flow Rate (m^3/h)	In SI Units (m/s)											
	Inside Diameter (mm)											
	100	150	200	250	300	338	394	445	483	584	737	889
50	1.77	0.79	0.44	0.28	0.20	0.15	0.11	0.09	0.08	0.05	0.03	0.02
60	2.12	0.94	0.53	0.34	0.24	0.19	0.14	0.11	0.09	0.06	0.04	0.03
80	2.83	1.26	0.71	0.45	0.31	0.25	0.18	0.14	0.12	0.08	0.05	0.04
100	3.54	1.57	0.88	0.57	0.39	0.31	0.23	0.18	0.15	0.10	0.07	0.04
200	7.07	3.14	1.77	1.13	0.79	0.62	0.46	0.36	0.30	0.21	0.13	0.09
500	17.68	7.86	4.42	2.83	1.96	1.55	1.14	0.89	0.76	0.52	0.33	0.22
600	21.22	9.43	5.31	3.40	2.36	1.86	1.37	1.07	0.91	0.62	0.39	0.27
800	28.29	12.58	7.07	4.53	3.14	2.48	1.82	1.43	1.21	0.83	0.52	0.36
1000	35.37	15.72	8.84	5.66	3.93	3.10	2.28	1.79	1.52	1.04	0.65	0.45

For different pipe inside diameter D, multiply table values by $(Dt/D)^2$ where Dt is the diameter in the table.

8.2 REYNOLDS NUMBER IN LIQUID FLOW

The flow in a liquid pipeline may be smooth, laminar flow, also known as viscous flow. In this type of flow, the liquid flows in layers or laminations without causing eddies or turbulence. If the pipe is transparent and we inject a dye into the flowing stream, it would flow smoothly in a straight line confirming smooth or laminar flow. As the liquid flow rate is increased, the velocity increases and the flow will change from laminar flow to turbulent flow with eddies and disturbances. This can be seen clearly when a dye is injected into the flowing stream. The point at which the flow changes from laminar to turbulent is not clearly defined. In fact, there is an intermediate zone called the critical flow region in which the flow is indeterminate. The flow type is characterized using a dimensionless parameter known as the Reynolds number.

The Reynolds number R is used to classify the type of flow in pipelines. It is calculated as follows:

$$R = VD\rho/\mu \text{ (USCS)} \tag{8.4}$$

where in USCS units

V – Average velocity, ft/s
D – Pipe inside diameter, ft
ρ – Liquid density, slug/ft^3
μ – Absolute viscosity, lb · s/ft^2
R – Reynolds number, a dimensionless value

Because the kinematic viscosity $\nu = \mu/\rho$, the Reynolds number can also be expressed as

$$R = VD/\nu \text{ (USCS)} \tag{8.5}$$

where

ν – Kinematic viscosity, ft^2/s

Care should be taken to ensure that proper units are used in Eqs (8.4) and (8.5) such that R is dimensionless.

In SI units, R is calculated from the following equation:

$$R = VD/\nu \text{ (SI)} \tag{8.6}$$

where

V – Average velocity, m/s
D – Pipe inside diameter, m
ν – Kinematic viscosity, m^2/s

Flow through pipes is classified into three main flow regimes, based on the value of the Reynolds number, R.

1. Laminar flow – $R \le 2000$
2. Critical flow – $R > 2000$ and $R \le 4000$
3. Turbulent flow – $R > 4000$

Depending on the Reynolds number, flow through pipes will fall in one of the above three flow regimes. Note that in some publications, an R value of 2100 is used as the limit of laminar flow.

Using customary units in the pipeline industry, the Reynolds number can be calculated using the following formula in USCS units:

$$R = 2214\, Q/(\nu D)\ \text{(USCS)} \tag{8.7}$$

where

Q – Flow rate, bbl/h
D – Pipe inside diameter, in.
ν – Kinematic viscosity, cSt

If the flow rate Q is in gal/min, R is calculated from

$$R = 3160\, Q/(\nu D)\ \text{(USCS)} \tag{8.8}$$

The corresponding equation in SI units is

$$R = 353{,}678\, Q/(\nu D)\ \text{(SI)} \tag{8.9}$$

where

Q – Flow rate, m^3/h
D – Pipe inside diameter, mm
ν – Kinematic viscosity, cSt

Consider a 16-in. pipeline, 0.250 in. wall thickness transporting a liquid of viscosity 250 cSt. At a flow rate of 50,000 bbl/day, the Reynolds number is, using Eq. (8.7),

$$R = 2214(50{,}000/24)/(250 \times 15.5) = 1190.32$$

Since R is less than 2000, this flow is in the laminar region. If the flow rate increases to 150,000 bbl/day, the Reynolds number becomes 3571, and the flow will be in the critical flow region (R between 2000 and 4000). It can be concluded that at flow rates more than approximately 169,000 bbl/day, the Reynolds number exceeds 4000; therefore, the flow will be in the turbulent region. Thus, for this 16-in. pipeline and a liquid viscosity of 250 cSt, flow will be fully turbulent at flow rates more than 169,000 bbl/day.

As the flow rate and velocity increase, the flow regime changes. With change in flow regime, the energy lost due to pipe friction increases. At laminar flow, there is less frictional energy lost when compared with turbulent flow.

Table 8.2 shows the Reynolds number for flow of water (viscosity = 1.0 cSt) in pipes of various diameters. For a liquid of viscosity ν cSt the Reynolds

TABLE 8.2 Reynolds Number for Flow of Water (Viscosity = 1.0 cSt)

Flow Rate (gal/min)	In USCS Units											
	Inside Diameter (in.)											
	4	6	8	10	12	13.5	15.5	17.5	19	23	29	35
100	79,000	52,667	39,500	31,600	26,333	23,407	20,387	18,057	16,632	13,739	10,897	9,029
200	158,000	105,333	79,000	63,200	52,667	46,815	40,774	36,114	33,263	27,478	21,793	18,057
500	395,000	263,333	197,500	158,000	131,667	117,037	101,935	90,286	83,158	68,696	54,483	45,143
1000	790,000	526,667	395,000	316,000	263,333	234,074	203,871	180,571	166,316	137,391	108,966	90,286
2000	1,580,000	1,053,333	790,000	632,000	526,667	468,148	407,742	361,143	332,632	274,783	217,931	180,571
3000	2,370,000	1,580,000	1,185,000	948,000	790,000	702,222	611,613	541,714	498,947	412,174	326,897	270,857
4000	3,160,000	2,106,667	1,580,000	1,264,000	1,053,333	936,296	815,484	722,286	665,263	549,565	435,862	361,143
5000	3,950,000	2,633,333	1,975,000	1,580,000	1,316,667	1,170,370	1,019,355	902,857	831,579	686,957	544,828	451,429
6000	4,740,000	3,160,000	2,370,000	1,896,000	1,580,000	1,404,444	1,223,226	1,083,429	997,895	824,348	653,793	541,714
Flow Rate (m^3/h)	**In SI Units**											
	Inside Diameter (mm)											
	100	**150**	**200**	**250**	**300**	**338**	**394**	**445**	**483**	**584**	**737**	**889**
10	35,368	23,579	17,684	14,147	11,789	10,464	8,977	7,948	7,323	6,056	4,799	3,978
20	70,736	47,157	35,368	28,294	23,579	20,928	17,953	15,896	14,645	12,112	9,598	7,957
50	176,839	117,893	88,420	70,736	58,946	52,319	44,883	39,739	36,613	30,281	23,994	19,892

(Continued)

TABLE 8.2 Reynolds Number for Flow of Water (Viscosity = 1.0 cSt)—cont'd

Flow Rate (m^3/h)	In SI Units											
	Inside Diameter (mm)											
	100	**150**	**200**	**250**	**300**	**338**	**394**	**445**	**483**	**584**	**737**	**889**
60	212,207	141,471	106,103	84,883	70,736	62,783	53,860	47,687	43,935	36,337	28,793	23,870
80	282,942	188,628	141,471	113,177	94,314	83,711	71,813	63,583	58,580	48,449	38,391	31,827
100	353,678	235,785	176,839	141,471	117,893	104,638	89,766	79,478	73,225	60,561	47,989	39,784
200	707,356	471,571	353,678	282,942	235,785	209,277	179,532	158,956	146,451	121,123	95,978	79,568
500	1,768,390	1,178,927	884,195	707,356	589,463	523,192	448,830	397,391	366,126	302,807	239,944	198,919
600	2,122,068	1,414,712	1,061,034	848,827	707,356	627,831	538,596	476,869	439,352	363,368	287,933	238,703
800	2,829,424	1,886,283	1,414,712	1,131,770	943,141	837,108	718,128	635,826	585,802	484,490	383,911	318,270
1000	3,536,780	2,357,853	1,768,390	1,414,712	1,178,927	1,046,385	897,660	794,782	732,253	605,613	479,889	397,838

For other liquids (visc = *v*) and other diameters (*D*), multiply table values by (Dt/*D*)(1/*v*), where Dt is the diameter in the table.

number can be easily obtained by multiplying the table value by ($1/\nu$), as explained in the footnote to the table.

8.3 PRESSURE AND HEAD OF A LIQUID

The pressure in a liquid is the force per unit area that acts in all directions at a point within the liquid. Imagine a large storage tank, 50 ft high containing water to a depth of 30 ft. At the liquid surface, the pressure is atmospheric or a gauge pressure of zero psig. At a depth of 20 ft below the liquid surface, the pressure is equal to that of the column of liquid 20 ft high. At various points along the 20 ft column, the liquid pressure increases from zero psig at the surface to some pressure P at a depth h.

In fact, the pressure P at a depth h is related to the specific gravity Sg of the liquid (in this case, water) as follows:

$$P = h \times \text{Sg}/2.31 \text{ (USCS)} \tag{8.10}$$

where P is in psig, h is in ft, and Sg is dimensionless.

Thus, for water at $h = 20$ ft, we have a pressure of

$$P = 20 \times 1.0/2.31 = 8.66 \text{ psig}$$

Accordingly at the bottom of the tank (depth of water = 30 ft), the pressure is

$$P = 30 \times 1.0/2.31 = 12.99 \text{ psig}$$

In comparison, if the tank contained diesel fuel (Sg = 0.85), the pressure at the same 30 ft depth is

$$P = 30 \times 0.85/2.31 = 11.04 \text{ psig}$$

The height h corresponding to a pressure P is called the liquid head.

Thus, pressure in psig can be easily converted to the head of liquid in ft, using Eq. (8.10).

In SI units, the pressure P in kPa may be related to the liquid head h in meters as follows:

$$P = h \times \text{Sg}/0.102 \text{ (SI)} \tag{8.11}$$

For example, a pressure of 500 kPa in a gasoline pipeline (Sg = 0.74) is equivalent to a head of h meters that can be found as follows, using Eq. (8.11):

$$500 = h \times 0.74/0.102$$

Solving for h, we get $h = 68.92$ m.

Table 8.3 shows the pressures in psi versus head in ft of water. For liquids other than water, the equivalent feet of head for a given pressure can be easily calculated by multiplying the table values by the factor (1/Sg) where Sg is the specific gravity of the liquid, as explained in the footnote to the table. For SI units, Table 8.4 provides the conversion values from pressure in kPa to head in meters of water.

In liquid pipelines, we refer to pressure in terms of gauge pressure (psig or kPa gauge). The absolute pressure (P_a) is equal to the gauge pressure (P_g) plus the local atmospheric pressure (P_{atm}):

$$P_a = P_g + P_{atm} \tag{8.12}$$

Thus, at a location where the atmospheric pressure is 14.7 psi, a liquid pressure of 800 psig is equal to an absolute pressure of 814.7 psia.

Note that, in USCS units, the gauge pressure and absolute pressure are denoted by psig and psia, respectively. In SI units, it is kPa(g) for gauge pressure and kPa(abs) for absolute pressure.

Similarly, for a liquid pressure of 3500 kPa (gauge) at a location where the atmospheric pressure is 101 kPa, the absolute pressure in the liquid is

$$P_a = 3500 + 101 = 3601\,\text{kPa (abs)}$$

Generally, the gauge pressure is used in liquid pipelines calculations. However, in gas pipelines, we almost always use the absolute pressure.

TABLE 8.3 Pressures in psi versus Head in Feet of Water (USCS)

Pressure (psig)	Head (ft)
100	231.00
200	462.00
500	1155.00
1000	2310.00
1200	2772.00
1400	3234.00
1600	3696.00
1800	4158.00
2000	4620.00

For liquids other than water, multiply table values of head by (1/Sg), where Sg is the specific gravity of the liquid.

TABLE 8.4 Pressures in kPa versus Head in Meters of Water (SI)

Pressure (kPa)	Head (m)
100	10.20
200	20.40
500	51.00
1000	102.00
1200	122.40
1400	142.80
1600	163.20
1800	183.60
2000	204.00
4000	408.00
6000	612.00
8000	816.00
10,000	1020.00

For liquids other than water, multiply table values of head by (1/Sg), where Sg is the specific gravity of the liquid.

8.4 PRESSURE DROP IN LIQUID FLOW

As liquid flows through a pipeline, depending on the flow rate, liquid gravity, viscosity and the internal roughness of the pipe, a certain amount of pressure loss occurs from the upstream end A to the downstream end B of the pipe. This pressure drop resulting from friction is also called the head loss (h) and can be calculated using the Darcy–Weisbach (or simply Darcy) equation, as follows:

$$h = f(L/D)\,(V^2/2g) \tag{8.13}$$

where the pressure drop due to friction, h, is in ft of liquid head and the other symbols are defined below in USCS units:

f – Darcy friction factor, a dimensionless number, 0.008–0.10
L – Pipe length, ft
D – Pipe inside diameter, ft
V – Average liquid velocity, ft/s
g – Acceleration due to gravity, 32.2 ft/s^2

For example, consider an NPS 20 pipeline, 0.500 in. wall thickness, 1000 ft long that has water (Sg = 1.0) flowing at 5000 gal/min. If we assume a friction factor $f = 0.02$, the head loss can be calculated as follows, from Eq. (8.10). First calculate the average flow velocity V using Eq. (8.1) as follows:

$$V = 0.4085 \times 5000/(20 - 2 \times 0.500)^2 = 5.6579\ \text{ft/s}$$

The head loss from Eq. (8.13) is

$$h = 0.02(1000 \times 12/19.0)(5.6579)^2/(2 \times 32.2) = 6.28\ \text{ft}$$

Converting this head to pressure from Eq. (8.10),

$$\text{Pressure loss} = 6.28 \times 1.0/2.31 = 2.72\ \text{psig}$$

Notice that we did not have to use the specific gravity of the water to calculate the head loss, h. But to convert head to pressure, we had to use specific gravity. This shows that the head loss (in ft or m) in a pipeline is independent of the specific gravity of the liquid.

The Darcy equation does not use the familiar pipeline units for flow rate, pipe diameter, and so on. So, we now present a pressure loss equation that uses the more customary pipeline units of measurement.

Pressure drop due to friction per unit length of pipe, in USCS units is

$$P_m = 0.0605 f\, Q^2 (\text{Sg}/D^5)\ (\text{USCS}) \tag{8.14}$$

where

P_m – Pressure drop resulting from friction, psi per mile of pipe (psi/mi)
Q – Liquid flow rate, bbl/day
f – Darcy friction factor, dimensionless
Sg – Liquid specific gravity
D – Pipe inside diameter, in.

Sometimes, the transmission factor F is used in place of the friction factor f. This factor F is directly proportional to the volume that can be transported through the pipeline and, therefore, has an inverse relationship with the friction factor f. The transmission factor F is calculated from the following equation:

$$F = 2/\sqrt{f} \tag{8.15}$$

Because the friction factor f ranges from 0.008 to 0.10 it can be seen from Eq. (8.15) that the transmission factor F approximately ranges from 6 to 22. As will be seen later, the transmission factor is used more commonly in gas pipelines than in liquid pipelines.

In SI units, the Darcy equation (in pipeline units) for the pressure drop in terms of the friction factor is represented as follows:

$$P_{km} = 6.2475 \times 10^{10} f\, Q^2 (\text{Sg}/\text{D}^5)\ (\text{SI}) \tag{8.16}$$

where

P_{km} – Pressure drop due to friction, kPa/km
Q – Liquid flow rate, m^3/h
f – Darcy friction factor, dimensionless
Sg – Liquid specific gravity, dimensionless
D – Pipe inside diameter, mm

Example Problem 8.1

A refined products pipeline, NPS 16, 0.250 in. wall thickness is used to ship diesel fuel (Sg = 0.85) at a flow rate of 4200 bbl/h. Assuming a friction factor of 0.015, calculate the frictional pressure drop in a 10-mi segment of the pipeline. What is the value of the transmission factor?

Solution

Using Eq. (8.14), we calculate the pressure loss as

$$P_m = 0.0605 \times 0.015 \times (4200 \times 24)^2 (0.85/(16 - 2 \times 0.25)^5) = 8.7605 \text{ psi/mi}$$

Therefore, the frictional pressure drop in a 10-mi segment is

$$\text{Pressure drop} = 10 \times 8.7605 = 87.61 \text{ psi}$$

The transmission factor is found from Eq. (8.15) as

$$F = 2/\sqrt{0.015} = 8.165$$

8.5 FRICTION FACTOR

So far, we have used assumed values of the Darcy friction factor f. The actual value of f depends on the Reynolds number, pipe diameter, and the internal roughness of the pipe. If the flow is laminar ($R < 2000$), the friction factor depends only on the Reynolds number and is not affected by the internal roughness as seen from the following equation:

$$f = 64/R \tag{8.17}$$

It can be seen from Eq. (8.17) that for laminar flow, the friction factor depends only on the Reynolds number and is independent of the internal condition of the pipe. Thus, regardless of whether the pipe is smooth or rough, the friction factor for laminar flow is a number that varies inversely as the Reynolds number. Therefore, if the Reynolds number $R = 1980$, then friction factor is $f = 64/1980 = 0.0323$.

From Eq. (8.17), it appears that for laminar flow, f decreases with increasing Reynolds number. Therefore, we are tempted to conclude from the pressure drop Eq. (8.14) that the pressure drop will decrease as the flow rate and hence R increases. However, this is not true. Because pressure drop is proportional to the flow rate Q squared per Eq. (8.14), the influence of Q is greater than that of f. Therefore, the pressure drop will increase with flow rate in the laminar region.

To illustrate, consider a 16-in. pipeline, 0.250 in. wall thickness transporting a liquid with Sg = 0.85 and viscosity = 250 cSt. At a flow rate of 50,000 bbl/day, the Reynolds number using Eq. (8.7) is

$$R = 2214(50{,}000/24)/(250 \times 15.5) = 1190.32$$

If the flow rate is increased from 50,000 to 80,000 bbl/day, the Reynolds number R will increase from 1190.32 to 1904.51 (laminar flow).

The friction factors at 50,000 bbl/day and 80,000 bbl/day flow rate are

$$f_1 = 64/1190.32 = 0.0538$$
$$f_2 = 64/1904.51 = 0.0336$$

The pressure drops due to friction using Eq. (8.14) are

$$P_{m_1} = 0.0605 \times 0.0538 \times (50{,}000)^2 \times (0.85/15.5^5) = 7.731 \text{ psi/mi}$$
$$P_{m_2} = 0.0605 \times 0.0336 \times (80{,}000)^2 \times (0.85/15.5^5) = 12.36 \text{ psi/mi}$$

Therefore, from above, it is clear that in laminar flow even though the friction factor decreases with flow rate increase, the pressure drop still increases with increase in flow rate.

As the flow rate increases and reaches the fully turbulent zone ($R > 4000$), the friction factor f will depend on the Reynolds number R, the inside diameter D, and the internal (absolute) roughness of the pipe, e. The ratio e/D is called the *relative roughness*, and is dimensionless, since both D and e are in the same units. Table 8.5 lists the pipe roughness for various pipe materials.

TABLE 8.5 Internal Pipe Roughness

Pipe Material	Roughness (in.)	Roughness (mm)
Riveted steel	0.0354–0.354	0.9–9.0
Commercial steel/welded steel	0.0018	0.045
Cast iron	0.0102	0.26
Galvanized iron	0.0059	0.15
Asphalted cast iron	0.0047	0.12
Wrought iron	0.0018	0.045
PVC, drawn tubing, glass	0.000059	0.0015
Concrete	0.0118–0.118	0.3–3.0

8.6 COLEBROOK–WHITE EQUATION

Various correlations exist for calculating friction factor f in the turbulent zone. These are based on experiments conducted by scientists and engineers during the last 70 years or more. A good all-purpose equation for the friction factor f is called the Colebrook–White equation, which is as follows:

$$1/\sqrt{f} = -2\,\mathrm{Log}_{10}[(e/3.7D) + 2.51/(R\sqrt{f})]\ \text{(USCS)} \tag{8.18}$$

and applies only for turbulent flow: $R > 4000$,
where

f – Darcy friction factor, dimensionless
D – Pipe internal diameter, in.
e – Absolute pipe roughness, in.
R – Reynolds number of flow, dimensionless

In SI units, the preceding equation for f remains the same, as long as the absolute roughness e and the pipe diameter D are both expressed in mm. All other terms in the equation are dimensionless.

In SI units, the Colebrook–White equation is as follows:

$$1/\sqrt{f} = -2\mathrm{Log}_{10}[(e/3.7D) + 2.51/(R\sqrt{f})]\ \text{(SI)} \tag{8.19}$$

where

f – Darcy friction factor, dimensionless
D – Pipe internal diameter, mm
e – Absolute pipe roughness, mm
R – Reynolds number of flow, dimensionless

In the critical zone, where the Reynolds number is between 2000 and 4000 there is no generally accepted formula for determining the friction factor. This is because the flow is unstable in this region and therefore the friction factor is indeterminate. Most users calculate the value of f based upon turbulent flow.

It can be seen from Eqs (8.18) and (8.19) that the calculation of f is not easy because it appears on both sides of the equation. A trial and error approach needs to be used. We assume a starting value of f (say 0.02), and substitute it in the right-hand side of Eq. (8.18). This will yield a second approximation for f, which can then be used to recalculate a better value of f, by successive iteration. Generally, three to four iterations will yield a satisfactory result for f correct to within 0.001. An example will illustrate the method. There are other correlations for friction factor that do not require such an iterative solution. Two such explicit correlations are the Churchill equation and Swamee–Jain equation, described in Appendix 5. The value of f using these equations is very close to that obtained using the Colebrook–White equation.

8.7 MOODY DIAGRAM

The friction factor equations discussed in the preceding section are also plotted on the Moody diagram as shown in Fig. 8.1. This diagram shows the friction factor on the left vertical axis, plotted against the Reynolds number on the horizontal axis, for various values of the relative roughness of the pipe (*e*/*D*).

The Moody diagram represents the complete friction factor map for laminar and all turbulent regions of pipe flows. It is commonly used in estimating friction factor in pipe flow. If the Moody diagram is not available, we must use a trial and error solution of Eq. (8.18) to calculate the friction factor.

To use the Moody diagram for determining the friction factor *f* we first calculate the relative roughness (*e*/*D*) and the Reynolds number *R* for the flow. Next, for this value of *R* on the horizontal axis, draw a vertical line that intersects with the appropriate relative roughness (*e*/*D*) curve. From this point of intersection on the (*e*/*D*) curve, we go horizontally to the left and read the value of the friction factor *f* on the vertical axis on the left.

Note that some publications refer to the Fanning friction factor. This is equal to one-fourth the value of the Darcy friction factor *f* discussed in this chapter. Unless otherwise specified, we will use the Darcy friction factor *f* throughout this book.

Table 8.6 shows the variation of the friction factor for various Reynolds numbers for three relative roughness values.

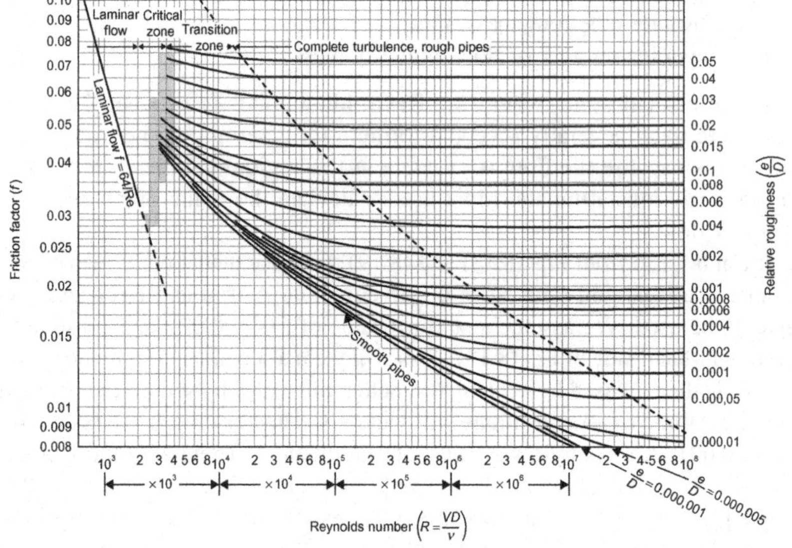

FIGURE 8.1 Moody diagram for friction factor.

TABLE 8.6 Darcy Friction Factors*

Reynolds Number	Friction Factor		
	e/D = 0.0001	e/D = 0.0002	e/D = 0.0003
1000	0.0664	0.0756	0.0799
2000	0.0512	0.0678	0.0762
4000	0.0407	0.0618	0.0733
5000	0.0380	0.0602	0.0725
10,000	0.0311	0.0559	0.0703
20,000	0.0261	0.0524	0.0684
30,000	0.0237	0.0506	0.0675
40,000	0.0222	0.0495	0.0669
50,000	0.0212	0.0487	0.0664
60,000	0.0204	0.0481	0.0661
70,000	0.0198	0.0476	0.0658
80,000	0.0192	0.0471	0.0656
90,000	0.0188	0.0468	0.0654
100,000	0.0185	0.0465	0.0652
125,000	0.0177	0.0458	0.0648
150,000	0.0172	0.0454	0.0646
200,000	0.0164	0.0447	0.0642
225,000	0.0161	0.0444	0.0640
250,000	0.0158	0.0442	0.0639
275,000	0.0156	0.0440	0.0638
300,000	0.0154	0.0438	0.0637
325,000	0.0153	0.0436	0.0636
350,000	0.0151	0.0435	0.0635
375,000	0.0150	0.0434	0.0634
400,000	0.0149	0.0433	0.0634
425,000	0.0147	0.0432	0.0633
450,000	0.0146	0.0431	0.0632

(*Continued*)

TABLE 8.6 Darcy Friction Factors*—cont'd

Reynolds Number	Friction Factor		
	e/D = 0.0001	e/D = 0.0002	e/D = 0.0003
500,000	0.0145	0.0429	0.0631
750,000	0.0139	0.0423	0.0628
1,000,000	0.0135	0.0419	0.0626

*Friction factor based on Swamee–Jain equation: $f = 0.25/[\mathrm{Log}_{10}(e/3.7D + 5.74/R^{0.9})]^2$.

Example Problem 8.2 (USCS)

Water (Sg = 1.0 and visc = 1.0 cSt) flows through an NPS 20, 0.375 in. wall pipe at 6000 gal/min. Calculate the friction factor using the Colebrook–White equation. Assume an absolute pipe roughness of 0.002 in. What is the head loss due to friction in 3500 ft of pipe?

Solution

First, we calculate the Reynolds number from Eq. (8.8) as follows:

$$R = 3160 \times 6000/(19.25 \times 1.0) = 984{,}935$$

Since $R > 4000$, the flow is fully turbulent and the friction factor f is calculated using Eq. (8.18) as follows:

$$1/\sqrt{f} = -2\,\mathrm{Log}_{10}[(0.002/(3.7 \times 19.25)) + 2.51/(984{,}935\sqrt{f})]$$

The above implicit equation for f must be solved by trial and error.

First assume a trial value of $f = 0.02$. Substituting in the equation above, we get successive approximations for f as follows:

$$f = 0.0133,\ 0.0135,\ \text{and}\ 0.0135$$

Therefore, the solution is $f = 0.0135$.

Converting the given flow rate from gal/min to bbl/day,

$$Q = 6000\ \text{gal/min} \times (60 \times 24\ \text{min/day}) \times (1/42\ \text{gal/bbl}) = 205{,}714.29\ \text{bbl/day}$$

Using Eq. (8.14), the pressure drop resulting from friction is

$$P_m = 0.0605 \times 0.0135 \times (205{,}714.29)^2(1.0/19.25^5) = 13.08\ \text{psi/mi}$$

Therefore, the pressure drop in 3500 ft of pipe is

$$\Delta P = 13.08 \times (3500/5280) = 8.67\ \text{psi}$$

This is converted to head in feet of water, using Eq. (8.10):

$$\text{Head loss due to friction} = 8.67 \times 2.31/1.0 = 20.03\ \text{ft}$$

Example Problem 8.3 (SI)

Diesel fuel (Sg = 0.85 and visc = 5.0 cSt) flows in a pipeline, DN 400, 8 mm wall thickness at 580 m^3/h. Calculate the average velocity, Reynolds number, and the friction factor. Assume an absolute pipe roughness of 0.05 mm. What is the pressure drop resulting from friction in 5 km length of the pipeline?

Solution

The inside diameter of the pipe, $D = 400 - 2 \times 8 = 384$ mm.

The average flow velocity is found using Eq. (8.3):

$$V = 353.6777(580)/384^2 = 1.39\,\text{m/s}$$

Next, we calculate the Reynolds number from Eq. (8.9) as follows:

$$R = 353{,}678\,(580)/(5 \times 384) = 106{,}840$$

Because $R > 4000$, the flow is fully turbulent and the friction factor f is calculated using Eq. (8.19) as follows:

$$1/\sqrt{f} = -2\text{Log}_{10}[(0.05/(3.7 \times 384)) + 2.51/(106{,}840\sqrt{f})]$$

This equation for f must be solved by trial and error.

First, assume a trial value of $f = 0.02$. Substituting in the equation above, and by successive iteration we get $f = 0.0612$.

The pressure drop due to friction is calculated from Eq. (8.16) as

$$P_{\text{km}} = 6.2475 \times 10^{10} \times 0.0612 \times (580)^2(0.85/384^5) = 130.94\,\text{kPa/km}$$

The pressure drop in 5 km length of the pipeline = 130.94 × 5 = 654.7 kPa

8.8 HAZEN–WILLIAMS EQUATION

The Hazen–Williams equation is commonly used in the design of water distribution lines and in calculation of frictional pressure drop in refined petroleum products such as gasoline, diesel, and so on. Instead of using a friction factor, this method uses the Hazen–Williams C-factor. The pipe roughness or liquid viscosity is not used in this approach.

The original form of the Hazen–Williams equation for head loss calculation takes into account the flow rate, pipe diameter, and length as follows:

$$h = 4.73 \times L \times (Q/C)^{1.852}/D^{4.87} \tag{8.20}$$

where

h – Head loss due to friction, ft
L – Length of pipe, ft
D – Inside diameter of pipe, ft

Q – Flow rate, ft^3/s
C – Hazen–Williams coefficient or C Factor, dimensionless. The C-factor ranges from 60 to 150. Typical values of Hazen–Williams C-factor are given in Appendix 5.

In customary pipeline units, the Hazen–Williams Eq. (8.20) can be rewritten as follows:

In USCS units,

$$Q = 0.1482(C)(D)^{2.63}(P_m/\text{Sg})^{0.54} \text{ (USCS)} \tag{8.21}$$

where

Q – Flow rate, bbl/day
D – Pipe inside diameter, in.
P_m – Pressure drop due to friction, psi/mi
Sg – Liquid specific gravity
C – Hazen–Williams C-factor

Another form of Hazen–Williams equation, when the flow rate is in gal/min and head loss due to friction is measured in feet of liquid per thousand feet of pipe is as follows:

$$\text{GPM} = 6.7547 \times 10^{-3}(C)(D)^{2.63}(H_L)^{0.54}\text{(USCS)} \tag{8.22}$$

where

GPM – Flow rate, gal/min
H_L – Friction loss, ft of liquid per 1000 ft of pipe

Other symbols remain the same.

In SI units, the Hazen–Williams formula is as follows:

$$Q = 9.0379 \times 10^{-8}(C)(D)^{2.63}(P_{km}/\text{Sg})^{0.54} \text{ (SI)} \tag{8.23}$$

where

Q – Flow rate, m^3/h
D – Pipe inside diameter, mm
P_{km} – Pressure drop due to friction, kPa/km
Sg – Liquid specific gravity
C – Hazen–Williams C-factor

Equations (8.21)–(8.23) are for calculating the flow rate from a given pressure drop. These equations can be rewritten to solve for the pressure drop or head loss from the given flow rate as shown next.

$$P_m = 34.32 \times (Q/C)^{1.852}(\text{Sg}/D^{4.87}) \text{ (USCS)} \tag{8.24}$$

where P_m is in psi/mi, Q in bbl/day, and D in inches.

$$P_m = 23{,}909 \times (Q/C)^{1.852}(Sg/D^{4.87}) \text{ (USCS)} \quad (8.25)$$

where P_m is in psi/mi, Q in gal/min, and D in inches.

$$P_{km} = 1.1101 \times 10^{13}(Q/C)^{1.852}(Sg/D^{4.87}) \text{ (SI)} \quad (8.26)$$

where P_{km} is in kPa/km, Q m^3/h, and D in mm.

Historically, many empirical formulas have been used to calculate frictional pressure drop in pipelines. The Hazen–Williams formula has been used widely in the analysis of pipeline networks and water distribution systems because of its simple form and ease of use. A review of the Hazen–Williams formula shows that the pressure drop due to friction depends on the liquid specific gravity, pipe diameter, and the Hazen–Williams coefficient or C-factor.

The Colebrook–White equation for friction factor is based on pipe roughness, pipe diameter, and the Reynolds number, which further depends on liquid specific gravity and viscosity. The Hazen–Williams C-factor does not appear to take into account the liquid viscosity or pipe roughness. It could be argued that the C-factor is in fact a measure of the pipe internal roughness. However, there does not seem to be any indication of how the C-factor varies from laminar flow to turbulent flow. Also, the flow rate appears to be directly proportional to the C-factor, whereas the pipe roughness and friction factor are negatively correlated with the flow rate.

It must be noted that the Hazen–Williams equation, although convenient from the standpoint of its explicit form, must be regarded as an empirical equation that is difficult to apply to all fluids under all conditions. Nevertheless, in real-world pipelines, with sufficient field data, we could determine specific C-factors for specific pipelines and fluids pumped.

Example Problem 8.4 (USCS)

A 6-in. (inside diameter) smooth pipeline is used to pump 630 gal/min of water. Using the Hazen–Williams formula, calculate the head loss in 3000 ft of this pipe. Assume C-factor = 140.

Solution

Using Eq. (8.22) and substituting given values, we get

$$630 = 6.7547 \times 10^{-3} \times 140(6.0)^{2.63}(H_L)^{0.54}$$

Solving for the head loss, we get

$$H_L = 27.49 \text{ ft per } 1000 \text{ ft}$$

Therefore, head loss for 3000 ft = 27.49 × 3 = 82.47 ft of water.

Example Problem 8.5 (SI)

A pipeline, DN 400 with 8 mm wall thickness is used to transport 750 m^3/h of gasoline (Sg = 0.745). Using the Hazen–Williams formula with $C = 145$, calculate the frictional pressure drop in a 5 km length of the pipeline.

Solution

Using Eq. (8.26) and substituting given values, we get

$$\text{The inside diameter } D = 400 - 2 \times 8 = 384\,\text{mm}$$

$$P_{km} = 1.1101 \times 10^{13} (750/145)^{1.852} (0.745/384^{4.87}) = 45.03\,\text{kPa/km}$$

Therefore, the pressure drop due to friction in 5 km = 45.03 × 5 = 225.17 kPa.

Table 8.7 shows the head loss using the Hazen–Williams equation in water pipelines for various pipe diameters and different flow rates for a C value of 120. The head loss for other liquids and C values can be easily calculated from the table values, as explained in the footnote to the table.

TABLE 8.7 Head Loss in Water Pipelines Using Hazen–Williams Equation

In USCS

Nominal Pipe Size (NPS)	Outside Diameter (in.)	Wall Thickness (in.)	Flow Rate (gpm)	Velocity (ft/s)	Head Loss* (ft/1000 ft)
½	0.840	0.109	15	15.84	2245.37
1	1.315	0.330	15	14.28	1745.63
1½	1.900	0.145	75	11.82	430.82
2	2.375	0.154	150	14.34	460.62
2½	2.875	0.203	200	13.40	330.26
3	3.500	0.216	250	10.85	173.35
3½	4.000	0.226	300	9.74	119.71
4	4.500	0.237	400	10.08	110.21
6	6.625	0.280	500	5.55	22.65
8	8.625	0.322	1000	6.41	21.47
10	10.75	0.365	1500	6.10	15.03
12	12.75	0.250	3000	8.17	20.39
	12.75	0.312	4000	11.11	36.50

TABLE 8.7 Head Loss in Water Pipelines Using Hazen–Williams Equation—cont'd

In USCS					
Nominal Pipe Size (NPS)	Outside Diameter (in.)	Wall Thickness (in.)	Flow Rate (gpm)	Velocity (ft/s)	Head Loss* (ft/1000 ft)
	12.75	0.344	4500	12.63	46.58
	12.75	0.375	5000	14.18	58.05
	12.75	0.406	5000	14.33	59.53
	12.75	0.500	5000	14.79	64.32
14	14.00	0.250	6000	13.45	45.85
	14.00	0.312	6000	13.70	47.96
	14.00	0.375	6000	13.96	50.22
	14.00	0.500	6000	14.50	55.10
16	16.00	0.250	8000	13.60	39.86
	16.00	0.281	8000	13.71	40.65
	16.00	0.312	8000	13.82	41.45
	16.00	0.375	8000	14.05	43.14
	16.00	0.500	8000	14.52	46.76
18	18.00	0.250	10,000	13.34	33.37
	18.00	0.281	10,000	13.43	33.95
	18.00	0.312	10,000	13.53	34.54
	18.00	0.500	10,000	14.13	38.43
	18.00	0.625	10,000	14.56	41.30
20	20.00	0.250	12,000	12.89	27.61
	20.00	0.375	12,000	13.23	29.40
	20.00	0.500	12,000	13.58	31.34
	20.00	0.625	12,000	13.94	33.42
	20.00	0.812	12,000	14.52	36.87
22	22.00	0.250	15,000	13.26	25.95
	22.00	0.375	15,000	13.57	27.47

(Continued)

TABLE 8.7 Head Loss in Water Pipelines Using Hazen–Williams Equation—cont'd

In USCS					
Nominal Pipe Size (NPS)	Outside Diameter (in.)	Wall Thickness (in.)	Flow Rate (gpm)	Velocity (ft/s)	Head Loss* (ft/1000 ft)
	22.00	0.500	15,000	13.89	29.10
	22.00	0.625	15,000	14.23	30.84
	22.00	0.750	15,000	14.58	32.72
24	24.00	0.218	20,000	14.71	28.29
	24.00	0.312	20,000	14.95	29.41
	24.00	0.500	20,000	15.44	31.83
	24.00	0.562	20,000	15.61	32.68
	24.00	0.625	20,000	15.79	33.57
26	26.00	0.312	20,000	12.69	19.72
	26.00	0.500	20,000	13.07	21.21
	26.00	0.625	20,000	13.34	22.27
	26.00	0.750	20,000	13.61	23.40
	26.00	0.875	20,000	13.89	24.60
28	28.00	0.312	25,000	13.63	20.60
	28.00	0.500	25,000	14.01	22.04
	28.00	0.625	25,000	14.27	23.06
	28.00	0.750	25,000	14.54	24.14
	28.00	0.875	25,000	14.82	25.28
30	30.00	0.312	30,000	14.20	20.48
	30.00	0.500	30,000	14.57	21.81
	30.00	0.625	30,000	14.83	22.75
	30.00	0.750	30,000	15.09	23.74
	30.00	0.875	30,000	15.36	24.78
32	32.00	0.312	30,000	12.45	14.86
	32.00	0.500	30,000	12.75	15.76

TABLE 8.7 Head Loss in Water Pipelines Using Hazen–Williams Equation—cont'd

In USCS					
Nominal Pipe Size (NPS)	Outside Diameter (in.)	Wall Thickness (in.)	Flow Rate (gpm)	Velocity (ft/s)	Head Loss* (ft/1000 ft)
	32.00	0.625	30,000	12.96	16.40
	32.00	0.750	30,000	13.17	17.06
	32.00	0.875	30,000	13.39	17.76
34	34.00	0.312	35,000	12.83	14.64
	34.00	0.500	35,000	13.13	15.47
	34.00	0.625	35,000	13.33	16.05
	34.00	0.750	35,000	13.54	16.66
	34.00	0.875	35,000	13.75	17.30
36	36.00	0.312	40,000	13.06	14.12
	36.00	0.500	40,000	13.34	14.87
	36.00	0.625	40,000	13.53	15.40
	36.00	0.750	40,000	13.73	15.95
	36.00	0.875	40,000	13.93	16.53
42	42.00	0.375	50,000	12.00	10.10
	42.00	0.500	50,000	12.15	10.40
	42.00	0.625	50,000	12.30	10.72
	42.00	0.750	50,000	12.45	11.04
	42.00	1.000	55,000	14.04	14.00
48	48.00	0.375	60,000	10.98	7.31
	48.00	0.500	60,000	11.10	7.50
	48.00	0.625	60,000	11.21	7.70
	48.00	0.750	60,000	11.34	7.90
	48.00	1.000	60,000	11.58	8.33

(*Continued*)

TABLE 8.7 Head Loss in Water Pipelines Using Hazen–Williams Equation—cont'd

In SI					
Nominal Pipe Size (NPS)	Outside Diameter (mm)	Wall Thickness (mm)	Flow Rate (m^3/h)	Velocity (m/s)	Head Loss* (m/km)
½	21.34	1.651	4	4.35	1587.44
1	33.40	1.651	8	3.12	473.55
1½	48.26	5.080	16	3.90	542.31
2	60.33	5.537	32	4.66	560.51
2½	73.03	5.156	50	4.50	394.92
3	88.9	5.486	75	4.37	290.66
4	114.3	8.560	100	3.75	168.97
6	168.28	10.973	200	3.30	83.10
8	219.08	10.312	400	3.59	68.03
10	273.05	15.062	600	3.60	53.85
12	323.85	3.962	1000	3.54	38.58
	323.85	6.350	1000	3.65	41.55
	323.85	9.525	1000	3.81	45.94
	323.85	12.700	1000	3.97	50.90
	323.85	17.450	1000	4.24	59.58
	323.85	25.400	1000	4.74	78.49
14	355.60	3.962	1200	3.51	33.92
	355.60	9.525	1200	3.75	39.74
	355.60	15.062	1200	4.01	46.77
	355.60	31.750	1200	4.97	79.22
16	406.40	4.775	1500	3.37	26.92
	406.40	9.525	1500	3.54	30.30
	406.40	14.275	1500	3.72	34.19
	406.40	17.450	1500	3.84	37.13
	406.40	36.500	1500	4.77	62.89

TABLE 8.7 Head Loss in Water Pipelines Using Hazen–Williams Equation—cont'd

In SI					
Nominal Pipe Size (NPS)	Outside Diameter (mm)	Wall Thickness (mm)	Flow Rate (m^3/h)	Velocity (m/s)	Head Loss* (m/km)
18	457.20	6.350	1800	3.22	21.72
	457.20	12.700	1800	3.41	25.02
	457.20	19.050	1800	3.62	28.93
	457.20	23.800	1800	3.79	32.35
	457.20	39.675	1800	4.46	47.92
20	508.00	6.350	2200	3.17	18.60
	508.00	12.700	2200	3.34	21.11
	508.00	19.050	2200	3.52	24.03
	508.00	38.100	2200	4.17	36.28
	508.00	49.987	2200	4.67	47.80
22	558.80	4.775	2600	3.05	15.32
	558.80	6.350	2600	3.08	15.75
	558.80	12.700	2600	3.23	17.66
	558.80	22.225	2600	3.48	21.09
	558.80	47.625	2600	4.28	34.99
24	609.60	4.775	3200	3.14	14.62
	609.60	6.350	3200	3.18	15.00
	609.60	12.700	3200	3.32	16.66
	609.60	26.187	3200	3.64	20.97
	609.60	49.987	3200	4.36	32.40
26	660.40	6.350	3800	3.20	13.86
	660.40	9.525	3800	3.27	14.54
	660.40	15.875	3800	3.40	16.03
	660.40	22.225	3800	3.54	17.70
	660.40	28.575	3800	3.69	19.59

(Continued)

TABLE 8.7 Head Loss in Water Pipelines Using Hazen–Williams Equation—cont'd

In SI					
Nominal Pipe Size (NPS)	Outside Diameter (mm)	Wall Thickness (mm)	Flow Rate (m^3/h)	Velocity (m/s)	Head Loss* (m/km)
28	711.20	6.350	4500	3.26	13.12
	711.20	9.525	4500	3.32	13.72
	711.20	15.875	4500	3.45	15.01
	711.20	22.225	4500	3.58	16.46
	711.20	28.575	4500	3.72	18.07
30	762.00	6.350	6000	3.78	15.88
	762.00	9.525	6000	3.84	16.55
	762.00	15.875	6000	3.98	18.00
	762.00	22.225	6000	4.12	19.61
	762.00	28.575	6000	4.27	21.39
32	812.80	6.350	7500	4.14	17.44
	812.80	9.525	7500	4.21	18.13
	812.80	15.875	7500	4.35	19.61
	812.80	22.225	7500	4.49	21.25
	812.80	28.575	7500	4.65	23.04
34	863.60	6.350	9000	4.40	18.12
	863.60	9.525	9000	4.46	18.79
	863.60	15.875	9000	4.60	20.23
	863.60	22.225	9000	4.74	21.80
	863.60	28.575	9000	4.89	23.53
36	914.40	6.350	10,200	4.44	17.22
	914.40	9.525	10,200	4.50	17.83
	914.40	15.875	10,200	4.63	19.11
	914.40	22.225	10,200	4.77	20.51
	914.40	28.575	10,200	4.91	22.03

TABLE 8.7 Head Loss in Water Pipelines Using Hazen–Williams Equation—cont'd

In SI					
Nominal Pipe Size (NPS)	Outside Diameter (mm)	Wall Thickness (mm)	Flow Rate (m^3/h)	Velocity (m/s)	Head Loss* (m/km)
42	1066.80	6.350	12,000	3.82	10.88
	1066.80	9.525	12,000	3.87	11.20
	1066.80	15.875	12,000	3.96	11.89
	1066.80	25.400	12,000	4.11	13.01
	1066.80	38.100	12,000	4.33	14.72
48	1219.20	9.525	15,000	3.68	8.74
	1219.20	15.875	15,000	3.76	9.21
	1219.20	25.400	15,000	3.89	9.96
	1219.20	38.100	15,000	4.06	11.09
	1219.20	50.800	15,000	4.25	12.37

For other values of C, multiply head loss by $(120/C)^{1.852}$.
**Head Loss: Based on Hazen–Williams Formula with C = 120.*

8.9 MINOR LOSSES

In most long distance pipelines, such as transmission and distribution lines, the pressure drop resulting from friction in the straight lengths of pipe forms the significant portion of the total pressure drop. Valves, fittings, and other devices installed on the pipeline contribute very little to the total pressure drop. Hence, in long pipelines, pressure losses through valves, fittings, and other restrictions are generally classified as *minor losses*. Minor losses include energy losses resulting from rapid changes in the direction or magnitude of liquid velocity in the pipeline. Thus pipe enlargements, contractions, bends, and restrictions such as check valves and gate valves are included in minor losses.

In short pipelines, such as terminal and plant piping, the pressure loss due to valves, fittings, and so on may be a substantial portion of the total pressure drop. In such cases, the term minor losses is a misnomer.

Therefore, in long pipelines, the pressure loss through bends, elbows, valves, fittings, and so on are classified as "minor" losses and in most instances may be neglected without significant error. However, in shorter pipelines, these losses must be included for accuracy. It has been found that at high Reynolds numbers, these minor losses varied approximately as the square of the velocity. Hence, minor losses can be represented as a multiple of the liquid velocity head or kinetic energy ($V^2/2g$).

Therefore, the pressure drop through valves and fittings can be expressed as $K(V^2/2g)$, where K is a dimensionless head loss coefficient. Comparing this with the Darcy–Weisbach equation for head loss in a pipe, we see that for straight pipe, the head loss is $V^2/2g$ multiplied by the factor (fL/D). Thus, the head loss coefficient for straight pipe is fL/D. We can also approximate the valve or fitting by its equivalent length, compared with the straight pipe. Thus, a gate valve may be said to have an L/D ratio to equate it to straight pipe with the same head loss. The L/D ratios for various valves are shown in Table 8.8.

TABLE 8.8 Equivalent Length to Diameter (L/D) Ratio for Valves and Fittings

Description	L/D
Gate valve	8
Globe valve	340
Angle valve	55
Ball valve	3
Plug valve straightway	18
Plug valve three-way thru-flow	30
Plug valve branch flow	90
Swing check valve	50
Lift check valve	600
Standard elbow – 90°	30
Standard elbow – 45°	16
Standard elbow long radius 90°	16
Standard Tee thru-flow	20
Standard Tee thru-branch	60
Mitre bends – $\alpha = 0$	2
Mitre bends – $\alpha = 30$	8
Mitre bends – $\alpha = 60$	25
Mitre bends – $\alpha = 90$	60

Using the head loss coefficient, the pressure drop in a valve or fitting is calculated as follows:

$$h = KV^2/2g \tag{8.27}$$

where

h – Head loss due to valve or fitting, ft
K – Head loss coefficient for the valve or fitting, dimensionless
V – Velocity of liquid through valve or fitting, ft/s
g – Acceleration due to gravity, 32.2 ft/s^2 in English units

A table of K values commonly used for valves and fittings is shown in Table 8.9.

8.10 FLOW OF GAS IN PIPELINES

The basis of pressure drop calculations in pipelines transporting gases is very similar to that of liquid pipelines. However, unlike liquids, gases are compressible; therefore, we have to consider the variation in gas density and the compressibility with the temperature and pressure of the flowing gas.

In a gas pipeline, the velocity of flow represents the speed at which the gas molecules move from one point to another. Unlike a liquid pipeline, due to compressibility, the gas velocity depends on the pressure and, hence, will vary along the pipeline even if the pipe diameter is constant. The highest velocity will be at the downstream end where the pressure is the least. Similarly, the least velocity will be at the upstream end, where the pressure is higher.

Consider a pipe transporting gas from point A to point B. Under steady state flow, at A, the mass flow rate of gas is designated as M and will be the same as the mass flow rate at point B if between A and B there is no injection or delivery of gas. The mass being the product of volume Q and density ρ, the following equation applies for point A:

$$M = Q \times \rho \tag{8.28}$$

The volume flow rate Q is the product of the average flow velocity u and cross-sectional area of the pipe A as follows:

$$Q = u \times \mathrm{A} \tag{8.29}$$

Therefore, combining the two equations and applying the conservation of mass to points A and B, we get

$$M_1 = u_1 \times \mathrm{A}_1 \times \rho_1 = M_2 = u_2 \times \mathrm{A}_2 \times \rho_2 \tag{8.30}$$

where subscript 1 and 2 refer to points A and B, respectively.

If the pipe is of uniform cross section between A and B, then $\mathrm{A}_1 = \mathrm{A}_2 = \mathrm{A}$.

TABLE 8.9 Head Loss Coefficient K for Valves and Fittings

Description	L/D	Nominal Pipe Size (in.)											
		½	¾	1	1¼	1½	2	2½ to 3	4	6	8 to 10	12 to 16	18 to 24
Gate valve	8	0.22	0.20	0.18	0.18	0.15	0.15	0.14	0.14	0.12	0.11	0.10	0.10
Globe valve	340	9.20	8.50	7.80	7.50	7.10	6.50	6.10	5.80	5.10	4.80	4.40	4.10
Angle valve	55	1.48	1.38	1.27	1.21	1.16	1.05	0.99	0.94	0.83	0.77	0.72	0.66
Ball valve	3	0.08	0.08	0.07	0.07	0.06	0.06	0.05	0.05	0.05	0.04	0.04	0.04
Plug valve straightway	18	0.49	0.45	0.41	0.40	0.38	0.34	0.32	0.31	0.27	0.25	0.23	0.22
Plug valve three-way thru-flow	30	0.81	0.75	0.69	0.66	0.63	0.57	0.54	0.51	0.45	0.42	0.39	0.36
Plug valve branch flow	90	2.43	2.25	2.07	1.98	1.89	1.71	1.62	1.53	1.35	1.26	1.17	1.08
Swing check valve	50	1.40	1.30	1.20	1.10	1.10	1.00	0.90	0.90	0.75	0.70	0.65	0.60
Lift check valve	600	16.20	15.00	13.80	13.20	12.60	11.40	10.80	10.20	9.00	8.40	7.80	7.22
Standard elbow – 90°	30	0.81	0.75	0.69	0.66	0.63	0.57	0.54	0.51	0.45	0.42	0.39	0.36
Standard elbow – 45°	16	0.43	0.40	0.37	0.35	0.34	0.30	0.29	0.27	0.24	0.22	0.21	0.19

TABLE 8.9 Head Loss Coefficient K for Valves and Fittings—cont'd

Description	L/D	Nominal Pipe Size (in.)											
		½	¾	1	1¼	1½	2	2½ to 3	4	6	8 to 10	12 to 16	18 to 24
Standard elbow long radius 90°	16	0.43	0.40	0.37	0.35	0.34	0.30	0.29	0.27	0.24	0.22	0.21	0.19
Standard Tee thru-flow	20	0.54	0.50	0.46	0.44	0.42	0.38	0.36	0.34	0.30	0.28	0.26	0.24
Standard Tee thru-branch	60	1.62	1.50	1.38	1.32	1.26	1.14	1.08	1.02	0.90	0.84	0.78	0.72
Mitre bends – $\alpha = 0$	2	0.05	0.05	0.05	0.04	0.04	0.04	0.04	0.03	0.03	0.03	0.03	0.02
Mitre bends – $\alpha = 30$	8	0.22	0.20	0.18	0.18	0.17	0.15	0.14	0.14	0.12	0.11	0.10	0.10
Mitre bends – $\alpha = 60$	25	0.68	0.63	0.58	0.55	0.53	0.48	0.45	0.43	0.38	0.35	0.33	0.30
Mitre bends – $\alpha = 90$	60	1.62	1.50	1.38	1.32	1.26	1.14	1.08	1.02	0.90	0.84	0.78	0.72

Therefore, the area term in the preceding equation may be dropped and the velocity at A and B are related by the following equation:

$$u_1 \times \rho_1 = u_2 \times \rho_2 \tag{8.31}$$

Because the flow of gas in a pipe results in variation of temperature from point A to point B, the gas density will also vary with temperature and pressure. If the density and velocity at one point are known the corresponding velocity at the other point may be calculated using Eq. (8.31).

At the pipe inlet, volume flow rate Q at standard conditions of 60°F and 14.7 psia are known, and we can calculate the velocity at any point along the pipeline at which the pressure and temperature of the gas are P and T, respectively.

Using Eq. (8.28) under steady state gas flow, the mass flow rate M at section 1 and 2 are the same.

Therefore,

$$M = Q_1\rho_1 = Q_2\rho_2 = Q_b\rho_b, \tag{8.32}$$

where Q_b is the gas flow rate at standard conditions (60°F and 14.7 psia), and ρ_b is the corresponding gas density.

Therefore, simplifying Eq. (8.32), we get

$$Q_1 = Q_b\left(\frac{\rho_b}{\rho_1}\right) \tag{8.33}$$

Applying the gas law Eq. (1.37) from Chapter 1, we get

$$\frac{P_1}{\rho_1} = Z_1RT_1$$

or

$$\rho_1 = \frac{P_1}{Z_1RT_1} \tag{8.34}$$

where P_1 and T_1 are the pressure and temperature at pipe section 1.

Similarly, at standard conditions,

$$\rho_b = \frac{P_b}{Z_bRT_b} \tag{8.35}$$

From Eqs (8.33)–(8.35), we get

$$Q_1 = Q_b\left(\frac{P_b}{T_b}\right)\left(\frac{T_1}{P_1}\right)\left(\frac{Z_1}{Z_b}\right) \tag{8.36}$$

Because $Z_b = 1.00$ approximately, we can simplify this to

$$Q_1 = Q_b\left(\frac{P_b}{T_b}\right)\left(\frac{T_1}{P_1}\right)Z_1 \tag{8.37}$$

Therefore, the gas velocity at section 1, using Eqs (8.31) and (8.37), is

$$u_1 = \frac{Q_b Z_1}{A}\left(\frac{P_b}{T_b}\right)\left(\frac{T_1}{P_1}\right) = \frac{4 \times 144}{\pi D^2} Q_b Z_1 \left(\frac{P_b}{T_b}\right)\left(\frac{T_1}{P_1}\right)$$
$$u_1 = 0.002122\left(\frac{Q_b}{D^2}\right)\left(\frac{P_b}{T_b}\right)\left(\frac{Z_1 T_1}{P_1}\right) \text{ (USCS units)} \tag{8.38}$$

where, in USCS units,

u_1 – Upstream gas velocity, ft/s
Q_b – Gas flow rate, measured at standard conditions, ft^3/day (SCFD)
D – Pipe inside diameter, in.
P_b – Base pressure, psia
T_b – Base temperature, °R (460 + °F)
P_1 – Upstream pressure, psia
T_1 – Upstream gas temperature, °R (460 + °F)
Z_1 – Gas compressibility factor at upstream conditions, dimensionless

Similarly, the gas velocity at section 2 is given by

$$u_2 = 0.002122\left(\frac{Q_b}{D^2}\right)\left(\frac{P_b}{T_b}\right)\left(\frac{Z_2 T_2}{P_2}\right) \text{ (USCS units)} \tag{8.39}$$

In general, the gas velocity at any point in a pipeline is given by

$$u = 0.002122\left(\frac{Q_b}{D^2}\right)\left(\frac{P_b}{T_b}\right)\left(\frac{ZT}{P}\right) \text{ (USCS units)} \tag{8.40}$$

where, in USCS units,

u – Gas velocity, ft/s
Q_b – Gas flow rate, measured at standard conditions, ft^3/day (SCFD)
D – Pipe inside diameter, in.
P_b – Base pressure, psia
T_b – Base temperature, °R (460 + °F)
P – Gas pressure, psia
T – Gas temperature, °R (460 + °F)
Z – Gas compressibility factor, dimensionless

Similarly, in SI units, the gas velocity at any point in a pipeline is given by

$$u = 14.7349\left(\frac{Q_b}{D^2}\right)\left(\frac{P_b}{T_b}\right)\left(\frac{ZT}{P}\right) \text{ (SI units)} \tag{8.41}$$

where

u – Gas velocity, m/s
Q_b – Gas flow rate, measured at standard conditions, m^3/day
D – Pipe inside diameter, mm

P_b – Base pressure, kPa
T_b – Base temperature, K (273 + °C)
P – Absolute pressure, kPa
T – Gas temperature, K (273 + °C)
Z – Gas compressibility factor, dimensionless

Because the right-hand side of Eq. (8.41) contains ratios of pressures, any consistent unit may be used such as kPa, MPa, or Bar.

8.11 EROSIONAL VELOCITY

We have seen from the preceding section that the gas velocity is directly related to the flow rate. As flow rate increases, so does the gas velocity. How high can the gas velocity be in a pipeline? As the velocity increases, vibration and noise are evident. In addition, higher velocities will cause erosion of the pipe interior over a long period of time. The upper limit of the gas velocity (erosional velocity) is calculated approximately from the following equation:

$$u_{max} = \frac{100}{\sqrt{\rho}} \text{ (USCS)} \tag{8.42}$$

where

u_{max} – Maximum or erosional velocity, ft/s
ρ – Gas density at flowing temperature, lb/ft^3

Because the gas density ρ may be expressed in terms of pressure and temperature, using the gas law Eq. (1.37), the maximum velocity Eq. (8.42) may be rewritten as

$$u_{max} = 100\sqrt{\frac{ZRT}{29GP}} \text{ (USCS)} \tag{8.43}$$

where

Z – Compressibility factor of gas, dimensionless
R – Gas constant = 10.73 ft^3 psia/lb · molR
T – Gas temperature, °R
G – Gas gravity (air = 1.00)
P – Gas pressure, psia

Similarly, in SI units, the erosional velocity in a gas pipeline is given by

$$u_{max} = 100\sqrt{\frac{ZRT}{29GP}} \text{ (SI)} \tag{8.44}$$

where

Z – Compressibility factor of gas, dimensionless
R – Gas constant = 8.314 J/mol·K
T – Gas temperature, K
G – Gas gravity (air = 1.00)
P – Gas pressure, kPa (absolute)

An acceptable operational velocity is usually considered to be 50% of the above erosional velocity.

Example Problem 8.6

A gas pipeline NPS 24 with 0.500 in. wall thickness transports natural gas (specific gravity = 0.6) at a flow rate of 300 MMSCFD at an inlet temperature of 80°F. Assuming isothermal flow, calculate the velocity of gas at the inlet and outlet of the pipe if the inlet pressure is 1200 psig and the outlet pressure is 850 psig. The base pressure and base temperature are 14.7 psia and 60°F, respectively. Assume compressibility factor $Z = 1.00$. What is the erosional velocity for this pipeline based on the above data and a compressibility factor $Z = 0.90$?

Solution

Using Eq. (8.39) and assuming the compressibility factor $Z = 1.00$, the velocity of gas at the inlet pressure of 1200 psig is

$$u_1 = 0.002122\left(\frac{300\times10^6}{23.0^2}\right)\left(\frac{14.7}{60+460}\right)\left(\frac{80+460}{1214.7}\right) = 15.12\,\text{ft/s}$$

The gas velocity at the outlet is by proportions

$$u_2 = 15.12\times\frac{1214.7}{864.7} = 21.24\,\text{ft/s}$$

The erosional velocity is found for $Z = 0.90$ using Eq. (8.43):

$$u_{\max} = 100\sqrt{\frac{0.9\times10.73\times540}{29\times0.6\times1214.7}} = 49.67\,\text{ft/s}$$

Example Problem 8.7 (SI)

A gas pipeline DN 500 with 12 mm wall thickness transports natural gas (specific gravity = 0.6) at a flow rate of 7.5 Mm^3/day at an inlet temperature of 15°C. Assuming isothermal flow, calculate the velocity of gas at the inlet and outlet of the pipe if the inlet pressure is 7 MPa and the outlet pressure is 6 MPa. The base pressure and base temperature are 0.1 MPa and 15°C. Assume compressibility factor $Z = 0.95$.

Solution

Inside diameter of pipe $D = 500 - (2 \times 12) = 476$ mm

Flow rate at standard conditions $Q_b = 7.5 \times 10^6$ m^3/day

Using Eq. (8.41) the velocity of gas at the inlet pressure of 7 MPa is

$$u_1 = 14.7349\left(\frac{7.5\times 10^6}{476^2}\right)\left(\frac{0.1}{15+273}\right)\left(\frac{0.95\times 288}{7.0}\right) = 6.62\,\text{m/s}$$

The gas velocity at the outlet is by proportions

$$u_2 = 6.62\times\frac{7.0}{6.0} = 7.72\,\text{m/s}$$

In the preceding Example problems, we have assumed the value of compressibility factor Z as constant. A more accurate solution will be to calculate the value of Z using one of the methods outlined in Chapter 1, such as the Standing-Katz method or the CNGA method.

For example, if we used the CNGA method, then the compressibility factor in Example Problem 8.6 will be

$$Z_1 = \frac{1}{\left[1+\dfrac{1200\times 344400\times (10)^{1.785\times 0.6}}{540^{3.825}}\right]}$$

$$= 0.8532 \text{ at inlet pressure of 1000 psig and temperature of } 80°\text{F}$$

and

$$Z_2 = \frac{1}{\left[1+\dfrac{850\times 344400\times (10)^{1.785\times 0.6}}{540^{3.825}}\right]}$$

$$= 0.8913 \text{ at outlet pressure of 850 psig and temperature of } 80°\text{F}$$

Using these values of compressibility factors, the corrected inlet and outlet gas velocities are as follows:

$$\text{Inlet velocity } u_1 = 0.8532\times 15.12 = 12.90\,\text{ft/s}$$
$$\text{Outlet velocity } u_2 = 0.8913\times 21.24 = 18.93\,\text{ft/s}$$

8.12 REYNOLDS NUMBER IN GAS FLOW

As discussed earlier in liquid pipelines, an important parameter in flow of fluids in a pipe is the nondimensional term, the Reynolds number. The Reynolds number is used to characterize the type of flow in a pipe, such as laminar, turbulent, or critical flow. For gas flow in a pipeline, the Reynolds number is a function of the gas flow rate, pipe inside diameter, and the gas density and viscosity. It is calculated from the following equation:

$$R = \frac{uD\rho}{\mu}\ \text{(USCS)} \qquad (8.45)$$

where, in USCS units,

R – Reynolds number, dimensionless
u – Average velocity of gas in pipe, ft/s
D – Inside diameter of pipe, ft
ρ – Gas density, slug/ft^3
μ – Gas viscosity, lb·s/ft^2

The corresponding equation for the Reynolds number in SI units is as follows:

$$R = \frac{uD\rho}{\mu} \text{ (SI)} \tag{8.46}$$

where

R – Reynolds number, dimensionless
u – Average velocity of gas in pipe, m/s
D – Inside diameter of pipe, m
ρ – Gas density, kg/m^3
μ – Gas viscosity, kg/m·s

Using customary pipeline units, a more suitable equation for Reynolds number is as follows:

$$R = 0.0004778\left(\frac{P_b}{T_b}\right)\left(\frac{GQ}{\mu D}\right) \text{ (USCS)} \tag{8.47}$$

where

P_b – Base pressure, psia
T_b – Base temperature, °R (460 + °F)
G – Specific gravity of gas (Air = 1.0)
Q – Gas flow rate, standard ft^3/day (SCFD)
D – Pipe inside diameter, in.
μ – Viscosity of gas, lb/ft·s

In SI units, the Reynolds number is

$$R = 0.5134\left(\frac{P_b}{T_b}\right)\left(\frac{GQ}{\mu D}\right) \text{ (SI)} \tag{8.48}$$

where

P_b – Base pressure, kPa
T_b – Base temperature, °K (273 + °C)
G – Specific gravity of gas (Air = 1.0)
Q – Gas flow rate, m^3/day (standard conditions)
D – Pipe inside diameter, mm
μ – Viscosity of gas, Poise

Laminar flow occurs in a pipeline when the Reynolds number is below a value of approximately 2000. Turbulent flow occurs when the Reynolds number is greater than 4000. For Reynolds numbers between 2000 and 4000, the flow is undefined and is referred to as critical flow.

Thus,

1. Laminar flow – $R \leq 2000$
2. Critical flow – $R > 2000$ and $R \leq 4000$
3. Turbulent flow – $R > 4000$

Most natural gas pipelines operate in the turbulent flow region. Therefore, the Reynolds number is greater than 4000. Turbulent flow is further divided into three regions known as smooth pipe flow, fully rough pipe flow, and transition flow.

Example Problem 8.8 (USCS Units)

A natural gas pipeline, NPS 24 with 0.500 in. wall thickness transports 150 MMSCFD. The specific gravity of gas is 0.6 and viscosity is 0.000008 lb/ft·s. Calculate the value of the Reynolds number of flow. Assume the base temperature and base pressure to be 60°F and 14.7 psia, respectively.

Solution

$$\text{Pipe inside diameter} = 24 - 2 \times 0.5 = 23.0\,\text{in.}$$
$$\text{The base temperature} = 60 + 460 = 520°\text{R}$$

Using Eq. (8.47), we get

$$R = 0.0004778\left(\frac{14.7}{520}\right)\left(\frac{0.6 \times 150 \times 10^6}{0.000008 \times 23}\right) = 6{,}606{,}704$$

Because R is greater than 4000, the flow is in the turbulent region.

Example Problem 8.9 (SI Units)

A natural gas pipeline, DN 500 with 12 mm wall thickness transports 4 Mm^3/day. The specific gravity of gas is 0.6 and viscosity is 0.00012 Poise. Calculate the value of the Reynolds number. Assume the base temperature and base pressure to be 15°C and 101 kPa, respectively.

Solution

$$\text{Pipe inside diameter} = 500 - 2 \times 12 = 476\,\text{mm}$$
$$\text{The base temperature} = 15 + 273 = 288\,\text{K}$$

Using Eq. (8.48), we get

$$R = 0.5134\left(\frac{101}{15 + 273}\right)\left(\frac{0.6 \times 4 \times 10^6}{0.00012 \times 476}\right) = 7{,}564{,}980$$

Since R is greater than 4000, the flow is in the turbulent region.

8.13 FRICTION FACTOR IN GAS FLOW

To calculate the pressure drop in a gas pipeline at a given flow rate, we must first determine the friction factor. The term friction factor is a dimensionless parameter that depends on the Reynolds number of flow and the pipe diameter and internal roughness of the pipe. The Darcy friction factor will be used throughout this book. Another friction factor known as the Fanning friction factor is preferred by some engineers. It is numerically equal to one-fourth the Darcy friction factor as follows:

$$f_{\mathrm{f}} = \frac{f_{\mathrm{d}}}{4} \tag{8.49}$$

where

f_{f} = Fanning friction factor
f_{d} = Darcy friction factor

To avoid confusion, in subsequent discussions the Darcy friction factor is used and will be represented by the symbol f.

For laminar flow, the friction factor is inversely proportional to the Reynolds number, as indicated below.

$$f = \frac{64}{R} \tag{8.50}$$

For turbulent flow, the friction factor is a function of Reynolds number, pipe inside diameter, and internal roughness of the pipe. Many empirical relationships for calculating f have been put forth by researchers. The more popular correlations include Colebrook–White and the AGA equations. These will be discussed shortly.

Refer to the Moody Diagram, Fig. 8.1 for variation of friction factor f with Reynolds number and the relative roughness of pipe (e/D). For turbulent flow in smooth pipes, the friction factor f depends only on the Reynolds number. For fully rough pipes, f depends more on the pipe internal roughness and less on the Reynolds number. In the transition zone between smooth pipe flow and flow in fully rough pipes, f depends on the pipe roughness, pipe inside diameter, and the Reynolds number.

Generally, the internal pipe roughness is expressed in micro inches (one-millionth of an inch). For example, an internal roughness of 0.0006 in. is referred to as 600 micro inches or 600 μ-in. If the pipe inside diameter is 15.5 in, the relative roughness, in this case, is calculated as

$$\text{Relative roughness} = \frac{0.0006}{15.5} = 0.0000387 = 3.87 \times 10^{-5}$$

From the Moody diagram, Fig. 8.1, for this relative roughness value and Reynolds number $R = 10$ million, we find the friction factor $f = 0.012$.

See Table 8.5 for the pipe roughness used for different pipe materials.

8.14 COLEBROOK–WHITE EQUATION FOR GAS FLOW

The Colebrook–White equation, discussed earlier in liquid flow, can also be used to calculate the friction factor in gas flow. The following form of the Colebrook equation is used to calculate the friction factor in gas pipelines in turbulent flow.

$$\frac{1}{\sqrt{f}} = -2\text{Log}_{10}\left(\frac{e}{3.7D} + \frac{2.51}{R\sqrt{f}}\right) \quad \text{for } R > 4000 \text{ (USCS)} \tag{8.51}$$

where, in USCS units,

f – Friction factor, dimensionless
D – Pipe inside diameter, in.
e – Absolute pipe roughness, in.
R – Reynolds number of flow, dimensionless

Because R and f are dimensionless, as long as consistent units are used for both e and D, the Colebrook equation is the same regardless of the units used. Therefore, in SI units, Eq. (8.51) is used with e and D expressed in mm.

It can be seen from Eq. (8.51) that, in order to calculate the friction factor f, we must use a trial and error approach. It is an implicit equation in f since f appears on both sides of the equation. We first assume a value of f (such as 0.01) and substitute it in the right-hand side of the equation. This will yield a second approximation for f that can then be used to calculate a better value of f and so on. Generally, three to four iterations are sufficient to converge on a reasonably good value of the friction factor.

Example 8.10 (USCS Units)

A natural gas pipeline, NPS 24 with 0.500 in. wall thickness transports 250 MMSCFD. The specific gravity of gas is 0.6 and viscosity is 0.000008 lb/ft·s. Calculate the friction factor using the Colebrook equation. Assume absolute pipe roughness = 600 μ-in. The base temperature and base pressure are 60°F and 14.7 psia, respectively.

Solution

$$\text{Pipe inside diameter} = 24 - 2 \times 0.5 = 23.0 \text{ in.}$$
$$\text{Absolute pipe roughness} = 600\mu\text{–in.} = 0.0006 \text{ in.}$$

First, we calculate the Reynolds number using Eq. (8.47):

$$R = 0.0004778\left(\frac{14.7}{60+460}\right)\left(\frac{0.6 \times 250 \times 10^6}{0.000008 \times 23.0}\right) = 11{,}011{,}173$$

Using the Colebrook–White Eq. (8.51), the friction factor is

$$\frac{1}{\sqrt{f}} = -2\text{Log}_{10}\left(\frac{0.0006}{3.7 \times 23.0} + \frac{2.51}{11{,}011{,}173\sqrt{f}}\right)$$

This equation will be solved by successive iteration. First, assuming $f = 0.01$ and substituting in the preceding equation, we get a better approximation as $f = 0.0099$. Repeating the iteration, we get the final value as $f = 0.00988$.

Therefore, the friction factor is 0.00988.

Example 8.11 (SI Units)

A natural gas pipeline, DN 500 with 12 mm wall thickness transports 6 Mm^3/day. The specific gravity of gas is 0.6 and viscosity is 0.00012 Poise. Calculate the friction factor using the Colebrook equation. Assume absolute pipe roughness = 0.03 mm and the base temperature and base pressure are 15°C and 101 kPa, respectively.

Solution

$$\text{Pipe inside diameter} = 500 - 2 \times 12 = 476\,\text{mm}$$

First, we calculate the Reynolds number using Eq. (8.48)

$$R = 0.5134\left(\frac{101}{15+273}\right)\left(\frac{0.6\times 6\times 10^6}{0.00012\times 476}\right) = 11{,}347{,}470$$

Using Eq. (8.51), the friction factor is

$$\frac{1}{\sqrt{f}} = -2\text{Log}_{10}\left(\frac{0.030}{3.7\times 476} + \frac{2.51}{11{,}347{,}470\sqrt{f}}\right)$$

This equation will be solved by successive iteration. Assume $f = 0.01$ initially and substituting above, we get a better approximation as $f = 0.0112$. Repeating the iteration, we get the final value as $f = 0.0112$.

Therefore, the friction factor is 0.0112.

8.15 TRANSMISSION FACTOR

In gas pipelines, the term transmission factor is used in conjunction with gas flow. The transmission factor, denoted by F, is considered the opposite of the friction factor f; whereas the friction factor indicates how difficult it is to move a certain quantity of gas through a pipeline, the transmission factor is a direct measure of how much gas can be transported through the pipeline. As the friction factor increases, the transmission factor decreases; therefore, the gas flow rate also decreases. Conversely, the higher the transmission factor, the lower the friction factor and, therefore, the higher will be the flow rate.

The transmission factor F is related to the friction factor f as follows:

$$F = \frac{2}{\sqrt{f}} \tag{8.52}$$

where

f = Friction factor
F = Transmission factor

It must be noted that the friction factor f in the above equation is the Darcy friction factor. Because some engineers prefer to use the Fanning friction factor, the relationship between the transmission factor F and the Fanning friction factor is given below for reference.

$$F = \frac{1}{\sqrt{f_f}} \tag{8.53}$$

where f_f is the Fanning friction factor.

For example, if the Darcy friction factor is 0.025, the transmission factor is calculated using Eq. (8.52) as

$$F = \frac{2}{\sqrt{0.025}} = 12.65$$

The Fanning friction factor, in this case, will be $\frac{0.025}{4} = 0.00625$.

Therefore, the transmission factor using Eq. (8.53) is $F = \frac{1}{\sqrt{0.00625}} = 12.65$, which is the same as calculated using the Darcy friction factor. Therefore, there is only one transmission factor while there are two different friction factors.

Having defined a transmission factor, we can rewrite the Colebrook Eq. (8.51) in terms of the transmission factor using Eq. (8.53) as follows:

$$F = -4\text{Log}_{10}\left(\frac{e}{3.7D} + \frac{1.255F}{R}\right) \tag{8.54}$$

Because R and F are dimensionless, as long as consistent units are used for both e and D, the transmission factor equation is the same regardless of the units used. Therefore, in SI units, Eq. (8.54) is used with e and D expressed in mm.

Similar to the calculation of the friction factor f from Eq. (8.51), the transmission factor F from Eq. (8.54) also requires an iterative approach.

Example Problem 8.12

For a gas pipeline, flowing 150 MMSCFD gas of specific gravity 0.65 and viscosity of 0.000008 lb/ft·s, calculate the friction factor and transmission factor considering an NPS 20 pipeline, 0.500 in. wall thickness and an internal roughness of 700 μ-in. Assume the base temperature and base pressure to be 60°F and 14.7 psia, respectively. If the flow rate increases by 50%, what is the impact on the friction factor and transmission factor?

Solution

$$\text{The base temperature} = 60 + 460 = 520°\text{R}$$
$$\text{Pipe inside diameter} = 20 - 2 \times 0.500 = 19.0\,\text{in.}$$

Using Eq. (8.47), we calculate the Reynolds number as

$$R = 0.0004778\left(\frac{14.7}{520}\right)\left(\frac{0.65 \times 150 \times 10^6}{0.000008 \times 19}\right) = 8{,}664{,}054$$

$$\text{The relative roughness} = \frac{700 \times 10^{-6}}{19} = 3.6842 \times 10^{-5}$$

Using Eq. (8.51), the friction factor is

$$\frac{1}{\sqrt{f}} = -2\text{Log}_{10}\left(\frac{0.00003684}{3.7} + \frac{2.51}{8{,}664{,}054\sqrt{f}}\right)$$

Solving by successive iteration, we get $f = 0.0104$

Therefore, the transmission factor, F, is found from Eq. (8.52) as follows:

$$F = \frac{2}{\sqrt{0.0104}} = 19.61$$

When flow rate is increased by 50%, the Reynolds number becomes by proportion

$$R = 1.5 \times 8{,}664{,}054 = 12{,}996{,}081$$

The new friction factor from Eq. (8.51) is

$$\frac{1}{\sqrt{f}} = -2\text{Log}_{10}\left(\frac{0.00003684}{3.7} + \frac{2.51}{12{,}996{,}081\sqrt{f}}\right)$$

Solving for f by successive iteration, we get $f = 0.0103$

The corresponding transmission factor is

$$F = \frac{2}{\sqrt{0.0103}} = 19.74$$

Compared with the previous values of 0.0104 for friction factor and 19.61 for transmission factor, we see the following changes:

$$\text{Decrease in friction factor} = \frac{0.0104 - 0.0103}{0.0104} = 0.0096 \text{ or } 0.96\%$$

$$\text{Increase in transmission factor} = \frac{19.74 - 19.61}{19.61} = 0.0066 \text{ or } 0.66\%$$

Thus increasing the flow rate by 50% reduces the friction factor by 0.96% and increases the transmission factor by 0.66%.

Example Problem 8.13

For a gas pipeline, flowing 3 Mm^3/day gas of specific gravity 0.6 and viscosity of 0.000119 Poise, calculate the friction factor and transmission factor considering a DN 400 pipeline, 10 mm wall thickness and an internal roughness of 0.02 mm. The base temperature and base pressure are 15°C and 101 kPa, respectively. If the flow rate is doubled, what is the impact on the friction factor and transmission factor?

Solution

$$\text{The base temperature} = 15 + 273 = 288\,\text{K}$$
$$\text{Pipe inside diameter} = 400 - 2 \times 10 = 380\,\text{mm}$$

Using Eq. (8.48), we calculate the Reynolds number as

$$R = 0.5134\left(\frac{101}{288}\right)\left(\frac{0.6 \times 3 \times 10^6}{0.000119 \times 380}\right) = 7{,}166{,}823$$

$$\text{The relative roughness} = \frac{0.02}{380} = 0.0000526$$

Using Eq. (8.51), the friction factor is

$$\frac{1}{\sqrt{f}} = -2\text{Log}_{10}\left(\frac{0.0000526}{3.7} + \frac{2.51}{7{,}166{,}823\sqrt{f}}\right)$$

Solving by iteration, we get $f = 0.0111$
Therefore, the transmission factor, F, is found from Eq. (8.52) as follows:

$$F = \frac{2}{\sqrt{0.0111}} = 18.98$$

When the flow rate is doubled, the Reynolds number becomes

$$R = 2 \times 7{,}166{,}823 = 14{,}333{,}646$$

The new value of the friction factor from Eq. (8.51) is

$$\frac{1}{\sqrt{f}} = -2\text{Log}_{10}\left(\frac{0.0000526}{3.7} + \frac{2.51}{14{,}333{,}646\sqrt{f}}\right)$$

Solving for f by successive iteration, we get $f = 0.0109$ and the transmission factor is from Eq. (8.52)

$$F = \frac{2}{\sqrt{0.0109}} = 19.16$$

Therefore, doubling the flow rate increased the transmission factor and decreased the friction factor as follows:

$$\text{Decrease in friction factor} = \frac{0.0111 - 0.0109}{0.0111} = 0.018 \text{ or } 1.8\%$$

$$\text{Increase in transmission factor} = \frac{19.16 - 18.98}{18.98} = 0.0095 \text{ or } 0.95\%$$

8.16 PRESSURE DROP IN GAS FLOW

The general flow equation, also called the fundamental flow equation, for the steady state isothermal flow in a gas pipeline is the basic equation for relating the pressure drop with flow rate. The most common form of this equation in

USCS units is given in terms of the pipe diameter, gas properties, pressures, temperatures, and flow rate as follows:

$$Q = 77.54\left(\frac{T_b}{P_b}\right)\left(\frac{P_1^2 - P_2^2}{GT_fLZf}\right)^{0.5} D^{2.5} \tag{8.55}$$

where, in USCS units,

Q – Gas flow rate, measured at standard conditions, ft³/day (SCFD)
f – Darcy friction factor, dimensionless
P_b – Base pressure, psia
T_b – Base temperature, °R (460 + °F)
P_1 – Upstream pressure, psia
P_2 – Downstream pressure, psia
G – Gas gravity (Air = 1.00)
T_f – Average gas flowing temperature, °R (460 + °F)
L – Pipe segment length, mi
Z – Gas compressibility factor at the flowing temperature, dimensionless
D – Pipe inside diameter, in.

Refer to Fig. 8.2 for an explanation of the symbols used. Note that for isothermal flow in the pipe segment, from section 1 to section 2, the average gas temperature T_f is assumed to be constant.

In SI units, the general flow equation is stated as follows:

$$Q = 1.1494 \times 10^{-3}\left(\frac{T_b}{P_b}\right)\left[\frac{(P_1^2 - P_2^2)}{GT_fLZf}\right]^{0.5} D^{2.5}\ \text{(SI)} \tag{8.56}$$

where

Q – Gas flow rate, measured at standard conditions, m³/day
f – Friction factor, dimensionless
P_b – Base pressure, kPa
T_b – Base temperature, K (273 + °C)
P_1 – Upstream pressure, kPa
P_2 – Downstream pressure, kPa
G – Gas gravity (Air = 1.00)

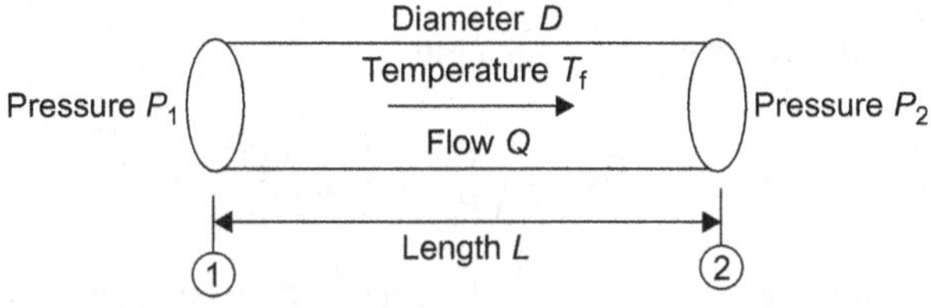

FIGURE 8.2 Steady flow in a gas pipeline.

T_f – Average gas flowing temperature, K (273 + °C)
L – Pipe segment length, km
Z – Gas compressibility factor at the flowing temperature, dimensionless
D – Pipe inside diameter, mm

Due to the format of Eq. (8.56), the pressures may also be in MPa or Bar as long as the same consistent unit is used.

Equation (8.55) relates the flow rate in a pipe segment of length L based on an upstream pressure of P_1 and a downstream pressure of P_2 as shown in Fig. 8.2. It is assumed that there is no elevation difference between the upstream and downstream points; therefore, the pipe segment is horizontal.

On examining the general flow Eq. (8.55), we see that for a pipe segment of length L and diameter D, the gas flow rate Q (at standard conditions) depends on several factors. Q depends on gas properties represented by the gravity G and the compressibility factor Z. If the gas gravity is increased (heavier gas), the flow rate will decrease. Similarly, as the compressibility factor Z increases, flow rate will decrease. Also as the gas flowing temperature T_f increases, throughput will decrease. Thus the hotter the gas, the lower will be the flow rate. Therefore, to increase flow rate, it helps to keep the gas temperature low. The impact of pipe length and inside diameter are also clear. As the pipe segment length increases for given pressure P_1 and P_2, the flow rate will decrease. However, the larger the diameter, the larger will be the flow rate. The term ${P_1}^2 - {P_2}^2$ represents the driving force that causes the flow rate from the upstream end to the downstream end. As the downstream pressure P_2 is reduced, keeping the upstream pressure P_1 constant, the flow rate will increase. It is obvious that when there is no flow rate, P_1 is equal to P_2. It is because of friction between the gas and pipe wall that the pressure drop ($P_1 - P_2$) occurs from the upstream point 1 to downstream point 2. As we have discussed before, the friction factor f depends on the internal pipe roughness, as well as the Reynolds number of flow. The general flow equation can be represented in terms of the transmission factor F instead of the friction factor f, as follows:

$$Q = 38.77F\left(\frac{T_b}{P_b}\right)\left(\frac{P_1^2 - P_2^2}{GT_fLZ}\right)^{0.5} D^{2.5} \tag{8.57}$$

where the transmission factor F and friction factor f are related by Eq. (8.52) as discussed earlier.

In SI units, the corresponding equation is

$$Q = 5.747\times 10^{-4}F\left(\frac{T_b}{P_b}\right)\left[\frac{(P_1^2 - P_2^2)}{GT_fLZ}\right]^{0.5} D^{2.5} \tag{8.58}$$

We will discuss several aspects of the general flow equation before moving on to the other formulas for pressure drop calculation.

8.17 EFFECT OF PIPE ELEVATIONS

When elevation difference between the ends of a pipe segment is included, the general flow equation is modified as follows:

$$Q = 38.77F\left(\frac{T_b}{P_b}\right)\left(\frac{P_1^2 - e^s P_2^2}{GT_f L_e Z}\right)^{0.5} D^{2.5} \quad (8.59)$$

In SI units,

$$Q = 5.747 \times 10^{-4} F\left(\frac{T_b}{P_b}\right)\left[\frac{(P_1^2 - e^s P_2^2)}{GT_f L_e Z}\right]^{0.5} D^{2.5} \quad (8.60)$$

and the equivalent length L_e is defined by

$$L_e = \frac{L(e^s - 1)}{s} \quad (8.61)$$

where the parameter s depends on the elevation difference between the ends of the pipe segment as follows:

$$s = 0.0375G\left(\frac{H_2 - H_1}{T_f Z}\right) \quad (8.62)$$

where, in USCS units,

s – Elevation adjustment parameter, dimensionless
H_1 – Upstream elevation, ft
H_2 – Downstream elevation, ft

Other symbols are as defined earlier.

The equivalent length L_e and the term e^s take into account the elevation difference between the upstream and downstream ends of the pipe segment. The parameter s depends on the gas gravity, gas compressibility factor, the flowing temperature, and the elevation difference.

In SI units, the elevation adjustment parameter s is defined as follows:

$$s = 0.0684G\left(\frac{H_2 - H_1}{T_f Z}\right) \quad (8.63)$$

where

H_1 – Upstream elevation, m
H_2 – Downstream elevation, m

Other symbols are as defined earlier.

In the calculation of L_e in Eq. (8.61), we have assumed that there is a single slope between the upstream point 1 and the downstream point 2 in Fig. 8.2. However, if the pipe segment of length L has a series of slopes, then we introduce a parameter j as follows, for each individual pipe subsegment that compose the pipe length from point 1 to point 2.

$$j = \frac{e^s - 1}{s} \tag{8.64}$$

The parameter j is calculated for each slope of each pipe subsegment of length L_1, L_2, and so on that make up the total length L. The equivalent length term L_e in Eqs (8.59) and (8.61) is calculated by summing the individual slopes as defined below:

$$L_e = j_1 L_1 + j_2 L_2 e^{s1} + j_3 L_3 e^{s2} + \cdots\cdots \tag{8.65}$$

Note that the general flow equation is the most commonly used equation to calculate the flow rate and pressure in a gas pipeline. In order to apply it correctly, we must use the correct friction factor or transmission factor. Several other equations such as Panhandle A, Panhandle B, and Weymouth calculate the flow rate for a given pressure without using a friction factor or transmission factor. These equations are listed in Appendix 5. However, an equivalent friction factor (or transmission factor) can be calculated using these methods as well.

8.18 THE AVERAGE GAS PRESSURE

In the general flow equation, the compressibility factor Z used must be calculated at the gas flowing temperature and average pressure in the pipe segment. Therefore, it is important to first calculate the average pressure in a pipe segment described in Fig. 8.2.

In Fig. 8.2, for the pipe segment, the upstream pressure is P_1 and downstream pressure is P_2. The average pressure for this segment must be used to calculate the compressibility factor of the gas at the average gas temperature T_f. As a first approximation, we may use an arithmetic average of $(P_1 + P_2)/2$. However, based on the variation of pressures in a gas pipeline, a more accurate value of the average gas pressure in a pipe segment is

$$P_{avg} = \frac{2}{3}\left(P_1 + P_2 - \frac{P_1 P_2}{P_1 + P_2}\right) \tag{8.66}$$

By mathematical manipulation, Eq. (8.66) may also be expressed in a slightly different form, as follows:

$$P_{avg} = \frac{2}{3}\left(\frac{P_1{}^3 - P_2{}^3}{P_1{}^2 - P_2{}^2}\right) \tag{8.67}$$

Note that all pressures used in the general flow equation are in absolute units. Therefore, gauge pressure units should be converted to absolute pressure by adding the base pressure.

For example, if the upstream and downstream pressures are 1000 and 900 psia, respectively, then from Eq. (8.66), the average gas pressure is

$$P_{avg} = \frac{2}{3}\left(1000 + 900 - \frac{1000 \times 900}{1900}\right) = 950.88\,\text{psia}$$

Compare this with the arithmetic average of

$$P_{avg} = \frac{1}{2}(1000 + 900) = 950\,\text{psia}$$

Example Problem 8.14 (USCS Units)

A gas pipeline NPS 20 with 0.500 in. wall thickness flows 200 MMSCFD gas of specific gravity 0.6 and viscosity of 0.000008 lb/ft·s. Using the Colebrook–White equation calculate the pressure drop in a 50-mi segment of pipe based on an upstream pressure of 1000 psig. Assume an internal pipe roughness of 600 μ-in and that the base temperature and base pressure are 60°F and 14.73 psia, respectively. Neglect elevation effects and use 60°F for gas flowing temperature and compressibility factor $Z = 0.88$.

Solution

$$\text{Inside diameter of pipe} = 20 - 2 \times 0.5 = 19.0\,\text{in.}$$
$$\text{The base temperature} = 60 + 460 = 520°\text{R}$$
$$\text{Gas flow temperature} = 60 + 460 = 520°\text{R}$$

First, we calculate the Reynolds number using Eq. (8.47):

$$R = 0.0004778\left(\frac{14.73}{520}\right)\left(\frac{0.6 \times 200 \times 10^6}{0.000008 \times 19}\right) = 10{,}685{,}214$$

The transmission factor F is calculated from Eq. (8.54) as follows:

$$F = -4\text{Log}_{10}\left(\frac{600 \times 10^{-6}}{3.7 \times 19} + \frac{1.255F}{10{,}685{,}214}\right)$$

Solving for F by successive iteration, we get

$$F = 19.86$$

Next, using general flow Eq. (8.55), we calculate the downstream pressure P_2 as follows:

$$200 \times 10^6 = 38.77 \times 19.86\left(\frac{60 + 460}{14.73}\right)\left[\frac{1014.73^2 - P_2{}^2}{0.6 \times 520 \times 50 \times 0.88}\right]^{0.5} \times 19^{2.5}$$

Solving for P_2, we get

$$P_2 = 854.12\,\text{psia} = 839.4\,\text{psig}$$

Therefore, the pressure drop = 1014.73 − 854.12 = 160.6 psi.

Example Problem 8.15 (SI Units)

A pipeline DN 500, 10 mm wall thickness is used to transport natural gas at a flow rate of 5 Mm^3/day in a 10-km long pipeline. Assuming isothermal flow at 20°C, calculate the inlet pressure required for a terminus delivery pressure of 5000 kPa(abs). The gas has a specific gravity 0.6 and viscosity of 0.00012 Poise. The pipe roughness is 0.015 mm and the base temperature and base pressure are 15°C and 101 kPa, respectively. Assume compressibility factor $Z = 0.895$.

Solution

$$\text{Inside diameter of pipe} = 500 - 2\times 10 = 480\,\text{mm}$$
$$\text{The base temperature} = 15 + 273 = 288\,\text{K}$$
$$\text{Gas flow temperature} = 20 + 273 = 293\,\text{K}$$

First, we calculate the Reynolds number using Eq. (8.48):

$$R = 0.5134\left(\frac{101}{288}\right)\left(\frac{0.6\times 5\times 10^6}{0.00012\times 480}\right) = 9{,}377{,}423$$

The transmission factor F is calculated from Eq. (8.54) as follows:

$$F = -4\text{Log}_{10}\left(\frac{0.015}{3.7\times 480} + \frac{1.255F}{9{,}377{,}423}\right)$$

Solving for F by successive iteration, we get

$$F = 19.82$$

Next, using general flow Eq. (8.56), we calculate the upstream pressure P_1 as follows:

$$5\times 10^6 = 5.747\times 10^{-4}\times 19.82\left(\frac{288}{101}\right)\left[\frac{(P_1^2 - 5000^2)}{0.6\times 293\times 10\times 0.895}\right]^{0.5} 480^{2.5}$$

Solving for the inlet pressure P_1, we get

$$P_1 = 5144\,\text{kPa(abs)}$$

SUMMARY

In this chapter, we reviewed the hydraulics of liquid and gas pipelines. First, we explained how velocity and Reynolds number of flow were calculated and how the types of flow (laminar, turbulent, and so on) were classified based on the Reynolds number. Next, the friction factor was calculated using the Moody diagram and the Colebrook–White equations. The Darcy equation for pressure drop calculation in liquids was introduced and a more practical version of the equation using the commonly used pipeline units was illustrated using examples. The Hazen–Williams equation for water flow was reviewed, together with the impact of the C-factor. Minor losses in valves and fittings were discussed. In gas pipelines, the friction factor, transmission factor, and the general flow equation were introduced and the method of calculating the pressure drop in a pipe segment was explained. The effect of the elevation difference between the upstream and the downstream ends of the pipeline was also discussed.

BIBLIOGRAPHY

A.E. Uhl, et al., Steady Flow in Gas Pipelines, AGA, AGA Report No. 10, New York, 1965.

Compressibility Factors, AGA, AGA Report No. 8, New York, 1992.

E.F. Brater, H.W. King, Handbook of Hydraulics, McGraw-Hill, New York, 1982.

Flow of Fluids through Valves, Fittings and Pipes, Crane Company, New York, 1976.

Gas Processors Suppliers Association, Engineering Data Book, tenth ed., Tulsa, OK, 1994.

E. Shashi Menon, Liquid Pipeline Hydraulics, Marcel-Dekker, New York, 2004.

E. Shashi Menon, Gas Pipeline Hydraulics, CRC Press, Boca Raton, FL, 2005.

D.L. Katz, et al., Handbook of Natural Gas Engineering, McGraw-Hill, New York, 1959.

W.D. McCain Jr., The Properties of Petroleum Fluids, Petroleum Publishing Company, Tulsa, OK, 1973.

M. Mohitpour, H. Golshan, A. Murray, Pipeline Design and Construction, second ed., ASME Press, New York, 2003.

E. Shashi Menon, Piping Calculations Manual, McGraw-Hill, New York, 2005.

Pipeline Design for Hydrocarbon Gases and Liquids, American Society of Civil Engineers, New York, 1975.

Chapter 9

Series and Parallel Piping and Power Required

E. Shashi Menon, Ph.D., P.E.

Chapter Outline

INTRODUCTION

In this chapter, we explain how to calculate the pressure required at the pipeline origin to safely transport a given throughput of liquid or gas. The effect of pipeline elevation profile, minimum delivery pressure required, in addition to friction losses, will be considered. The pumping power and how many pumps or compressor stations are required will be reviewed.

We will also discuss pipes configured in series and parallel. System head curves for liquid pipelines will be introduced along with how they interact with pump curves. Flow injection and delivery, as well as looping pipelines to increase the pipeline throughput will be explained.

9.1 TOTAL PRESSURE REQUIRED TO TRANSPORT LIQUIDS

The total pressure P_T required at the beginning of a pipeline to transport liquids at a given flow rate from point A to point B will depend on the following:

- Flow rate
- Liquid specific gravity and viscosity
- Pipe diameter, wall thickness, and roughness
- Pipe length from A to B
- Pipeline elevation changes from A to B

From the discussions in Chapter 8, increasing flow rate will result in higher pressure drop and increased total pressure P_T. Higher specific gravity and viscosity of the liquid transported will cause increased pressure drop and hence increase in P_T. If we increase the pipe diameter, keeping all other items constant, the frictional pressure drop will decrease and hence P_T will also decrease. Similarly, increasing pipe wall thickness or pipe roughness will increase pressure drop due to friction and P_T. On the other hand, if only the pipe length was increased, the pressure drop as well as P_T will increase. If the elevation changes from A to B are small, its effect on P_T is small, compared with large elevation changes, which will increase P_T.

Let us review how the pipeline elevation profile affects the total pressure P_T. If the pipelines were installed along a fairly flat terrain, with no appreciable elevation difference between the beginning of the pipeline A and the terminus B, the total pressure P_T will not be affected. But, if the elevation difference between A and B was substantial and B was at a higher elevation than A, P_T will be higher than that for the flat terrain pipeline.

The total pressure required (P_T) is composed of three main parts:

1. Frictional head
2. Elevation head
3. Delivery pressure at terminus

For example, a 50-mi-long pipeline from point A to point B transports diesel (Sg = 0.85) fuel at a flow rate of 5000 bbl/h. If the total pressure drop due to friction in the pipeline is 900 psi, the elevation difference from point A to point B is 600 ft (uphill flow), and the minimum delivery pressure required at the terminus B is 50 psi, we can determine the total pressure required at A as the sum of the three components as follows:

$$\text{Total pressure at A} = 900\,\text{psi} + 600\,\text{ft} + 50\,\text{psi}$$

In liquid pipelines, psi and psig are used interchangeably. If the pressure is in absolute units, it will be denoted by psia. In gas pipelines, psia is used more frequently than psig.

Converting to consistent units of psi, the total pressure becomes

$$\text{Total pressure} = 900 + (600)(0.85/2.31) + 50 = 1171\,\text{psi}$$

This assumes that there are no controlling peaks or high elevation points between point A and point B. Suppose that an intermediate point C located half way between A and B had an elevation of 1600 ft, compared with the elevations of point A (100 ft) and point B (700 ft). In this case, the elevation of point C becomes a controlling factor. The calculation of the total pressure required at A is as follows:

Let us assume that the 50-mi pipeline is of uniform diameter and thickness throughout its entire length. Therefore, the pressure drop per mile will be a constant value for the entire pipeline calculated as

$$900/50 = 18\,\text{psi/mi}$$

The total frictional pressure drop for the pipe segment from point A to C (peak), located 25 mi away, is given below.

$$\text{Frictional pressure drop from A to C} = 18 \times 25 = 450\,\text{psi}$$

Since C is the midpoint of the pipeline, the frictional pressure drop between C and B is also equal to 450 psi.

Consider now the portion of the pipeline from A to C, with a frictional pressure drop of 450 psi and an elevation difference between A and C of

$$1600\,\text{ft} - 100\,\text{ft} = 1500\,\text{ft}$$

The total pressure required at A to get over the peak at C is the sum of the friction and elevation components as follows:

$$\text{Total pressure} = 450 + (1500)(0.85/2.31) = 1001.95\,\text{psi}$$

It must be noted that this pressure of 1001.95 psi at A will just about get the liquid over the peak at point C with zero gauge pressure. Sometimes, it is desired that the liquid at the top of the hill be at some minimum pressure above the liquid vapor pressure at the flowing temperature. If the transported liquid was LPG, we would require a minimum pressure in the pipeline of 250–300 psi. On the other hand with crude oils and refined products with low vapor pressure, the minimum pressure required may be only 10–20 psi. In this example, we assume that we are dealing with low vapor pressure liquids and a minimum pressure of 10 psi is adequate at the high points in a pipeline.

The revised total pressure at A is as follows:

$$\text{Total pressure} = 450 + (1500)(0.85/2.31) + 10 = 1012\,\text{psi, approximately.}$$

Therefore, starting with a pressure of 1012 psi at A will result in a pressure of 10 psi at the highest point C after accounting for frictional pressure drop and elevation difference between A and the peak C.

Once the liquid reaches the high point C at 10 psi, it flows downhill from point C to the terminus B, assisted by gravity. Therefore, for the pipe segment from C to B, the elevation difference (1600 – 700 ft) helps the flow, whereas the friction impedes the flow. The delivery pressure at the terminus B can be calculated by considering the elevation difference and frictional pressure drop between C and B as follows:

$$\begin{aligned}\text{Delivery pressure at terminus B} &= 10 + (1600 - 700)(0.85/2.31) - 450 \\ &= -109 \text{ psi, approximately.}\end{aligned}$$

Since we require a positive pressure of 50 psi at B, we need to adjust the starting pressure at A as follows:

$$\text{Pressure at A} = 1012 + 50 + 109 = 1171 \text{ psi}$$

The above value is incidentally the same pressure we calculated at the beginning of this section, without considering the point C.

Based on this, the pressure at peak C = 10 + 50 + 109 = 169 psi.

Thus, a pressure of 1171 psi at A, the beginning of the pipeline, results in a delivery pressure of 50 psi at the terminus B and clears the peak at C with a pressure of 169 psi. This is shown in Fig. 9.1 where the hydraulic pressure gradient has been plotted for the pipeline and the pressures are

$$P_A = 1171 \text{ psi} \quad P_C = 169 \text{ psi} \quad \text{and} \quad P_B = 50 \text{ psi}$$

All pressures have been rounded off to the nearest one psi.

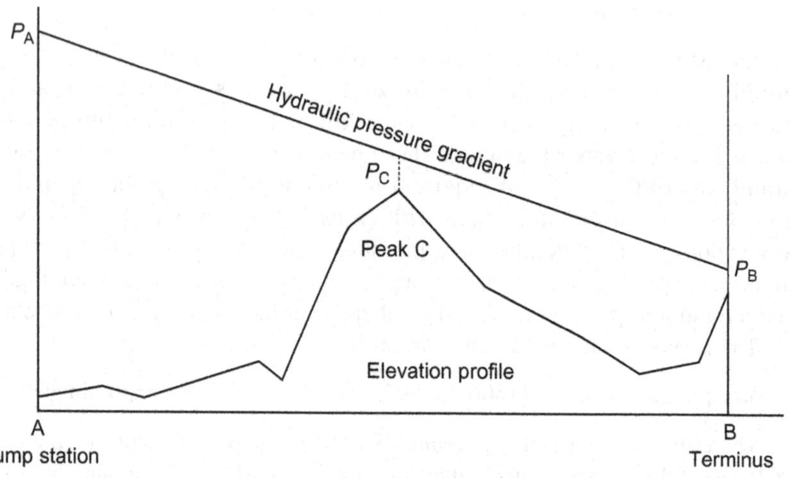

FIGURE 9.1 Hydraulic gradient.

Although the higher elevation point at C appeared to be controlling, our calculation showed that the pressure required at A depended more on the required delivery pressure at the terminus B. In many cases, this may not be true. The pressure required may be dictated more by the controlling peak, and therefore the arrival pressure of the liquid at the pipeline terminus may be higher than the minimum required.

9.2 HYDRAULIC PRESSURE GRADIENT IN LIQUIDS

Due to friction losses, the liquid pressure in a pipeline decreases continuously from the pipe inlet to the pipe delivery terminus. If there is no significant elevation difference between the two ends of the pipeline, the inlet pressure at the beginning of the pipeline will decrease continuously by the friction loss at a particular flow rate. When there is elevation difference along the pipeline, the decrease in pipeline pressure along the pipeline will be due to the combined effect of pressure drop due to friction and the algebraic sum of pipeline elevations. Thus, starting at 1000 psi pressure at the beginning of the pipeline, assuming 15 psi/mi pressure drop due to friction in a flat pipeline (no elevation difference) with constant diameter, at a distance of 20 mi, the pressure would drop to

$$1000 - 15 \times 20 = 700\,\text{psi}$$

If the pipeline is 60 mi long, the total pressure drop due to friction is

$$15 \times 60 = 900\,\text{psi}$$

The pressure at the end of the pipeline is

$$1000 - 900 = 100\,\text{psi}$$

Thus, the liquid pressure in the pipeline has uniformly dropped from 1000 psi in the beginning to 100 psi at the end of the 60 mi length. This pressure profile is referred to as the hydraulic pressure gradient in the pipeline. The hydraulic pressure gradient is a graphical representation of the variation of pressures along the pipeline. It is shown along with the pipeline elevation profile. Since elevation is plotted in feet, the pipeline pressures are also represented in feet of liquid head. This is shown in Figs 9.1 and 9.2.

In the example problem discussed in Section 9.1, we calculated the pressure required at the beginning of the pipeline to be 1171 psi for pumping diesel fuel at a flow rate of 5000 bbl/h. This pressure requires one pump station at the origin of the pipeline at point A.

Suppose the pipe length is 100 mi and the maximum allowable operating pressure (MAOP) is limited to 1200 psi. Let us assume the total pressure required at A is calculated to be 1800 psi at a flow rate of 5000 bbl/h. Obviously we cannot pump the liquid from A with a discharge pressure of 1800 psi, since this will exceed the 1200 psi MAOP. To be within the MAOP limit, we would require an intermediate pump station between A and B. The total pressure

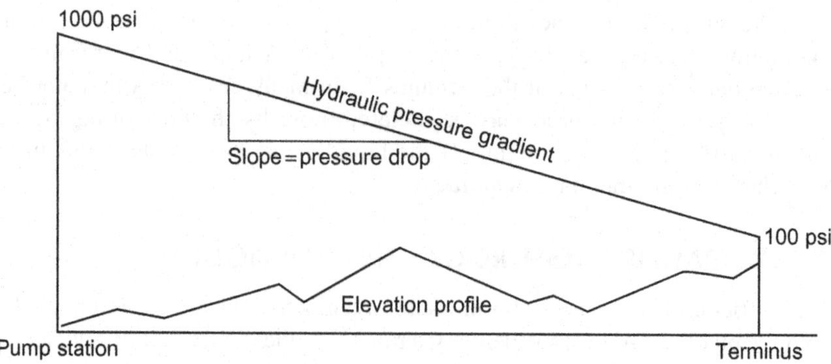

FIGURE 9.2 Hydraulic pressure gradient.

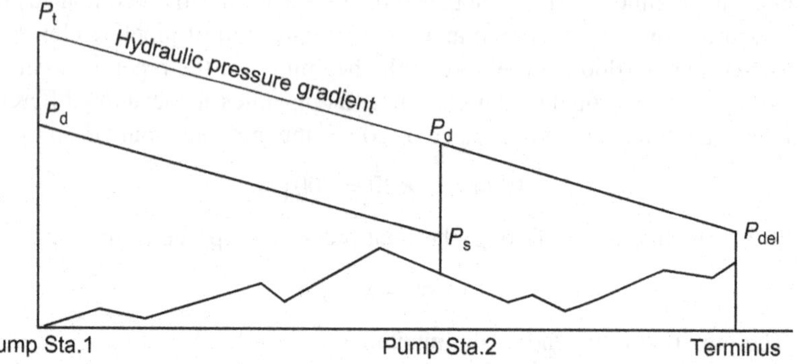

FIGURE 9.3 Hydraulic pressure gradient – two pump stations.

required to pump the diesel from A to B will be provided in steps. The first pump station at A will provide approximately half the pressure followed by the second pump station located at some intermediate point providing the other half. This results in a sawtooth-like hydraulic gradient as shown in Fig. 9.3.

The actual discharge pressures at each pump station can be calculated considering pipeline elevations between A and B and the required minimum suction pressures at each of the two pump stations, as explained next.

Let P_s and P_d represent the common suction and discharge pressure, respectively, for each pump station and P_{del} is the required delivery pressure at the pipe terminus B. The total pressure P_t required at A can be written as follows:

$$P_t = P_{friction} + P_{elevation} + P_{del} \tag{9.1}$$

where

P_t – Total pressure required at A
$P_{friction}$ – Total frictional pressure drop between A and B

$P_{elevation}$ – Elevation head between A and B
P_{del} – Required delivery pressure at B

Based on geometry, from Fig. 9.3, we get the following

$$P_t = P_d + P_d - P_s \tag{9.2}$$

Solving for P_d, we get

$$P_d = (P_t + P_s)/2 \tag{9.3}$$

where

P_d – Pump station discharge pressure
P_s – Pump station suction pressure

For example, if the total pressure calculated is 1800 psi and the pipeline MAOP is 1200 psi, we will need two pump stations. Assuming a minimum suction pressure of 50 psi, from Eq. (9.3), each pump station would have a discharge pressure of

$$P_d = (1800 + 50)/2 = 925\,\text{psi}$$

Each pump station operates at 925 psi discharge pressure and the pipeline MAOP is 1200 psi. Therefore, based on pipeline pressures, we have the capability of increasing pipeline flow rate further to fully utilize the 1200 psi MAOP at each pump station. This would also require upgrading the pumping equipment at each pump station, since more pump power will be required at the higher flow rate. We can estimate how much additional flow rate can be obtained if we increase P_d at each pump station to 1200 psi MAOP.

Suppose the pipeline elevation head between A and B accounts for 300 psi, converted to pressure units. This component of the total pressure required (P_t = 1800 psi) depends only on the pipeline elevation and liquid specific gravity and therefore does not vary with flow rate. Similarly, the delivery pressure of 50 psi at B is also independent of flow rate. We can then calculate the frictional component (which depends on flow rate) of the total pressure P_t using Eq. (9.1) as follows:

$$\text{Frictional pressure drop} = 1800 - 300 - 50 = 1450\,\text{psi}$$

Assuming 100 mi length of the pipeline, the frictional pressure drop per mile of pipe is

$$P_m = 1450/100 = 14.5\,\text{psi/mi}$$

This is the pressure drop at a diesel flow rate of 5000 bbl/h. From our discussions in Chapter 8, the pressure drop per mile, P_m, varies approximately as the square of the flow rate Q, as long as the liquid properties and pipe size do not change. Using Eq. (8.14), we get

$$P_m = K(Q)^2 \tag{9.4}$$

where K is a constant for this pipeline that depends on liquid properties and pipe diameter.

Note that the K value in Eq. (9.4) is not the same as the head loss coefficient discussed in Chapter 8.

Therefore, using Eq. (9.4), we get Eq. (9.5) at the initial flow rate of 5000 bbl/h

$$14.5 = K(5000)^2 \tag{9.5}$$

When flow rate is increased to some value Q to fully utilize the 1200 psi MAOP of the pipeline, the total pressure required at A is calculated using Eq. (9.3), by setting $P_d = 1200$.

$$1200 = (P_t + 50)/2$$

or

$$P_t = 2400 - 50 = 2350\,\text{psi}$$

As mentioned before, this total pressure is the sum of friction, elevation, and delivery pressure components at the higher flow rate Q.

From Eq. (9.1), we can write

$$2350 = P_{\text{friction}} + P_{\text{elevation}} + P_{\text{del}}$$

or

$$2350 = P_{\text{friction}} + 300 + 50$$

Therefore,

$$P_{\text{friction}} = 2350 - 300 - 50 = 2000\,\text{psi at the higher flow rate } Q$$

Therefore, the frictional pressure drop per mile at the higher flow rate Q is

$$P_m = 2000/100 = 20\,\text{psi/mi}$$

From Eq. (9.4), we can write

$$20 = K(Q)^2 \tag{9.6}$$

where Q is the unknown higher flow rate in bbl/h.

By dividing Eq. (9.6) by Eq. (9.5), we get the following:

$$20/14.5 = (Q/5000)^2$$

Solving for Q we get

$$Q = 5000(20/14.5)^{1/2} = 5872\,\text{bbl/h}$$

Therefore, with the two pump stations each discharging at the 1200 psi MAOP, we can increase the flow rate to 5872 bbl/h. As previously mentioned, this will definitely require additional or modified pumps at both pump stations to provide the higher discharge pressure. Pumps will be discussed in Chapter 11.

In the preceding sections, we have considered a pipeline to be of uniform diameter and wall thickness for its entire length. In reality, pipe diameter and wall thickness change, depending on the service requirements, design code, and the local regulatory requirements. Pipe wall thickness may have to be increased due to different specified minimum yield strength (SMYS) of pipe because a higher or lower grade of pipe was used at some locations. Also, sometimes, cities or counties through which the pipeline traverses may dictate different design factors (0.66 instead of 0.72) to be used, thus requiring a different wall thickness. If there are significant elevation changes along the pipeline, the low elevation points may require higher wall thickness to withstand the higher pipe operating pressures. In all these cases, the pressure drop due to friction (slope of the hydraulic gradient) will not be a constant value throughout the entire pipeline length.

Therefore, when pipe diameter and wall thickness change along a pipeline, the slope of the hydraulic gradient, as shown in Fig. 9.2, will no longer be uniform. Due to varying frictional pressure drop (because of pipe diameter and wall thickness change), the slope of the hydraulic gradient will vary along the pipe length.

9.3 SERIES PIPING IN LIQUID PIPELINES

When different lengths of pipes are joined end to end, the pipes are said to be in series. The entire flow passes through all pipes, without any branching, from the beginning to the end, as shown in Fig. 9.4.

Consider a simpler pipeline consisting of two different lengths and pipe diameters joined together in series. This series piping system is composed of a 1000-ft long segment of NPS 16 pipe, connected with a pipe segment NPS 14, 500 ft long. At the common junction of the two pipe segments, a fitting, known as a reducer, will be used to join the larger NPS 16 pipe with the smaller NPS 14. This fitting will be a 16-in. × 14-in. reducer. We can calculate the total pressure drop through this series piping by adding the individual pressure drops in the 16-in. and the 14-in. pipe segment and accounting for the pressure loss in the 16-in. × 14-in. reducer.

In series piping, we can also use the equivalent length method to convert the length of one of the two pipes in terms of the other pipe diameter and calculate the pressure drop in the equivalent pipeline. A pipe segment is equivalent to

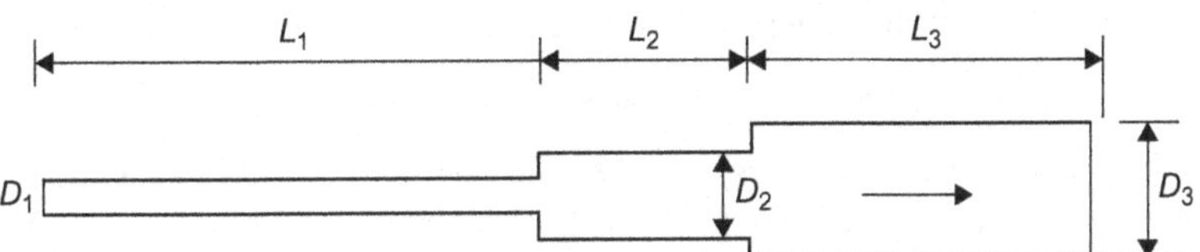

FIGURE 9.4 Pipes in series.

another pipe segment if the same head loss due to friction occurs in the equivalent pipe compared with that of the other pipe. Thus, we can find a certain length of NPS 16 pipe that will have the same pressure drop as the given NPS 14 pipe. In our example, we have 500 ft of NPS 14 pipe that we want to convert to an equivalent length of NPS 16 pipe. Obviously, the NPS 16 pipe has less friction loss per unit length compared with the NPS 14 pipe. Therefore, the 500 ft of NPS 14 will be equivalent to a larger length of NPS 16 pipe to have the same total pressure drop. We will explain a method of finding this equivalent length for different pipe diameters.

A pipe A of length L_A and inside diameter D_A is connected in series with pipe B of length L_B and inside diameter D_B. If we replace this two-pipe system with a single pipe of length L_E and diameter D_E, we have what is known as the equivalent length of pipe. This equivalent length of pipe may be based on either D_A, or D_B, or a different diameter D_E.

From Chapter 8, using Eq. (8.14) for pressure drop, the equivalent length L_E in terms of pipe diameter D_E is as follows:

$$L_E/(D_E)^5 = L_A/(D_A)^5 + L_B/(D_B)^5 \tag{9.7}$$

If we set $D_E = D_A$, we get

$$L_E = L_A + L_B(D_A/D_B)^5 \tag{9.8}$$

Thus, we have converted the two pipe segments into a single pipe segment of equivalent length L_E, based on diameter D_A. This length L_E of pipe diameter D_A will have the same total frictional pressure drop as the two lengths L_A and L_B in series. Note that this equivalent length method discussed above is only approximate. If elevation changes are involved in some cases, errors may result unless there are no controlling elevations along the pipeline system. Appendix 6 lists the equivalent lengths of pipes based on Eq. (9.8).

Example Problem 9.1

A series pipeline, similar to the one shown in Fig. 9.4, consists of NPS 16 pipe, 0.281 in. wall thickness, 7 mi long connected with an NPS 14 pipe, 0.250 in. wall thickness, 5 mi long. Calculate the equivalent length of this pipeline based on NPS 14 pipe and the total head loss due to friction at 6000 gal/min of water. Use Hazen–Williams C = 120.

Solution

The equivalent length of the pipeline from Eq. (9.8) is

$$5 + 7 \times (14 - 0.50)^5/(16 - 0.562)^5 = 8.58 \text{ mi of 14 in. pipe}$$

Based on equivalent pressure drop, the actual length of 7 mi of 16-in. and 5 mi of 14-in. pipe, totaling 12 mi, is replaced with a single equivalent 14-in. pipe, 8.58 mi long. For now, we will neglect the effect of the 16-in. × 14-in. reducer, which will be very small.

From Table 8.7, Chapter 8, we determine the head loss in NPS 14 pipe at 6000 gal/min as 45.85 ft/1000 ft.

The total head loss in the entire pipeline is then

$$\text{Total head loss} = 45.85 \times 8.58 \times 5280/1000 = 2077\,\text{ft of water} = 899\,\text{psi},$$

converting feet of head to psi using Eq. (8.10)

For more accuracy, we should include the minor loss in the 16-in. × 14-in. reducer by determining the equivalent length of the reducer and adding it to the 8.58 mi to obtain the total equivalent length, including the fitting.

9.4 PARALLEL PIPING IN LIQUID PIPELINES

Pipes are said to be in parallel if they are connected so that the liquid flow splits into two or more separate pipes and rejoins downstream into another pipe as shown in Fig. 9.5

The liquid initially flows through pipe AB, and at B, a portion of the flow branches off into pipe BCE, while the remainder flows through the pipe BDE. At point E, the flows recombine to the original value and the liquid flows through the pipe EF.

To solve for the pressures and flow rates in a parallel piping system shown in Fig. 9.5, we use the following basis:

1. Conservation of total flow
2. Common pressure loss across each parallel pipe.

According to 1, the total flow entering each junction of pipe must equal the total flow leaving the junction.

Thus, if the flow into the junction B is Q and the flow in branch BCE is Q_{BC} and flow in the branch BDE is Q_{BD}, we can state that

$$Q = Q_{BC} + Q_{BD} \tag{9.9}$$

According to 2, the pressure drop due to friction in branch BCE must equal the pressure drop in the branch BDE. Referring to pressure as P_B at point B and P_E at point E, we see that

$$\text{Pressure drop in branch BCE} = P_B - P_E \tag{9.10}$$

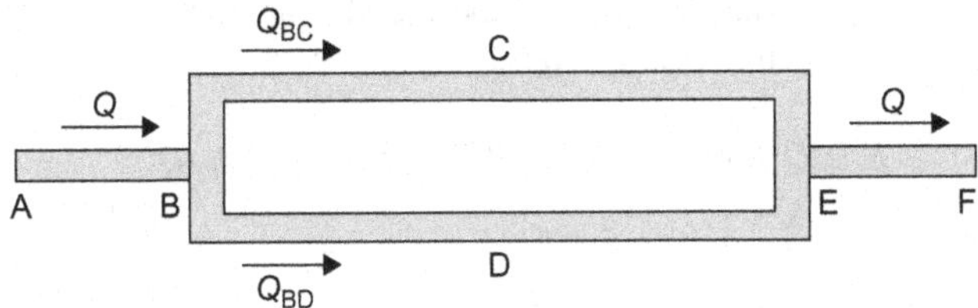

FIGURE 9.5 Parallel pipes.

$$\text{Pressure drop in branch BDE} = P_B - P_E \tag{9.11}$$

We assume that the flows Q_{BC} and Q_{BD} are in the direction of BCE and BDE, respectively.

For three pipes in parallel, we can rewrite Eqs (9.9) and (9.11) as follows:

$$Q = Q_{BC} + Q_{BD} + Q_{BE} \tag{9.12}$$

and

$$\Delta P_{BCE} = \Delta P_{BDE} = \Delta P_{BE} \tag{9.13}$$

where ΔP is the pressure drop in respective parallel pipes.

From Eqs (9.12) and (9.13), we can solve for flow rates and pressures in any parallel piping system. Another approach, using equivalent pipe diameters for pipes in parallel is as follows. Similar to the equivalent length in series piping, we can calculate an equivalent pipe diameter for pipes connected in parallel. In this method, we replace the two or more pipe branches by a single pipe with an equivalent diameter that has the same head loss as the individual pipes. Thus, if we had a NPS 16 pipe and NPS 14 pipe in parallel, we could determine an equivalent pipe with a diameter D_E that has the same pressure drop as each of the two pipes. This equivalent pipe will carry the full flow, which is the sum of the flows in the NPS 16 and NPS 14 pipes.

Assuming now that we have the two parallel pipes BCE and BDE in Fig. 9.5,

$$Q = Q_{BC} + Q_{BD}$$

and

$$\Delta P_{EQ} = \Delta P_{BCE} = \Delta P_{BDE}$$

where ΔP is the common pressure drop in the branches.

The pressure ΔP_{EQ} for the equivalent pipe (diameter D_E) can be written using Eq. (8.14) as follows:

$$\Delta P_{EQ} = K(L_E)(Q)^2/D_E^5$$

where K is a constant that depends on the liquid properties.

Equating the branch pressure drops,

$$K\,L_E Q^2/D_E^5 = K\,L_{BC} Q_{BC}^2/D_{BC}^5 = K\,L_{BD} Q_{BD}^2/D_{BD}^5$$

Simplifying we get

$$L_E Q^2/D_E^5 = L_{BC} Q_{BC}^2/D_{BC}^5 = L_{BD} Q_{BD}^2/D_{BD}^5 \tag{9.14}$$

If we assume that the pipe branches are all equal lengths,

$$L_{BC} = L_{BD} = L_E$$

This results in

$$Q^2/D_E^5 = Q_{BC}^2/D_{BC}^5 = Q_{BD}^2/D_{BD}^5$$

Substituting for Q_{BD} in terms for Q_{BC} from Eq. (9.9), we get

$$Q^2/D_E^5 = Q_{BC}^2/D_{BC}^5 \tag{9.15}$$

and

$$Q_{BC}^2/D_{BC}^5 = (Q - Q_{BC})^2/D_{BD}^5 \tag{9.15a}$$

From the above Eqs (9.15) and (9.15a), we can solve for the two flows Q_{BC}, Q_{BD}, and the equivalent diameter D_E in terms of the known quantities Q, D_{BC}, and D_{BD}. Appendix 6 lists the equivalent diameter for pipes in parallel.

Example Problem 9.2

A parallel pipe system, similar to the one shown in Fig. 9.5 is located in a horizontal plane with the following data.

Flow rate Q = 4000 gal/min of water
Pipe branch BCE = NPS 12, 0.375 in. wall thickness, 7500 ft
Pipe branch BDE = NPS 10, 0.375 in. wall thickness, 4000 ft

Determine the flow rate through each of the two branch pipes and calculate the equivalent pipe diameter for a single pipe 5000 ft long between B and E to replace the two parallel pipes.

Solution

$$Q_1^2 L_1/D_1^5 = Q_2^2 L_2/D_2^5$$

where suffix 1 and 2 refer to the two branches BCE (NPS 12) and BDE (NPS 10), respectively.

$$\begin{aligned}(Q_2/Q_1)^2 &= (D_2/D_1)^5\,(L_1/L_2)\\ &= (10/12)^5 \times (7500/4000)\\ Q_2/Q_1 &= 0.8681\end{aligned}$$

Since $Q_1 + Q_2 = 4000$, solving for the flow rates, we get

$$Q_1 = 2141 \text{ gal/min}$$
$$Q_2 = 1859 \text{ gal/min}$$

The equivalent pipe diameter for a single pipe 5000 ft long is calculated as follows:

$$(4000)^2(5000)/D_E^5 = (2141)^2 \times 7500/(12)^5$$

or

$$D_E = 14.21 \text{ in.}$$

Therefore, a 14.21 in. inside diameter pipe, 5000 ft long between B and E will replace the two parallel pipes for the same head loss.

9.5 TRANSPORTING HIGH VAPOR PRESSURE LIQUIDS

Transportation of high vapor pressure liquids such as liquefied petroleum gas (LPG) requires that a certain minimum pressure be maintained throughout the pipeline. This minimum pressure must be greater than the liquid vapor pressure at the flowing temperature to prevent vaporization that could damage the pumps. If the vapor pressure of LPG at the flowing temperature is 250 psi, the minimum pressure in the pipeline must be greater than 250 psi. At high elevation points or peaks along the pipeline, the pipeline pressure must not fall below 250 psi. Also, at the delivery terminus, the pressure must satisfy the minimum pressure requirements. Accordingly, for an LPG pipeline, the terminus delivery pressure may be 300 psi or higher to account for meter manifold piping losses at the delivery point. Sometimes, with high vapor pressure liquids, the delivery may be into a pressure vessel or a pressurized bullet or sphere maintained at 500 to 600 psi. This will therefore require even higher minimum pressures compared with the vapor pressure of the liquid. Hence, both the delivery pressure and the minimum pressure must be considered when analyzing pipelines transporting high vapor pressure liquids.

9.6 PUMPING POWER REQUIRED IN LIQUID PIPELINES

In the preceding pages, we have discussed how to calculate the pressure required to transport a given amount of liquid through a pipeline system. Based upon the flow rate and the MAOP of the pipeline, we may require one or more pump stations to safely transport the liquid. In a liquid pipeline, the pressure required will be provided by centrifugal or positive displacement pumps at each pump station. The selection and performance of pumps are discussed in Chapter 11. However, in the following pages, we will estimate the pumping power required to transport a given flow rate of a liquid through the pipeline regardless of the type of pumping equipment used.

9.6.1 Hydraulic Horsepower

Power required is defined as energy or work performed per unit time. In USCS units, energy is measured in ft·lb and power is stated as ft·lb/min and in the units of horsepower (HP). One HP is defined as 33,000 ft·lb/min or 550 ft·lb/s. In SI units, energy is measured in Joules and power in Joules/second (watts). The larger unit kilowatts (kW) is more commonly used. One HP is equal to 0.746 kW.

Consider a situation where 150,000 gal of water per hour is to be raised 600 ft to supply the needs of a small community. The work done in lifting 150,000 gal of water by 600 ft in 1 hour is

$$(150{,}000/7.48)\times 62.34\times 600 = 750{,}080{,}214 \text{ ft lb/h}$$

assuming specific weight of water = 62.34 lb/ft^3 and 1 ft^3 = 7.48 gal

The power required is then

$$\text{HP} = \frac{7.50\times 10^8}{60\times 33{,}000} = 379.0$$

This is also known as the hydraulic horsepower (HHP), since we have not considered pumping efficiency. If the pumping efficiency is 80%, the actual HP = 379.0/0.8 = 474.

As a liquid flows through a pipeline, pressure loss occurs due to friction. The pressure needed at the beginning of the pipeline to account for friction and any elevation changes can be used to calculate the amount of energy required to transport the liquid. Factoring in the time element, we get the power required to transport the liquid.

Example Problem 9.3

Consider 5000 bbl/h of diesel fuel (Sg = 0.85) transported through a pipeline with one pump station operating at 1200 psi discharge pressure. If the pump station suction pressure is 50 psi, the pump has to produce (1200 – 50) or 1150 psi differential pressure to pump the diesel fuel at the given flow rate. Calculate the HP required at this flow rate. The specific weight of water is 62.34 lb/ft^3.

Solution

Flow rate of liquid in lb/min is calculated as follows:

$$M = 5000 \text{ bbl/h } (5.6146 \text{ ft}^3\text{/bbl})(1 \text{ h/60 min})(0.85)(62.34 \text{ lb/ft}^3)$$

or

$$M = 24{,}793 \text{ lb/min}$$

The head developed by the pump is

$$(1200 - 50) = 1150 \text{ psi} = 1150\times 2.31/0.85 \text{ ft of diesel head} = 3126 \text{ ft}$$

Therefore, the power required is

$$\text{HP} = \frac{(\text{lb/min})(\text{ft. head})}{33{,}000}$$

$$\text{HP} = 24{,}793\times 3126/33{,}000 = 2349$$

In the above calculation, no efficiency has been considered. In other words, we have assumed 100% pumping efficiency. Therefore, the above HP calculated is referred to as hydraulic horsepower (HHP), based on 100% efficiency.

9.6.2 Brake Horsepower

The brake horsepower takes into account the pump efficiency. If a pump efficiency of 75% is used, we can calculate the brake horsepower (BHP) in the above example as follows:

Brake horsepower = hydraulic horsepower/pump efficiency
$\text{BHP} = \text{HHP}/0.75 = 2349/0.75 = 3132$

In addition, if an electric motor is used to drive the pump, the motor horsepower required is calculated as follows:

$$\text{Motor HP} = \text{BHP/motor efficiency}$$

Generally, induction motors used for driving pumps have fairly high efficiencies, ranging from 95% to 98%. Using 95% for motor efficiency, we can calculate the motor HP required as follows:

$$\text{Motor HP} = 3132/0.95 = 3297$$

Pump companies state pump flow rates in gal/min and pump pressures in feet of liquid head, whereas in the petroleum industry, flow rates may be stated in bbl/h or gal/min in USCS units.

The formula for pumping power (BHP) required in terms of customary pipeline units is as follows:

$$\text{BHP} = Q \times P/(2449 \times E) \quad \text{(USCS)} \tag{9.16}$$

where

Q – Flow rate, bbl/h
P – Differential pressure, psi
E – Efficiency expressed as a decimal value less than 1.0

Other formulas for BHP expressed in terms of flow rate in gal/min and pressure in psi or ft of liquid are as follows:

$$\text{BHP} = Q \times H \times \text{Spgr}/(3960 \times E) \quad \text{(USCS)} \tag{9.17}$$

and

$$\text{BHP} = Q \times P/(1714 \times E) \quad \text{(USCS)} \tag{9.18}$$

where

Q – Flow rate, gal/min
H – Differential head, ft
P – Differential pressure, psi
E – Efficiency (less than 1.0)
Spgr – Liquid specific gravity, dimensionless

In SI units, the brake power in kW is expressed as follows:

$$\text{Power (kW)} = Q \times H \times \text{Spgr}/(367.460 \times E) \quad \text{(SI)} \tag{9.19a}$$

where

Q – Flow rate, m^3/h
H – Differential head, m
E – Efficiency (less than 1.0)
Spgr – Liquid specific gravity, dimensionless

$$\text{Power (kW)} = Q \times P/(3600 \times E) \quad \text{(SI)} \tag{9.19b}$$

where

Q – Flow rate, m^3/h
P – Pressure, kPa
E – Efficiency (less than 1.0)

Appendix 6 provides tables to determine the brake power required at various flow rates and pressures based on water. These values can be corrected easily for other liquids.

9.7 SYSTEM HEAD CURVES – LIQUID PIPELINES

A system head curve, also sometimes known as system curve, for a pipeline is a graphic representation of how the pressure to pump a liquid in a pipeline varies with the flow rate. Figure 9.6 shows a typical system head curve. As the flow rate increases, the head required increases. From Chapter 8, Eq. (8.14) we know that the pressure drop due to friction varies as the square of the flow rate. If the head loss at 1000 gal/min is 10 psi/mi, at 2000 gal/min the head loss is approximately $(2000/1000)^2 \times 10 = 40$ psi/mi. Doubling the flow rate causes the head loss to be 4 times as before.

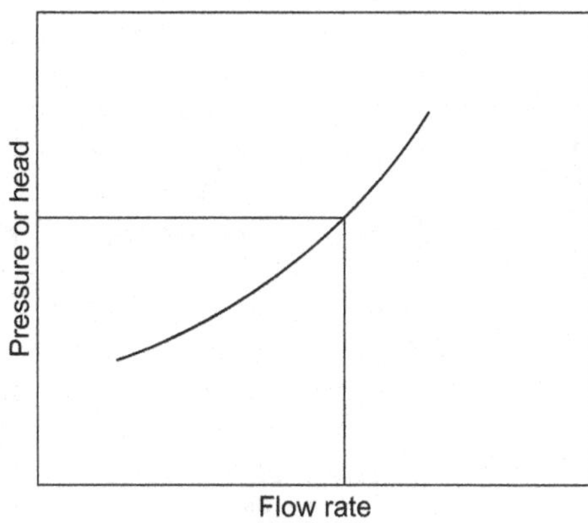

FIGURE 9.6 System head curve.

Therefore, if we neglect the elevation profile, and consider 50 psi terminus delivery pressure, the total pressure P_T required at the origin of a 30-mi pipeline is 350 psi at 1000 gal/min and 1250 psi at 2000 gal/min. Thus, we can calculate the head loss and hence the total pressure P_T required for various flow rates and plot a curve representing the system curve. We can also take into account the elevations along the pipeline using the method described in Section 9.1 to obtain the values of P_T and plot the system curve. Usually, the pressures are converted to feet of head as indicated in Fig. 9.6. However, sometimes the system head curve may be plotted in psi as well. In USCS units, the horizontal scale is the flow rate in gal/min or bbl/h. In SI units, the head is plotted in meters of liquid and the flow rates in m^3/h.

To recap, consider a pipeline of inside diameter D and length L that transports a liquid of specific gravity Sg and viscosity v from a pump station at A to a delivery terminus located at B. We calculate the pressure required at A to transport the liquid at a particular flow rate Q. By varying flow rate Q, we can determine the pressure required at A for each flow rate such that a given delivery pressure at B is maintained. For each flow rate Q, we calculate the pressure drop due to friction for the length L of the pipeline, add the head required to account for elevation difference between A and B and finally, add the delivery pressure required at B as follows, using Eq. (9.1):

Pressure at A = Friction pressure drop + Elevation head + Delivery pressure

Once the pressure at A is calculated for each flow rate, we can plot a system head curve as shown in Fig. 9.6.

In Chapter 11, system head curves along with pump head curves will be reviewed in detail. We will see how the system head curve in conjunction with the pump head curve will determine the operating point for a particular pump–pipeline configuration. Since a system head curve represents the pressure required to pump various flow rates through a given pipeline, we can plot a

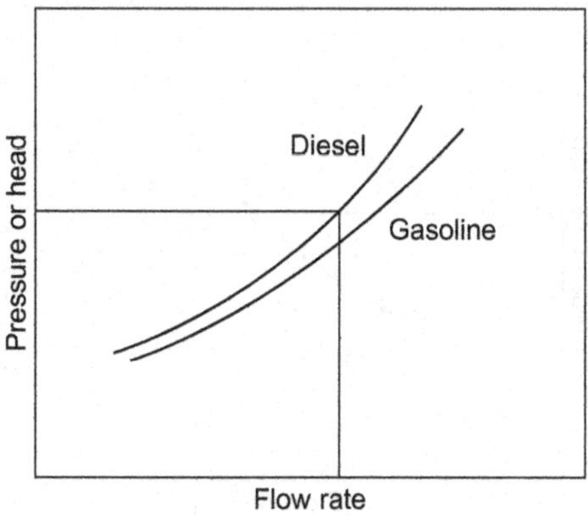

FIGURE 9.7 System head curve – different products.

family of such curves for different liquids as shown in Fig. 9.7. The higher specific gravity and viscosity of diesel fuel requires greater pressures (psi or kPa) compared with gasoline. Hence, the diesel system head curve is located above that of gasoline as shown in Fig. 9.7.

The shape of the system curve varies depending upon the amount of friction head component compared with the elevation head. This is illustrated in Figs 9.8

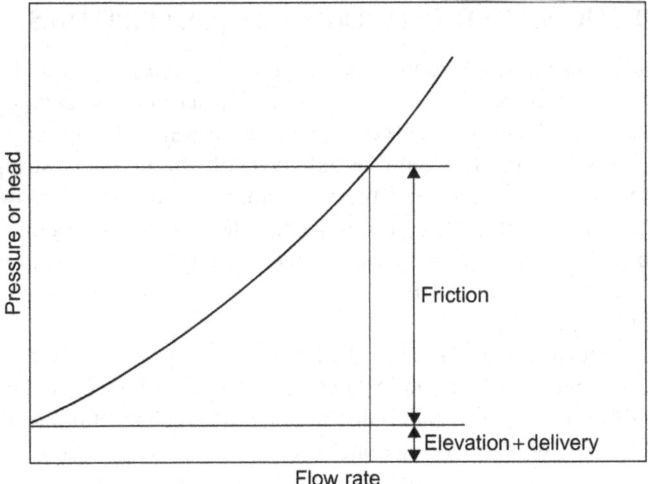

FIGURE 9.8 System head curve – high friction.

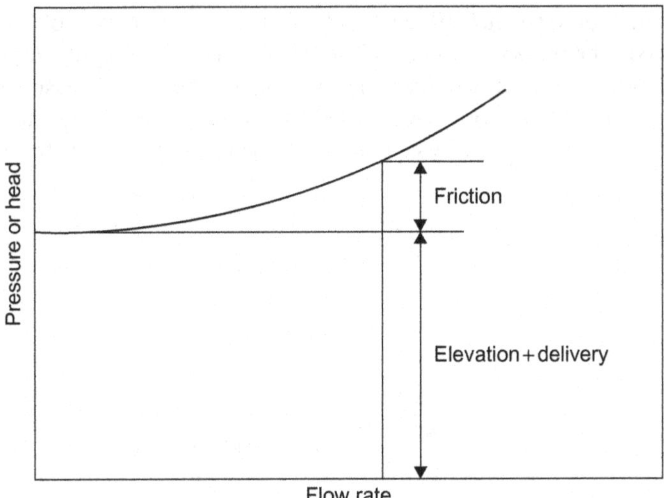

FIGURE 9.9 System head curve – high elevation.

and 9.9 that show two system curves. In Fig. 9.8, the friction component is higher than the elevation component. Most of the system head required is due to the friction in the pipe.

Alternatively, if the elevation differences are much higher than the frictional head loss, then the system curve is less sensitive to flow rate changes.

Figure 9.9 shows a system head curve that consists mostly of the static head due to the pipe elevations.

9.8 INJECTIONS AND DELIVERIES – LIQUID PIPELINES

In most pipelines liquid enters the pipeline at the beginning and would continue toward the end of the pipeline to be delivered at the terminus, with no deliveries or injection at any intermediate point along the pipeline. However, there are situations in which liquid would be delivered off the pipeline (stripping) at some intermediate location and the remainder would continue toward the pipeline terminus. Similarly, liquid may enter the pipeline (injection) at some intermediate location thereby adding to the existing volume in the pipeline. These are called deliveries off the pipeline and injection into the pipeline. This is illustrated in Fig. 9.10.

Consider the pipeline AB in Fig. 9.10 that shows liquid entering the pipeline at A at a flow rate of 6000 bbl/h. At a point C, there is a flow injection into the pipeline at the rate of 1000 bbl/h. Further along the pipeline, at point D, there is a delivery of 3000 bbl/h off the pipeline. Therefore, a resultant volume of (6000 + 1000 – 3000) or 4000 bbl/h is delivered to the pipeline terminus at B. To calculate the pressure required at A for such a pipeline with injection and deliveries, we have to consider separately each pipe segment that has a constant flow rate, as explained next.

First, the pipe segment between A and C that has a uniform flow of 6000 bbl/h is analyzed. The pressure drop in AC is calculated considering the 6000 bbl/h flow rate, pipe diameter, and liquid properties. Similarly, the pressure drop in the pipe segment CD at a flow rate of 7000 bbl/h is calculated taking into account the blended liquid properties by combining the incoming stream at C (6000 bbl/h)

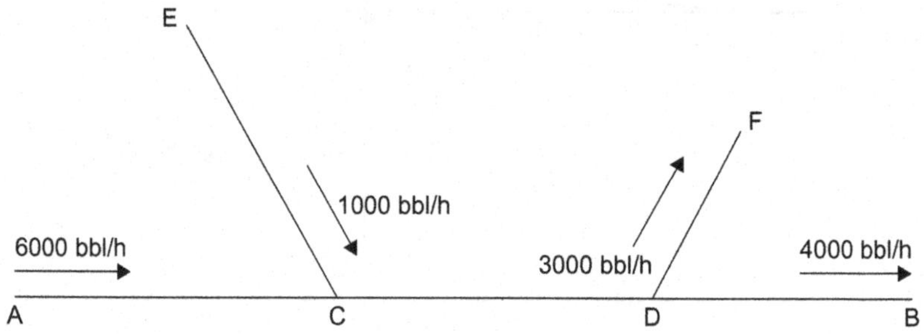

FIGURE 9.10 Injection and delivery.

along the main line with the injection stream (1000 bbl/h) at C. Finally, the pressure drop in the pipe segment DB is calculated considering a volume of 4000 bbl/h and the liquid properties in that segment, which would be the same as that of pipe segment CD. The total frictional pressure drop between A and B will be the sum of the three pressure drops calculated above. After adding any elevation head between A and B and accounting for the required delivery pressure at B, the total pressure required at point A is calculated.

9.9 PIPE LOOPS IN LIQUID PIPELINES

A pipe loop is a length of pipe installed in parallel between two points on a main pipeline as shown in Fig. 9.11. Earlier in this chapter, we discussed parallel pipes and equivalent diameters. We will now discuss how looping an existing pipeline will reduce the pressure drop due to friction and thus require less pumping power.

The purpose of the pipe loop is to split the flow through a parallel segment of the pipeline between the two locations resulting in a reduced pressure drop in that segment of the pipeline.

Example Problem 9.4

A 50-mi pipeline from point A to point B is constructed of NPS 16, 0.250 in. wall thickness pipe. The flow rate at the inlet A is 4000 bbl/h. The crude oil properties are specific gravity of 0.85 and viscosity of 10 cSt at a flowing temperature of 70°F.

1. Calculate the pressure required at A without any pipe loop. Assume 50 psi delivery pressure at the terminus B and a flat pipeline elevation profile.
2. A 10-mi portion, CD, starting at milepost 10 is looped with an identical length of NPS 16-in. pipeline. Calculate the revised pressure at A.
3. What is the difference in pump BHP required at A between cases (1) and (2) above? Use 80% pump efficiency and 50 psi pump suction pressure.

Use Chapter 8 friction factor tables with a relative roughness $e/D = 0.0001$.

Solution

First, calculate the Reynolds number R, using Eq. (8.7):

$$R = 2214 \times 4000/(15.5 \times 10) = 57{,}135$$

From Table 8.6, for $e/D = 0.0001$ and $R = 57{,}135$, we obtain the friction factor f by interpolation as $f = 0.0207$ and the pressure drop from Eq. (8.14) as

$$P_m = 0.0605 \times 0.0207 \times (4000 \times 24)^2 (0.85/15.5^5) = 10.97\ \text{psi}$$

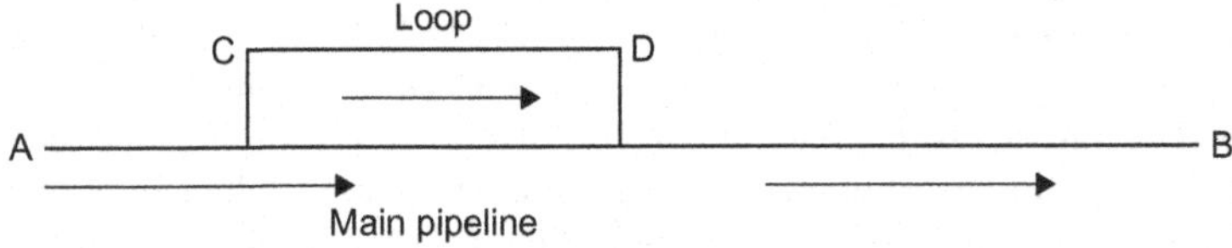

FIGURE 9.11 Pipe loop.

Appendix 5 also provides tables from which the frictional pressure drop can be interpolated for various flow rates, pipe sizes, and friction factors.

1. The total pressure required at A = (50 ×10.97) + 50 = 599 psi.
2. With 10 mi of pipe loop, the flow rate through each parallel pipe is 2000 bbl/h. The revised pressure drop at the reduced flow is calculated as follows:

$$R = 2214 \times 2000/(15.5 \times 10) = 28{,}568$$

From Table 8.6, for e/D = 0.0001 and R = 28,568, we get f = 0.024 and

$$P_m = 0.0605 \times 0.024 \times (2000 \times 24)^2 (0.85/15.5^5) = 3.18\,\text{psi/mi}$$

The pressure drop in the 10-mi loop = 10 × 3.18 = 31.8 psi
The pressure drop in the 10-mi section AC = 10 × 10.97 = 109.7 psi
The pressure drop in the last 30-mi section DB = 10 × 10.97 = 109.7 psi
Therefore, the total pressure required at A in the looped system is

$$\text{Total pressure} = 109.7 + 31.8 + 109.7 + 50 = 301.2\,\text{psi, with the pipe loop}$$

Therefore, due to the 10-mi loop, the total pressure required at A reduces from 599 to 301.2 psi, or a reduction of approximately 298 psi.
3. Pumping power required without the loop is, from Eq. (9.16)

$$\text{BHP} = 4000 \times (599 - 50)/(2449 \times 0.8) = 1121$$

The power required considering the loop is

$$\text{HP} = 4000 \times (301.2 - 50)/(2449 \times 0.8) = 513$$

Therefore, installing the 10-mi pipe loop results in a reduction of

$$(1121 - 513)/1121 = 54.2\% \text{ in pumping power}$$

9.10 GAS PIPELINES

9.10.1 Total Pressure Required to Transport Gases

Similar to liquid pipelines, we can calculate the total pressure required in a gas pipeline for a specific gas flow rate, using the pressure drop equations discussed for gas pipelines in Chapter 8. Unlike the liquid flow equation for pressure drop, the formulas for gas pipelines, such as the general flow equation, include the length of the pipe segment and the elevation effect. Therefore, gas pipelines require a slightly different approach to determine the total pressure required at the origin of the pipeline. Equations (8.56) and (8.59), discussed in Chapter 8, consider a certain pipe segment of length L, with an upstream pressure P_1 and downstream pressure P_2. The elevation difference $(H_2 - H_1)$ between the upstream and downstream ends of the pipe segment is taken into account using the elevation adjustment parameter s that includes the elevation difference as well as the gas gravity, flow temperature, and the gas compressibility factor.

Because of the nonlinear variation in gas pressure along the pipeline, the compressibility factor, and the elevation effect, for accuracy, we must subdivide the gas pipeline into short segments, calculate the pressure drops in stages, and add them up to obtain the total pressure required at the origin.

9.11 HYDRAULIC PRESSURE GRADIENT IN GAS PIPELINE

The hydraulic pressure gradient is a graphical representation of the gas pressures along the pipeline as shown in Fig. 9.12. Unlike liquids, due to the gas compressibility and nonlinear variation of gas pressure along the pipeline, the hydraulic pressure gradient in a gas pipeline is a curved line, as shown in Fig. 9.12. The horizontal axis shows the distance along the pipeline starting at the upstream end. The vertical axis shows the pipeline pressures.

The slope of the hydraulic gradient at any point represents the pressure loss due to friction per unit length of pipe. The slope is more pronounced as we move toward the downstream end of the pipeline, since the pressure drop is larger toward the end of the pipeline. If the flow rate through the pipeline is a constant value (no intermediate injections or deliveries) and pipe size is uniform throughout, the hydraulic gradient appears to be a slightly curved line as shown in Fig. 9.12 with no perceptible breaks. If there are intermediate deliveries or injections along the pipeline, the hydraulic gradient will be a series of broken lines as indicated in Fig. 9.13.

In a long distance gas pipeline, due to pipe MAOP limits, intermediate compressor stations may be required to boost the gas pressure to the required value so that the gas can be delivered at the minimum contract delivery pressure at the pipeline terminus. This is illustrated in Fig. 9.14, where a 200-mi pipeline from Compton to Beaumont requires 1600 psig at Compton to deliver gas to Beaumont at 800 psig. Since the MAOP is limited to 1350 psig, Compton will discharge at 1350 psig and the gas pressure will drop to 900 psig at the intermediate compressor station located at Sheridan. At Sheridan, the gas pressure will be boosted to 1350 psig from the suction pressure of 900 psig, which will eventually drop to the required delivery pressure of 800 psig at Beaumont.

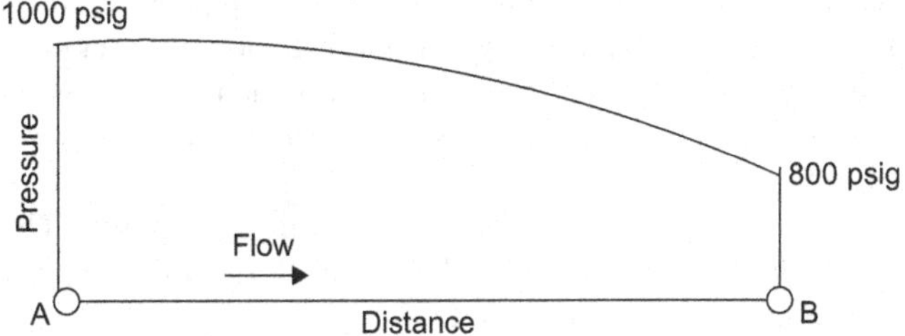

FIGURE 9.12 Hydraulic pressure gradient in gas pipeline.

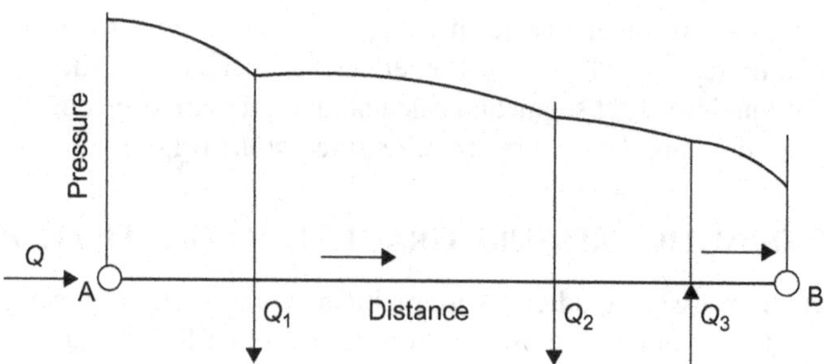

FIGURE 9.13 Hydraulic pressure gradient – deliveries and injections.

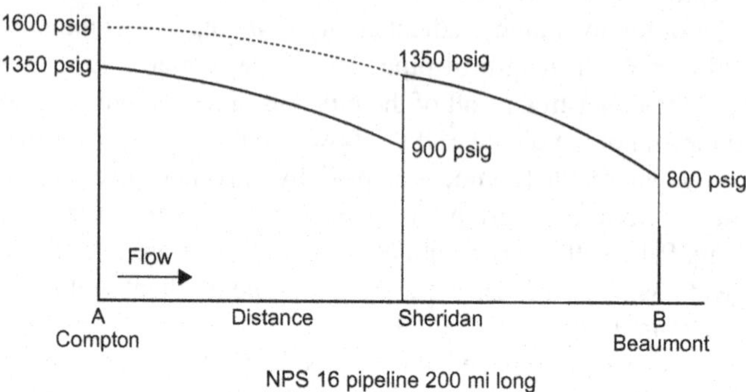

FIGURE 9.14 Compton to Beaumont pipeline.

9.12 SERIES PIPING IN GAS PIPELINES

Similar to liquid pipelines, gas pipelines may be configured in series. In Fig. 9.15, a pipe segment AB with a diameter of 16 in. is used to transport a volume of 100 MMSCFD. After making a delivery of 20 MMSCFD at B, the remainder of 80 MMSCFD flows through the 14-in. diameter pipe BC. At C, a delivery of 30 MMSCFD is made and the balance volume of 50 MMSCFD is delivered to the terminus D through a 12-in. pipeline CD.

The pressure required to transport gas in a series pipeline from point A to point D in Fig. 9.15 is calculated by considering each pipe segment such as AB, BC, etc., and applying the appropriate flow equation, such as the general flow equation for each segment as was described earlier in series piping in liquid pipelines.

We can also use the equivalent length concept in series piping. This method can be applied when the same uniform flow exists throughout the pipeline, with no intermediate deliveries or injections.

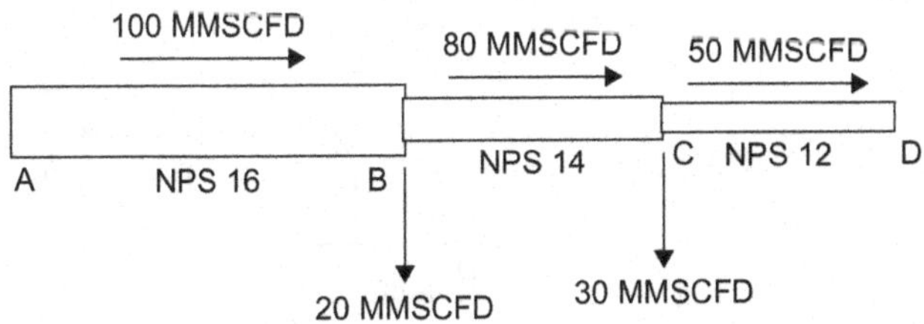

FIGURE 9.15 Series piping in gas pipeline.

Similar to liquid pipelines in series, Eqs (9.7) and (9.8), the total equivalent length L_e for three pipe segments in terms of diameter D_1 is

$$L_e = L_1 + L_2\left(\frac{D_1}{D_2}\right)^5 + L_3\left(\frac{D_1}{D_3}\right)^5 \tag{9.20}$$

9.13 PARALLEL PIPING IN GAS PIPELINES

Sometimes two or more pipes are connected in parallel such that the gas flow splits among the branch pipes and eventually combine downstream into a single pipe as illustrated in Fig. 9.16. This is also called a looped piping system in which each parallel pipe is called a loop. The reason for installing parallel pipes or loops is to reduce the pressure drop in a certain section of the pipeline due to pipe MAOP limits or to increase the flow rate in a bottleneck section. By installing a pipe loop from B to E as shown in Fig. 9.16, we are effectively reducing the overall pressure drop in the pipeline from A to F, since between B and E the total flow Q is split into Q_1 and Q_2 through two pipes.

We can use the same method for parallel pipes we discussed in liquid pipelines to calculate the flow rates and the pressure drops in gas pipelines with parallel pipe segments as shown in Fig. 9.16.

Using the general flow equation, the pressure drop due to friction in branch BCE can be calculated from

$$(P_B^2 - P_E^2) = \frac{K_1 L_1 Q_1^2}{D_1^5} \tag{9.21}$$

where

K_1 – A parameter that depends on gas properties, gas temperature, etc.
L_1 – Length of pipe branch BCE
D_1 – Inside diameter of pipe branch BCE
Q_1 – Flow rate through pipe branch BCE

Other symbols are as defined earlier.

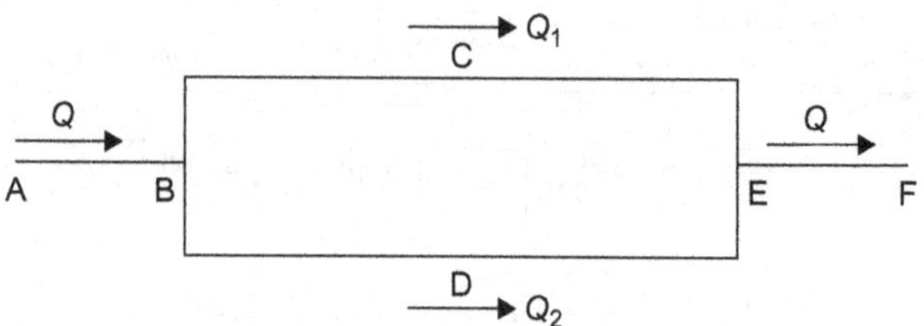

FIGURE 9.16 Parallel piping in gas pipeline.

K_1 is a parameter that depends on the gas properties, gas temperature, base pressure, and base temperature. Similarly, the pressure drop due to friction in branch BDE is calculated from

$$(P_B^2 - P_E^2) = \frac{K_2 L_2 Q_2^2}{D_2^5} \tag{9.22}$$

where

K_2 – A constant like K_1
L_2 – Length of pipe branch BDE
D_2 – Inside diameter of pipe branch BDE
Q_2 – Flow rate through pipe branch BDE

Other symbols are as defined earlier.

In Eqs (9.21) and (9.22), the constants K_1 and K_2 are equal because they do not depend on the diameter or length of the branch pipes BCE and BDE. Combining both equations, we get the following for common pressure drop through each branch:

$$\frac{L_1 Q_1^2}{D_1^5} = \frac{L_2 Q_2^2}{D_2^5} \tag{9.23}$$

This is similar to parallel pipe analysis in liquid pipelines, such as Eq. (9.14). Simplifying further, we get the following relationship between the two flow rates Q_1 and Q_2:

$$\frac{Q_1}{Q_2} = \left(\frac{L_2}{L_1}\right)^{0.5} \left(\frac{D_1}{D_2}\right)^{2.5} \tag{9.24}$$

We can also use the equivalent diameter approach, as we did with liquid pipelines in parallel. Because the pressure drop in the equivalent diameter pipe, which flows the full volume Q, is the same as that in any of the branch pipes, from Eq. (9.22), we have

$$(P_B^2 - P_E^2) = \frac{K_e L_e Q^2}{D_e^5} \tag{9.25}$$

where

$$Q = Q_1 + Q_2 \tag{9.26}$$

And K_e is a constant for the equivalent diameter pipe of length L_e flowing the full volume Q. Equating the value of $(P_B^2 - P_E^2)$ to the corresponding values considering each branch separately, we get

$$\frac{K_1 L_1 Q_1^2}{D_1^5} = \frac{K_2 L_2 Q_2^2}{D_2^5} = \frac{K_e L_e Q^2}{D_e^5} \tag{9.27}$$

Simplifying further, we get

$$D_e = D_1 \left[\left(\frac{1 + \text{Const 1}}{\text{Const 1}} \right)^2 \right]^{1/5} \tag{9.28}$$

where

$$\text{Const 1} = \sqrt{\left(\frac{D_1}{D_2}\right)^5 \left(\frac{L_2}{L_1}\right)} \tag{9.29}$$

The individual flow rates Q_1 and Q_2 are calculated from

$$Q_1 = \frac{Q\,\text{Const 1}}{1 + \text{Const 1}} \tag{9.30}$$

and

$$Q_2 = \frac{Q}{1 + \text{Const 1}} \tag{9.31}$$

SUMMARY

In this chapter, we extended the pressure drop concepts developed in Chapter 8 to calculate the total pressure required to transport a liquid or gas through a pipeline taking into account the elevation profile of the pipeline and the required delivery pressure at the terminus. The concept of hydraulic pressure gradient was explained. We analyzed pipes in series and parallel and introduced the concept of equivalent pipe length and equivalent pipe diameter. System head curve calculations for liquid pipelines were introduced. Flow injection and delivery in pipelines were studied and the impact on the hydraulic gradient discussed. We also compared the advantage of looping a pipeline to reduce overall pressure drop and pumping horsepower. For a pipeline system, the hydraulic horsepower, brake horsepower, and motor horsepower calculations were illustrated using examples. A more comprehensive analysis of pumps and compressors and power required will be discussed in Chapters 11 and 12.

BIBLIOGRAPHY

A.E. Uhl, et al., Steady Flow in Gas Pipelines, AGA, AGA Report No. 10, New York, 1965.

Compressibility Factors, AGA, AGA Report No. 8, New York, 1992.

A. Benaroya, Fundamentals and Application of Centrifugal Pumps, The Petroleum Publishing Company, Tulsa, OK, 1978.

E.F. Brater, H.W. King, Handbook of Hydraulics, McGraw-Hill, New York, 1982

Flow of Fluids through Valves, Fittings and Pipes, Crane Company, New York, 1976.

G.M. Jones, R.L. Sanks, G. Tchobanoglous, B. Bossermann, Pumping Station Design, Revised third ed., Elsevier, Inc., 2008.

Gas Processors Suppliers Association, Engineering Data Book, tenth ed., Tulsa, OK, 1994.

E. Shashi Menon, Liquid Pipeline Hydraulics, Marcel-Dekker, New York, 2004.

E. Shashi Menon, Gas Pipeline Hydraulics, CRC Press, Boca Raton, FL, 2005.

E. Shashi Menon, Piping Calculations Manual, McGraw-Hill, New York, 2005.

Pipeline Design for Hydrocarbon Gases and Liquids, American Society of Civil Engineers, New York, 1975.

V.S. Lobanoff, R.R. Ross, Centrifugal Pumps Design & Application, Gulf Publishing, Burlington, MA, 1985.

Chapter 10

Valve Stations

Barry G. Bubar, P.E.

Chapter Outline

INTRODUCTION

Pipelines require a large number of valves installed along the length of their route, also at supporting facilities necessary for the pipeline to function. These valves work like gateways, where they are usually fully open on the mainline, trunk line, or transmission line to allow hazardous liquids or natural gas to flow to their delivery destination. Valves are also installed at pipeline branch connections along the route, for delivery or receipts of product being shipped on the mainline. These branch block valves or lateral/takeoff valves, as they are sometimes called, may be open or closed depending on shipping orders. It should be recognized that these mainline and lateral block valves are the first line of protection for the pipeline. With these block valves, the operator can isolate any segment of the line or lateral for repair work or to isolate some part of the pipeline in case of a leak, damage, or rupture.

Numerous other valves, not on the mainline, which are needed for pipeline supporting operation, are located at pump and compressor stations, pressure regulator stations, meter runs, launcher/receiver stations, breakout tankage, and at distribution manifolds. Most of these valves are controlled by the pipeline control center (CC) and the CC operator needs to know each valve location and its status, and he or she will remotely control some of these valves by design. Otherwise, local facility operators open and close the valves, as required, and provide valve status to the CC, which are recorded, including time and date, after which that valve position shows on the CC operating board. Because each valve in the pipeline system is extremely important, all valves are assigned a number and that number must be tagged on the hand wheel or on the motor operator valve (MOV), if so equipped, for valve identification.

The federal codes[1–2] have different requirements for block valve placement and spacing on natural gas pipeline versus the placement and spacing of valves on hazardous liquid pipelines, due to the different nature of risk for the two commodities. The required block valve placement, spacing, and valve usage for each pipeline type will be covered here in detail.

10.1 WHAT TO EXPECT

First, we will address the design and placement of pipeline valves, and the valve specifications required, and then we will cover pipeline code requirements for safety and the maintenance of pipeline valves, and lastly pipeline valve selection, Ball or Gate.

Pipeline block valves are either placed above the ground, or placed below the ground in concrete vaults, or directly buried using valve stem risers. In any case, these valve stations must allow access to the valve or valves for operation and maintenance; and they should provide structural support for the valve to prevent the pipeline from settling and also to provide protection from unauthorized personnel.

10.2 VALVE USAGE

First, we will address valve usage and valve facility locations for hazardous liquid pipeline systems:

- *Block valves*: Block valves on the mainline must be full port gate valves, designed with a block-and-bleed feature to insure leak-tight performance.

The full port feature is necessary to allow smooth pipeline pigging. Having leak-tight valves in the mainline will be extremely useful during pipeline hydrostatic testing and beneficial to owners and customers by minimizing product contamination.

- Some block valves will be equipped with MOVs, which allow the control center to operate the valve remotely and monitor valve status (open, closed, or in-transit), via the SCADA system.
- *Valves at laterals*: Valves installed at lateral takeoffs on the mainline should also be leak tight, having block-and-bleed features, and they should meet API 6D requirements. They may be equipped with MOVs, where electric power can be obtained, in order to allow remote operation and provide valve status. The MOV-equipped valve will provide the control center (CC) with the ability to quickly close it, in emergencies, thus isolating the lateral without interrupting flow on the mainline. This is a code requirement as per CFR Title 49, Part 195.260(d).[1] A full port design is not necessary for the lateral valve, unless a pig launcher station is included there, to allow pigging of that lateral.
- *Breakout tanks*: Inlet and outlet valves on breakout tanks or storage tanks must be installed in a manner that will permit isolating each tank from other pipeline facilities, in case of an upset condition or pipeline hazard; this is a code requirement as per Part 195.260(b). Some pipelines operate by floating on tankage. This means that the inlet valve to the tank is left open to the pipeline, so that pipeline pressure at that location is regulated to 10–15 psi (69–103 kPa). In this case, the storage tank valves must be equipped with a MOV, allowing remote valve operation when needed.
- *Pump station valves*: Pump stations must have valves on the suction side of the station and also on the discharge side to the pipeline, which will permit isolation and shutdown of the pumps in the event of an emergency situation; this is a code requirement as per Part 195.260(a). These required isolation valves should be equipped with MOVs to allow remote operation of the valves and provide valve status. If the pump station has operating personal available on a 24-hour basis, the valves with MOVs would be set to local control and will provide quicker valve operation when necessary, compared to manual operation.
- *Launcher/Receiver valves*: A launcher/receiver station has three main valves and of course a number of small valves on related vents, drains, and instruments. Valve A is the mainline valve, which can be either a ball or a gate valve (preferably ball valve), meeting API 6D specs with MOV installed, when needed for remote operation, but it does not need to be of full port design. Valve A is normally open unless pipeline pigging is underway. Valve B, located at the inlet to the scraper trap, must be a full port API 6D spec valve with MOV (optional) to allow pig passage into the chamber; finally, Valve C, the kicker/bypass valve that makes the launcher/receiver station work, should be an API 6D ball valve with MOV (optional). Refer to Chapter 15, Fig. 15.1, showing valve placement and typical piping for a bidirectional launcher/receiver station.

- *Pipeline pressure relief valves*: Each pipeline may have a pressure relief valve (PRV), sometimes called a pressure safety valve, usually located, but not necessarily, at the pump station, which is independent of the programmed high-pressure shutdown switches and SCADA devices. These pipeline pressure-limiting valves are set to prevent pressure from exceeding 110% of the pipeline's MAOP, in accordance with the code as per Part 195.406(b). The PRV is connected directly to the pipeline through a nozzle with an isolation valve between the PRV and the pipeline. This isolation valve is needed to take the PRV out of service during pipeline hydrostatic testing and also for removal of this pressure limiting device and replacement as required by code as per Part 195.428(a), for annual testing of high pipeline pressure safety devices. During normal pipeline operation, the isolation valve, which is a full port ball or gate valve, is locked in the open position so that the PRV is fully operational. In some PRV installations, a flow switch is installed in the discharge piping, to alarm if the PRV opens to control pipeline pressure.
- *Valves at distribution manifold*: Distribution manifolds are found in many petroleum-pipeline facilities and have a variety of piping configurations. They are designed with valves above ground at pump stations and tank farms and in some cases, with valves below grade in concrete containments in places like truck- and ship-loading facilities.
- Figure 10.1 shows a 15-valve distribution manifold being constructed in a crude oil tank farm in California. This manifold connects with three crude oil pipelines coming into the facility, and with the five crude oil storage tanks located at that site.
- An example of this manifold's flexibility is where two separate pipelines can be delivering different crudes from different gathering locations, into two

FIGURE 10.1 The construction of a distribution manifold.

selected storage tanks at this tank farm, and at the same time, a local shipping pump can be shipping stored crude oil out of a third storage tank, by using the third pipeline, connected to the manifold, to ship to some other location, miles away. This is only one example of the manifold's flexibility in using pipeline and storage tank capacity that can be configured and controlled by the pipeline CC, also located miles away from this tank farm location, via pipeline SCADA. The 15 valves installed in the manifold were Class 150 cast steel, rising-steam gate valves modified for MOVs. The MOVs provide the CC with the ability to remotely operate each valve as necessary to make these pipeline line-ups and provide the status of each valve. Twin seal-type plug valves, with leak-tight sealing and available with MOV from other suppliers, could have been selected as well for this manifold. But they are more expensive and it was believed that they were not needed for this crude oil service.

- *Pipeline control valves*: Pipeline control valves work in concert with centrifugal pumps to control desired pipeline pressure and flow rate. Pipeline control valves are usually vee-ball design, with a controller that senses pipeline pressure, pump suction pressure, and flow rate. Based on instantaneous pipeline data, the controller modulates the control valve by throttling. If pipeline pressure is too high, the control valve will throttle the flow, forcing the pump to move back on the pump curve, which lowers the flow rate and pushes down pipeline pressure. If pipeline flow rate is too low, the control valve will open up, thus reducing the amount of throttling and letting the pump move out on its curve, which will cause an increase in flow rate.

Generally, the control valve is installed downstream of the pump discharge and sandwiched between two isolation valves, with bypass piping with ball valve in parallel to the control valve. This piping arrangement allows the pipeline to be operated manually, using the bypass ball valve for critical adjustment, while the malfunctioning control valve is being repaired or replaced.

- *Valves at pipeline meter stations*: When possible, pipeline meters piping should be designed to locate the meters on the low-pressure side of the shipping pumps because low-pressure positive displacement meters are much cheaper than their high-pressure cousins. The low-pressure meter (say with Class 150 flanges) will bolt up with a low-pressure filter, which is needed to protect the meter, the low pressure meter-prover valves, and the isolation valves. Sometimes PD meters, or even turbine meters, are piped in parallel, at stations where variable shipping rates on the pipeline can often occur. One meter is needed for low rates and both meters in parallel are needed at the highest flow rates. Each meter run should have block-and-bleed type isolation valves, which can easily verify no-leak sealing, including the valves at the prover connections, and also for the bypass-line block valve, when a bypass line is part of the meter piping design.

This is required because leak-tight isolation valves are necessary and required when proving pipeline meters.

Selection of twin seal type plug valves work well for meter runs since they are leak tight, and have low-pressure drop, when in the full-open position, and because a full port valve is not needed in a meter run. These twin seal plug valves have two sealing slips with elastemer seals, which are not in contact with the valve-seating surface during plug rotation, but both slips are wedged closed at valve shutoff, providing no-leak sealing. When valve seal maintenance is required, the slips can be replaced from the bottom of the valve without removing the valve from the meter piping.

Each valve installed in the pipeline system, especially block valves, must comply with CFR Title 49, Part 195.116, which provides a list of valve requirements needed for procurement and is listed below:

- The valves must be of sound engineering design.
- Valve material must be compatible with all anticipated liquids that will be shipped through the pipeline.
- Each valve must be hydrostatically tested, without leakage, in accordance with API-6D.
- All valves other than check valves must be equipped with a means to clearly indicate valve position.
- Each valve must be marked at the factory, using a name plate.
- The valve must show class designation and the maximum working pressure.

The types of pipeline valves covered by API-6D are as follows: double disc with wedge gate (like WKM Pow-R-Seal) and single-wedge conduit gate valve, ball valves, check valves, and plug valves (such as General Twin Seal).

10.3 SOME OTHER VALVES NOT LISTED BY API-6D

Some other valves include but are not limited to cast steel, rising-steam, wedge gate valves, pressure relief valves, and pipeline control valves, but they must meet ASME B 16.5 flange requirements. The popular cast steel, rising-stem wedge gate valves are covered by Valve Specifications API-600 and BS-1414.

They are available in pressure Class 150 through 2500, and sizes 2 in. through 24 in. (25 mm through 600 mm), with face to face dimensions specified by ASME B 16.10 and BS-2080. These cast steel valves are frequently used in refineries, pipeline pump stations, and tank farms and are popular in the oil fields.

The rising steam threads are separated from the pipeline fluid and are easy to lubricate. The position of the rising steam clearly shows the amount of valve opening. The back-seating feature, with valve fully open, allows repacking the stuffing box with the valve on line. Seat rings screwed into the body are replaceable. Hard-faced wedge gate and seat rings can be supplied with the valve, when specially ordered.

Pressure relief valves can be procured with flanged or screwed ends. The size, capacity, set point, and pressure rating must be specified on a data sheet.

The pipeline control valves must be sized for the target pipeline fluid, flow rate, and pressure range. The size and type of control valve that will be selected will be based on the data sheet information. The data sheet specification should include the type and size of the control valve actuator to be used.

10.4 VALVE PRESSURE CLASS

The maximum service pressure and test pressure requirements for pipeline valves, in accordance with API-6D, for liquid and natural gas service at ambient temperatures, ranging from −20°F (−29°C) to 100°F (38°C) are presented in Table 10.1 below:

TABLE 10.1 Valve Pressure Criteria by ASME Class

ASME Class Rating (Pressure Psi)	Pipeline Working (Max. Pressure)	Hydrostatic Shell (Test Pressure)	Hydrostatic Seat (Test Pressure)
150	275	425	300
300	720	1100	800
600	1440	2175	1600
900	2160	3250	2400
1500	3600	5400	4000

10.5 PIPELINE DESIGN AND VALVE SELECTION

It is important during the design of a new pipeline system that the valves and related equipment selected have a working pressure that will not limit future pipeline performance upgrades. In other words, a pipeline designer should select pipeline equipment with working pressure that will not limit the future strength of the pipeline. For example, if the pipeline design pressure is 700 psi (4830 kPa) and the pipe stress at that pressure is only 50% of the specified minimum yield strength (SMYS), rather than selecting Class 300 steel flanged equipment that can operate at 720 psi (4970 kPa), it would be smart to select Class 600 flanged equipment, such as valves, flanged equipment, pig launchers, meters, strainers, etc., for original construction. Then the pipeline can be upgraded to, say, 70% SMYS sometime in the future by simply retesting. This future upgrade could very well be necessary to increase pipeline throughput, due to expanding market demand; see Table 10.1 for valve selection.

10.6 MAINLINE VALVE LOCATIONS

Pipeline block valves shall be located and spaced along the mainline as necessary for safe operation and pipeline maintenance purposes. These block valves or isolation valves are needed in case of a natural disaster, such as a landslide or an earthquake, which could break the pipeline, or for third party damage to the pipeline caused by road graders or backhoes accidentally hitting the pipeline and causing rupture. Once this severe leak is reported, or detected by the pipeline leak detection system, closing adjacent block valves to isolate the spill will significantly reduce damage to the environment and minimize product loss.

Pipeline code as per Part 195.260(c) requires that block valves shall be placed along the pipeline route, at locations that will minimize damage or pollution caused by accidental hazardous liquid discharge, as appropriate for terrain in open country or in populated areas. Block valves should be placed at the edge of environmentally sensitive locations such as marshes and wetlands. When selecting block valve locations, in open country, a 10 mi (16 km) or longer spacing may be a good starting point. But the actual valve placement must be located in a place that is accessible to pipeline crews, for valve operation and pipeline maintenance.[3] In hilly or mountainous terrain where pipeline hydrostatic test section length is short, due to large elevation differences, it is helpful to locate the near block valve in between two of these test sections so that time and money can be saved when hydrostatically testing against this properly positioned block valve.

In populated areas, and where elevation differences are generally small, block valve spacing may be 5 mi (8 km). It may be a good design idea to place a block valve right next to a pipeline lateral valve, for packaging reasons. Most likely, in populated areas, these two valves would be placed together in an underground concrete vault, for valve protection and to limit exposure.

Pipeline code as per Part 195.260(e & f) requires that block valves must be installed on each sides of a water crossing, such as a river, that is more than 100 ft (30 m) wide from high-water mark to high-water mark. Also, block valves must be installed on each side of a reservoir located on, or next to the pipeline right-of-way, which holds water for human consumption.

10.7 VALVE STATION DESIGN

Pipeline block valve and lateral takeoff valves along the mainline, in open country, are usually placed above the ground to allow full valve access for operation and maintenance. The buried pipeline is brought above ground, using two 45-degree-long radius bends to move the center line of the pipe to 3 or 5 ft (1 m +) above grade, depending on pipe diameter, for block valve placement and bolt-up. If only one valve is involved at this valve station, the pipe will then dive back into the ground using two more

45-degree-long radius bends so that the pipeline goes back in its right-of-way, with about the same amount of cover, as it had when coming into the valve station. With 24 in. (610 mm) diameter pipeline and larger, 45-degree 3R bends will work fine for both utility pigging and smart pig passing. It is a good idea to have at least 2-ft (610 mm) tangents on each long radius bend, going into and out of the block valve, to allow for smooth pig passing. For smaller diameter pipelines, it is a good idea to rely on a smart pig supplier for the recommended bend radius needed for pig passing. This is because smaller size pipelines need longer radius bends to accommodate the longer and smaller diameter internal inspection devices. Some older and smaller diameter pipelines were designed using 90-degree elbows to bring pipelines up at valve stations. This was done before designers were thinking much about pipeline/pig compatibility.

To support above grade pipe span and valve weight, a pipe support placed next to the valve, using a concrete spread footing, is recommended. The block valve may or may not have a MOV but the valve hand wheel must be locked so that only authorized persons can operate the valve. This above-ground block valve must be protected from third-party damage and vandalism, by erecting a 6-ft to 8-ft-high (2–2.7 m) chain-link fence, with double-door opening to allow service truck access for valve maintenance.

The fenced facility must have proper signage, stating a warning, such as *Danger – Do Not Enter. In Emergency, Call Dispatcher*, with telephone number.

When the block valve is motor operated or electric power is available, the fenced area should be protected by using a motion detector, which would alarm at the pipeline control center (CC) if unauthorized persons gained entry.

Some pipeline block valve locations will include the following: lateral pipeline valves with or without MOVs and also locations where there is a need for above-ground pig receiver/launcher stations, where a block valve (Valve A) is common to both the receiver trap and to the pig launcher barrel. There will also be locations where a booster pump station is constructed in open country on the pipeline easement. The booster pump station may be located next to a block valve, but it will have motor-operated suction and discharge valves, on the lines going and coming from the pipeline, as explained earlier in this chapter, in Section 10.2 under *Pump station valves*. This booster station might consist of one or more electric motor driven pumps, designed for outside service, with pressure control valve and a control building with switch gear, and voice and SCADA transmission. This unmanned station with above-ground valves and with pump and pipeline instruments must be protected from vandalism by using a high chain-link fence, and as explained above the pipeline security would be enhanced by using motion detectors inside the fence and motion-sensing cameras, along with a control building break-in alarm system, to notify the pipeline CC operators should unauthorized entry occur.

10.8 BURIED VALVE VAULTS

In urban areas, including residential and commercial, it is not safe or practical to design valve stations above the ground.

In most cases, local building codes would not allow placing pipeline block valves above the ground for safety reasons and also because of neighborhood beautification concerns. So, in most cases, pipeline block valves and valves at lateral connections are placed under the ground, in concrete vaults. These precast concrete vaults can be purchased in different sizes, and they can be selected depending on the main pipeline diameter and the number of valves to be placed in the vault and also the number of pipe penetrations through the vault walls needed for laterals.

Figure 10.2 shows the bottom half of a large precast concrete vault being lowered down into the excavated location at the side of a paved road. In this case, the size of the excavation was approximately 12 ft (4 m) deep by 12 ft (4 m) wide by 20 ft (6.7 m) long. Figure 10.2 also shows some of the shoring necessary to allow pipefitters and laborers to work at that depth, assembling the vault, completing valve piping, and making all the necessary pipeline tie-ins. This shoring design used heavy steel plates to hold back the earth, and the plates were held in position by the wide flange vertical columns. After the large hole was excavated, the edge of the hole was drilled as required for the wide flange columns and the columns were driven deep enough into the earth to provide support for the heavy plates. Once all the columns were driven in place, the heavy plates were slid between the flanges, down to the bottom of the hole, thus completing the shoring.

Figure 10.3 shows part of the piping assembled in the bottom half of the vault, with four twin seal plug valves, supplied with Limitorque motor operators. Notice that there are two products pipelines going into the vault and two coming out of the vault, with a crossover MOV that has a slip blind flanged-up, in the open position. These twin seal plug valves were an excellent

FIGURE 10.2 Lowering the cast concrete vault bottom into excavation.

FIGURE 10.3 Assembling valves and piping in the bottom half of vault.

choice for this light product valve box, since these light product lines do not get pigged, and leak-tight shutoff is required to minimize product contamination. One major goal with these product pipelines is not allowing jet fuel to be contaminated with gasoline or diesel fuel. If jet fuel is contaminated, it must be downgraded and shipped back to the refinery for processing.

Also notice that the plug valves have valve position indicators that clearly show the open/closed position of the valve. The concrete vault that was selected is large enough to allow manual operation of each valve, using the hand wheel. When the top half of the vault is installed, the vault is sealed, using an elastic sealing compound applied to the two mating surfaces designed to make the vault water tight. With the top half in place, the inside height is approximately 8 ft (2.7 m), providing plenty of head room to climb over pipes to allow access to a desired valve.

Since the vault is located at the side of a roadway, the aluminum manway/closure was selected for heavy truck traffic loading and equipped with a suitable locking device for vault security. The design calls for a ladder to be attached, to allow access to the bottom the vault, which is easily reached from the open manway. The pipes through the vault must also be water tight. This can be accomplished by using link seals between the vault pipe openings and the pipes and then sealing the spaces between the links with silicone sealer.

Because water tight designs will eventually leak, a sump pump should be installed, which will pump rain water to a stand pipe, next to the above-ground power/control cabinet on the side of the road. Also, a high water level switch should be installed to alarm the pipeline control center.

With electricity available for the MOVs at the site, the vault will have adequate lighting inside to allow operators and maintenance persons to perform their duties, without the use of handheld lighting. A vault is a confined space. Company personnel or operators entering the vault must follow the rules for confined space entry.

10.9 DIRECT BURIAL OF VALVES

Buried pipeline block valves with risers have been used on liquid and gas pipelines for years. They do have some benefits but also some disadvantages. The benefits are that they are usually butt welded to the pipeline and supported at the bottom of the pipeline trench and require no additional supporting. Since the valves are welded directly to the pipeline, there is no concern regarding flange leakage. The block valves are usually full port gate valves with block-and-bleed design. The valve body drain and seat lube tubing are brought above the ground, generally 12 in (300 mm) or so under the hand wheel for maintenance access. The valve stem raiser with gear box and hand wheel is about 4 ft (1.4 m) longer than standard. The valve and buried stem must be totally covered with a coldtar corrosion preventive coating and backfilled and compacted with good quality pipeline-shading material. The disadvantage, of course, is that you cannot get to the valve for repair. If there is a malfunction, it must be dug out and the line drained, just like if it were above the ground. Then, a valve technician may be able to make the repair in place, by removing the valve stem riser and bonnet, but if that fails, the bad valve must be cut out of the pipeline and replaced, by welding a new or refurbished valve in its place.

The fact that the pipeline and the block valve are both buried, and only the hand wheel and maintenance tubing come up above the ground, may make the installation less susceptible to vandalism. An above-ground enclosure or "dog house" could be placed over the valve operator, large enough to allow an operator to use the hand wheel. The enclosure door would be locked and of course the hand wheel would be locked in place to prevent unauthorized use. Fencing in the enclosure and attaching proper warning signs would provide additional security.

10.10 NATURAL GAS PIPELINE VALVES

Valve selection and placement for natural gas pipelines and related facilities, including compressor stations, are similar to that of liquid pipeline line valves in that they both use the API-6D specifications. The main difference is that gas pipelines primarily use ball valves for pipeline blocking and isolation purposes. They also use ball valves in compressor stations, gas meter runs, pressure regulator stations, pig launcher/receiver stations, etc.

The smaller size ball valves, up to 24 in. (600 mm) diameter, have a two-piece cast steel shell design with floating ball and are available with full port and reduced port design and with ASME Class 150 through 600. For transmission lines sizes of 24, 30, 36, and 48 in. (600, 750, 900, and 1200 mm) diameter, trunnion mounted, side-entry ball valves with three-piece steel shell designs are commonly used for pipeline block valves, see Fig. 10.4. These compact ball valves are also available in two-piece top-entry for easy in-line maintenance and three-piece welded body to eliminate leak path through the body. These large, heavy-ball valves need upper and lower trunnion support to keep the ball in place during rotation under the intense gas pressure and ball weight, and to allow the floating valve seats to maintain proper position. These soft-seated valves have a resilient sealing material inserted into the metal seat holder to provide soft seating-seal action, in addition to the metal to metal seating between the ball and seat ring.

The valves are available with full port design, necessary for smart pigging, and in ASME pressure Class 150 up through 2500. The larger size ball valves, greater than 36 in. (900 mm) diameter, are not available in pressure Class 900 through 2500. All these ball valves have bubble-tight shutoff, and body-bleed for leak testing. They also come with a secondary stem sealing injection system as a backup for environmental reasons. We will talk more about selecting ball valves versus gate valves in both natural gas and liquid pipeline service.

10.11 VALVE PLACEMENT ON GAS PIPELINES

A high-pressure gas transmission line is not much different from the hazardous liquid pipeline, when it comes to valve placement. The block valves will be located above ground in open right-of-way or buried in concrete vaults in populated locations. Refer to sections 10.7 Valve Station Design and 10.8 Buried Valve Vaults, for above-ground valve design and valve vault design. Gas companies have rigorous requirements for venting concrete vaults before entering. Some valve vaults are equipped with exhaust fans. This is because of the high likelihood of gas leaking from flanged piping and valves located in the confined space. The gas company enforces strict vault testing and venting procedures, before entering any vault.

10.12 BLOCK VALVE SPACING ON GAS TRANSMISSION LINES

Because of the nature of pipeline hazards, which can occur with natural gas, in comparison with the random nature of hazards with liquid pipelines, the code can specify the placement of block valves, in the following natural gas pipeline class locations:[2]

Class 1 locations: Each point on the pipeline must be within 10 mi (16 km) of a valve, or in other words allowing a maximum of 20 mi (32 km) valve spacing. (Class 1 requirements: 10 or fewer buildings intended for human occupancy, in a 1 mi (1.6 km) length of pipeline right-of-way.)

Class 2 locations: Each point on the pipeline must be within 7 mi (12 km) of a valve, or in other words allowing a maximum of 15 mi (24 km) valve spacing. (Class 2 requirements: more than 10 but fewer than 46 buildings, in a 1 mi (1.6 km) pipeline length.)

Class 3 locations: Each point on the pipeline must be within 4 mi (6.4 km) of a valve, or in other words allowing a maximum of 8 mi (13 km) valve spacing. (Class 3 requirements: 46 or more buildings or when the pipeline is located 100 yards from a playground, etc., in a 1 mi pipeline length.

Class 4 locations: Each point on the pipeline must be within 2½ mi (4 km) of a valve, or in other words allowing a maximum of 5 mi (8 km) valve spacing. (Class 4 requirements: an area where buildings with four or more stories above the ground are prevalent, over a 1 mi pipeline length.)

These valves and their operating devices must be readily accessible and protected from tampering and damage. Also, each gas transmission block valve must be supported to prevent settling of the valve or movement of the pipe to which it is attached.

Each section of transmission line, between mainline block valves, must have a blowdown valve with enough capacity to allow the section to blow down as rapidly as possible. Each blowdown discharge point must be located so that the gas can be blown to atmosphere without hazard. If the transmission line is adjacent to overhead electric lines, the gas must be directed away from the electrical conductors.

10.13 VALVE MAINTENANCE FOR LIQUID AND GAS PIPELINES AS PER CODE

10.13.1 Hazardous Liquid Pipeline Valves

Each pipeline operator shall maintain all valves that are necessary for safe operation of the pipeline system. This is to insure that the valves are in good working order at all times. Each operator shall at intervals not exceeding 7 months, but at least twice each calendar year, inspect each mainline block valve and lateral valve by partially operating the valve, to determine that it is functioning properly as per CFR Title 49, Part 195.420(a & b). It may be a good idea to lubricate the valve and valve operator, if so equipped, during one of these required valve inspections.

A valve maintenance record must be kept, listing all valves that were inspected by valve number and explaining the inspection method used. This valve inspection record should state the inspection results, and what needs to be repaired or changed as a result of the inspection, and the date that the valve repair was made. Also, it is a good idea to list the date that valve lubrication was done.

10.14 OVERPRESSURE SAFETY VALVES AND PRESSURE LIMITING DEVICES FOR HAZARDOUS LIQUID PIPELINES

Each operator shall at intervals not exceeding 15 months, but at least once each calendar year, or in the case of pipelines used to carry highly volatile liquids at intervals not exceeding 7 months, but at least twice each calendar year, inspect and test each pressure relief valve (PRV), pressure limiting device, or pressure regulator to determine that it is functioning properly, and that it is in good working condition, and is adequate from the standpoint of capacity and reliability of operation, as per Part 195.428.

10.15 NATURAL GAS PIPELINE VALVES MAINTENANCE

Each gas transmission line valve and gas distribution line valve, which might be required for operation during any emergency, must be inspected and partially operated at intervals not exceeding 15 months, but at least once each calendar year.

Each operator must take prompt remedial action to correct any valve found inoperable, unless the operator can designate an alternative valve for that purpose, as per CFR Title 49, Part 192.745 and Part 192.747.

10.16 PRESSURE LIMITING AND REGULATING STATIONS FOR GAS PIPELINES

Each pressure limiting station, pressure relief valve, and pressure regulator station and its equipment must be subjected at intervals not exceeding 15 months, but at least once each calendar year, to inspection and testing to determine that it is in good working condition and is adequate from the standpoint of capacity and reliability, and that it is set to control or relieve at the correct pressure, consistent with the pressure limits. For steel pipelines where the maximum allowable operating pressure (MAOP) produces a hoop stress equal or greater than 72% of SMYS, the set point shall be MAOP plus 4%. The above pressure limiting code requirements are as per Part 192.739.

10.17 GENERAL VALVE STATION PROTECTION

All petroleum and natural gas pipelines must be protected from tampering by unauthorized persons. Since most pipeline valves expose the pipeline to risk when located either above the ground or in vaults or even directly buried, unauthorized access to the valves must be prevented. Pipelines can be a target of vandalism, sabotage, or even terrorist attacks. In war, pipelines are often the target of military attacks, as destruction of pipelines can seriously disrupt logistics and the resulting pipeline leakage, fire, and explosions will terrorize the local population.

10.18 PIPELINE VALVE SELECTION – BALL OR GATE?

Why use ball valves on gas pipelines for code-required block valves or isolation purposes when one also has a choice of gate valves, globe valves, and butterfly valves?

Well, right off the bat, it is now necessary to run utility pigs and smart pigs in natural gas pipelines and transmission lines for operation, maintenance, and internal inspection reasons, and since you cannot push a pig through a globe and butterfly valves, those two choices are out. This brings us back to the question, why are ball valves preferred over gate valves in natural gas pipeline service?

First, ball valves are more modern; they were invented in the late 1940s. This was made possible by the availability of polymers such as PTFE and advancement in machining capability and accuracy. On the other hand, the gate valve was invented many years before, and it cannot compete with the ease of operation and the compactness of a ball valve.

A ball valve has a much smaller envelope than a gate valve and does not need collection zones at the top and bottom of the valve, for a sliding gate. With a gate valve, there must be room for the gate to move from open to closed and it could have difficulty displacing what has collected there. The ball valve enjoys ease of operation: 90-degree rotation to go from full open to close, not requiring multiples turns that are necessary to push the gate valve slide from open to close. The sealing member (ball) rotates in its own volume, instead of having to displace the volume of the gate. Ball valves shut off tight, bubble-tight. This is because the resilient soft-seats provide a sealing action, in addition to the metal to metal seating between the ball and the seat ring. This is especially true in a relatively clean environment such as natural gas service.

Refer to Fig. 10.4 showing a cutaway of a side-entry, full port ball valve. The valve seat-seals stay in contact with the ball, and for natural gas transmission service, they are not affected by paraffin, abrasives, and multiproduct flow that can cause seal damage. So, a natural gas pipeline is a good environment for using ball valves, since natural gas is treated and relatively free of abrasives. Using ball valves in crude oil or heavy oil pipeline service could eventually expose the soft-seats to contamination and possible damage.

Many pipeline engineers have selected the WKM Pow-R-Seal gate valves for crude oil and heavy oil and even sour oil and gas service because the wedging action of the gate provides positive shutoff in both the open and close positions. The sliding wedge feature allows the gate surfaces to pull away from both upstream and downstream valve seats, to minimize or eliminate seat damage resulting from abrasives, as the valve is operated from the open position to the closed position and of course from closed to the open position.

Then, there is the matter of valve size. In large diameter pipelines, such as 30–48 in. (750–1200 mm) diameter, the gate valves get enormous! The 48-in. gate valve is over five-diameters tall or more than 20 feet (6.3 m) in overall height. In service, the valve body would probably be rotated into the horizontal position to allow access to the valve and actuator. I have noticed that many valve suppliers limit gate valve size, to 24 in. (600 mm) diameter.

With ball valves, you are dealing with a valve size slightly larger than the pipe diameter itself. The gate valve actuator must be more robust than an actuator, sized for a ball valve of the same diameter and service. The actuator must be able to provide the "hammer blow" torque necessary to seat and unseat the gate under full differential pressure and with a buildup of debris.

From the environmental standpoint, it is much easier sealing a ball valve steam packing-gland that only rotates 90 degrees than dragging one pipe diameter length of gate valve stem through a packing box when closing, and

Bolted body side-entry

Parts List			
1	Body	13	Body closure nut
2	Closure	14	Socket screw
3	Ball	15	Spring
4	Seat	15a	Antistatic spring
5	Stem	16	Seat greaser
6	Bearing retainer	17	Check valve
7	Gland plate	18	Stem greaser
8	Operator flange	19	Vent bleeder valve
9	Stem key	20	Drain plug
10	Body gasket	21	Ball bushing
10a	Stem gasket	22	Stem thrust washer
10b	Gland plate gasket	23	Ball thrust washer
11	Body O-ring	24	Pin
11a	Seat O-ring	24a	Pin
11b	Seat O-ring	30	Lifting plug
11c	Stem O-ring	31	Valve support
11d	Gland plate O-ring	90	Stop seat socket screw[(p)]
12	Body closure stud	91	Stop seat washer[(p)]

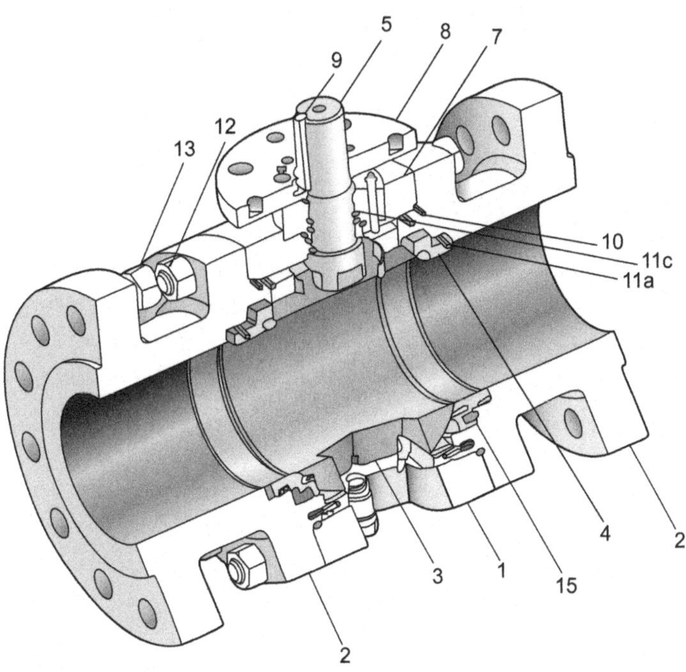

FIGURE 10.4 *Full port, side-entry* ball valve cutaway. *Source: Courtesy of Valbart-Flowserve Company.*

then dragging that diameter's length of contaminated stem back into the atmosphere, on the opening cycle.

In summary, we can say that the ball valve is more efficient in natural gas pipeline service, and especially in all larger size pipelines, starting with 24 inch (600 mm) diameter and larger.

REFERENCES

[1] Code of Federal Regulations, Title 49, Part 195, Transportation of Hazardous Liquids by Pipeline, 2010.
[2] Code of Federal Regulations, Title 49, Part 192, Transportation of Natural and Other Gas by Pipeline, 2010.
[3] ASME Code for Pressure Piping, B31.4, Addenda, 1991.

BIBLIOGRAPHY

California Code of Regulations, Title 2, Division 3, Chapter 1, Article 5.5 (2 CCR 2560-2571) Marine Oil Terminal Pipelines, 1997.

Chapter 11

Pump Stations

E. Shashi Menon, Ph.D., P.E.

Chapter Outline

INTRODUCTION

In this chapter, we will discuss the pump stations and pumping configurations used in liquid pipelines. The optimum locations of pump stations for hydraulic balance will be reviewed. Centrifugal pumps and positive displacement pumps will be discussed along with their performance characteristics. We will introduce affinity laws for centrifugal pumps, the importance of NPSH and how to calculate power requirements when pumping different liquids. Also, viscosity-corrected pump performance using the Hydraulic Institute chart will be explained. The performance of two or more pumps in series or parallel configuration will be examined and how to estimate the operating point for a pump in conjunction with the pipeline system head curve. Next, we will look at some of the major components in a pump station, such as pumps, drivers, and control valves. The use of variable speed pumps to save pumping power under different operating conditions, such as batching, will be reviewed.

11.1 MULTIPUMP STATION PIPELINES

In Chapter 9, we explained how the total pressure required to pump a liquid through a pipeline from point A to point B at a specified flow rate was determined. We found that depending on the flow rate and MAOP, one or more pump stations located along the pipeline may be required to handle the throughput. Suppose calculations show that the total pressure required at the pipeline origin, taking into account the frictional pressure loss, elevation profile of the pipeline, and the terminus delivery pressure is 1950 psig. If the MAOP of the pipeline is 1200 psig, we conclude that we need two pump stations to handle the throughput, without exceeding the MAOP. The first pump station located at the beginning of the pipeline will discharge at the maximum pressure of 1200 psig, and as the liquid flows through the pipeline, the pressure will reduce because of friction as well as decrease (or increase) because of the elevation profile and, finally, reach some point at the minimum pressure, such as 50 psig. The second pump station at this location will boost the liquid pressure to some value (less than MAOP), which will then be sufficient to take the product all the way to the terminus.

The question is then, where do we locate this intermediate booster pump station? If the pipeline is of uniform diameter and wall thickness, the pipeline elevation profile is fairly flat and there are no intermediate injections or deliveries, the second pump station will be located at the midpoint along the pipeline. If the pipeline is 100 mi long, the first pump station will be at the origin (milepost 0.0), and the second booster pump station will be at milepost 50.0.

11.2 HYDRAULIC BALANCE AND PUMP STATIONS REQUIRED

Suppose calculations indicate that at a flow rate of 5000 gal/min, a 100-mi pipeline requires a pressure of 1950 psig at the beginning of the pipeline to deliver the liquid to the pipeline terminus at some minimum pressure, 50 psig. This 1950 psig pressure may be provided in two steps of 975 psig each or three steps of approximately 650 psig each. In fact, as a result of the MAOP limit of the pipe, we may not be able to operate with just one pump station at the beginning of the pipeline, discharging at 1950 psig. Therefore, in long pipelines, the total pressure required to pump the liquid is provided in two or more stages by installing intermediate booster pumps along the pipeline.

In the example case with 1950 psig requirement, and an MAOP of 1200 psig, we would provide this pressure as follows. The pump station at the beginning of the pipeline will provide a discharge pressure of 975 psig, which would be consumed by friction loss in the pipeline, and at some point (roughly halfway) along the pipeline, the pressure will drop to zero. At this location, we boost the liquid pressure to 975 psig using an intermediate booster pump station. We have, of course, neglected the elevation profile of the pipeline and assumed an essentially flat profile.

This pressure of 975 psig will be sufficient to take care of the friction loss in the second half of the pipeline length. The liquid pressure will reduce to zero at the end of the pipeline. However, the liquid pressure at any point along the pipeline must be more than the vapor pressure of the liquid at the flowing temperature. In addition, the intermediate pump station requires certain minimum suction pressure. Therefore, we cannot allow the pressure at any point to drop to zero. Accordingly, we will locate the second pump station at a point where the pressure has dropped to a suitable minimum suction pressure, such as 50 psig. The minimum suction pressure required is also dictated by the particular pump and may have to be higher than 50 psi to account for any restrictions and suction piping losses at the pump station. For the present, we will assume 50 psig suction pressure is adequate for each pump station. Hence, starting with a discharge pressure of 1025 psig (975 + 50), we will locate the second pump station along the pipeline where the pressure has dropped to 50 psig. This pump station will then boost the liquid pressure back up to 1025 psig and will deliver the liquid to the pipeline terminus at 50 psig. Thus, each pump station provides 975 psig differential pressure (discharge pressure less suction pressure) to the liquid, together matching the total pressure requirement of 1950 psig at 5000 gal/min flow rate.

Note that in the above analysis, if we considered the pipeline elevations, the location of the intermediate booster pump will be different from that of a pipeline along a flat terrain.

When each pump station supplies the same amount of energy to the liquid, we say that the pump stations are in hydraulic balance. This will result in the same power added to the liquid at each pump station. For a single flow rate at the inlet of the pipeline (no intermediate injections or deliveries), the hydraulic balance will also result in identical discharge pressures at each pump station. Because of the topography of the pipeline route, it may not be possible to locate the intermediate pump station at the theoretical locations for hydraulic balance. It is possible that the hydraulically balanced pump station location of milepost 50 may actually be in the middle of a swamp or a river. Hence, we will have to relocate this pump station to a more suitable location after field investigation. If the revised location of the second pump station were at milepost 52, the hydraulic balance would no longer be valid. Recalculating the hydraulics with the revised pump station location will show that the pump stations are not in hydraulic balance and will not be operating at the same discharge pressure or providing the same amount of power. Therefore, while it is desirable to have all pump stations balanced, it may not be practical. Hydraulically balanced pump station locations afford the advantage of using identical pumps and motors and the convenience of maintaining a common set of spare parts (pump rotating elements, mechanical seal, and so on) at a central operating district location.

In Chapter 9, we introduced the hydraulic pressure gradient and how to locate an intermediate pump station. We presented a method to calculate the discharge pressure for a two pump station pipeline system, knowing the total pressure required for a particular flow rate. We will now discuss a method to calculate the pump station pressures for hydraulic balance.

Figure 11.1 shows a pipeline with varying elevation profile, but no significant controlling peaks along the pipeline. First, the total pressure P_T is calculated for a given flow rate and liquid properties. If we used only one pump station at the beginning, the hydraulic gradient with the pump station discharging at pressure P_T is as shown, delivering the liquid at the terminus delivery pressure of P_{del}. Because P_T may be higher than the MAOP of the pipeline, let's assume that three pump stations are required to provide the pressures needed without exceeding the MAOP. Each pump station will discharge at some common pressure P_D that is just below the MAOP. If P_S represents the common pump station suction pressure, from the geometry of the hydraulic gradient, we see that

$$P_D + (P_D - P_S) + (P_D - P_S) = P_T \tag{11.1}$$

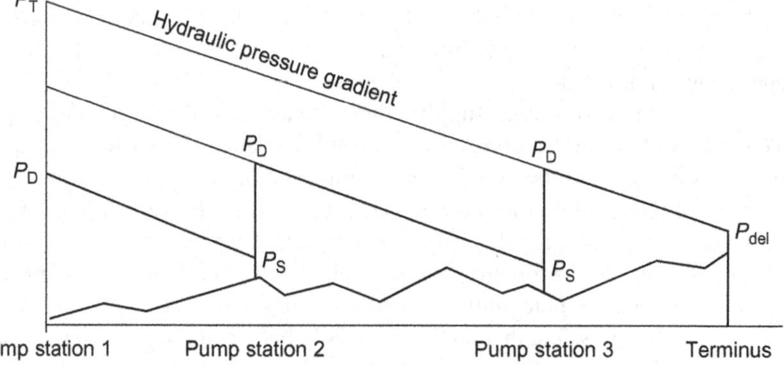

FIGURE 11.1 Hydraulic gradient – multiple pump stations.

Because this is based on one origin pump station and two intermediate pump stations, we can extend the above equation for N pump stations as follows:

$$P_D + (N-1)(P_D - P_S) = P_T \tag{11.2}$$

Solving for the number of pump stations, N, we get

$$N = (P_T - P_S)/(P_D - P_S) \tag{11.3}$$

Equation (11.3) can be used to estimate the number of pump stations required for hydraulic balance, knowing the discharge pressure limit P_D at each pump station and the calculated total pressure P_T.

Solving Eq. (11.3) for the pump station discharge pressure, we get

$$P_D = (P_T - P_S)/N + P_S \tag{11.4}$$

For example, suppose we calculated the total pressure required as 2950 psig and the pump station suction pressure is 50 psig. The number of pump stations required with 1200 psig discharge pressure is, using Eq. (11.3)

$$N = \frac{(2950-50)}{(1200-50)} = 2.52$$

Rounding up to the nearest whole number, $N=3$.

Therefore, three pump stations are required, to limit the maximum discharge pressure to 1200 psig. From Eq. (11.4), with $N=3$, each pump station will operate at a discharge pressure of

$$P_D = (2950-50)/3 + 50 = 1017\,\text{psig}$$

After calculating the discharge pressure required for hydraulic balance, as explained above, the pump stations can be graphically located along the pipeline profile. Let's assume we need three pump stations, as shown in Fig. 11.1. First, the pipeline profile (milepost versus elevation) is plotted, and the hydraulic gradient superimposed on it by drawing the sloped line starting at P_T at A and ending at P_{del} at D as shown in Fig. 11.1. Because the pipeline elevations are in feet, the pressures must also be plotted in feet of liquid head. Thus, P_T and P_{del} must be converted to feet of head first. Next, starting at the first pump station A at discharge pressure P_D, a line is drawn parallel to the hydraulic gradient. The location B of the second pump station will be established at a point where the hydraulic gradient between A and B meet the vertical line at the required suction pressure P_S. The process is continued to determine the location C of the third pump station.

In the preceding analysis, we made some simplifying assumptions. We assumed that the pressure drop per mile (slope of the hydraulic gradient) was constant throughout the pipeline, based on constant pipe inside diameter and flow rate throughout the entire pipeline. With variable pipe diameter or wall thickness, the slope of the hydraulic gradient may not be constant, as will be explained shortly.

11.3 TELESCOPING PIPE WALL THICKNESS

Reviewing a typical hydraulic gradient as shown in Fig. 11.2, we note that as a result of friction, the liquid pressure decreases from the pump station to the terminus in the direction of flow. The pipeline segment immediately downstream of a pump station will be at a higher pressure such as 1200 psig or more, depending on MAOP, whereas the tail end of that segment before the next pump station (or the terminus) will be subject to lower pressures in the range of 50–100 psig. If we use the same wall thickness throughout the pipeline, we will be underutilizing the downstream portion of the piping that is subject to a lower pressure. Therefore, a more efficient approach would be to reduce the pipe wall thickness as we move away from a pump station toward the suction side of the next pump station or the delivery terminus. For example, the pipe wall thickness immediately downstream of the pump station may be 0.375 in. for some length and then the wall thickness gradually reduces to 0.250 in. closer to the suction of the next pump station.

From the discussions on pipe strength in Chapter 7, the higher pipe wall thickness immediately adjacent to the pump station will be able to withstand the higher discharge pressure. As the pressure reduces down the length of the pipeline, the wall thickness would be reduced to some value just enough to withstand the lower pressures as we approach the next pump station or delivery terminus. This process of varying the wall thickness to compensate for reduced pipeline pressures is referred to as telescoping pipe wall thickness.

However, telescoping pipe wall thickness must be done cautiously. Suppose a pipeline has two pump stations, and the second pump station is shut down for some reason, then the flow rate will reduce and the hydraulic gradient will be flatter as shown in Fig. 11.3.

It can be seen that portions of the pipeline on the upstream side of the second pump station will be subject to higher pressure than when the second pump

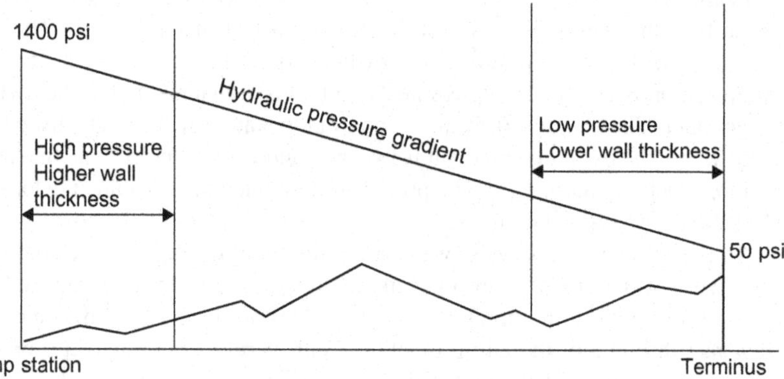

FIGURE 11.2 Telescoping pipe wall thickness.

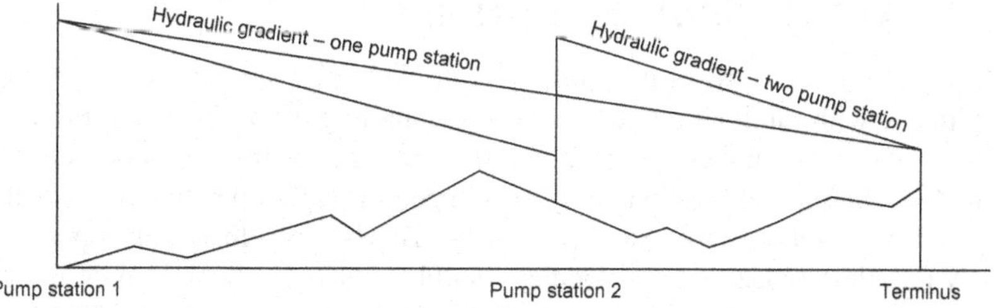

FIGURE 11.3 Hydraulic gradient – pump station shutdown.

station was online. Therefore, wall thickness reductions (telescoping) implemented upstream of a pump station must be able to handle the higher pressures resulting from the shutdown of an intermediate pump station.

11.4 CHANGE OF PIPE GRADE – GRADE TAPERING

Similar to reducing the pipe wall thickness to compensate for lower pressures as we approach the next pump station or delivery terminus, the pipe material grade may also be varied. Thus, the high pressure sections immediately downstream of a pump station may be constructed of API 5L X-52 grade steel, whereas the lower pressure section may be constructed of API 5L X-42 grade pipe, thereby reducing the total cost. This process of varying the pipe grade is referred to as grade tapering. Sometimes, a combination of telescoping and grade tapering is used to reduce pipe material cost. Note that such wall thickness variation and pipe grade reduction to match the requirements of steady state pressures may not always be feasible. Consideration must be given to increased pipeline pressures under intermediate pump station shut down and upset conditions such as pump start up, valve closure, and so on. These transient conditions cause surge pressures in a pipeline and therefore must be taken into account when selecting optimum wall thickness and pipe grade.

11.5 SLACK LINE AND OPEN CHANNEL FLOW

Most pipelines typically flow full, with no vapor space or a free liquid surface. However, under certain topographic conditions with drastic elevation changes, we may encounter pipeline sections that are partially full, resulting in open channel flow or slack line flow. Slack line operation may be unavoidable in some liquid pipelines. However, when pumping high vapor pressure liquids and in batched pipelines, slack line flow cannot be allowed. When pumping high vapor pressure products, the liquid must not vaporize so as to prevent pump damage. In batched pipelines, slack line flow will cause intermingling of batches, which is unacceptable for product quality reasons.

11.6 BATCHING DIFFERENT LIQUIDS

Batching is the process of transporting multiple liquids simultaneously through a pipeline with minimal mixing. Some commingling of the batches is unavoidable at the boundary or interface between contiguous batches. For example, gasoline, diesel, and kerosene may be shipped through a pipeline in a batched mode from a refinery to a storage terminal. Batched pipelines must operate in fully turbulent mode or the velocities should be sufficiently high to ensure the Reynolds number is more than 4000. If the flow were laminar ($R < 2000$), the product batches would intermingle, thereby contaminating or degrading the products. Also, batched pipelines must be run in packed conditions (no slack line or open channel flow) to avoid contamination or intermingling of batches, in pipelines that have significant elevation changes.

In a batched pipeline, the total frictional pressure drop for a given flow rate is calculated by adding the individual pressure drops for each product, based on its specific gravity, viscosity, and the batch length.

In batched pipelines, we must first calculate the line fill of the pipeline as follows, in USCS units:

$$\text{Line fill volume, bbl} = 5.129\,L(D)^2 \quad \text{(USCS)} \tag{11.5a}$$

where D is pipe inside diameter in in., and L is pipe length in mi.

In SI units, line fill is calculated from

$$\text{Line fill volume, m}^3 = 7.855 \times 10^{-4} L(D)^2 \quad \text{(SI)} \tag{11.5b}$$

where, D is pipe inside diameter in mm, and L is pipe length in km.

Appendix 7 shows the line fill volumes for various pipe diameters and wall thickness in USCS and SI units.

Example Problem 11.1 (USCS)

A refined products pipeline, NPS 12 in., 0.250 in. wall thickness, 120 mi long from Douglas Refinery to Hampton Terminal is used to ship three products at a flow rate of 1500 gal/min as shown in Fig. 11.4. The instantaneous condition shows the three batches within the pipeline.

Assuming the following physical properties and batch sizes for the three products, calculate the pressure drop for each liquid batch at the given flow rate. Use Hazen–Williams head loss with the C-factors shown. The batch sizes for diesel and kerosene are 50,000 and 30,000 bbl, respectively. Gasoline fills the remainder of the pipeline.

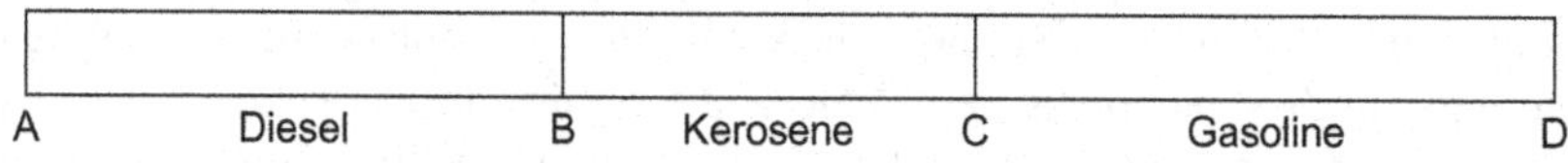

FIGURE 11.4 Batched pipeline.

Product	Specific Gravity	C-Factor
Diesel	0.85	125
Kerosene	0.82	135
Gasoline	0.74	140

Solution

First, calculate the total liquid volume in the pipeline, also known as the line fill volume, using Eq. (11.5a)

$$\text{Line fill} = 5.129(120)(12.25)^2 = 92{,}361\ \text{bbl}$$

The gasoline batch size = 92,361 bbl – 80,000 bbl = 12,361 bbl.

Using Eq. (11.5a), the batch lengths are calculated as 64.96 mi for diesel, 38.98 mi for kerosene, and 16.06 mi for gasoline.

The frictional head loss in the different batch segments are calculated from Table 8.7, Chapter 8, adjusting for the *C* values, and flow rates.

For example, for water with $C = 120$, at 3000 gal/min in NPS 12 pipe, Table 8.7 shows a head loss of 20.39 ft per 1000 ft of pipe.

For diesel, multiplying by $\left(\frac{120}{C} \times \frac{Q}{3000}\right)^{1.852}$ results in the head loss as

$$\text{Diesel head loss} = 20.39 \times \left(\frac{120}{125} \times \frac{1500}{3000}\right)^{1.852} = 5.24\ \text{ft per 1000 ft of pipe}$$

Converting to psi/mi,

$$\text{Diesel pressure loss} = (5.24 \times 0.85/2.31) \times 5.28\ \text{psi/mi} = 10.18\ \text{psi/mi}$$

Similarly, the pressure loss in kerosene and gasoline are calculated as

Diesel: 10.18 psi/mi	Batch length: 64.96 mi
Kerosene: 9.82 psi/mi	Batch length: 38.98 mi
Gasoline: 8.86 psi/mi	Batch length: 16.06 mi

In the preceding, for each liquid, the pipeline length that represents the batch volume is shown. Thus, the diesel batch will start at milepost 0.0 and end at milepost 64.96. Similarly, the kerosene batch will start at milepost 64.96 and end at milepost (64.96 + 38.98) = 103.94. Finally, the gasoline batch will start at milepost 103.94 and end at milepost (103.94 + 16.06) = 120.0 for the snapshot configuration shown in Fig. 11.4.

The batch lengths calculated above are based on 92,361/120 = 769.68 bbl per mile of 12-inch pipe calculated using Eq. (11.5a). The total frictional pressure drop for the entire 100-mi pipeline is obtained by adding up the individual frictional pressure drops for each product as follows:

$$\text{Total pressure drop} = 10.18(64.96) + 9.82(38.98) + 8.86(16.06) = 1186.4\ \text{psig}$$

In addition, the elevation head and the delivery pressure at Hampton are added to the frictional pressure loss of 1186.4 psig to determine the total pressure required at Douglas Refinery.

When batching different products such as gasoline and diesel, flow rates vary as the batches move through the pipeline because of the changing composition of liquid in the pipeline. In order to economically operate the batched pipeline, by minimizing pumping cost, there exists an optimum batch size for the various products in a pipeline system. An analysis needs to be made over a finite period, such as a week or a month, to determine the flow rates and pumping costs considering various batch sizes. The combination of batch sizes that result in the least total pumping cost, consistent with shipper and market demands will then be the optimum batch sizes for the particular pipeline system.

11.7 CENTRIFUGAL PUMPS VERSUS RECIPROCATING PUMPS

Centrifugal and reciprocating pumps are used to pump liquids through a pipeline from an originating point to the delivery terminus at the required flow rate and pressure. To increase the flow rate, more pump pressure will be required. The majority of liquid pipelines use centrifugal pumps because of their flexibility and lower operating cost compared with reciprocating pumps.

Reciprocating pumps belong to the category of positive displacement (PD) pumps and are typically used for liquid injection lines in oil pipeline gathering systems.

A centrifugal pump increases the kinetic energy of a liquid because of centrifugal velocity from the rotation of the pump impeller. This kinetic energy is converted to pressure energy in the pump volute. The higher the impeller speed, the higher the pressure developed. Larger impeller diameter increases the velocity and hence the pressure generated by the pump. Compared with PD pumps, centrifugal pumps have a lower efficiency. However, centrifugal pumps can operate at higher speeds to generate higher flow rates and pressures. Centrifugal pumps also have lower maintenance requirements than PD pumps.

PD pumps, such as reciprocating pumps, operate by forcing a fixed volume of liquid from the inlet to outlet of the pump. These pumps operate at lower speeds than centrifugal pumps. Reciprocating pumps cause intermittent flow. Rotary screw pumps and gear pumps are also PD pumps, but operate continuously compared with reciprocating pumps.

Modern liquid pipelines are mostly designed with centrifugal pumps because of their flexibility in volumes and pressures. In petroleum pipeline installations where liquid from a field gathering system is injected into a main pipeline, PD pumps may be used. Figures 11.5 and 11.6 show typical centrifugal pumps and reciprocating pumps used in the pipeline industry.

Centrifugal pumps are generally classified as radial flow, axial flow, and mixed flow pumps. Radial flow pumps develop head by centrifugal force. Axial flow pumps, however, develop the head due to a propelling or lifting action of the impeller vanes on the liquid. Radial flow pumps are used when high heads are required, while the axial flow and mixed flow pumps are mainly used for low head–high capacity applications.

FIGURE 11.5 Typical centrifugal pump.

FIGURE 11.6 Reciprocating pump.

The performance of a centrifugal pump is represented by a series of curves, collectively called the pump characteristic curves. These show how the pump head, efficiency, and pump power varies with flow rate (aka capacity) as shown in Fig. 11.7.

The head curve shows the pump head on the left vertical axis, while the flow rate is shown on the horizontal axis. This curve may be referred to as the H–Q curve or the head–capacity curve. The term capacity is used interchangeably

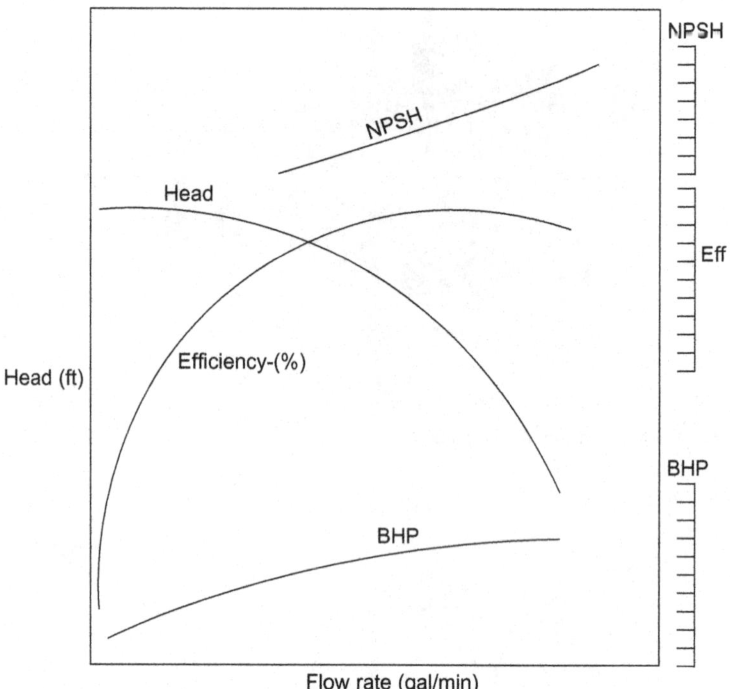

FIGURE 11.7 Centrifugal pump performance.

with flow rate when dealing with pumps. The efficiency curve is called the E–Q curve and shows how the pump efficiency varies with capacity. The power curve, such as the brake horsepower (BHP) versus capacity curve, indicates the power required to run the pump at various flow rates. Another important pump characteristic is the NPSH versus flow rate curve. NPSH or net positive suction head is important when pumping high vapor pressure liquids and will be discussed later in this chapter.

The performance curves for a particular model pump are typically plotted for a particular pump impeller size and speed (example: 10 in. impeller, 3560 RPM). The manufacturer's pump curves are always based on water as the liquid pumped. When pumping liquids other than water, these curves may require adjustments for specific gravity and viscosity of the liquid. In USCS units, the pressure generated by the pump is measured in feet of water and flow rate is shown in gal/min. In SI units, the head is stated in meters of water and the flow rate may be in m^3/h or *L*/s. In USCS units, pump power is always stated as BHP, while kW is used in SI units.

In addition to the four characteristic curves, pump vendors provide pump head curves drawn for different impeller diameters and iso-efficiency curves.

An example of this is seen in Fig. 11.8.

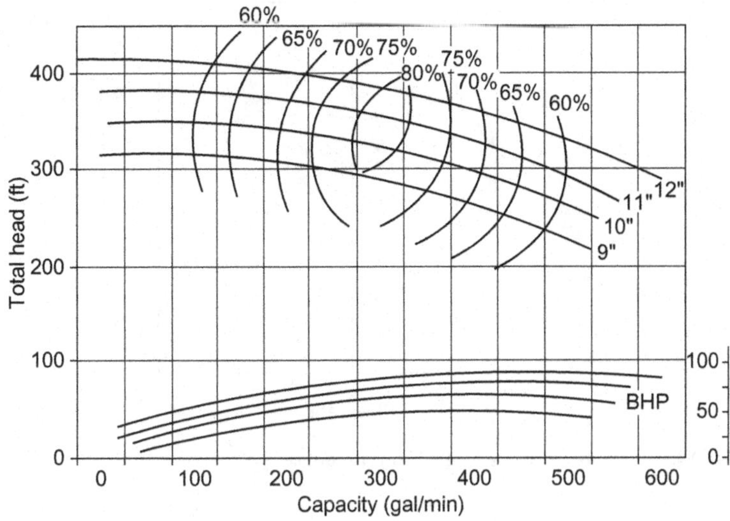

FIGURE 11.8 Centrifugal pump performance for different impeller sizes.

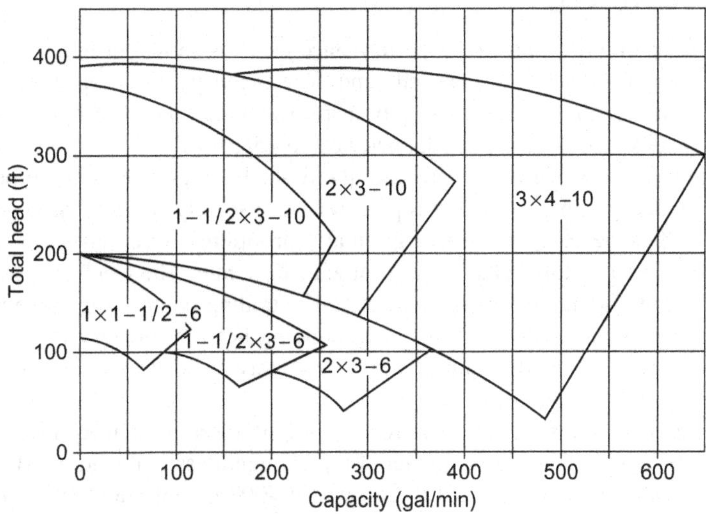

FIGURE 11.9 Centrifugal pump composite rating chart.

Another set of curves provided by centrifugal pump vendors is called a composite rating chart and is shown in Fig. 11.9.

A PD pump continuously pumps a fixed volume at various pressures. It is able to provide any pressure required at a fixed flow rate, within the limits of structural design. This fixed flow rate depends on the geometry of the pump such as bore, stroke, and so on. A typical PD pump pressure volume curve is shown in Fig. 11.10.

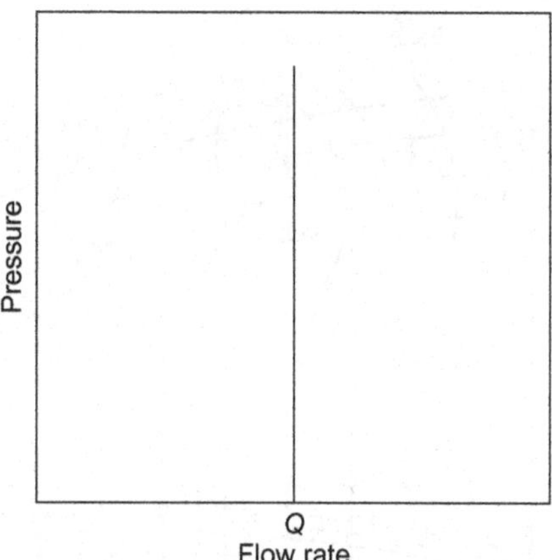

FIGURE 11.10 Positive displacement pump performance.

11.8 CENTRIFUGAL PUMP HEAD AND EFFICIENCY VERSUS FLOW RATE

A typical pump manufacturer's performance curve is shown in Fig. 11.11.

It shows the head versus capacity and the efficiency versus capacity as well as the BHP versus capacity curves for a specific model pump at a pump speed of 3560 RPM and an impeller diameter of 16-15/16 in.

The solid curves are for water and the dashed curves represent the performance when pumping a heavy liquid, with a viscosity of 1075 SSU. It can be seen that the H–Q curve is a gradually drooping curve that starts at the highest value (shutoff head) of head at zero flow rate. The head decreases as the flow rate through the pump increases. The trailing point of the curve represents the maximum flow and the corresponding head the pump can generate. The shutoff head for this pump is approximately 1320 ft and the maximum capacity is 7000 gal/min.

Pump head is always plotted in feet of head of water. Therefore, the pump is said to develop the same head in feet of liquid, regardless of the liquid. Because this particular pump develops 1280 ft head at 4000 gal/min flow rate, the corresponding pump pressure in psig depends on the specific gravity of the liquid. Using Eq. (8.10),

$$\text{When pumping water, pump pressure} = 1280 \times 1/2.31 = 554\,\text{psi}$$

$$\text{When pumping gasoline, pump pressure} = 1280 \times 0.74/2.31 = 410\,\text{psi}$$

The head curve will be the same for water or a petroleum liquid, if its viscosity is less than 10 cSt. At higher viscosities, the pump does not produce the same head as that produced when pumping water and the H–Q curve will

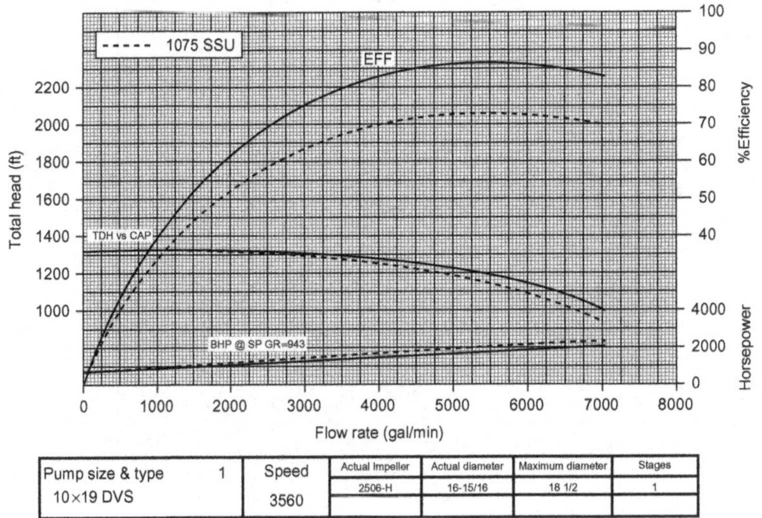

FIGURE 11.11 Typical centrifugal pump performance curves.

have to be corrected downward, as explained later. In fact, Fig. 11.11 shows that when pumping a crude oil with a viscosity of 1075 SSU (236 cSt), the head generated at 4000 gal/min flow rate drops to 1260 ft. The effect of the viscosity is more pronounced on the pump efficiency. With water, the pump efficiency is approximately 82.5% at 4000 gal/min. However, this drops to 70% with the viscous liquid. The maximum efficiency of this pump is 86% when pumping water and occurs at a capacity of 5500 gal/min. The flow rate and head corresponding to the maximum pump efficiency is called the best efficiency point (BEP). In this case at the BEP, $Q = 5500$ gal/min, $H = 1180$ ft, and $E = 86\%$.

The head versus flow rate curve is shown for this particular pump at an impeller diameter of 16-15/16 in. operating at 3560 RPM. Within the same pump body, a certain range of impeller sizes can be accommodated. The maximum impeller size for this pump is 18½ in. Similarly, the minimum impeller size will be specified by the vendor. The pump head versus capacity curve in Fig. 11.12 shows a series of head curves at different impeller sizes for a fixed pump speed. It can be seen that the 10-in. and 12-in. curves are parallel to the 11-in. curve. The variation of the pump H–Q curves with impeller diameter follow the affinity laws, discussed later in this chapter.

Similar to H–Q variation with pump impeller diameter, we can generate curves for different impeller speeds. If the impeller diameter is kept constant at 10 in. and the initial H–Q curve was based on a pump speed of 3560 RPM, then by varying the pump speed, we can generate a family of parallel curves as shown in Fig. 11.13.

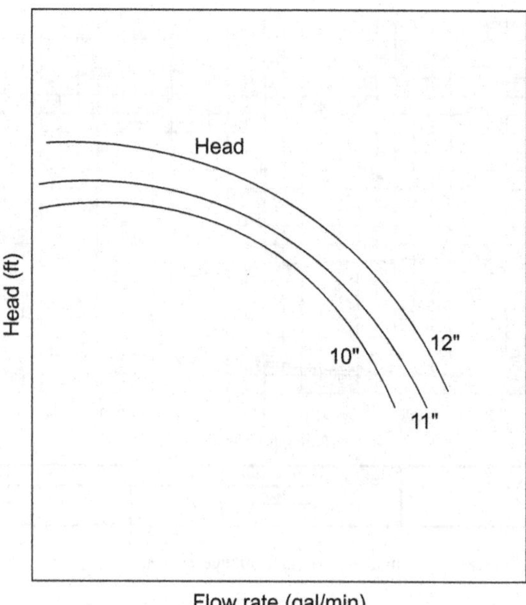

FIGURE 11.12 Head versus flow rate at different impeller sizes.

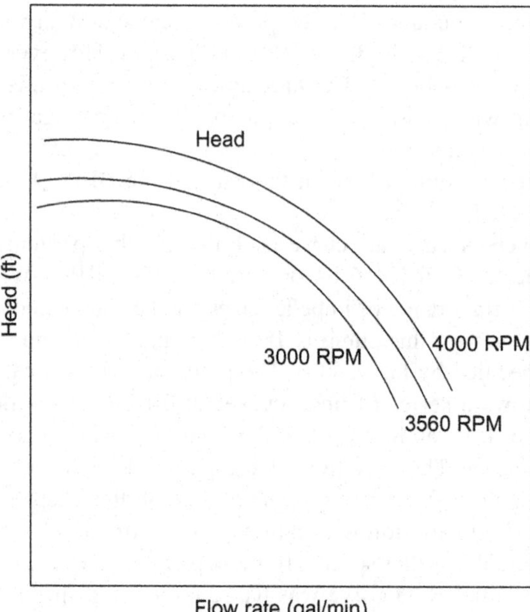

FIGURE 11.13 Head versus flow rate at different speeds.

A pump may develop the head in stages. A single-stage pump may generate 200 ft head at 2500 gal/min. A three-stage pump of this design will generate 600 ft head at same flow rate and pump speed. An application that requires 2400 ft head at 3500 gal/min will be handled by a six-stage pump with each stage providing 400 ft of head. Destaging is the process of reducing the active number of stages in a pump to reduce the total head developed. The six-stage pump discussed above may be destaged to four stages if we need only 1600 ft head at 3500 gal/min.

When choosing a centrifugal pump for a particular application, we try to get the operating point as close as possible to the BEP. In order to allow for future increase in flow rates, the initial operating point is chosen slightly to the left of the BEP. This will ensure that with increase in pipeline throughput, the operating point on the pump curve will move to the right, which would result in a slightly better efficiency.

11.9 BHP VERSUS FLOW RATE

From the head versus flow rate (H–Q) curve and efficiency versus flow rate curves, we can calculate the brake horsepower (BHP) required at every flow rate as follows:

$$\text{Pump BHP} = \frac{Q \times H \times Sg}{3960 \times E} \quad \text{(USCS)} \tag{11.6}$$

where Q is pump flow rate, gal/min; H is pump head, ft; E is pump efficiency as a decimal value, less than 1.0; and Sg is liquid specific gravity (for water Sg = 1.0).

In SI units, power in kW can be calculated as follows:

$$\text{Power kW} = \frac{Q \times H \times Sg}{367.46 \times E} \quad \text{(SI)} \tag{11.7}$$

where Q is pump flow rate, m^3/h; H is pump head, m; E is pump efficiency as a decimal value, less than 1.0; and Sg is liquid specific gravity (for water Sg = 1.0).

For example, from Fig. 11.11, at the BEP the flow rate, head, and efficiency for the water curve are $Q = 5500$ gal/min, $H = 1180$ ft, and $E = 86\%$.

The BHP at this flow rate for water is calculated from Eq. (11.6) as

$$\text{Pump BHP} = \frac{5500 \times 1180 \times 1.0}{3960 \times 0.86} = 1906$$

Similarly, BHP can be calculated at various flow rates from 0 to 7000 gal/min by reading the corresponding head and efficiency values from the H–Q curve and E–Q curve and using Eq. (11.6) as above. The BHP versus flow rate curve can be seen below the H–Q curve in Fig. 11.11.

Note that there is a dashed curve above the solid BHP curve. The dashed curve is based on crude oil with Sg = 0.943 and viscosity = 1075 SSU. Because of the higher viscosity, the BHP required for crude oil is higher than that for water.

As we discussed in Chapter 9, the BHP calculated above is the brake horsepower demanded by the pump. An electric motor driving the pump will have an efficiency that ranges from 95 to 98%. Therefore, the motor HP required is equal to the pump BHP divided by the motor efficiency as follows:

$$\text{Motor HP} = \text{Pump BHP/Motor efficiency} = 1906/0.95$$
$$= 2006\ \text{HP, based on 95\% motor efficiency}$$

11.10 NPSH VERSUS FLOW RATE

In addition to the H–Q, E–Q, and BHP versus capacity curves discussed in the preceding sections, the pump performance data will include a fourth curve for net positive suction head (NPSH) versus capacity. This curve is generally located above the head, efficiency, and BHP curves as shown in Fig. 11.14.

The NPSH curve shows the variation of the minimum net positive suction head at the impeller suction versus the flow rate. The NPSH increases as the flow rate increases. NPSH represents the resultant positive pressure at pump

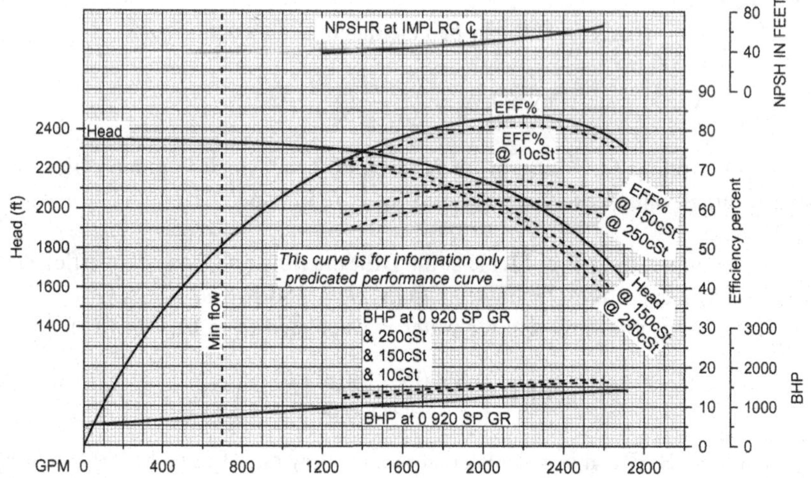

FIGURE 11.14 NPSH versus flow rate.

suction after accounting for frictional loss and liquid vapor pressure. NPSH is discussed in detail later in this chapter.

11.11 SPECIFIC SPEED

The specific speed of a centrifugal pump is used for comparing geometrically similar pumps and for classifying the different types of centrifugal pumps.

Specific speed may be defined as the speed at which a geometrically similar pump must be run such that it produces a head of 1 ft at a flow rate of 1 gal/min. It is calculated as follows:

$$N_S = N Q^{1/2} / H^{3/4} \tag{11.8}$$

where N_S is pump specific speed; N is pump impeller speed, RPM; Q is flow rate or capacity, gal/min; and H is head, ft.

Both Q and H are measured at the best efficiency point (BEP) for the maximum impeller diameter. The head H is measured per stage for a multistage pump.

Another related term, called the suction specific speed is defined as follows:

$$N_{SS} = N Q^{1/2} / (\mathrm{NPSH_R})^{3/4} \tag{11.9}$$

where N_{SS} is suction specific speed; N is pump impeller speed, RPM; Q is flow rate or capacity, gal/min; and $\mathrm{NPSH_R}$ is NPSH required at BEP.

When applying the above equations to calculate the pump-specific speed and suction-specific speed, use the full Q value for single- or double-suction pumps for N_S calculation. For N_{SS} calculation, use one-half the Q value for double suction pumps. Appendix 7 lists specific speeds for various pump speeds, capacities, heads and NPSH.

Table 11.1 lists the specific speed range for centrifugal pumps.

TABLE 11.1 Specific Speeds of Centrifugal Pumps

Description	Application	Specific Speed (N_S)
Radial vane	Low capacity/high head	500–1000
Francis–screw type	Medium capacity/medium head	1000–4000
Mixed–flow type	Medium to high capacity, low to medium head	4000–7000
Axial–flow type	High capacity/low head	7000–20,000

Example Problem 11.2

Calculate the specific speed of a five-stage, double-suction centrifugal pump, 12-in. diameter impeller that when operated at 3560 RPM generates a head of 2200 ft at a capacity of 3000 gal/min at the BEP on the head capacity curve. If the NPSH required is 25 ft, calculate the suction-specific speed.

Solution

$$N_S = N\,Q^{1/2}/H^{3/4} = 3560\,(3000)^{1/2}/(2200/5)^{3/4} = 2030$$

The suction-specific speed is

$$N_{SS} = N\,Q^{1/2}/\text{NPSH}_R^{3/4} = 3560(3000/2)^{1/2}/(25)^{3/4} = 12{,}332$$

11.12 AFFINITY LAWS FOR CENTRIFUGAL PUMPS

The affinity laws for centrifugal pumps are used to predict pump performance for changes in impeller diameter and impeller speed.

The affinity laws are represented as follows:

For impeller diameter change,

$$Q_2/Q_1 = D_2/D_1 \tag{11.10}$$

$$H_2/H_1 = (D_2/D_1)^2 \tag{11.11}$$

where Q_1 and Q_2 are initial and final flow rates, respectively; H_1 and H_2 are initial and final heads, respectively; and D_1 and D_2 are initial and final impeller diameters, respectively.

Similarly, affinity laws state that for the same impeller diameter, if pump speed is changed, then flow rate is directly proportional to the speed, while the head is directly proportional to the square of the speed. As with diameter change, the BHP is proportional to the cube of the impeller speed. This is represented mathematically as follows:

For impeller speed change

$$Q_2/Q_1 = N_2/N_1 \tag{11.12}$$

$$H_2/H_1 = (N_2/N_1)^2 \tag{11.13}$$

where Q_1 and Q_2 are initial and final flow rates, respectively; H_1 and H_2 are initial and final heads, respectively; and N_1 and N_2 are initial and final impeller speeds, respectively.

Note that the affinity laws for speed change are exact. However, the affinity laws for impeller diameter change are only approximate and valid for small changes in impeller sizes. The pump vendor must be consulted to verify that the predicted values using affinity laws for impeller size changes are accurate or if any correction factors are needed. With speed and impeller size changes, the efficiency versus flow rate can be assumed to be the same.

Example Problem 11.3

The head and efficiency versus capacity data for a centrifugal pump with a 10-in. impeller is as shown below.

Q (gal/min)	0	800	1600	2400	3000
H (ft)	3185	3100	2900	2350	1800
E (%)	0	55.7	78	79.3	72

The pump is driven by a constant speed electric motor at a speed of 3560 RPM.

1. Determine the performance of this pump with a 11-in. impeller, using affinity laws.
2. If the pump drive were changed to a variable frequency drive (VFD) motor with a speed range of 3000–4000 RPM, calculate the new H–Q curve for the maximum speed of 4000 RPM with the original 10-in. impeller.

Solution

1. Using affinity laws for impeller diameter changes, the multiplying factor for flow rate is

$$\text{factor} = 11/10 = 1.1 \text{ and the multiplier for head is } (1.1)^2 = 1.21$$

Therefore, we will generate a new set of Q and H values for the 11-in. impeller by multiplying the given Q values by the factor 1.1 and the H values by the factor 1.21 as follows:

Q (gal/min)	0	880	1760	2640	3300
H (ft)	3854	3751	3509	2844	2178

The above flow rate and head values represent the predicted performance of the 11-in. impeller. The efficiency versus flow rate curve for the 11-in. impeller will approximately be the same as that of the 10-in. impeller.

2. Using affinity laws for speeds, the multiplying factor for the flow rate is

$$\text{factor} = 4000/3560 = 1.1236 \text{ and the multiplier for head is } (1.1236)^2 = 1.2625$$

Therefore, we will generate a new set of Q and H values for the pump at 4000 RPM by multiplying the given Q values by the factor 1.1236 and the H values by the factor 1.2625 as follows:

Q (gal/min)	0	899	1798	2697	3371
H (ft)	4021	3914	3661	2967	2273

The above flow rates and head values represent the predicted performance of the 10-in. impeller at 4000 RPM. The new efficiency versus flow rate curve will approximately be the same as the given curve for 3560 RPM.

11.13 EFFECT OF SPECIFIC GRAVITY AND VISCOSITY ON PUMP PERFORMANCE

As mentioned earlier, the pump vendor's performance curves are always based on water as the pumped liquid. When a pump is used to pump a viscous liquid with a viscosity greater than 10 cSt, the H–Q and E–Q curves must be corrected for viscosity, using the Hydraulic Institute standards. Because BHP is a function of the specific gravity, the BHP calculated for water must also be adjusted for the specific gravity of the viscous liquid. Therefore for a viscous liquid, capacity, head, and efficiency must be corrected as well as the specific gravity used to adjust the water BHP. The Hydraulic Institute has published viscosity correction charts that can be applied to correct the water performance curves to produce viscosity-corrected curves. Figure 11.15 shows a typical chart from the Hydraulic Institute Engineering Data book. For any application involving high-viscosity liquids, the pump vendor should be given the liquid properties. The viscosity-corrected performance curves will be supplied by the vendor as part of the pump proposal. For preliminary analysis, you may also use these charts to generate the viscosity-corrected pump curves.

The Hydraulic Institute method of viscosity correction requires determining the best efficiency point (BEP) values for Q, H, and E from the water performance curve. This is called the 100% BEP point. Three additional sets of Q, H, and E values are obtained at 60%, 80%, and 120% of the BEP flow rate from the water performance curve. From these four sets of data, the Hydraulic Institute chart can be used to obtain the correction factors C_q, C_h, and C_e for flow, head, and efficiency for each set of data. These factors are used to multiply the Q, H, and E values from the water curve, thus generating corrected values of Q, H, and E for 60%, 80%, 100%, and 120% BEP values. Example Problem 11.4 illustrates the Hydraulic Institute method of viscosity correction. Note that for multistage pumps, the values of H must be per stage.

Example Problem 11.4

The water performance of a single-stage centrifugal pump for 60%, 80%, 100%, and 120% of the BEP is as shown below:

Q (gal/min)	450	600	750	900
H (ft)	114	108	100	86
E (%)	72.5	80.0	82.0	79.5

Calculate the viscosity-corrected pump performance when pumping oil with a specific gravity of 0.90 and a viscosity of 1000 SSU at pumping temperature.

Solution

By inspection, the BEP for this pump curve is

$$Q = 750 \quad H = 100 \quad \text{and} \quad E = 82$$

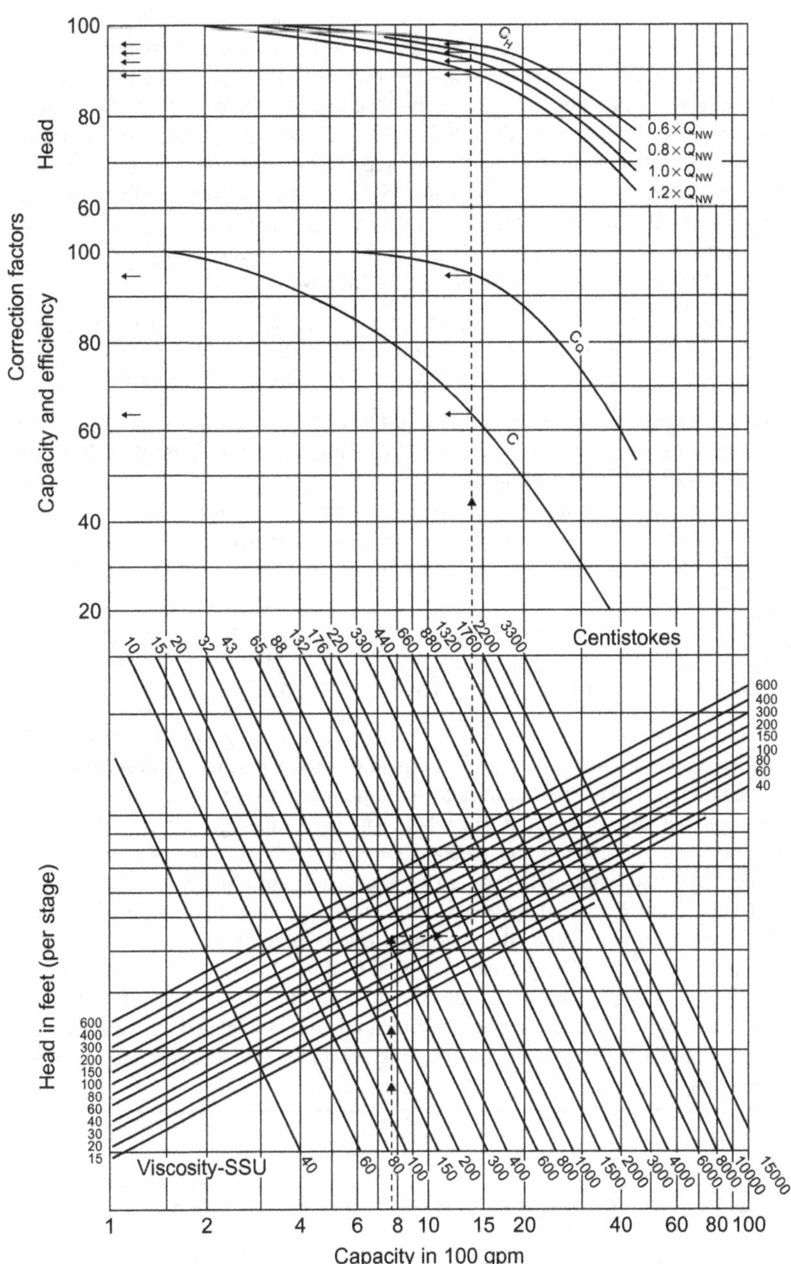

FIGURE 11.15 Viscosity correction chart. *Source: Courtesy of the Hydraulic Institute, Parsippany, NJ, www.Pumps.org.*

We first establish the four sets of capacities to correspond to 60%, 80%, 100%, and 120%. These have already been given as 450, 600, 750, and 900 gal/min. Because the head values are per stage, we can directly use the BEP value of head along with the corresponding capacity to enter the Hydraulic Institute Viscosity Correction chart at 750 gal/min on the lower horizontal scale. We go vertically from 750 gal/min to the intersection point on the line representing the 100 ft head curve, and then horizontally to intersect the 1000 SSU viscosity line, and finally, vertically up to intersect the three correction factor curves C_e, C_q, and C_h.

From the Hydraulic Institute chart, Fig. 11.15, we obtain the values of C_q, C_h, and C_e for flow rate, head, and efficiency as follows:

C_q	0.95	0.95	0.95	0.95
C_h	0.96	0.94	0.92	0.89
C_e	0.635	0.635	0.635	0.635

These correspond to Q values of 450 (60% of Q_{NW}), 600 (80% of Q_{NW}), 750 (100% of Q_{NW}), and 900 (120% of Q_{NW}). The term Q_{NW} is the BEP flow rate from the water performance curve.

Using these correction factors, we generate the Q, H, and E values for the viscosity-corrected curves by multiplying the water performance value of Q by C_q, H by C_h, and E by C_e and obtain the following result.

Q_V	427	570	712	855
H_V	109.5	101.5	92.0	76.5
E_V	46.0	50.8	52.1	50.5
BHP_V	23.1	25.9	28.6	29.4

The last row of values for viscous BHP was calculated using Eq. (9.17) in Chapter 9. The Hydraulic Institute chart used to obtain the correction factors consists of two separate charts. One chart applies to small pumps of up to 100 gal/min capacity and head per stage of 6–400 ft. The other chart applies to larger pumps with a capacity between 100 gal/min and 10,000 gal/min and a head range of 15–600 ft per stage. Note that when data is taken from a water performance curve, the head has to be corrected per stage because the Hydraulic Institute charts are based on head in ft per stage rather than the total pump head. Therefore, if a six-stage pump has a BEP at 2500 gal/min and 3000 ft of head with an efficiency of 85%, the head per stage to be used with the chart will be $3000/6 = 500$ ft. The total head (not per stage) from the water curve can then be multiplied by the correction factors from the Hydraulic Institute charts to obtain the viscosity-corrected head for the six-stage pump.

11.14 PUMP CONFIGURATION – SERIES AND PARALLEL

In the preceding section, we discussed the performance of a single pump. To transport liquid through a pipeline, we may need to use more than one pump at a pump station to provide the necessary flow rate or head requirement. These pumps may be operated in series or parallel configurations. Series pumps are generally used for higher heads and parallel pumps for increased flow rates. When pumps are operated in series, the same flow rate goes through each pump and the resultant head is the sum of the heads generated by each pump. In parallel operation, the flow rate is split between the pumps, while each pump produces the same head. Series and parallel pump configurations are illustrated in Fig. 11.16a,b.

The choice of series or parallel pumps for a particular application depends on many factors, including pipeline elevation profile, as well as operational flexibility. Figure 11.17 shows the combined performance of two identical pumps in series and parallel configurations. It can be seen that parallel pumps are used when we need larger flows. Series pumps are used when we need higher heads than each individual pump.

If the pipeline elevation profile is essentially flat, the pump pressure required is mainly to overcome the pipeline friction. However, if the pipeline has drastic elevation changes, the pump head generated is mainly for the static lift and to a lesser extent for pipe friction. In the latter case, if two pumps are used in series and one shuts down, the remaining pump alone will only be able to provide half the head and, therefore, will not be able to provide the necessary head for the static lift at any flow rate. If the pumps were configured in parallel, then shutting down one pump will still allow the other pump to provide the necessary head for the static lift at half the previous flow rate. Thus, parallel pumps are generally used when elevation differences are considerable. Series pumps are used where pipeline elevations are not significantly high.

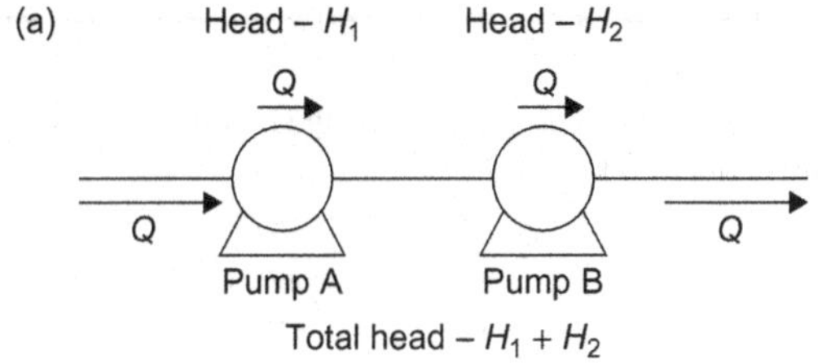

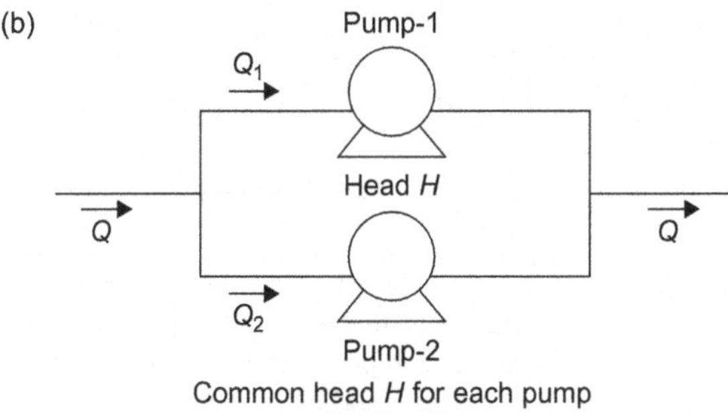

FIGURE 11.16 (a) Pumps in series. (b) Pumps in parallel.

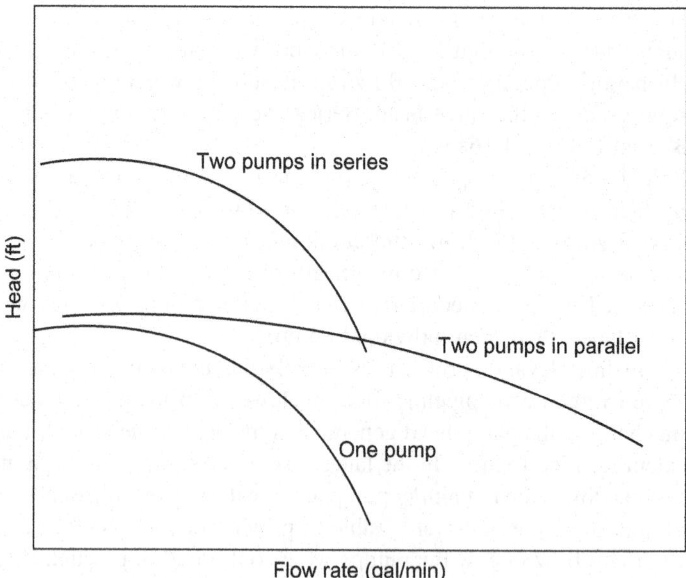

FIGURE 11.17 Pump performance in series and parallel.

Example Problem 11.5

One large pump and one small pump are operated in series. The H–Q characteristics of the pumps are defined as follows:

Pump-1					
Q (gal/min)	0	800	1600	2400	3000
H (ft)	2389	2325	2175	1763	1350
Pump-2					
Q (gal/min)	0	800	1600	2400	3000
H (ft)	796	775	725	588	450

1. Calculate the combined performance of pump-1 and pump-2 in series configuration.
2. Can these pumps be configured to operate in parallel?

Solution

1. Pumps in series have the same flow through each pump and the heads are additive. We can, therefore, generate the total head produced in series configuration by adding the head of each pump for each flow rate given as follows: Combined performance of pump-1 and pump-2 in series:

Q (gal/min)	0	800	1600	2400	3000
H (ft)	3185	3100	2900	2351	1800

2. To operate satisfactorily, in a parallel configuration, the two pumps must have a common range of heads so that at each common head, the corresponding flow rates can be added to determine the combined performance. Pump-1 and pump-2 are mismatched for parallel operation. Therefore, they cannot be operated in parallel.

11.15 PUMP HEAD CURVE VERSUS SYSTEM HEAD CURVE

In Chapter 9, we discussed the development of pipeline system head curves. In this chapter, we will see how the system curve and the pump head curve together determine the operating point (Q, H) on the pump curve.

The system head curve for the pipeline represents the pressure required to pump a liquid through the pipeline at various flow rates (increasing pressure with increasing flow rate) and the pump H–Q curve shows the pump head available at various flow rates. When the head requirements of the pipeline match the available pump head, we have a point of intersection of the system head curve with the pump head curve as shown in Fig. 11.18. This is the operating point for this pipeline and pump combination.

This figure shows system head curves for diesel and gasoline. The point of intersection of the pump head curves and the system head curve for diesel (point A) indicate the operating point for this pump when transporting diesel. Similarly, when pumping gasoline, the corresponding operating point (point B) is as shown in the Fig. 11.18. Therefore, with 100% diesel in the pipeline, the flow rate would be Q_A and the corresponding pump head would be H_A as

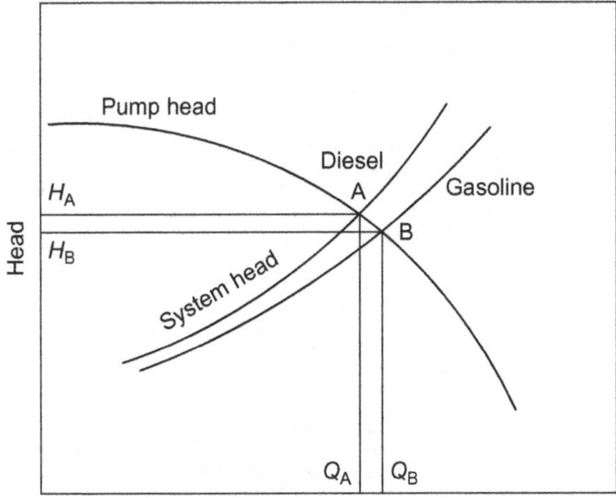

FIGURE 11.18 Pump curve – system head curve.

shown. Similarly, with 100% gasoline in the pipeline, the flow rate would be Q_B and the corresponding pump head would be H_B as shown in the Fig. 11.18.

11.16 MULTIPLE PUMPS VERSUS SYSTEM HEAD CURVE

Figure 11.19 shows the pipeline system head curve superimposed on the pump head curves to show the operating point with one pump, two pumps in series, and the same two pumps in parallel configurations. The operating points are shown as A, C, and B with flow rates of Q_A, Q_C, and Q_B, respectively.

In certain pipeline systems, depending on the flow requirements, we may be able to obtain higher throughput by switching from a series pump configuration to a parallel pump configuration. From Fig. 11.19, it can be seen that a steep system curve would be better with series pumps. If the system curve is relatively flat, parallel pumps operation will result in increased flow.

11.17 NPSH REQUIRED VERSUS NPSH AVAILABLE

The NPSH of a centrifugal pump is defined as the net positive suction head required at the pump impeller suction to prevent pump cavitation at any flow rate. Cavitation occurs when the suction pressure at the impeller reduces

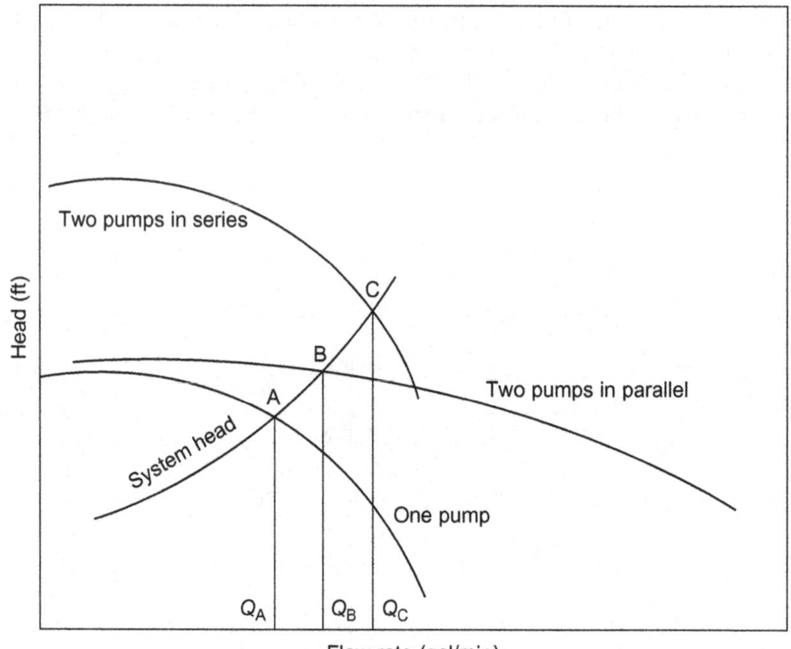

FIGURE 11.19 Multiple pumps and system head curve.

below the liquid vapor pressure. This will damage the pump impeller and render it useless. NPSH required for a pump at any flow rate is given by the pump vendor's NPSH curve, as in Fig. 11.14. For that pump curve, it can be seen that the NPSH required ranges from 38 ft at 1200 gal/min to 64 ft at 2600 gal/min. To obtain satisfactory performance with this pump, the actual *available* NPSH ($NPSH_A$) must be more than the *required* NPSH ($NPSH_R$). The $NPSH_A$ is calculated for a piping system by taking into account the positive tank head, including atmospheric pressure and subtracting the pressure drop resulting from friction in the suction piping and the liquid vapor pressure at the pumping temperature. The resulting value of NPSH for this piping configuration will represent the net pressure of the liquid at pump suction, above its vapor pressure. Shortly, we will review this calculation for a typical pump and piping system.

Consider a centrifugal pump with the suction and delivery tanks and interconnecting piping as shown in Fig. 11.20.

The vertical distance from the liquid level on the suction side of the pump center line is defined as the static suction head. More correctly, it is the static suction lift (H_S) when the center line of the pump is above that of the liquid supply level as in Fig. 11.20. If the liquid supply level were higher than the pump center line, it will be called the static suction head on the pump. Similarly, the vertical distance from the pump center line to the liquid level on the delivery side is the static discharge head (H_d) as shown in Fig. 11.20. The total static head on a pump is defined as the sum of the static suction head and the static discharge head. It represents the vertical distance between the liquid supply level and the liquid discharge level. The static suction head, static discharge head, and the total static head on a pump are all measured in feet of liquid in USCS or meters of liquid in the SI system.

The friction head, measured in feet of liquid, is the head loss resulting from friction in both suction and discharge piping. It represents the pressure required to overcome the frictional resistance of all piping, fittings, and valves on the suction and discharge side of the pump as shown in Fig. 11.20.

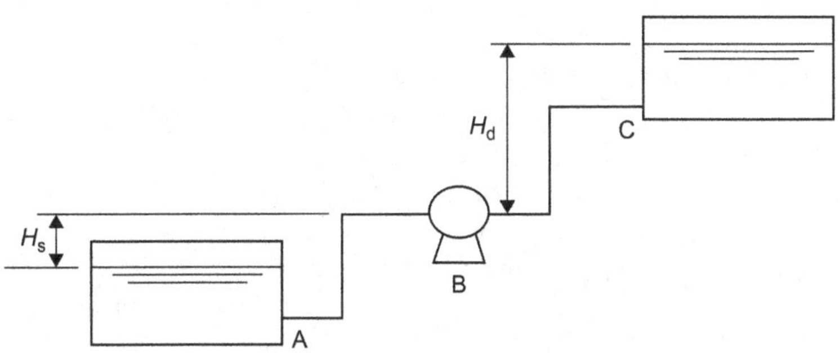

FIGURE 11.20 Suction and discharge heads for a centrifugal pump.

On the suction side of the pump, the available suction head H_S will be reduced by the friction loss in the suction piping. This net suction head on the pump will be the available suction head at the pump center line.

$$\text{Suction head} = H_S - H_{fs} \tag{11.14}$$

where H_{fs} is the friction loss in suction piping.

The NPSH available is calculated by adding the suction head to the atmospheric pressure on the liquid surface in the suction tank and subtracting the vapor pressure of the liquid at the flowing temperature, as follows:

$$\text{NPSH}_A = (P_a - P_v)(2.31/\text{Sg}) + H + E1 - E2 - h_f \tag{11.15}$$

where

P_a – Atmospheric pressure, psi
P_v – Liquid vapor pressure at flowing temperature, psi
Sg – Liquid specific gravity at flowing temperature
H – Tank head, ft
$E1$ – Elevation of tank bottom, ft
$E2$ – Elevation of pump suction, ft
h_f – Friction loss in suction piping, ft

Example Problem 11.6

A centrifugal pump is used to pump a liquid from a storage tank through 500 ft of NPS 16 suction piping as shown in Fig. 11.21. The head loss in the suction piping is estimated to be 12.5 ft (using equations in Chapter 8).

1. Calculate the NPSH available at a flow rate of 3500 gal/min.
2. The pump data indicate $\text{NPSH}_R = 24$ ft at 3500 gal/min and 52 ft at 4500 gal/min. Can this piping system handle the higher flow rate without the pump cavitating?
3. If cavitation is a problem in (2) above, what changes must be made to the piping system to prevent pump cavitation at 4500 gal/min?

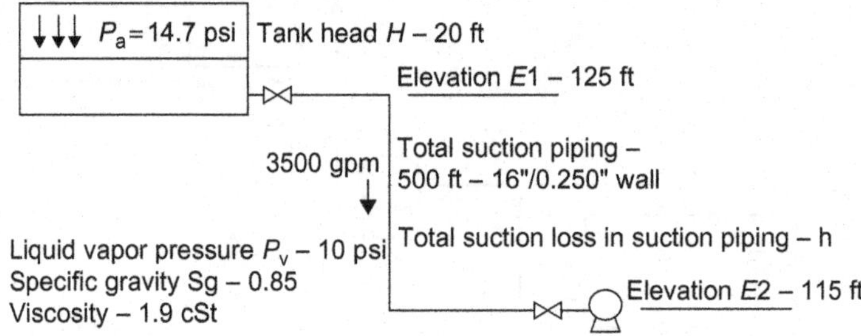

FIGURE 11.21 NPSH calculation.

Solution

1. NPSH available in ft of liquid head is from Eq. (11.15):

$$NPSH_A = (P_a - P_v)(2.31/Sg) + H + E1 - E2 - h_f$$

Substituting given values, we get

$$NPSH_A = (14.7 - 10) \times 2.31/0.85 + 20 + 125 - 115 - 12.5 = 30.27\ \text{ft}$$

2. Since $NPSH_R = 24$ ft at 3500 gal/min and $NPSH_A > NPSH_R$ the pump will not cavitate at this flow rate. Next, we will check for the higher flow rate.
At 4500 gal/min flow rate, the head loss needs to be estimated.
Using Chapter 8 concepts, at 4500 gal/min

$$h_f = (4500/3500)^2 \times 12.5 = 20.7\ \text{ft}$$

Recalculating $NPSH_A$ at the higher flow rate, we get

$$NPSH_A = (14.7 - 10) \times 2.31/0.85 + 20 + 125 - 115 - 20.7 = 22.07\ \text{ft}$$

Because the pump data indicates $NPSH_R = 52$ ft at 4500 gal/min, the pump does not have adequate NPSH ($NPSH_A < NPSH_R$) and therefore will cavitate at the higher flow rate.

3. The extra head required to prevent cavitation = 52 − 22.07 = 29.9 ft

One solution is to locate the pump suction at an additional 30 ft or more below the tank. This may not be practical. Another solution is to provide a small vertical can type pump that can serve as a booster between the tank and the main pump. This booster pump will provide the additional head required to prevent cavitation.

11.18 PUMP STATION CONFIGURATION

A typical piping layout within a pump station in a simplified form is shown in Fig. 11.22. In addition to the components shown, a pump station will contain other ancillary equipment such as strainers, flow meters, and so on.

The pipeline enters the station boundary at point A, where the station block valve MOV-101 is located. The pipeline leaves the station boundary on the discharge side of the pump station at point B, where the station block valve MOV-102 is located.

Station bypass valves designated as MOV-103 and MOV-104 are used for bypassing the pump station in the event of pump station maintenance or other reasons where the pump station must be isolated from the pipeline. Along the main pipeline there is located a check valve, CKV-101, that prevents reverse flow through the pipeline. This typical station layout shows two pumps configured in series. Each pump pumps the same flow rate and the total pressure generated is the sum of the pressures developed by each pump. On the suction side of the pump station, the pressure is designated as P_s while the discharge pressure on the pipeline side is designated as P_d. With constant speed motor driven pumps, there

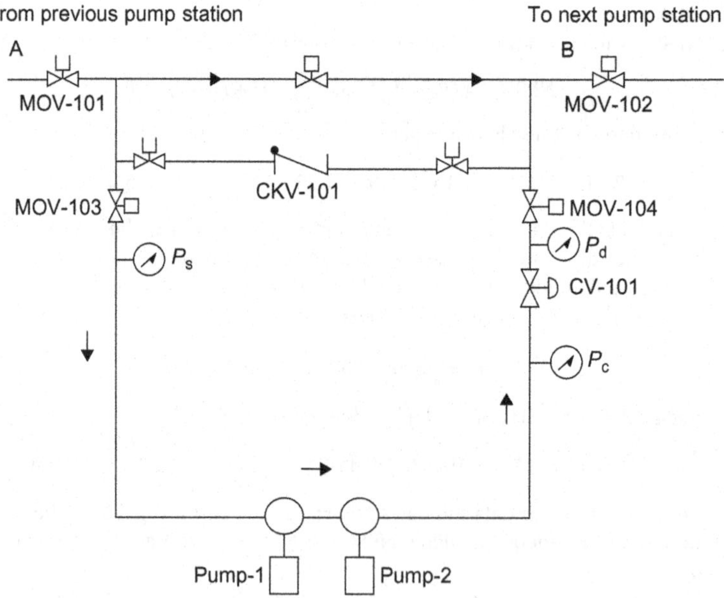

FIGURE 11.22 Typical pump station layout.

is always a control valve on the discharge side of the pump station shown as CV-101 in Fig. 11.22. This control valve controls the pressure to the required value P_d by creating a pressure drop across it between the pump discharge pressure P_c and the station discharge pressure P_d. Because the pressure within the case of the second pump represents the sum of the suction pressure and the total pressure generated by both pumps, it is referred to as the case pressure P_c.

If the pump is driven by a variable speed drive (VSD) motor or an engine, the control valve is not needed as the pump may be slowed down or speeded up as required to generate the exact pressure P_d. In such a situation, the case pressure will equal the station discharge pressure P_d.

In addition to the valves shown, there will be additional valves on the suction and discharge of the pumps. Also, not shown in the figure is a check valve located immediately after the pump discharge that prevents reverse flow through the pumps.

11.19 CONTROL PRESSURE AND THROTTLE PRESSURE

Mathematically, if ΔP_1 and ΔP_2 represent the differential head produced by pump-1 and pump-2 in series, we can write

$$P_c = P_s + \Delta P_1 + \Delta P_2 \qquad (11.16)$$

where P_c is the case pressure in pump-2 or upstream pressure at the control valve.

The pressure throttled across the control valve is defined as

$$P_{\text{thr}} = P_{\text{c}} - P_{\text{d}} \tag{11.17}$$

where

P_{thr} – Control valve throttle pressure
P_{d} – Pump station discharge pressure

The throttle pressure represents the mismatch that exists between the pump and the system pressure requirements at a particular flow rate. P_{d} is the pressure at the pump station discharge needed to transport liquid to the next pump station or delivery terminus. The case pressure P_{c} is the available pressure due to the pumps. If the pumps were driven by variable speed motors, P_{c} would exactly match with P_{d} and there would be no throttle pressure. In this case, no control valve is needed. The case pressure is also referred to as control pressure because it is the pressure upstream of the control valve. The throttle pressure represents unused pressure developed by the pump and hence results in wasted HP and dollars. The objective should be to reduce the amount of throttle pressure in any pumping situation. With VSD pumps, there is no HP wasted because the pump case pressure exactly matches with the station discharge pressure.

11.20 VARIABLE SPEED PUMPS

If there are two or more pumps in series configuration, one of the pumps may be driven by a variable speed drive (VSD) motor or an engine. With parallel pumps, all pumps will have to be VSD pumps because parallel configurations require matching heads at the same flow rate. In the case of two pumps in series, we could convert one of the two pumps to be driven by a variable speed motor. This pump can then slow down to the required speed that would develop just the right amount of head, which when added to the head developed by the constant speed pump would provide exactly the total head required to match the pipeline system requirement.

11.21 VSD PUMP VERSUS CONTROL VALVE

In a single pump station pipeline with one pump, a control valve is used to regulate the pressure for a given flow rate.

The pipeline from Essex pump station to the Kent delivery terminal is 120 mi long, NPS 16, 0.250 in. wall thickness pipe with an MAOP of 1440 psi, as shown in Fig. 11.23. The pipeline is designed to operate at 1400 psi discharge pressure, pumping 4000 bbl/h of liquid (specific gravity: 0.89 and viscosity: 30 cSt at 60°F) on a continuous basis. The delivery pressure required at Kent is 50 psi. The pump suction pressure at Essex is 50 psi,

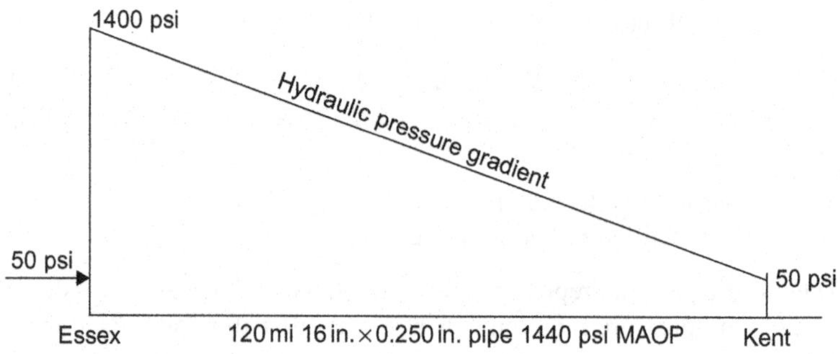

FIGURE 11.23 Essex to Kent pipeline.

and the pump differential pressure is 1400 − 50 = 1350 psi. Let's assume the single pump at Essex has the following design point:

$$Q = 2800\,\text{gal/min} \quad H = 3800\,\text{ft}$$

Converting the head of 3800 ft into psi, the pump pressure developed is

$$3800 \times 0.89/2.31 = 1464\,\text{psi}$$

We can see that this pressure combined with the 50-psi pump-suction pressure would produce a pump discharge pressure (and hence a case pressure) of 1464 + 50 = 1514 psi

Since the MAOP is 1440 psi, this will overpressure the pipeline by 74 psi. The control valve located just downstream of the pump discharge will be used to reduce the discharge pressure to the required pressure of 1400 psi. This is shown by the modified system curve (2) in Fig. 11.24.

The system head curve (1) in Fig. 11.24 represents the pressure versus flow rate variation for our pipeline from Essex to Kent. At $Q = 2800$ gal/min (4000 bbl/h) flow rate, point C on the pipeline system head curve shows the operating point that requires a pipeline discharge pressure of 1400 psi at Essex. We have superimposed the pump H–Q curve and pressures are in ft of liquid head, so the pressure at C is

$$1400 \times 2.31/0.89 = 3634\,\text{ft}$$

Because the pump suction pressure is 50 psi, the vertical axis includes this suction head, and the pump head curve in Fig. 11.24 includes this suction head as well. The pump H–Q curve shows that at a flow rate of 2800 gal/min, the differential head generated is 3800 ft. Adding the 50 psi suction pressure point B on the H–Q curve corresponds to

$$3800 + 50 \times 2.31/0.89 = 3930\,\text{ft}$$

Therefore, the point of intersection of the pump H–Q curve and the pipeline system curve is the operating point A, which is at a higher flow rate than

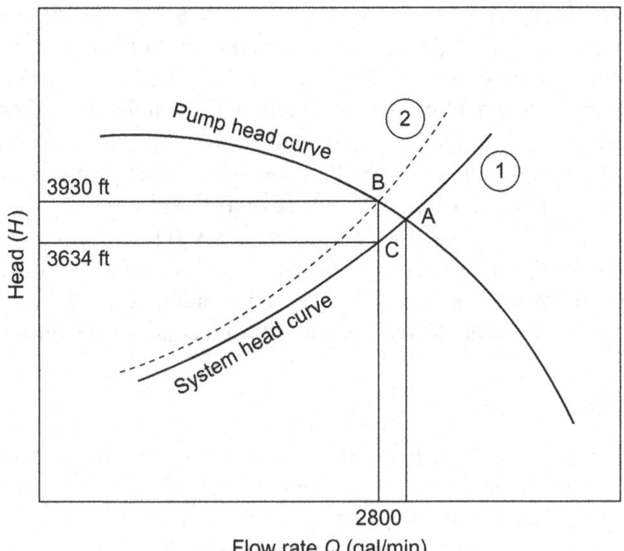

FIGURE 11.24 System curve and control valve.

2800 gal/min. Also, the pump discharge pressure at A will be higher than that at C. To limit this pressure within the MAOP, some form of pressure control is required. Thus, utilizing the control valve on the pump discharge will move the operating point from A to B on the pump curve corresponding to 2800 gal/min and a higher head of 3930 ft as calculated above. The control valve pressure drop is the head difference BC as follows:

$$\text{Control valve pressure drop, BC} = 3930 - 3634 = 296\ \text{ft}$$

This pressure drop across the control valve, also called the pump throttle pressure, is

$$\text{Throttle pressure} = 296 \times 0.89/2.31 = 114\ \text{psi}$$

Therefore, due to the slightly oversized pump and the pipe MAOP limit, we use a control valve on the pump discharge to limit the pipeline discharge pressure to 1400 psi at the required flow rate of 4000 bbl/h (2800 gal/min).

A dashed hypothetical system head curve designated as (2), passing through the point B on the pump head curve is the artificial system head curve due to the restriction imposed by the control valve.

The preceding analysis applies to a pump driven by a constant speed electric motor. If we had a VSD pump that can vary the pump speed from 60% to 100% rated speed and if the rated speed were 3560 RPM, then the pump speed range will be 3560 × 0.60 = 2136 to 3560 RPM.

Since the pump speed can be varied from 2136 to 3560 RPM, the pump head curve will correspondingly vary according to the centrifugal pump affinity laws. Therefore, we can find some speed (less than 3560 RPM) at which the pump will generate the required head corresponding to point C in Fig. 11.24.

From the preceding discussion, we can conclude that use of VSD pumps can provide the right amount of pressure required for a given flow rate, thus avoiding pump throttle pressures (and hence wasted HP) common with constant speed motor-driven pumps with control valves. However, VSD pumps are expensive to install and operate compared with the use of a control valve. A typical control valve installation may cost $150,000, whereas a VSD may require $300,000–$500,000 incremental cost compared with the constant speed motor-driven pump. We will have to factor in the increased operating cost of the VSD pump compared with the dollars lost in wasted HP from control valve throttling.

SUMMARY

In this chapter, we discussed pump stations and pumping configurations used in liquid pipelines. The optimum locations of pump stations for hydraulic balance were analyzed. Centrifugal pumps and positive displacement pumps were compared along with their performance characteristics. We explained the affinity laws for centrifugal pumps, the importance of NPSH, and how to calculate power requirements when pumping different liquids. Viscosity-corrected pump performance using the Hydraulic Institute chart was explained. The performance of two or more pumps in series or parallel configuration and the operating point for a pump in conjunction with the pipeline system head curve were examined. We discussed some of the major components in a pump station, such as pumps and drivers and control valves. The use of variable speed pumps to save pumping power under different operating conditions was reviewed.

BIBLIOGRAPHY

A. Benaroya, Fundamentals and Application of Centrifugal Pumps, The Petroleum Publishing Company, Tulsa, OK, 1978.

Flow of Fluids through Valves, Fittings and Pipes, Crane Company, New York, 1976.

G.M. Jones, R.L. Sanks, G. Tchobanoglous, B. Bossermann, Pumping Station Design, Revised third ed., Elsevier, Inc., 2008.

E. Shashi Menon, Liquid Pipeline Hydraulics, Marcel-Dekker, New York, 2004.

E. Shashi Menon, Gas Pipeline Hydraulics, CRC Press, Boca Raton, FL, 2005.

Hydraulic Institute Engineering Data Book, Hydraulic Institute, 1979.

I.J. Karassik, et al., Pump Handbook, McGraw-Hill, New York, NY, 1976.

M. Volk, Pump Characteristics and Applications, second ed., CRC Press, Taylor and Francis Group, 2005.

E. Shashi Menon, Piping Calculations Manual, McGraw-Hill, New York, 2005.

E. Shashi Menon, P.S. Menon, Working Guide to Pumps and Pumping Stations, Gulf Professional Publishing, Burlington, MA, 2010.

V.S. Lobanoff, R.R. Ross, Centrifugal Pumps Design & Application, Gulf Publishing, Burlington, MA, 1985.

Chapter 12

Compressor Stations

E. Shashi Menon, Ph.D., P.E.

Chapter Outline

INTRODUCTION

In this chapter, we will discuss the number and size of compressor stations required to transport gas in a pipeline. The optimum locations and pressures at which compressor stations operate will be determined based on the pipeline flow rate, allowable pipe operating pressures, and pipeline topography. Centrifugal and positive displacement compressors used in natural gas transportation will be compared with reference to their performance characteristics and cost. Typical compressor station design and equipment used will be discussed. Isothermal, adiabatic, and polytropic compression processes and the compressor power required will be discussed. The discharge temperature of compressed gas and its impact on pipeline throughput and gas cooling will also be reviewed.

12.1 COMPRESSOR STATION LOCATIONS

Compressor stations are installed on long distance gas pipelines to provide the pressure needed to transport gas from one location to another. Due to the maximum allowable operating pressure (MAOP) limitations for pipelines, more than a single compressor station may be needed on a long pipeline. The locations and pressures at which these compressor stations operate are determined by the MAOP, compression ratio, power available, and environmental and geotechnical factors.

Consider a pipeline that is designed to transport 200 million standard ft^3/day (MMSCFD) of natural gas from Dover to an industrial plant, which is 70 mi away, at Grimsby. According to methods outlined in Chapter 9, we would calculate the pressure required at Dover to ensure delivery of the gas at a minimum contract pressure of 500 psig at Grimsby. Suppose the MAOP of the pipeline is 1200 psig and we calculated the required pressure at Dover to be 1130 psig. It is clear that there is no violation of MAOP and hence a single compressor station at Dover would suffice to deliver gas to Grimsby at the required 500 psig delivery pressure. If the pipeline length had been 150 mi instead, calculations would show that in order to deliver the same quantity of gas to Grimsby at the same 500 psig terminus pressure, the pressure required at Dover would have to be 1580 psig. Obviously, because this is greater than the MAOP, we would need more than the origin compressor station at Dover.

Having determined that we need two compressor stations, we need to locate the second compressor station between Dover and Grimsby. Where would this compressor station be installed? A logical site would be the midpoint between Dover and Grimsby, if the hydraulic characteristics (pipe size and ground profile) are uniform throughout the pipeline.

For simplicity, let us assume the pipeline elevation profile is fairly flat and therefore, elevation differences can be ignored. Based on this, we will locate the intermediate compressor station at the midpoint, Kent, as shown in Fig. 12.1. Next, we will determine the pressures at the two compressor stations.

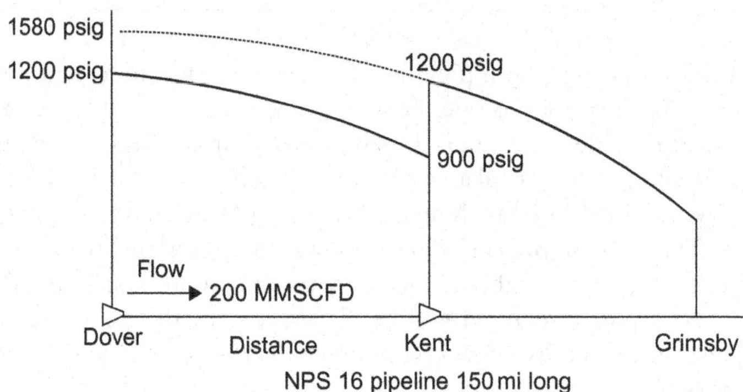

FIGURE 12.1 Gas pipeline with two compressor stations.

Because the MAOP is limited to 1200 psig, assume that the compressor at Dover discharges at this pressure. Because of friction, the gas pressure drops from Dover to Kent as indicated in Fig. 12.1. Suppose the gas pressure reaches 900 psig at Kent and is then boosted back to 1200 psig by the compressor at Kent. Therefore, the compressor station at Kent has a suction pressure of 900 psig and a discharge pressure of 1200 psig. The gas continues to move from Kent to Grimsby starting at 1200 psig at Kent. As the gas reaches Grimsby, the pressure may or may not be equal to the desired terminus delivery pressure of 500 psig. Therefore, if the desired pressure at Grimsby is to be maintained, the discharge pressure of the Kent compressor station may have to be adjusted. Alternatively, Kent could discharge at the same 1200 psig but its location along the pipeline may have to be adjusted. We assumed the 900 psig suction pressure at the Kent compressor station quite arbitrarily. It could well have been 700 or 1000 psig. The actual value depends on the so-called compression ratio desired. The compression ratio is simply the ratio of the compressor discharge pressure to its suction pressure, both pressures being expressed in absolute units.

$$\text{Compression ratio } r = \frac{P_\text{d}}{P_\text{s}} \tag{12.1}$$

where the compressor suction and discharge pressures P_s and P_d are in absolute units.

In the present case, using the assumed values, the compression ratio for Kent is

$$r = \frac{1200 + 14.7}{900 + 14.7} = 1.33$$

In the above calculation, we assumed the base pressure to be 14.7 psia. If instead of the 900 psig, we had chosen a suction pressure of 700 psig, the compression ratio would be

$$r = \frac{1200 + 14.7}{700 + 14.7} = 1.7$$

A typical compression ratio for centrifugal compressors is around 1.5. A larger value of the compression ratio will mean more compressor power, whereas a smaller compression ratio means less power required. In gas pipelines, it is desirable to keep the average pipeline pressure as high as possible to reduce the total compression power. Therefore, if the suction pressure at Kent is allowed to reduce to 700 psig or lower, the average pressure in the pipeline would be lower than if we had used 900 psig. Obviously, there is a trade-off between the number of compressor stations, the suction pressure, and compression power required. We will discuss this in more detail later in this chapter.

Going back to the example problem above, we concluded that we may have to adjust the location of the Kent compressor station or adjust its discharge pressure to ensure the 500 psig delivery pressure at Grimsby. Alternatively, we could leave the intermediate compressor station at the halfway point discharging at

1200 psig, and eventually deliver gas to Grimsby at some pressure such as 600 psig. If calculations showed that discharging out of Kent results in 600 psig at Grimsby, we have satisfied the minimum contract delivery pressure (500 psig) requirement at Grimsby. However, there is extra energy associated with the extra 100 psig delivery pressure. If the power plant can use this extra energy then there is no waste. However, if the power plant requirement is 500 psig maximum, then a pressure regulator must be installed at the delivery point. The extra 100 psig would be reduced through the pressure regulator or pressure control valve at Grimsby and some energy will be wasted. Another option would be to keep the Kent compressor at the midpoint but reduce its discharge pressure such that it will result in the requisite 500 psig at Grimsby. Because pressure drop in gas pipelines is nonlinear, remembering our discussion in Chapter 8, Kent discharge pressure may have to be reduced by less than 100 psig to provide the fixed 500 psig delivery pressure at Grimsby. This would mean that Dover will operate at 1200 psig discharge while Kent would discharge at 1150 psig. This will, of course, not ensure maximum average pressure in the pipeline. However, this is still a solution and in order to choose the best option, we must compare two or more alternative approaches, taking into account the total compressor power required as well as the cost involved. By moving the Kent compressor station slightly upstream or downstream, there will be changes in the suction and discharge pressures and hence the compressor power required. The change in the capital cost may not be significant. However, the compressor power variation will result in change in energy cost and therefore in annual operating cost. We must, therefore, take into account the capital cost and annual operating cost in order to come up with the optimum solution. An example will illustrate this method.

Example Problem 12.1 (USCS)

A natural gas pipeline 140 mi long from Danby to Leeds is constructed of NPS 16, 0.250 in. wall thickness pipe with an MAOP of 1400 psig. The gas specific gravity and viscosity are 0.6 and 8×10^{-6} lb/ft · s, respectively. The pipe roughness may be assumed to be 700 μ in., and the base pressure and base temperature are 14.7 psia and 60°F, respectively. The gas flow rate is 175 MMSCFD at 80°F and the delivery pressure required at Leeds is 800 psig. Determine the number and locations of compressor stations required, neglecting elevation difference along the pipeline. Assume $Z = 0.85$.

Solution

We will use Colebrook–White equation to calculate the pressure drop.

The Reynolds number is calculated from Eq. (8.47) as follows:

$$R = \frac{0.0004778 \times 175 \times 10^6 \times 0.6 \times 14.7}{15.5 \times 8 \times 10^{-6} \times 520} = 11{,}437{,}412$$

$$\text{Relative roughness} = e/D = \frac{700 \times 10^{-6}}{15.5} = 4.5161 \times 10^{-5}$$

Using the Colebrook–White Eq. (8.51), we get the friction factor as

$$\frac{1}{\sqrt{f}} = -2\,\text{Log}_{10}\left[\frac{4.516\times 10^{-5}}{3.7} + \frac{2.51}{11{,}437{,}412\sqrt{f}}\right]$$

Solving for f by successive iteration, we get

$$f = 0.0107$$

Using the general flow Eq. (8.55), neglecting elevation effects, we calculate the pressure required at Danby as follows:

$$175\times 10^6 = 77.54\left(\frac{1}{\sqrt{0.0107}}\right)\left(\frac{520}{14.7}\right)\left(\frac{P_1^2 - 814.7^2}{0.6\times 540\times 140\times 0.85}\right)^{0.5}\times (15.5)^{2.5}$$

Solving for the pressure at Danby, we get

$$P_1 = 1594\,\text{psia} = 1594 - 14.7 = 1579.3\,\text{psig}$$

It can be seen from Fig. 12.2 that because the MAOP is 1400 psig, we cannot discharge at 1579.3 psig at Danby.

We will need to reduce the discharge pressure at Danby to 1400 psig and install an intermediate compressor station between Danby and Leeds as shown in Fig. 12.3.

Initially, let's assume that the intermediate compressor station will be located at Hampton, halfway between Danby and Leeds. For the pipe segment from Danby to Hampton, we calculate the suction pressure at the Hampton compressor station as follows:

Using the general flow Eq. (8.55),

$$175\times 10^6 = 77.54\left(\frac{1}{\sqrt{0.0107}}\right)\left(\frac{520}{14.7}\right)\left(\frac{1414.7^2 - P_2^2}{0.6\times 540\times 70\times 0.85}\right)^{0.5}\times (15.5)^{2.5}$$

Solving for the pressure at Hampton (suction pressure)

$$P_2 = 1030.95\,\text{psia} = 1016.25\,\text{psig}$$

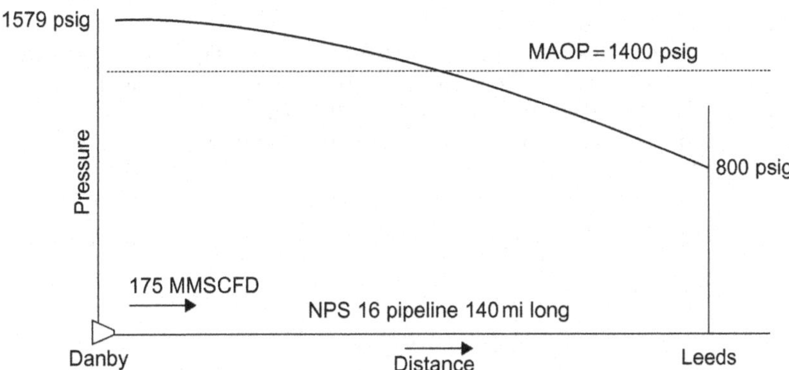

FIGURE 12.2 Pipeline with 1400 psig MAOP.

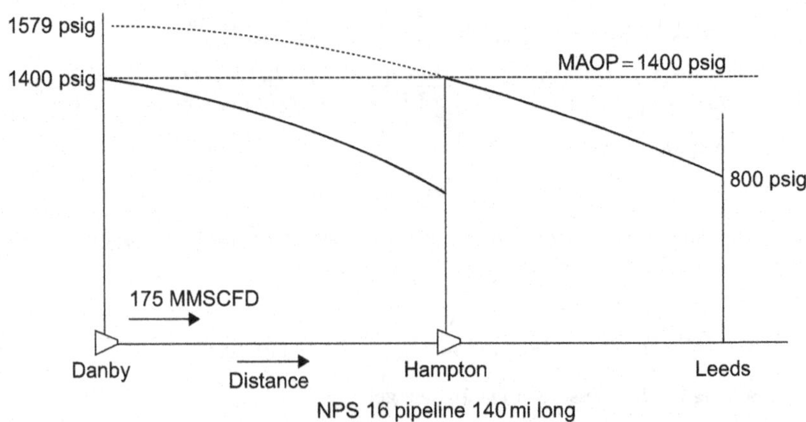

FIGURE 12.3 Danby to Leeds with intermediate compressor stations.

At Hampton, if we boost the gas pressure from 1016.25 to 1400 psig (MAOP), then the compression ratio at Hampton is $\frac{1414.7}{1030.95} = 1.37$

This is a reasonable compression ratio for a centrifugal compressor. Next, we will calculate the delivery pressure at Leeds, starting with the 1400 psig discharge pressure at Hampton.

For the 70-mi pipe segment from Hampton to Leeds, using the general flow equation, we get

$$175 \times 10^6 = 77.54\left(\frac{1}{\sqrt{0.0107}}\right)\left(\frac{520}{14.7}\right)\left(\frac{1414.7^2 - P_2^2}{0.6 \times 540 \times 70 \times 0.85}\right)^{0.5} \times (15.5)^{2.5}$$

Solving for P_2, the pressure at Leeds, we get

$$P_2 = 1030.95\,\text{psia} = 1016.25\,\text{psig}$$

This is exactly the suction pressure at Hampton we calculated earlier. This is because, hydraulically, Hampton is at the midpoint of the 140-mi pipeline discharging at the same pressure as Danby.

The calculated pressure at Leeds is higher than the 800 psig desired. Hence, we must move the location of Hampton compressor station slightly toward Danby so that the calculated delivery at Leeds will be 800 psig. We will calculate the distance L required between Hampton and Leeds. To achieve this, we use the general flow Eq. (8.55) as follows:

$$175 \times 10^6 = 77.54\left(\frac{1}{\sqrt{0.0107}}\right)\left(\frac{520}{14.7}\right)\left(\frac{1414.7^2 - 814.7^2}{0.6 \times 540 \times L \times 0.85}\right)^{0.5} \times (15.5)^{2.5}$$

Solving for length L, we get

$$L = 99.77\,\text{mi}$$

Therefore, the Hampton compressor station must be located at approximately 99.8 mi from Leeds, or 40.2 mi from Danby. Relocating Hampton from the

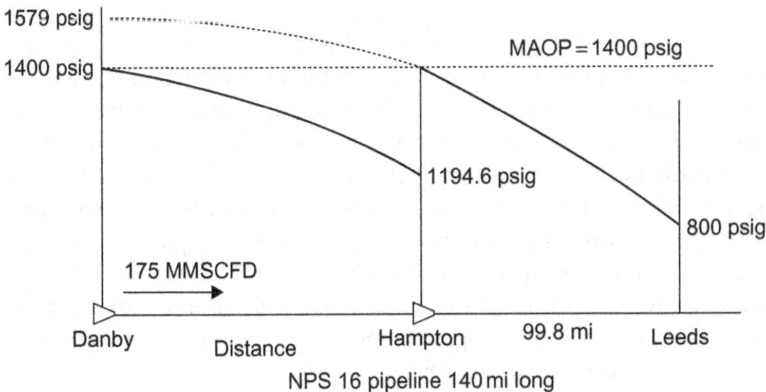

FIGURE 12.4 Danby to Leeds with relocated Hampton compressor station.

midpoint (70 mi) results in a higher suction pressure at Hampton, and hence a different compression ratio. This can be calculated as follows:

Using the general flow Eq. (8.55) for the 40.2-mi pipe segment between Danby and Hampton, we get

$$175 \times 10^6 = 77.54\left(\frac{1}{\sqrt{0.0107}}\right)\left(\frac{520}{14.7}\right)\left(\frac{1414.7^2 - P_2^2}{0.6 \times 540 \times 40.2 \times 0.85}\right)^{0.5} \times (15.5)^{2.5}$$

Solving for P_2, we get

$$P_2 = 1209.3 \text{ psia} = 1194.6 \text{ psig}$$

Therefore, the suction pressure at Hampton = 1194.6 psig

$$\text{The compression ratio at Hampton} = \frac{1414.7}{1209.3} = 1.17$$

Figure 12.4 shows the revised location of the Hampton compressor station.

12.2 HYDRAULIC BALANCE

In the preceding discussions, we considered each compressor station operating at the same discharge pressure and also considered the same compression ratio. However, in order to provide the desired delivery pressure at the terminus, we had to relocate the intermediate compressor station, thus changing its compression ratio. From the definition of compression ratio, Eq. (12.1), we can say that each compressor station operates at the same suction and discharge pressures if doing so results in adequate delivery pressure at the terminus. If there are no intermediate gas flows in or out (injection or delivery) of the pipeline, other than at the beginning and end, as in Example 12.1, each compressor station is required to compress the same amount of gas. Therefore, with pressures and flow rates being the same, each compressor station will require the same amount of power. This is known as hydraulic balance. In a long pipeline with multiple compressor stations, if each compressor station adds the same amount of energy to the gas, this is a hydraulically balanced pipeline.

One of the advantages of a hydraulically balanced pipeline is that all compression equipment may be identical and result in minimum inventory of spare parts, thus reducing the maintenance cost. Also in order to pump the same gas volume through a pipeline, hydraulically balanced compressor stations will require less total power than if the stations were not in hydraulic balance.

Next, we will review the different gas compression processes, such as isothermal, adiabatic, and polytropic compression. This will be followed by an estimation of the compressor power required.

12.3 ISOTHERMAL COMPRESSION

Isothermal compression occurs when the gas pressure and volume vary such that the temperature remains constant. Isothermal compression requires the least amount of work compared with other forms of compression. This process is of theoretical interest because, in practice, maintaining the temperature constant in a gas compressor is not practical.

Figure 12.5 shows the pressure versus volume variation for isothermal compression. Point 1 represents the inlet conditions of pressure P_1, volume V_1, and at temperature T_1. Point 2 represents the final compressed conditions of pressure P_2, volume V_2, and at constant temperature T_1.

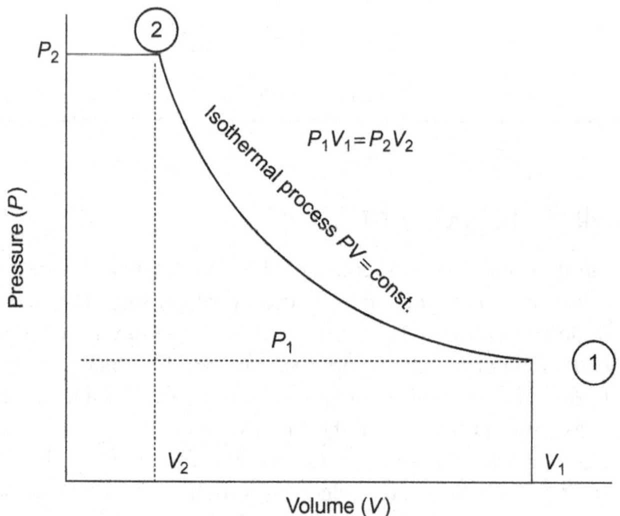

FIGURE 12.5 Isothermal compression.

The relationship between pressure P and volume V in an isothermal process is as follows:

$$PV = C \tag{12.2}$$

where C is a constant. Recall from Chapter 1 that this is basically Boyle's law. Applying subscripts 1 and 2, we can state that

$$P_1 V_1 = P_2 V_2 \tag{12.3}$$

The work done in compressing 1 lb of natural gas isothermally is given by

$$W_i = \frac{53.28}{G} T_1 \,\mathrm{Log_e}\left(\frac{P_2}{P_1}\right) \text{ (USCS units)} \tag{12.4}$$

where

W_i – Isothermal work done, ft · lb/lb of gas
G – Gas gravity, dimensionless
T_1 – Suction temperature of gas, °R
P_1 – Suction pressure of gas, psia
P_2 – Discharge pressure of gas, psia
$\mathrm{Log_e}$ – Natural logarithm to base e (e = 2.718)

The ratio $\left(\frac{P_2}{P_1}\right)$ is also called the compression ratio.

In SI units, the work done in isothermal compression of one kg of gas is

$$W_i = \frac{286.76}{G} T_1 \,\mathrm{Log_e}\left(\frac{P_2}{P_1}\right) \text{ (SI units)} \tag{12.5}$$

where

W_i – Isothermal work done, J/kg of gas
T_1 – Suction temperature of gas, K
P_1 – Suction pressure of gas, kPa absolute
P_2 – Discharge pressure of gas, kPa absolute

Other symbols are as defined earlier.

Example Problem 12.2 (USCS)

Natural gas is compressed isothermally at 80°F from an initial pressure of 800 psig to a pressure of 1000 psig. The gas gravity is 0.6. Calculate the work done in compressing 4 lb of gas. Use 14.7 psia and 60°F for the base pressure and temperature, respectively.

Solution

Using Eq. (12.4), the work done per lb of gas is

$$W_i = \frac{53.28}{0.6}(80+460)\,\mathrm{Log_e}\left(\frac{1000+14.7}{800+14.7}\right) = 10{,}527 \text{ ft} \cdot \text{lb/lb}$$

The total work done in compressing 4 lb of gas is

$$W_T = 10{,}527 \times 4 = 42{,}108\ \text{ft} \cdot \text{lb}$$

12.4 ADIABATIC COMPRESSION

Adiabatic compression occurs when there is no heat transfer between the gas and the surroundings. The term adiabatic and isentropic are used synonymously, although isentropic really means constant entropy. Actually, an adiabatic process, without friction, is called an isentropic process. In an adiabatic compression process, the gas pressure and volume are related as follows:

$$PV^{\gamma} = \text{Const} \tag{12.6}$$

where

γ – Ratio of specific heats of gas, $\frac{C_p}{C_v}$

C_p – Specific heat of gas at constant pressure
C_v – Specific heat of gas at constant volume
Const – A constant

γ is sometimes called the adiabatic or isentropic exponent for the gas and has a value ranging between 1.2 and 1.4.

Therefore, considering subscript 1 and 2 as the beginning and ending conditions, respectively, of the adiabatic compression process, we can write

$$P_1 V_1^{\gamma} = P_2 V_2^{\gamma} \tag{12.7}$$

Figure 12.6 contains an adiabatic compression diagram showing the variation of pressure versus volume of the gas.

The work done in compressing 1 lb of natural gas adiabatically can be calculated as follows:

$$W_a = \frac{53.28}{G} T_1 \left(\frac{\gamma}{\gamma - 1}\right) \left[\left(\frac{P_2}{P_1}\right)^{\frac{\gamma-1}{\gamma}} - 1\right] \text{(USCS units)} \tag{12.8}$$

where

W_a – Adiabatic work done, ft · lb/lb of gas
G – Gas gravity, dimensionless
T_1 – Suction temperature of gas, °R
γ – Ratio of specific heats of gas, dimensionless
P_1 – Suction pressure of gas, psia
P_2 – Discharge pressure of gas, psia

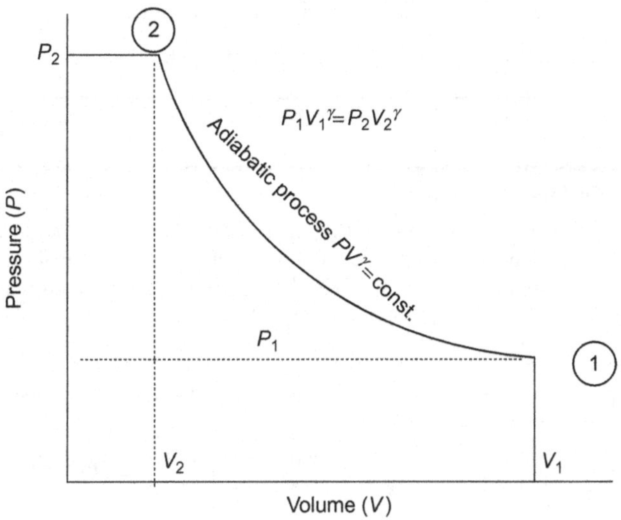

FIGURE 12.6 Adiabatic compression.

In SI units, the work done in adiabatic compression of 1 kg of gas is

$$W_a = \frac{286.76}{G} T_1 \left(\frac{\gamma}{\gamma - 1}\right) \left[\left(\frac{P_2}{P_1}\right)^{\frac{\gamma-1}{\gamma}} - 1 \right] \text{(SI units)} \tag{12.9}$$

where

W_a – Adiabatic work done, J/kg of gas
T_1 – Suction temperature of gas, K
P_1 – Suction pressure of gas, kPa absolute
P_2 – Discharge pressure of gas, kPa absolute

Other symbols are as defined earlier.

Example Problem 12.3 (USCS)

Natural gas is compressed adiabatically from an initial temperature and pressure of 60°F and 500 psig, respectively, to a final pressure of 1000 psig. The gas gravity is 0.6 and the ratio of specific heat is 1.3. Calculate the work done in compressing 5 lb of gas. Use 14.7 psia and 60°F for the base pressure and temperature, respectively.

Solution

Using Eq. (12.8), the work done in adiabatic compression is

$$W_a = \frac{53.28}{0.6}(60 + 460)\left(\frac{1.3}{0.3}\right)\left[\left(\frac{1014.7}{514.7}\right)^{\frac{0.3}{1.3}} - 1\right] = 33{,}931 \text{ ft} \cdot \text{lb/lb}$$

Therefore, the total work done in compressing 5 lb of gas is

$$W_T = 33{,}931 \times 5 = 169{,}655 \text{ ft} \cdot \text{lb}$$

Example Problem 12.4 (SI)

Calculate the work done in compressing 2 kg of gas (gravity = 0.65) adiabatically from an initial temperature of 20°C and pressure of 700 kPa to a final pressure of 2000 kPa. The specific heat ratio of gas is 1.4, and the base pressure and base temperature are 101 kPa and 15°C, respectively.

Solution

Using Eq. (12.9), the work done in adiabatic compression of 1 kg of gas is

$$W_a = \frac{286.76}{0.65}(20+273)\left(\frac{1.4}{0.4}\right)\left[\left(\frac{2000+101}{700+101}\right)^{\frac{0.4}{1.4}} - 1\right] = 143{,}512 \text{ J/kg}$$

Therefore, the total work done in compressing 2 kg of gas is

$$W_T = 143{,}512 \times 2 = 287{,}024 \text{ J}$$

12.5 POLYTROPIC COMPRESSION

Polytropic compression is similar to adiabatic compression, but there can be heat transfer, unlike adiabatic compression. In a polytropic process, the gas pressure and volume are related as follows:

$$PV^n = \text{Const} \tag{12.10}$$

where

n – Polytropic exponent
Const – A constant different than the one in Eq. (12.6)

As before from initial to final conditions, we can state that

$$P_1 V_1^n = P_2 V_2^n \tag{12.11}$$

Because polytropic compression is similar to adiabatic compression, we can easily calculate the work done in polytropic compression by substituting n for γ in Eqs (12.8) and (12.9).

Example Problem 12.5 (USCS)

Natural gas is compressed polytropically from an initial temperature and pressure of 60°F and 500 psig, respectively, to a final pressure of 1000 psig. The gas gravity is 0.6 and the base pressure and base temperature are 14.7 psia and 60°F, respectively. Calculate the work done in compressing 5 lb of gas using a polytropic exponent of 1.5.

Solution

Polytropic compression is similar to adiabatic compression and, therefore, the same equation can be used for work done, substituting the polytropic exponent n for the adiabatic exponent γ (the ratio of specific heat).

Using Eq. (12.8), the work done in polytropic compression of 1 lb of gas is

$$W_p = \frac{53.28}{0.6}(60+460)\left(\frac{1.5}{0.5}\right)\left[\left(\frac{1014.7}{514.7}\right)^{\frac{0.5}{1.5}} - 1\right] = 35{,}168 \text{ ft} \cdot \text{lb/lb}$$

Therefore, the total work done in compressing 5 lb of gas is

$$W_T = 35{,}168 \times 5 = 175{,}840 \text{ ft} \cdot \text{lb}$$

12.6 DISCHARGE TEMPERATURE OF COMPRESSED GAS

In adiabatic or polytropic compression of natural gas, the final temperature of the gas can be calculated, given the initial temperature, and the initial and final pressures, as follows. The initial conditions are called suction conditions and the final state is the discharge condition.

From Eq. (12.6) for adiabatic compression and the perfect gas law from Chapter 1, we eliminate the volume V to get a relationship between pressure, temperature, and the compressibility factor:

$$\left(\frac{T_2}{T_1}\right) = \left(\frac{Z_1}{Z_2}\right)\left(\frac{P_2}{P_1}\right)^{\frac{\gamma-1}{\gamma}} \tag{12.12}$$

where

T_1 – Suction temperature of gas, °R
T_2 – Discharge temperature of gas, °R
Z_1 – Gas compressibility factor at suction, dimensionless
Z_2 – Gas compressibility factor at discharge, dimensionless

Other symbols are as defined earlier.

Replacing the adiabatic exponent γ in Eq. (12.12) with the polytropic exponent n, we can similarly calculate the discharge temperature for polytropic compression from the following equation:

$$\left(\frac{T_2}{T_1}\right) = \left(\frac{Z_1}{Z_2}\right)\left(\frac{P_2}{P_1}\right)^{\frac{n-1}{n}} \tag{12.13}$$

where all symbols are as defined earlier.

Example Problem 12.6 (USCS)

Gas is compressed adiabatically ($\gamma = 1.4$) from 60°F suction temperature and a compression ratio of 2.0. Calculate the discharge temperature, assuming $Z_1 = 0.99$ and $Z_2 = 0.85$.

Solution

Using Eq. (12.12),

$$\left(\frac{T_2}{60+460}\right) = \left(\frac{0.99}{0.85}\right)(2.0)^{\frac{0.4}{1.4}} = 1.4198$$

$$T_2 = 1.4198 \times 520 = 738.3°R = 278.3°F$$

Therefore, the discharge temperature of the gas is 278.3°F.

12.7 COMPRESSION POWER REQUIRED

We have seen from the preceding calculations that the amount of energy input to the gas by the compression process will depend on the gas pressure and the quantity of gas. The latter is proportional to the gas flow rate. The power (HP in USCS units and kW in SI units) that represents the energy expended per unit time will also depend on the gas pressure and the flow rate. As the flow rate increases, the pressure also increases and hence the power needed will also increase. For a gas compressor, the power required can be stated in terms of the gas flow rate and the discharge pressure of the compressor station as explained next.

If the gas flow rate is Q and the suction and discharge pressures of the compressor station are P_1 and P_2, respectively, the compressor station provides the differential pressure of $(P_2 - P_1)$ to the gas flowing at Q. This differential pressure is called the head. The discharge pressure P_2 will depend on the type of compression (adiabatic, polytropic, and so on).

The head developed by the compressor is also defined as the amount of energy supplied to the gas per unit mass of gas (ft · lb/lb). Therefore, by multiplying the mass flow rate of gas by the compressor head, we can calculate the total energy supplied to the gas per unit time, or the power supplied to the gas. Dividing this power by the efficiency of the compressor, we get the compressor power input. The equation for the compressor power can be expressed as follows, in USCS units:

$$\text{HP} = 0.0857\left(\frac{\gamma}{\gamma-1}\right)QT_1\left(\frac{Z_1+Z_2}{2}\right)\left(\frac{1}{\eta_a}\right)\left[\left(\frac{P_2}{P_1}\right)^{\frac{\gamma-1}{\gamma}}-1\right] \quad \text{(USCS)} \qquad (12.14)$$

where

HP – Compression horsepower

γ – Ratio of specific heats of gas, dimensionless

Q – Gas flow rate, MMSCFD
T_1 – Suction temperature of gas, °R
P_1 – Suction pressure of gas, psia
P_2 – Discharge pressure of gas, psia
Z_1 – Compressibility of gas at suction conditions, dimensionless
Z_2 – Compressibility of gas at discharge conditions, dimensionless
η_a – Compressor adiabatic (isentropic) efficiency, less than 1.0

In SI units, the compressor power is as follows:

$$\text{Power} = 4.0639\left(\frac{\gamma}{\gamma-1}\right)QT_1\left(\frac{Z_1+Z_2}{2}\right)\left(\frac{1}{\eta_a}\right)\left[\left(\frac{P_2}{P_1}\right)^{\frac{\gamma-1}{\gamma}}-1\right] \text{ (SI)} \tag{12.15}$$

where

Power – Compression power, kW
γ – Ratio of specific heats of gas, dimensionless
Q – Gas flow rate, million m^3/day (Mm^3/day)
T_1 – Suction temperature of gas, K
P_1 – Suction pressure of gas, kPa
P_2 – Discharge pressure of gas, kPa
Z_1 – Compressibility of gas at suction conditions, dimensionless
Z_2 – Compressibility of gas at discharge conditions, dimensionless
η_a – Compressor adiabatic (isentropic) efficiency, decimal value

The adiabatic efficiency η_a generally ranges from 0.75 to 0.85. By considering a mechanical efficiency η_m of the compressor driver, we can calculate the brake horsepower (BHP) required to run the compressor as follows:

$$\text{BHP} = \frac{\text{HP}}{\eta_m} \text{ (USCS)} \tag{12.16}$$

where HP is the horsepower calculated from the preceding equations taking into account the adiabatic efficiency η_a of the compressor.

Similarly in SI units, the brake power is calculated as follows:

$$\text{Brakepower} = \frac{\text{Power}}{\eta_m} \text{ (SI)} \tag{12.17}$$

The mechanical efficiency η_m of the driver may range from 0.95 to 0.98. Therefore, the overall efficiency η_o is defined as the product of the adiabatic efficiency η_a and the mechanical efficiency η_m.

$$\eta_o = \eta_a \times \eta_m \tag{12.18}$$

Appendix 8 shows compressor power required for various values of compression ratio, suction temperature and number of stages.

From the adiabatic compression Eq. (12.6), using the perfect gas law, we eliminate the volume V, and the discharge temperature of the gas is then related to the suction temperature and the compression ratio as follows:

$$\left(\frac{T_2}{T_1}\right) = \left(\frac{P_2}{P_1}\right)^{\frac{\gamma-1}{\gamma}} \tag{12.19}$$

The adiabatic efficiency, η_a, may also be defined as the ratio of the adiabatic temperature rise to the actual temperature rise. Thus, if the gas temperature because of compression increases from T_1 to T_2, the actual temperature rise is $(T_2 - T_1)$.

The theoretical adiabatic temperature rise is obtained from the relationship between adiabatic pressure and temperature as follows, considering the gas compressibility factors similar to Eq. (12.12):

$$\left(\frac{T_2}{T_1}\right) = \left(\frac{Z_1}{Z_2}\right)\left(\frac{P_2}{P_1}\right)^{\frac{\gamma-1}{\gamma}} \tag{12.20}$$

Simplifying and solving for T_2,

$$T_2 = T_1\left(\frac{Z_1}{Z_2}\right)\left(\frac{P_2}{P_1}\right)^{\frac{\gamma-1}{\gamma}} \tag{12.21}$$

Therefore, the adiabatic efficiency is

$$\eta_a = \frac{T_1\left(\frac{Z_1}{Z_2}\right)\left(\frac{P_2}{P_1}\right)^{\frac{\gamma-1}{\gamma}} - T_1}{T_2 - T_1} \tag{12.22}$$

Simplifying, we get

$$\eta_a = \left(\frac{T_1}{T_2 - T_1}\right)\left[\left(\frac{Z_1}{Z_2}\right)\left(\frac{P_2}{P_1}\right)^{\frac{\gamma-1}{\gamma}} - 1\right] \tag{12.23}$$

where T_2 is the actual discharge temperature of the gas.

If the inlet gas temperature is 80°F and the suction and discharge pressures are 800 and 1400 psia, respectively, and if the discharge gas temperature is given as 200°F, we can calculate the adiabatic efficiency from Eq. (12.23). Using $\gamma = 1.4$, the adiabatic efficiency is, assuming compressibility factors to be equal to 1.0 (approximately),

$$\eta_a = \left(\frac{80 + 460}{200 - 80}\right)\left[\left(\frac{1400}{800}\right)^{\frac{1.4-1}{1.4}} - 1\right] = 0.7802 \tag{12.24}$$

Thus, the adiabatic compression efficiency is 0.7802.

Example Problem 12.7 (USCS)

Calculate the compressor power required for an adiabatic compression of 100 MMSCFD gas with inlet temperature of 70°F and 725 psia pressure. The discharge pressure is 1305 psia. Assume compressibility factors at suction and discharge condition to be Z_1 = 1.0 and Z_2 = 0.85 and adiabatic exponent γ = 1.4 with the adiabatic efficiency η_a = 0.8. If the mechanical efficiency of the compressor driver is 0.95, what is the compressor BHP? Also estimate the discharge temperature of the gas.

Solution

From Eq. (12.15), the horsepower required is

$$\text{HP} = 0.0857 \times 100\left(\frac{1.40}{0.40}\right)(70+460)\left(\frac{1+0.85}{2}\right)\left(\frac{1}{0.8}\right)\left[\left(\frac{1305}{725}\right)^{\frac{0.40}{1.40}} - 1\right] = 3362$$

Using Eq. (12.17), the compressor driver BHP is calculated based on a mechanical efficiency of 0.95.

$$\text{BHP required} = \frac{3362}{0.95} = 3539$$

The outlet temperature of the gas is estimated using Eq. (12.23) and with some simplification as follows:

$$T_2 = (70+460) \times \left[\frac{\left(\frac{1}{0.85}\right)\left(\frac{1305}{725}\right)^{\frac{0.4}{1.4}} - 1}{0.8}\right] + (70+460)$$

$$= 789.44\,\text{R} = 329.44°\text{F}$$

Therefore, the discharge temperature of the gas is 329.44°F.

Example Problem 12.8 (SI)

Natural gas at 4 million m^3/day (Mm^3/day) and 20°C is compressed isentropically (γ = 1.4) from a suction pressure of 5 MPa absolute to a discharge pressure of 9 MPa absolute in a centrifugal compressor with an isentropic efficiency of 0.82. Calculate the compressor power required, assuming the compressibility factors at suction and discharge condition to be Z_1 = 0.95 and Z_2 = 0.85, respectively. If the mechanical efficiency of the compressor driver is 0.95, what is the driver power required? Calculate the discharge temperature of the gas.

Solution

From Eq. (12.16), the power required is

$$\text{Power} = 4.0639 \times 4\left(\frac{1.40}{0.40}\right)(20+273)\left(\frac{0.95+0.85}{2}\right)\left(\frac{1}{0.82}\right)\left[\left(\frac{9}{5}\right)^{\frac{0.40}{1.40}} - 1\right] = 3346\,\text{kW}$$

$$\text{Compressor power} = 3346\,\text{kW}$$

Using Eq. (12.17), we calculate the driver power required as follows:

$$\text{Driver power required} = \frac{3346}{0.95} = 3522\,\text{kW}$$

The discharge temperature of the gas is estimated from Eq. (12.23) as

$$T_2 = \frac{20+273}{0.82} \times \left[\left(\frac{0.95}{0.85}\right)\left(\frac{9}{5}\right)^{\frac{0.4}{1.4}} - 1 \right] + (20+273) = 408.07\,\text{K} = 135.07°\text{C}$$

12.8 OPTIMUM COMPRESSOR LOCATIONS

In earlier sections, we discussed a pipeline with two compressor stations to deliver gas from Danby to Leeds power plant. In this section, we will explore how to locate the intermediate compressor stations at the optimum location taking into account the overall horsepower required. In Section 12.2, we discussed hydraulic balance. In the next example, we will analyze optimum compressor locations by considering both hydraulically balanced and unbalanced compressor station locations.

Example Problem 12.9 (USCS)

A gas transmission pipeline is 240 mile long, NPS 30, 0.500 in. wall thickness and runs from Payson to Douglas. The origin compressor station is at Payson and two intermediate compressor stations tentatively located at Williams (milepost 80) and Snowflake (milepost 160) as shown in Fig. 12.7. There are no intermediate flow deliveries or injections and the inlet flow rate at Payson is 900 MMSCFD.

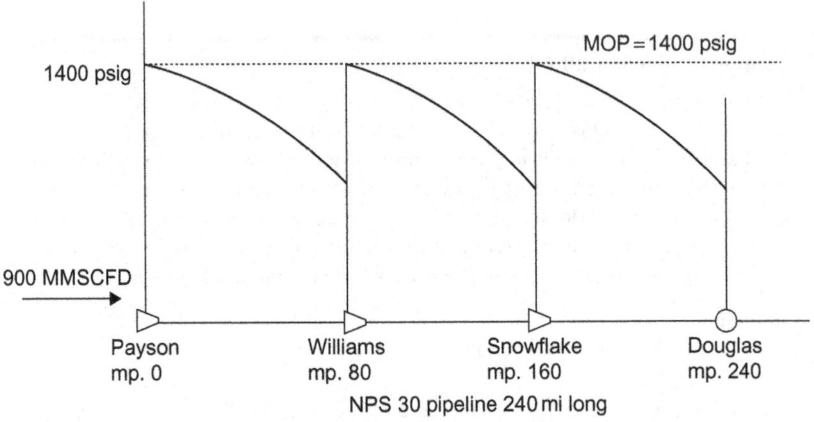

FIGURE 12.7 Gas pipeline with three compressor stations.

The terminus delivery pressure required at Douglas is 600 psig and the MAOP of the pipeline is 1400 psig throughout. Neglect the effects of elevation and assume constant gas flow temperature of 80°F and constant values of transmission factor $F = 20$ and compressibility factor $Z = 0.85$ throughout the pipeline. The gas gravity is 0.6, base pressure = 14.7 psia, and base temperature = 60°F. Use a polytropic compression coefficient of 1.38 and a compression efficiency of 0.9. The objective is to determine the locations of the intermediate compressor stations.

Solution

First, we calculate for each of the three segments, Payson to Williams, Williams to Snowflake, and Snowflake to Douglas, the downstream pressure starting with an upstream pressure of 1400 psig (MAOP). Accordingly, using the general flow equation for the Payson to Williams segment, we calculate the downstream pressure at Williams starting with a pressure of 1400 psig at Payson. This downstream pressure becomes the suction pressure at the Williams compressor station. Next, we repeat the calculations for the second segment from Williams to Snowflake to calculate the downstream pressure at Snowflake, based on an upstream pressure of 1400 psig at Williams. This downstream pressure becomes the suction pressure at the Snowflake compressor station. Finally, we calculate the downstream pressure at Douglas, for the third and final pipe segment from Snowflake to Douglas, based on an upstream pressure of 1400 psig at Snowflake. This final pressure is the delivery pressure at the Douglas terminus. We have, thus, calculated the suction pressures at each of the two intermediate compressor stations at Williams and Snowflake and also calculated the final delivery pressure at Douglas. This pressure calculated at Douglas may or may not be equal to the desired delivery pressure of 600 psig because we performed a *forward calculation* going from Payson to Douglas. Therefore, because the delivery pressure is usually a desired or contracted value, we will have to adjust the location of the last compressor station at Snowflake to achieve the desired delivery pressure at Douglas, as we did before in the Danby to Leeds pipeline.

Another approach would be to perform a backward calculation starting at Douglas and proceeding toward Payson. In this case, we would start with segment 3 and calculate the location of the Snowflake compressor station that will result in an upstream pressure of 1400 psig at Snowflake. Thus, we locate the Snowflake compressor station that will cause a discharge pressure of 1400 psig at Snowflake and a delivery pressure of 600 psig at Douglas. Having located the Snowflake compressor station, we can now recalculate the suction pressure at Snowflake by considering pipe segment 2 and using an upstream pressure of 1400 psig at Williams. We will not have to repeat calculations for segment 1 because the location of Williams has not changed and, therefore, the suction pressure at Williams will remain the same as the previously calculated value. We have thus been able to determine the pressures along the pipeline with the given three compressor station configurations such that the desired delivery pressure at Douglas has been achieved and each compressor station discharges at the MAOP value of 1400 psig. But are these the optimum locations of the intermediate compressor stations Williams and Snowflake? In other words, are all compressor stations in hydraulic balance? We can say that these compressor stations are optimized and are in hydraulic balance only if each compressor station operates at the same compression ratio and therefore adds the same amount of horsepower to the gas at each compressor station. The locations of Williams and Snowflake may not result in the same suction pressures even though the discharge pressures are the same. Therefore, chances are that Williams may be operating at a lower compression ratio than Snowflake or Payson or vice versa, which will not result in hydraulic balance. However, if the compression ratios are close enough that the required compressor sizes are the same, then we could still be in hydraulic balance and the stations could be at optimum locations.

Next, let's perform the calculations and determine how much tweaking of the compressor station locations is required to optimize these stations. First, we will perform the backward calculations for segment 3, starting with a downstream pressure of 600 psig at Douglas and an upstream pressure of 1400 psig at Snowflake. With these constraints, we will calculate the pipe length L miles between Snowflake and Douglas.

Using the general flow Eq. (8.55), neglecting elevations,

$$900\times10^6 = 38.77\times20.0\left(\frac{520}{14.7}\right)\left(\frac{1414.7^2-614.7^2}{0.6\times540\times L\times0.85}\right)^{0.5}(29)^{2.5}$$

Solving for pipe length we get

$$L = 112.31\ \text{mi}$$

Therefore, in order to discharge at 1400 psig at Snowflake and deliver gas at 600 psig at Douglas, the Snowflake compressor station will be located at a distance of 112.31 mi upstream of Douglas or at mile post (240 – 112.31) = 127.69 measured from Payson.

Next, keeping the location of the Williams compressor station at milepost 80, we calculate the downstream pressure at Snowflake for pipe segment 2 starting at 1400 psig at Williams. This calculated pressure will be the suction pressure of the Snowflake compressor station.

Using the general flow Eq. (8.55), neglecting elevations,

$$900\times10^6 = 38.77\times20.0\left(\frac{520}{14.7}\right)\left(\frac{1414.7^2-P_2^2}{0.6\times540\times47.69\times0.85}\right)^{0.5}(29)^{2.5}$$

where the pipeline segment length between Williams and Snowflake was calculated as 127.69 – 80 = 47.69 mi.

Solving for suction pressure at Snowflake, we get,

$$P_2 = 1145.42\ \text{psia} = 1130.72\ \text{psig}$$

Therefore, the compression ratio at Snowflake $= \frac{1414.7}{1145.42} = 1.24$

Next, for pipe segment 1 between Payson and Williams, we will calculate the downstream pressure at Williams starting at 1400 psig at Payson. This calculated pressure will be the suction pressure of the Williams compressor station.

Using the general flow Eq. (8.55), neglecting elevations,

$$900\times10^6 = 38.77\times20.0\left(\frac{520}{14.7}\right)\left(\frac{1414.7^2-P_2^2}{0.6\times540\times80\times0.85}\right)^{0.5}(29)^{2.5}$$

Solving for suction pressure at Williams, we get,

$$P_2 = 919.20\,\text{psia} = 904.5\,\text{psig}$$

Therefore, the compression ratio at Williams = $\frac{1414.7}{919.2} = 1.54$

Therefore, from the preceding calculations, the compressor station at Williams requires a compression ratio r = 1.54 while the compressor station at Snowflake requires a compression ratio r = 1.24. Obviously, this is not a hydraulically balanced compressor station system. Further, we do not know what the suction pressure is at the Payson compressor station. If we assume that Payson receives gas at, approximately, the same suction pressure as Williams (905 psig), both the Payson and Williams compressor stations will have the same compression ratio of 1.54. In this case, the Snowflake compressor will be the odd one operating at the compression ratio of 1.24. How do we balance these compressor stations? One way would be to obtain the same compression ratios for all three compressor stations by simply relocating the Snowflake compressor station toward Douglas such that its suction pressure will reduce from 1131 to 905 psig, while keeping the discharge at Snowflake at 1400 psig. This will then ensure that all three compressor stations will be operating at the following suction and discharge pressures and compression ratios:

$$\text{Suction pressure, } P_s = 904.5\,\text{psig}$$

$$\text{Discharge pressure, } P_d = 1400\,\text{psig}$$

$$\text{Compression ratio, } r = \frac{1400 + 14.7}{904.5 + 14.7} = 1.54$$

However, because the Snowflake compressor is now located closer to Douglas than before (127.69), the discharge pressure of 1400 psig at Snowflake will result in a higher delivery pressure at Douglas than the required 600 psig as shown in Fig. 12.8.

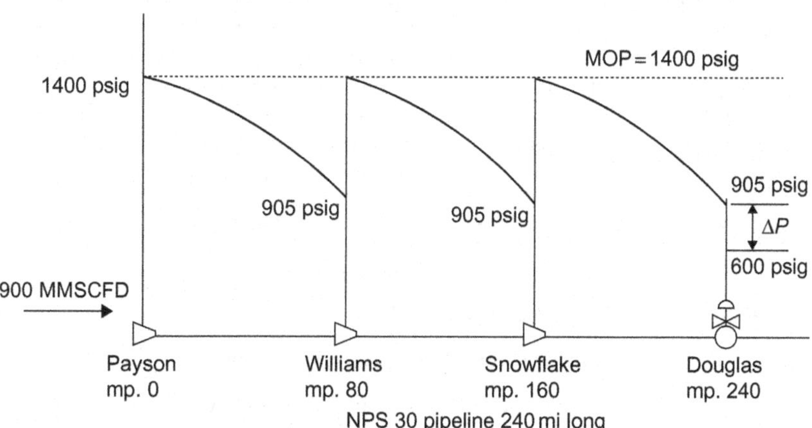

FIGURE 12.8 Pressure regulation at Douglas.

If the additional pressure at Douglas can be tolerated by the customer, then there will be no problem. But If the customer requires no more than 600 psig, we have to reduce the delivery pressure to 600 psig by installing a pressure regulator at Douglas as shown in Fig. 12.8. Therefore, by balancing the compressor station locations we have also created a problem of getting rid of the extra pressure at the delivery point. Pressure regulation means wasted horsepower. The advantage of the balanced compressor stations versus the negative aspect of the pressure regulation must be factored in to the decision process.

To illustrate this pressure regulation scenario, we will now determine the revised location of the Snowflake compressor station for hydraulic balance. We will calculate the length of pipe segment 2 by assuming 1400 psig discharge pressure at Williams and a suction pressure of 904.5 psig at Snowflake.

Using the general flow Eq. (8.55), neglecting elevations,

$$900 \times 10^6 = 38.77 \times 20.0 \left(\frac{520}{14.7}\right) \left(\frac{1414.7^2 - 919.2^2}{0.6 \times 540 \times L \times 0.85}\right)^{0.5} (29)^{2.5}$$

Solving for pipe length for segment 2, we get

$$L = 80\,\text{mi}$$

Therefore, the Snowflake compressor station should be located at a distance of 80 mi from Williams or at milepost 160. We could have arrived at this without the above calculations because elevations are neglected and the Payson to Williams pressure profile will be the same as the pressure profile from Williams to Snowflake. With the Snowflake compressor station located at milepost 160 and discharging at 1400 psig, we conclude that the delivery pressure at Douglas will also be 904.5 psig because all three pipe segments are hydraulically the same. We see that the delivery pressure at Douglas is approximately 305 psig more than the desired pressure. As indicated earlier, a pressure regulator will be required at Douglas to reduce the delivery pressure down to 600 psig. We can compare the hydraulically balanced scenario with the previously calculated case where Payson and Williams operate at a compression ratio of 1.54 and Snowflake operates at lower compression ratio of 1.24. By applying approximate cost per installed horsepower, we can compare these two cases. First, using Eq. (12.14), calculate the horsepower required at each compressor station, assuming polytropic compression and a compression ratio of 1.54 for a balanced compressor station.

$$\text{HP} = 0.0857 \times 900 \times \left(\frac{1.38}{0.38}\right)(80 + 460)\left(\frac{1 + 0.85}{2}\right)\left(\frac{1}{0.9}\right)\left[(1.54)^{\frac{0.38}{1.38}} - 1\right] = 19{,}627$$

Therefore, the total horsepower required in the hydraulically balanced case is

$$\text{Total HP} = 3 \times 19{,}627 = 58{,}881$$

At a cost of $2000 per installed HP,

$$\text{Total HP cost} = \$2000 \times 58{,}881 = \$117.76\,\text{million}$$

In the hydraulically unbalanced case, the Payson and Williams compressor stations will operate at a compression ratio of 1.54 each while the Snowflake compressor station will require a compression ratio of 1.24.

Using Eq. (12.14), the horsepower required at the Snowflake compressor station is

$$HP = 0.0857 \times 900 \times \left(\frac{1.38}{0.38}\right)(80+460)\left(\frac{1+0.85}{2}\right)\left(\frac{1}{0.9}\right)\left[(1.24)^{\frac{0.38}{1.38}} - 1\right] = 9487$$

Therefore, the total horsepower required in the hydraulically unbalanced case is

$$\text{Total HP} = (2 \times 19{,}627) + 9487 = 48{,}741$$

At a cost of \$2000 per installed HP,

$$\text{Total HP cost} = \$2000 \times 48{,}741 = \$97.48 \text{ million}$$

The hydraulically balanced case requires (58,881 – 48,741) or 10,140 HP more and will cost approximately (\$117.76 – \$97.48) or \$20.28 million more. In addition to the extra HP cost, the hydraulically balanced case will require a pressure regulator that will waste energy and result in extra equipment cost. Therefore, the advantages of using identical components, by reducing spare parts and inventory in the hydraulically balanced case must be weighed against the additional cost. It may not be worth spending the extra \$20 million to obtain this benefit. The preferred solution in this case is for the Payson and Williams compressor stations to be identical (compression ratio = 1.54) and the Snowflake compressor station to be a smaller one (compression ratio = 1.24) requiring the lesser compression ratio and horsepower to provide the required 600 psig delivery pressure at Douglas.

12.9 COMPRESSORS IN SERIES AND PARALLEL

When compressors operate in series, each unit compresses the same amount of gas but at different compression ratios such that the overall pressure increase of the gas is achieved in stages as shown in Fig. 12.9.

It can be seen from Fig. 12.9 that the first compressor compresses gas from a suction pressure of 900 to 1080 psia at a compression ratio of 1.2. The second

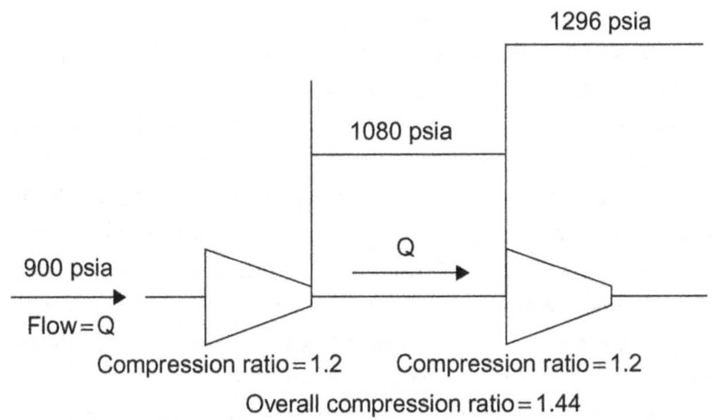

FIGURE 12.9 Compressors in series.

compressor takes the same volume and compresses it from 1080 psia to a discharge pressure of 1080 × 1.2 = 1296 psia. Thus, the overall compression ratio of the two identical compressors in series is 1296/900 = 1.44. We have thus achieved the increase in pressure in two stages. At the end of each compression cycle, the gas temperature would rise to some value calculated in accordance with Eq. (12.23). Therefore, with multiple stages of compression, unless the gas is cooled between stages, the final gas temperature may be too high. High gas temperatures are not desirable because the throughput capability of a gas pipeline decreases with gas flow temperature. Therefore, with compressors in series, the gas is cooled to the original suction temperature between each stage of compression such that the final temperature at the end of all compressors in series is not exceedingly high. Suppose the calculated discharge temperature of a compressor is 232°F, starting at 70°F suction temperature and with a compression ratio of 1.4. If two of these compressors were in series and there was no cooling between compressions, the final gas temperature would reach approximately

$$\frac{(232+460)(232+460)}{70+460} = 903.5\,R = 443.5°\text{F}$$

This is too high a temperature for pipeline transportation. However, if we cool the gas back to 70°F before compressing it through the second compressor, then the final temperature of the gas coming out of the second compressor will be approximately 232°F.

Compressors are installed in parallel so that the large volumes necessary can be provided by multiple compressors each producing the same compression ratio. Three identical compressors with a compression ratio of 1.4 may be used to provide 900 MMSCFD gas flow from a suction pressure of 900 psia. In this example, each compressor will compress 300 MMSCFD from 900 psia to a discharge pressure of

$$P_2 = 900 \times 1.4 = 1260\,\text{psia}$$

This is illustrated schematically in Fig. 12.10.

Unlike compressors in series, the discharge temperature of the gas coming out of the parallel bank of compressors will not be high because the gas does not undergo multiple compression ratios. The gas temperature on the discharge side of each parallel compressor will be approximately the same as that of a single compressor with the same compression ratio. Therefore, three parallel compressors, each compressing the same volume of gas at a compression ratio of 1.4, will have a final discharge temperature of 232°F starting from a suction temperature of 70°F. Gas cooling is required at these temperatures in order to achieve efficient gas transportation and also to operate at a temperature not exceeding the limits of the pipe-coating material. Generally, pipe coating requires gas temperature not to exceed 140–150°F.

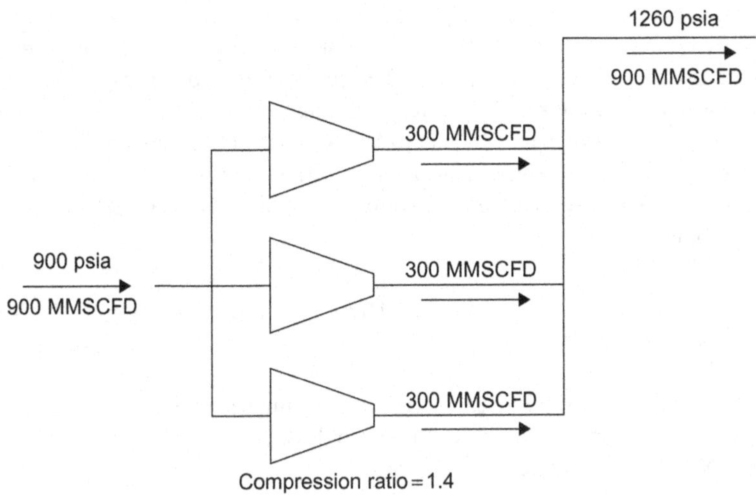

FIGURE 12.10 Compressors in parallel.

The compression ratio was defined earlier as the ratio of the discharge pressure to the suction pressure. The higher the compression ratio, the higher will be the gas discharge temperature in accordance with Eq. (12.23).

Consider a suction temperature of 80°F and the suction and discharge pressures of 900 and 1400 psia, respectively. The compression ratio is 1400/900 = 1.56. Using Eq. (12.19), the approximate discharge temperature will be

$$\left(\frac{T_2}{80+460}\right) = \left(\frac{1400}{900}\right)^{\frac{1.3-1}{1.3}} = 598.36\,R \text{ or } 138.36°\text{F}$$

If the compression ratio is increased from 1.56 to 2.0, the discharge temperature will become 173.67°F. It can be seen that the discharge temperature of the gas increases considerably with the compression ratio. Because the throughput capacity of a gas pipeline decreases with increase in gas temperature, we must cool the discharge gas from a compressor to ensure maximum pipeline throughput.

Typically, centrifugal compressors are used in gas pipeline applications to have a compression ratio of 1.5–2.0; there may be instances where higher compression ratios are required because of lower gas receipt pressures and higher pipeline discharge pressures to enable a given volume of gas to be transported through a pipeline. For this, reciprocating compressors are used. Manufacturers limit the maximum compression ratio of reciprocating compressors to a range of 5–6. This is because of high forces that are exerted on the compressor components, which require expensive materials, as well as complicated safety needs.

Suppose a compressor is required to provide gas at 1500 psia from a suction pressure 200 psia. This requires an overall compression ratio of 7.5. Because this is beyond the acceptable range of compression ratio, we will have to provide this compression in stages. If we provide the necessary pressure by using two compressors in series, each compressor will require a compression ratio of $\sqrt{7.5}$ or approximately 2.74. The first compressor raises the pressure from 200 psia to 200 × 2.74 = 548 psia. The second compressor will then boost the gas pressure from 548 psia to 548 × 2.74 = 1500 psia, approximately. In general, if n compressors are installed in series to achieve the required total compression ratio r_t, we can, then, calculate the individual compression ratio of each compressor to be

$$r = (r_t)^{\frac{1}{n}} \tag{12.25}$$

where

r – Compression ratio of each compressor, dimensionless
r_t – Overall compression ratio, dimensionless
n – Number of compressors in series

It has been found that by providing the overall compression ratio by means of identical compressors in series, power requirements will be minimized. Thus, in the preceding example, we assumed that two identical compressors in series each providing a compression ratio of 2.74 resulting in an overall compression ratio of 7.5 will be a better option than if we had a compressor with a compression ratio of 3.0 in series with another compressor with a compression ratio of 2.5. To illustrate this further if an overall compression ratio of 20 is required and we were to use three compressors in series, the most economical option will be to use identical compressors each with a compression ratio of $(20)^{\frac{1}{3}} = 2.71$.

12.10 TYPES OF COMPRESSORS – CENTRIFUGAL AND POSITIVE DISPLACEMENT

Compressors used in a natural gas transportation system are either positive displacement (PD) type or centrifugal (CF) type. Positive displacement compressors generate the pressure required by trapping a certain volume of gas within the compressor and increasing the pressure by reduction of volume. The high pressure gas is then released through the discharge valve into the pipeline. Piston-operated reciprocating compressors fall within the category of positive displacement compressors. These compressors have a fixed volume and are able to produce high-compression ratios. Centrifugal compressors, however, develop the pressure required by the centrifugal force due to rotation of the compressor wheel that translates the kinetic energy into pressure energy of the gas. Centrifugal compressors are more commonly used in gas transmission systems due to their flexibility. Centrifugal compressors have lower capital cost and lower maintenance expenses. They can handle larger volumes within a small area compared with positive displacement compressors. They also operate at high speeds and are of balanced construction. However, centrifugal compressors have less efficiency than positive displacement compressors.

Positive displacement compressors have flexibility in pressure range and have higher efficiency and can deliver compressed gas at a wide range of pressures. They are also not very sensitive to the composition of the gas. Positive displacement compressors have pressure ranges up to 30,000 psi and range from very low HP to more than 20,000 HP per unit. Positive displacement compressors may be single stage or multistage depending on the compression ratio required. The compression ratio per stage for positive displacement compressors is limited to 4.0 because higher ratio causes higher discharge pressures that affect the valve life of positive displacement compressors. Heat exchangers are used between stages of compression so that the compressed heated gas is cooled to the original suction temperature before being compressed in the next stage. The HP required in a positive displacement compressor is usually estimated from charts provided by the compressor manufacturer. The following equation may be used for large slow-speed compressors with compression ratios more than 2.5 and for gas specific gravity of 0.65:

$$BHP = 22rNQF \tag{12.26}$$

where

BHP – Brake horsepower
r – Compression ratio per stage
N – Number of stages
Q – Gas flow rate, MMSCFD at suction temperature and 14.4 psia
F – Factor that depends on the number of compression stages.
F – 1.0 for single-stage compression, 1.08 for two-stage compression, and 1.10 for three-stage compression

In Eq. (12.26), the constant 22 is changed to 20 when gas gravity is between 0.8 and 1.0. Also for compression ratios between 1.5 and 2.0, the constant 22 is replaced with a number between 16 and 18.

Example Problem 12.10 (USCS)

Calculate the BHP required to compress 5 MMSCFD gas at 14.4 psia and 70°F, with an overall compression ratio of 7 considering two-stage compression.

Solution

Consider two identical stages, compression ratio per stage $= \sqrt{7.0} = 2.65$.

Using Eq. (12.26), we get

$$BHP = 22 \times 2.65 \times 2 \times 5 \times 1.08 = 629.64$$

Centrifugal compressors may be a single-wheel or single-stage compressor or multiwheel or multistage compressor. Single-stage centrifugal compressors have a volume range of 100–150,000 ft^3/min at actual conditions (ACFM). Multistage centrifugal compressors handle volume range of 500–200,000 ACFM. The operational speeds of centrifugal compressors range from 3000 to 20,000 rev/min (rpm). The upper limit of speed will be limited by the wheel tip speed and stresses induced in the impeller. Advancement in technology has produced compressor wheels operating at speeds in excess of 30,000 rpm. Centrifugal compressors are driven by electric motors, steam turbines, or gas turbines. Sometimes, speed increasers are used to increase the speeds necessary to generate the pressure.

12.11 COMPRESSOR PERFORMANCE CURVES

The performance curve of a centrifugal compressor that can be driven at varying speeds typically shows a graphic plot of the inlet flow rate in actual cubic feet per minute (ACFM) against the head or pressure generated at various percentages of the design speed. Figure 12.11 shows a typical centrifugal compressor performance curve or performance map.

The limiting curve on the left-hand side is known as the surge line, and the corresponding curve on the right side is known as the stonewall limit. The performance of a centrifugal compressor at different rotational speeds follows the so-called affinity laws. According to the affinity laws, as the rotational speed of centrifugal compressor is changed, the inlet flow and head vary as the speed and the square of the speed, respectively, as indicated in the following equation.

For compressor speed change,

$$\frac{Q_2}{Q_1} = \frac{N_2}{N_1} \tag{12.27}$$

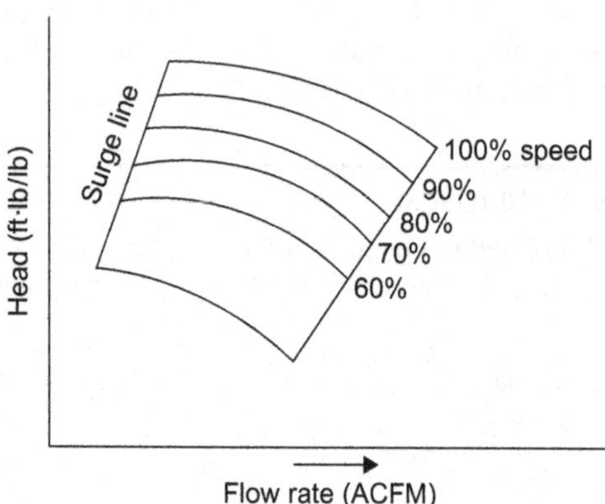

FIGURE 12.11 Typical centrifugal compressor performance curve.

$$\frac{H_2}{H_1} = \left(\frac{N_2}{N_1}\right)^2 \tag{12.28}$$

where

Q_1, Q_2 = initial and final flow rates
H_1, H_2 = initial and final heads
N_1, N_2 = initial and final compressor speeds

In addition, the horsepower for compression varies as the cube of the speed change as follows:

$$\frac{HP_2}{HP_1} = \left(\frac{N_2}{N_1}\right)^3 \tag{12.29}$$

Example Problem 12.11 (USCS)

The compressor head and volume flow rate for a centrifugal compressor at 18,000 rpm are as follows:

Flow Rate, Q (ACFM)	Head, H (ft·lb/lb)
360	10,800
450	10,200
500	9700
600	8200
700	5700
730	4900

Using the affinity laws, determine the performance of this compressor at a speed of 20,000 rpm.

Solution

The ratio of speeds is

$$\frac{20{,}000}{18{,}000} = 1.11$$

The multiplier for the flow rate is 1.11 and the multiplier for the head is $(1.11)^2$ or 1.232.

Using the affinity laws, the performance of the centrifugal compressor at 20,000 rpm is as follows:

Flow Rate, Q (ACFM)	Head, H (ft·lb/lb)
399.6	13,306
499.5	12,566
555.0	11,950
666.0	10,102
777.0	7022
810.0	6037

12.12 COMPRESSOR HEAD AND GAS FLOW RATE

The head developed by a centrifugal compressor is calculated from the suction and discharge pressures, the compressibility factor, and the polytropic or adiabatic exponent. We will explain how the actual or inlet flow rate (ACFM) is calculated from the standard gas flow rate. Knowing the maximum head that can be generated per stage, the number of stages needed can be calculated.

Consider a centrifugal compressor used to raise the gas pressure from 800 to 1440 psia starting at a suction temperature of 70°F and gas flow rate of 80 MMSCFD. The average compressibility factor from the suction to the discharge side is 0.95. The compressibility factor at the inlet is assumed to be 1.0, the polytropic exponent is 1.3, and the gas gravity is 0.6. The head generated by the compressor is calculated from the previously introduced equation

$$H = \frac{53.28}{0.6} \times 0.95 \times (70+460)\left(\frac{1.3}{0.3}\right)\left[\left(\frac{1440}{800}\right)^{\frac{0.3}{1.3}} - 1\right] = 28{,}146\ \text{ft}\cdot\text{lb/lb}$$

The actual flow rate at inlet conditions is calculated using the gas law as

$$Q_{act} = \frac{80 \times 14.7 \times 1.0}{800} \times \frac{70+460}{60+460} \times \frac{10^6}{24 \times 60} = 1040.5\ \text{ft}^3/\text{min (ACFM)}$$

If this particular compressor, according to vendor data, can produce a maximum head per stage of 10,000 ft · lb/lb, then the number of stages required to produce the required head is

$$n = \frac{28146}{10000} = 3$$

rounding off to the nearest whole number. Next, suppose that this compressor has a maximum design speed of 16,000 rpm. The actual operating speed necessary for the three-stage compressor is, according to the affinity laws

$$N_{act} = 16{,}000\sqrt{\frac{28{,}146}{3 \times 10{,}000}} = 15{,}498$$

Therefore, in order to generate 28,146 ft · lb/lb of head at a gas flow rate of 1040.5 ACFM, this three-stage compressor must run at a speed of 15,498 rpm.

12.13 COMPRESSOR STATION PIPING LOSSES

As the gas enters the suction side of the compressor, it flows through some complex piping system within the compressor station. Similarly, the compressed gas leaving the compressor traverses the compressor station discharge piping system that consists of valves and fittings before entering the main pipeline on its way to the next compressor station or delivery terminus. This is illustrated in Fig. 12.12.

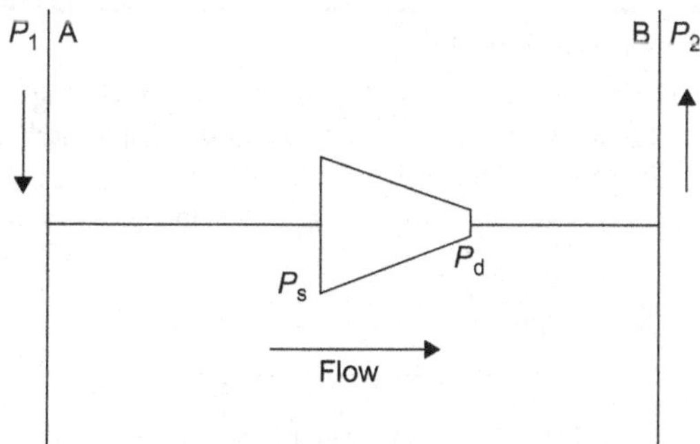

FIGURE 12.12 Compressor station suction and discharge piping.

It can be seen from Fig. 12.12 that at the compressor station boundary A on the suction side, the gas pressure is P_1. This pressure drops to a value P_s at the compressor suction, as the gas flows through the suction piping from A to B. This suction piping consisting of valves, fittings, filters, and meters causes a pressure drop of ΔP_s to occur. Therefore, the actual suction pressure at the compressor is

$$P_s = P_1 - \Delta P_s \tag{12.30}$$

where

P_s – Compressor suction pressure, psia
P_1 – Compressor station suction pressure, psia
ΔP_s – Pressure loss in compressor station suction piping, psi

At the compressor, the gas pressure is raised from P_s to P_d through a compression ratio r as follows:

$$r = \frac{P_d}{P_s} \tag{12.31}$$

where

r – Compression ratio, dimensionless
P_d – Compressor discharge pressure, psia

The compressed gas then flows through the station discharge piping, and loses pressure until it reaches the station discharge valve at the boundary B of the compressor station. If the station discharge pressure is P_2, we can write

$$P_2 = P_d - \Delta P_d \tag{12.32}$$

where

P_2 – Compressor station discharge pressure, psia
ΔP_d – Pressure loss in compressor station discharge piping, psi

Typically, the values of ΔP_s and ΔP_d range from 5 to 15 psi.

Example Problem 12.12 (USCS)

A compressor station on a gas transmission pipeline has the following pressures at the station boundaries. Station suction pressure = 850 psia and station discharge pressure = 1430 psia. The pressure losses in the suction piping and discharge piping are 5 and 10 psi, respectively. Calculate the compression ratio of this compressor station.

Solution

From Eq. (12.30), the compressor suction pressure is

$$P_s = 850 - 5 = 845\,\text{psia}$$

Similarly, the compressor discharge pressure is

$$P_d = 1430 + 10 = 1440\,\text{psia}$$

Therefore, the compression ratio is

$$r = \frac{1440}{845} = 1.70$$

12.14 COMPRESSOR STATION SCHEMATIC

A typical compressor station schematic showing the arrangement of the valves, piping, and the compressor itself is shown in Fig. 12.13.

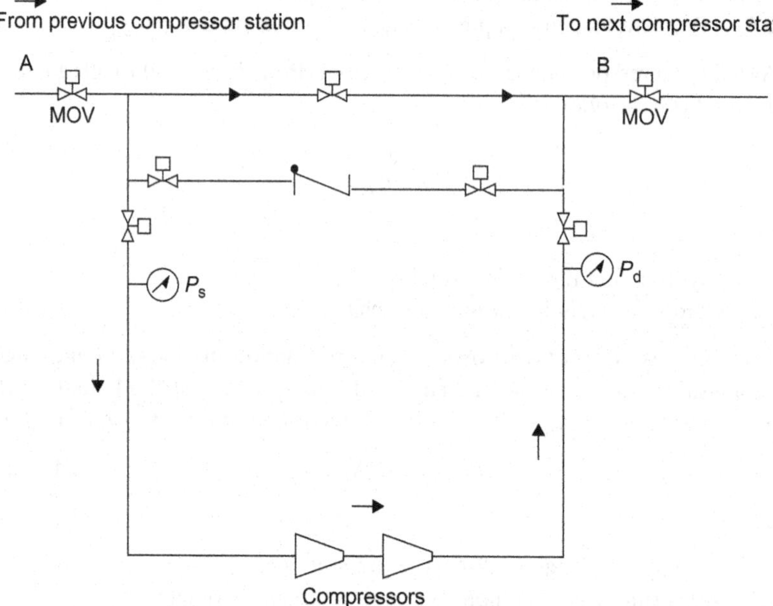

FIGURE 12.13 Compressor station schematic.

SUMMARY

In this chapter, we discussed compressing a gas to generate the pressure needed to transport the gas from one point to another along a pipeline. The compression ratio and the power required to compress a certain volume of gas, as well as the discharge temperature of the gas exiting the compressor, were explained and illustrated using examples. In a long distance gas transmission pipeline, locating intermediate compressor stations and minimizing energy loss were discussed. Hydraulically balanced and optimized compressor station locations were reviewed. Various compression processes, such as isothermal, adiabatic, and polytropic were explained. The different types of compressors such as positive displacement and centrifugal were explained along with their advantages and disadvantages. The need for configuring compressors in series and parallel was explored. The centrifugal compressor performance curve was discussed and the effect of rotational speed on the flow rate and head using the affinity laws was illustrated with examples. Finally, the impact of the compressor station yard piping pressure drops and how they affect the compression ratio and horsepower were discussed.

BIBLIOGRAPHY

Flow of Fluids through Valves, Fittings and Pipes, Crane Company, New York, 1976.

Gas Processors Suppliers Association, Engineering Data Book, tenth ed., Tulsa, OK, 1994.

E. Shashi Menon, Gas Pipeline Hydraulics, CRC Press, Boca Raton, FL, 2005.

D.L. Katz, et al., Handbook of Natural Gas Engineering, McGraw-Hill, New York, 1959.

W.D. McCain Jr., The Properties of Petroleum Fluids, Petroleum Publishing Company, Tulsa, OK, 1973.

E. Shashi Menon, Piping Calculations Manual, McGraw-Hill, New York, 2005.

V.S. Lobanoff, R.R. Ross, Centrifugal Pumps Design & Application, Gulf Publishing, Burlington, MA, 1985.

M. Mohitpour, H. Golshan, A. Murray, Pipeline Design and Construction, second ed., ASME Press, New York, 2003.

Chapter 13

Corrosion Protection

E. Shashi Menon, Ph.D., P.E.

Chapter Outline

INTRODUCTION

In this chapter, we discuss what pipeline corrosion is, how corrosion occurs, and the method employed to protect liquid and gas pipelines and associated facilities from corrosion damage.

13.1 CORROSION IN PIPELINES

Pipelines that are buried in the ground or situated above the ground are subject to the surrounding environment that over a period causes corrosion. This could result in reduced pipe wall thickness, strength, and integrity. Unprotected pipelines, whether above the ground or buried in the ground, or submerged in water, are susceptible to corrosion. Unless these pipelines are properly installed and maintained, they will eventually deteriorate. This degradation of pipelines as a result of environmental effects on pipe material, coatings, or welding is called corrosion. Corrosion can weaken the structural integrity of pipelines and render them unsafe for transporting liquids or gases. Over the years, pipeline failures due to corrosion have been documented by regulatory agencies and in newspapers, trade journals, and on the Internet.

13.2 CAUSES OF PIPELINE FAILURE

The US Department of Transportation's Research and Special Programs Administration, Office of Pipeline Safety (RSPA/OPS) has compiled data on pipeline accidents and their causes.

The combined data for the period 2002–2003 show that "outside force" damage contributes to a larger number of pipeline accidents and incidents than any other category of causes, when you consider all accidents involving liquid and gas pipelines together. When liquid pipeline data is considered separately, corrosion is the major factor for accidents.

Outside force damage includes the effects of natural causes such as earth movement, lightning, heavy rains and flood, temperature, high winds, as well as third-party damage due to excavation by the operator or fire or explosion external to the pipeline, being struck by vehicles not related to excavation, rupture of previously damaged pipe, and vandalism. Excavation damage ranges from damage to the external coating of the pipe, which can lead to accelerated corrosion and the potential for future failure, to cutting directly into the line and causing leaks or, in some cases, catastrophic failure.

Tables 13.1–13.3 show the RSPA/OPS statistics, for the period 2002–2003. They show the various categories of pipeline accidents. It can be seen that in liquid pipelines, corrosion accounted for 25.4% of all accidents, representing 23.8% of total property damage. In comparison, for natural gas transmission pipeline accidents, corrosion caused 25.6% of all accidents and 36.6% of property damage. In natural gas distribution pipelines, these numbers were 1.2% and 0.1%, respectively.

Corrosion is highly complex and requires extensive expertise and significant resources to control. The 2001 US Federal Highway Administration-funded cost of corrosion study, "Corrosion Costs and Preventive Strategies in the United States," which was initiated by NACE International and conducted by CC Technologies, Inc., determined the annual direct cost of corrosion to be $276 billion, or approximately 3.1% of the gross domestic product. This was based on an analysis of direct costs by industry sector. The study also found that $121 billion is attributed to corrosion control methods, services, research and development, education, and training. Corrosion cannot be completely eliminated, and there will always be a cost associated with controlling corrosion.

TABLE 13.1 Hazardous Liquid Pipeline Accident Summary by Cause 1/1/2002–12/31/2003

Reported Cause	Number of Accidents	% of Total Accidents	Barrels Lost	Property Damages	% of Total Damages	Fatalities	Injuries
Excavation	40	14.7	35,075	$8,987,722	12.0	0	0
Natural forces	13	4.8	5,045	$2,646,447	3.5	0	0
Other outside force	12	4.4	3,068	$2,062,535	2.8	0	0
Materials or weld failure	45	16.5	42,606	$30,681,741	41.0	0	0
Equipment failure	42	15.4	5,717	$2,761,068	3.7	0	0
Corrosion	69	25.4	55,610	$17,775,629	23.8	0	0
Operations	14	5.1	8,332	$817,208	1.1	0	4
Other	37	13.6	20,022	$9,059,811	12.1	1	1
Total	272		175,475	$74,792,161		1	5

TABLE 13.2 Natural Gas Transmission Pipeline Incident Summary by Cause 1/1/2002–12/31/2003

Reported Cause	Number of Incidents	% of Total Incidents	Property Damages	% of Total Damages	Fatalities	Injuries
Excavation damage	32	17.8	$4,583,379	6.9	2	3
Natural force damage	12	6.7	$8,278,011	12.5	0	0
Other outside force damage	16	8.9	$4,688,717	7.1	0	3
Corrosion	46	25.6	$24,273,051	36.6	0	0
Equipment	12	6.7	$5,337,364	8.0	0	5
Materials	36	20.0	$12,130,558	18.3	0	0
Operation	6	3.3	$2,286,455	3.4	0	2
Other	20	11.1	$4,773,647	7.2	0	0
Total	180		$66,351,182		2	13

TABLE 13.3 Natural Gas Distribution Pipeline Incident Summary by Cause 1/1/2002–12/31/2003

Reported Cause	Number of Incidents	% of Total Incidents	Property Damages	% of Total Damages	Fatalities	Injuries
Construction/operation	20	8.1	$3,086,000	6.7	0	16
Corrosion	3	1.2	$60,000	0.1	2	9
Outside force	153	62.2	$32,334,352	70.1	6	48
Other	70	28.5	$10,617,683	23.0	13	31
Total	246		$46,098,035		21	104

The study also estimated that 25%–30% of annual corrosion costs could be saved by employing good corrosion management practices.

Corrosion in a pipeline is a naturally occurring phenomenon defined as the gradual deterioration of the pipe material or its properties due to a reaction of the pipe with its environment. The environment consists of the fluid in the pipe, the surrounding soil, and the atmosphere. There are proved methods that properly applied can mitigate and control corrosion of pipelines ensuring safe and economic operation.

Pipelines are protected from corrosion by coating the pipe externally using fusion-bonded epoxy (FBE) and other types of coating as well as internally lining pipes and using cathodic protection (CP) techniques described in section 13.4. Figure 13.1 shows a failure in pipe coating that resulted in corrosion of pipelines due to improperly applied coating and heat shrink sleeves.

13.3 TYPES OF CORROSION

There are several types of pipeline corrosion as listed below. Most corrosion failures occur due to a combination of more than one of these types.

1. General attack corrosion
2. Localized corrosion
3. Galvanic corrosion
4. Environmental cracking
5. Flow assisted corrosion
6. Intergranular corrosion
7. Dealloying
8. Fretting corrosion
9. High-temperature corrosion

13.3.1 General Attack Corrosion

This type of corrosion occurs more or less uniformly over an exposed surface compared to localized corrosion. This results in fairly uniform thinning of pipe material. The mechanism is an electrochemical process that occurs at the surface of the material.

13.3.2 Localized Corrosion

Localized corrosion occurs at a specific area of the pipe surface. This includes pitting, crevice, and filiform corrosion. Pitting results in deep, narrow attack causing rapid penetration of the pipe wall thickness. Crevice corrosion occurs at localized sites such as crevices where materials meet. Filiform corrosion is a form of oxygen cell corrosion that occurs underneath organic or metallic coatings on pipe materials.

Blistering of FBE joint field coating

Corrosion under a 3-year-old 3-layer PE coating

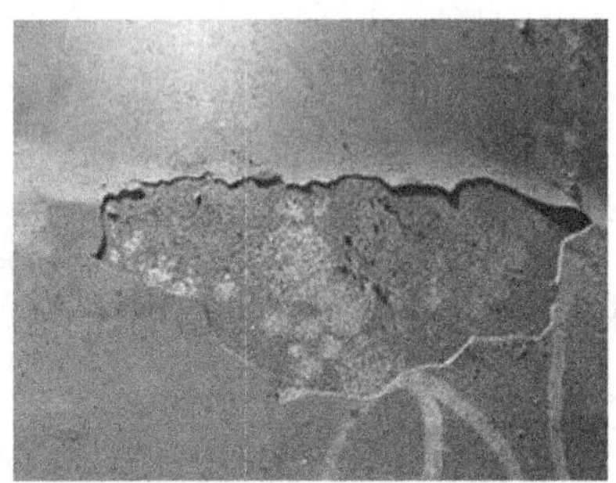

Foaming of an FBE coating

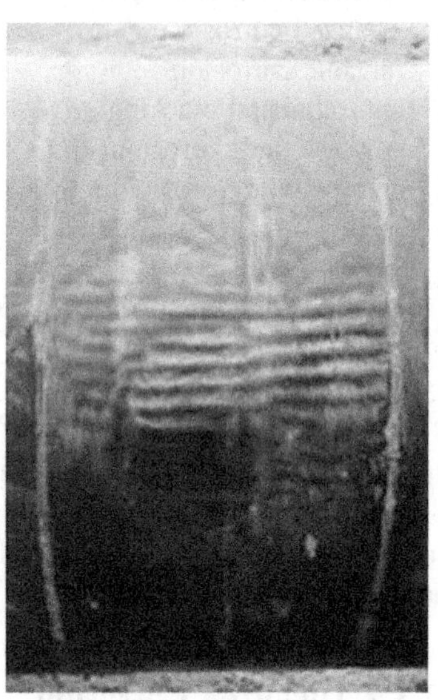

Heat shrink failure

FIGURE 13.1 Pipeline coating corrosion damage.

13.3.3 Galvanic Corrosion

Galvanic corrosion occurs due to the potential differences between metals when they are in electrical contact and exposed to an electrolyte. Dissimilar metals accelerate the process. This is due to electrochemical action that causes one metal to be the anode and the other the cathode. The anode corrodes and the cathode is protected. Table 13.4 shows the galvanic series for metals in seawater. The more active metal is the anode that corrodes and the less active one is the cathode.

13.3.4 Environmental Cracking

Environmental cracking occurs very rapidly and can be catastrophic. It is the brittle failure of a ductile material due to the combined effect of corrosion

TABLE 13.4 Galvanic Series for Metals in Seawater

Active (More Negative) End
Magnesium
Zinc
Aluminum alloys
Carbon steel
Cast iron
13% Cr (type 410) SS (active)
18-8 (type 304) SS (active)
Naval brass
Copper
70-30 Copper-nickel alloy
13% Cr (type 410) SS (passive)
Titanium
18-8 (type 304) SS (passive)
Graphite
Gold
Noble (More Positive) End
Platinum

Source: NACE International Basic Corrosion Course Handbook.

and tensile stress. This type includes stress corrosion cracking (SCC), hydrogen-induced cracking (HIC), liquid metal embrittlement (LME), and corrosion fatigue (CF).

13.3.5 Flow-Assisted Corrosion

Flow-assisted corrosion occurs due to the combined action of fluid flow and corrosion. It includes erosion-corrosion, impingement, and cavitation.

13.3.6 Intergranular Corrosion

Intergranular corrosion occurs at the grain boundaries of a metal when the grain boundaries or areas directly adjacent to them are anodic to the surrounding grain materials, caused by differences in impurity levels or the strain energy of the misalignment of atoms in the grain boundaries.

13.3.7 Dealloying

Dealloying occurs when one constituent of an alloy is removed preferentially, thus creating an altered residual structure. Since alloys consist of mixtures of elements, one element acts as an anode compared to another. Thus, by galvanic action, selective corrosion occurs.

13.3.8 Fretting Corrosion

Fretting corrosion occurs due to metal deterioration resulting from repetitive slip at the interface between two surfaces in contact. The movement between surfaces can remove protective films or, combined with the abrasive action of corrosive products, mechanically remove material from surfaces.

13.3.9 High-Temperature Corrosion

This type of corrosion occurs due to chemical reactions resulting in the deterioration of metals at high temperatures.

13.4 CORROSION CONTROL

Modern pipelines are protected from corrosion using CP techniques that include external pipe coatings combined with impressed current or sacrificial anodes. Proper use of today's technology can extend the life of the pipeline considerably, by proper application and consistent maintenance. Furthermore, employing persons trained in corrosion control is crucial to the success of any corrosion mitigation program. When pipeline operators assess risk, corrosion control must be an integral part of their evaluation.

As a first step, the environment in which the pipeline is located must be investigated. These include soil to pipe potential measurements at various points along the length of the pipeline as well as determination of the resistivity of the soil. Sometimes, just changing the environment surrounding a pipeline, such as reducing moisture by improving drainage, can be a simple and effective way to reduce corrosion.

Four common methods used to control corrosion on pipelines are protective coatings and linings, cathodic protection, materials selection, and inhibitors.

13.4.1 Protective Coatings

External pipe coatings are used for CP of the exterior of the pipeline. Internally coated pipe (lining) is used to protect the interior of the pipe from caustic and abrasive fluids transported through the pipeline. Internally coated pipe also reduces friction in the pipeline and hence improves the possible flow rate. Protective coatings are used in conjunction with CP systems to provide the most cost-effective protection for pipelines.

External coatings, applied on the outside of the pipe to form a barrier between the pipe material and the environment consist of the following:

1. Bitumastic materials such as coal tar, asphalt, or bitumen used with steel, cast iron, and concrete pipes
2. Epoxies such as FBE and other polyester materials
3. Wrappings consisting of tape applied around the pipe used on coated and uncoated pipes
4. Shrink sleeves used to protect welded joints of pipe

Common coating application methods include brush or roller, spray, and dipping. In addition to proper coating selection and application methods, substrate preparation is critical to the success of the coating. The majority of coating failures are caused either completely or partially by faulty surface preparation, such as leaving contaminants on the surface or having an inadequate anchor (sand blast) pattern. See Fig. 13.1 for examples of corrosion due to improperly applied external coating.

13.4.2 Cathodic Protection (CP)

The concept of cathodic protection goes back to 1824, in London, when the copper sheeting on British ships was being corroded. Experiments were conducted that resulted in protecting copper against corrosion from seawater utilizing iron anodes. Since then, CP has grown to have many uses in marine and underground structures, water storage tanks, gas pipelines, oil platform supports, and many other facilities exposed to corrosive environments.

CP is a technique by which a buried pipeline is made as the cathode and another metal is used as the anode that gets corroded by electrochemical process instead of the pipe material, as shown in Fig. 13.2.

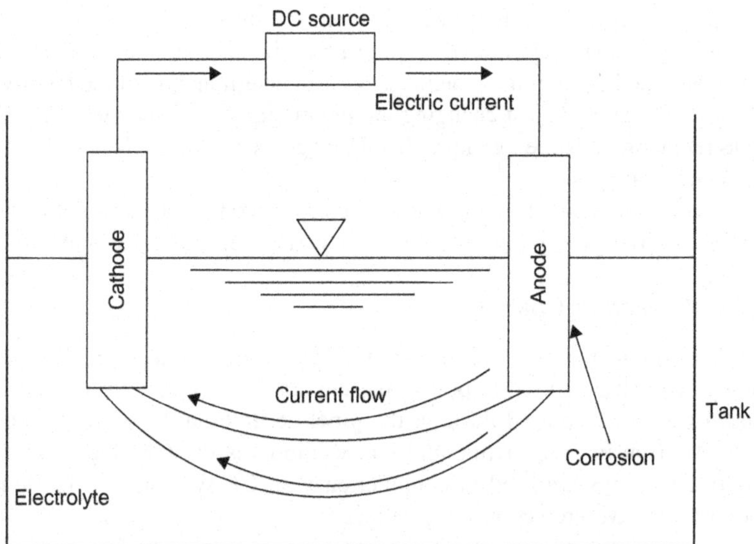

FIGURE 13.2 Electrolytic cell.

The electric circuit requires a source of protective current. This is provided by either an active (impressed current) or a passive (sacrificial) system of galvanic anodes (usually magnesium, aluminum, or zinc).

A direct current (DC) system is used to counteract the normal external corrosion of a metal pipeline resulting from the potential difference between the pipe and the soil.

Cathodic protection basically reduces the corrosion rate of a metallic structure by reducing its corrosion potential, bringing the metal closer to an immune state. The two main methods of achieving this goal are by either sacrificial anodes or the impressed current method:

- **Sacrificial anodes:** This involves connecting the pipe to a zinc or magnesium electrode and the surrounding soil in a buried pipeline. This creates a galvanic cell shown in Fig. 13.3 in which the pipe becomes the cathode and the zinc or magnesium electrode becomes the anode that gets corroded. Hence, the reason for the term sacrificial anodes.
- **Impressed current method:** This method requires the use of an external DC source (rectifier). The negative terminal of the rectifier is connected to the steel pipe and the positive terminal is connected to the ground through an electrode. The pipe thus becomes the cathode as shown in Fig. 13.4. When a copper sulfate electrode is used, the impressed current rectifier is adjusted so that the pipe to soil potential is maintained at −0.85 V. This potential should be maintained at various points along the pipeline.

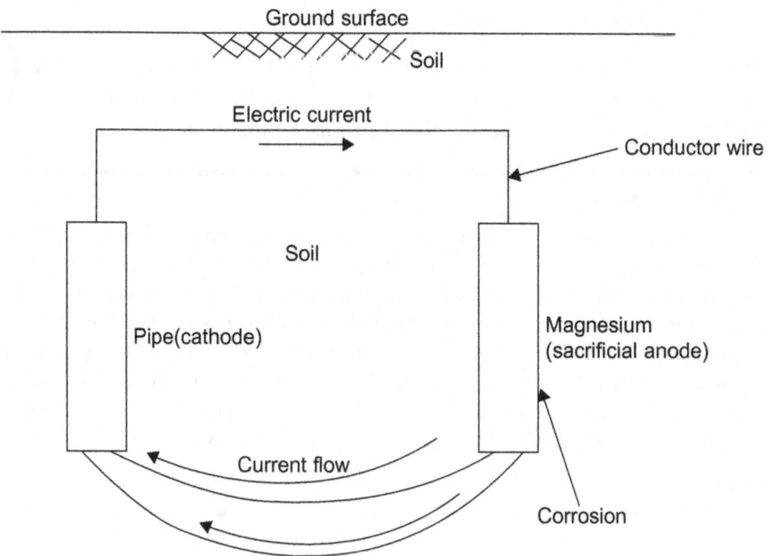

FIGURE 13.3 Galvanic cell.

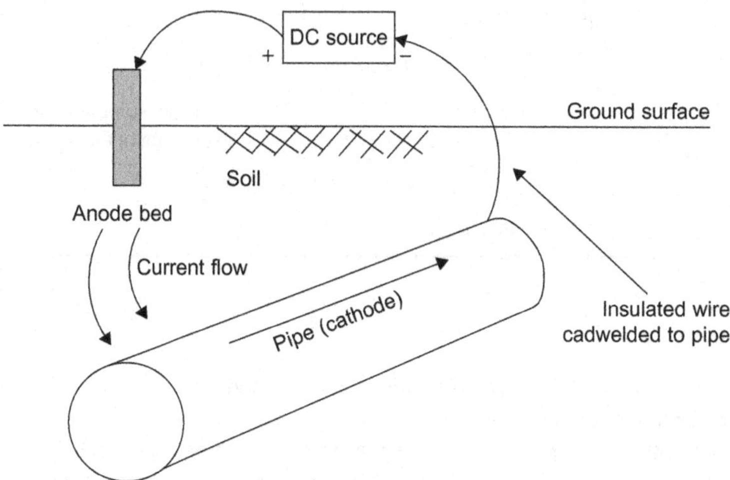

FIGURE 13.4 Impressed current system.

13.4.3 Materials Selection and Design

Corrosion can be mitigated by proper selection of materials for the specific application. Corrosion-resistant materials include stainless steels, plastics, and special alloys used in conjunction with pipelines. Materials selection must take into account the design life of the pipeline and the environment surrounding it.

13.4.4 Corrosion Inhibitors

Corrosion inhibitors such as acids, cooling waters, and steam are used to decrease the rate of corrosion by adding them to the environment. These inhibitors control corrosion of the metal by forming thin films at the metal surface.

Example Problem 13.1

An impressed current CP system is to be designed for a 24 mi pipeline NPS 20 with FBE coating. The maximum protective range for the CP installation is 30 mi. The ground bed is proposed to be installed at milepost 12. The soil resistivity at the anode burial depth is 5000 ohm·cm. The current required after 25 years of service is estimated to be 6.5×10^{-6} A/ft^2. (1) Determine the total current requirement for the 24 mi pipeline. (2) If silicone-chromium cast iron anodes are used assuming 45 lb weight each, estimate the number of anodes required for the 25-year life span. Assume a consumption rate of 1.2 lb/A·yr and 70% consumption.

Solution

1. Total current required = current density × total surface area:

$$I_T = 6.5 \times 10^{-6} \times \pi \times (20/12) \times 24 \times 5280 = 4.3\ \text{A}$$

2. Since the standard rectifier size is 8 A, the number of anodes can be calculated from the consumption rate, percent volume consumption, rectifier size, anode weight, and life span as follows:

$$N = L \times I \times C/(V \times W)$$

where L is the life in years, I is the rectifier size, in A, C is the consumption rate, lb/A·yr, V is the percent volume consumption, and W is the weight of the anode, lb.

$$N = 25 \times 8 \times 1.2/(0.7 \times 45) = 7.6$$

Therefore, eight anodes are required.

SUMMARY

In this chapter, we discussed corrosion in general and how it affects buried and above-ground pipelines. The various types of corrosion were reviewed, and the methods employed to mitigate corrosion damage in pipelines were discussed. The importance of externally coating pipelines for corrosion mitigation and internally coating pipelines to reduce friction and abrasion were explained. The two main methods of cathodic protection used in modern-day pipelines, sacrificial anodes and impressed current systems, were discussed.

BIBLIOGRAPHY

L.S. Van Delinder, Corrosion Basics, An Introduction, ed., NACE, Houston, TX, 1984.

NACE International Basic Corrosion Course Handbook, NACE, Houston, TX, 2000.

Chapter 14

Leak Detection

Hal S. Ozanne

Chapter Outline

INTRODUCTION

The installation and operation of a leak detection and leak prevention system on pipeline systems is normally a standard part of a pipeline system and operations. In some cases, leak-detection systems are mandated by regulations. The purpose of the system is to be able to prevent a leak from occurring and to detect a leak of product from the line, of significant volume, and take the necessary measures to shut the line down to minimize the impact of a leak and to repair.

Leaks can be hazardous, causing an explosion or an environmental situation such as crude oil or refined products leaking from a line and flowing into a waterway. In addition, a leak equates to a loss of product in the system, which equates to a loss of revenue to the pipeline operator. There is a potential for loss of life and damage to property with any size leak but even more so with larger leaks.

The cost to repair damage to property and the environment can be very significant. There can be fines accessed to the pipeline operator and in some instances, managers of the pipeline operating companies have been tried, found guilty of neglect, and sentenced to prison. Therefore, it is imperative that a reliable leak-detection system be installed and maintained on all pipeline systems.

When a leak in a pipeline results in significant damages to surrounding property and/or death, the pipeline industry comes under severe public scrutiny. The result may mean that the United States Department of Transportation (USDOT) Office of Pipeline Safety (OPS) will be called before the US Congress to explain how these catastrophic events can take place. In many cases, the result is that more stringent regulations will be imposed on the pipeline industry. The operators would do better to self-police themselves to take the necessary actions to install more robust leak prevention and detection systems. This is a better solution than the OPS imposing very costly regulations on the industry, which can be avoided.

In April of 2008, the USDOT/Pipeline and Hazardous Materials Safety Administration issued a Leak Detection Report to the US Congress. Figures 14.1 and 14.2 are excerpts from that report. The first figure shows statistical data as to volumes lost due to leaks in pipelines. The second one lists systems that can be used to prevent or lessen the impacts from a leak. Both figures help emphasize the importance of leak detection on pipeline systems.

Leaks in a pipeline can be caused by many things including internal or external corrosion of the pipe, third party damage such as construction equipment striking the pipe, equipment failure such as a valve or flange failure, operating the line above its rated operating pressure or temperature, which may result in a failure, and intentional damage or sabotage by outside parties.

Some commodities that are transported by pipe are corrosive by their nature. The correct metallurgy of the pipe material must be selected to protect it from the corrosive properties of the product being transported. This may require

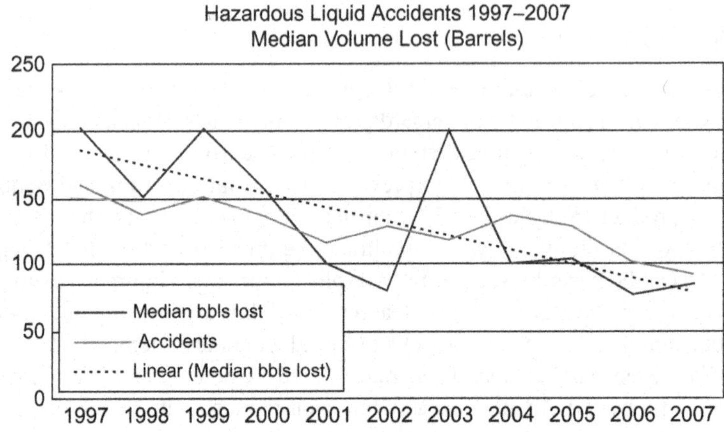

FIGURE 14.1 Hazardous liquid accidents 1997–2007.

Emphasis on prevention
Congressional report

- Surveillance
- Cathodic protection
- Pressure control
- Relief systems
- Damage prevention
  - Line marking
  - One-call systems
  - Public awareness
  - Common ground alliance
- Pipelines and informed planning alliance – PIPA

FIGURE 14.2 Emphasis on prevention.

extensive testing and research of the product to be transported and the metallurgy of the pipe material.

14.1 PREVENTION

The most effective way to prevent a leak from occurring is to take measures to prevent a leak from developing. Prevention methods include the following:

- Pressure regulation
- Cathodic protection
- Corrosion coupons
- Pipeline markers
- Smart pigging
- Pipeline security
- In-line valves
- Patrolling

14.2 PRESSURE REGULATION

The discharge pressure of the product entering the pipeline should be monitored and regulated to prevent the pressure from exceeding the maximum allowable operating pressure (MAOP) of the pipeline. This can be accomplished by installing overpressure devices, such as high pressure detection switches, on pumps or compressors discharging into the pipeline. The switches are set to a pressure just below the pipeline MAOP. When the pressure reaches the set point, the pumping units will be shut down to prevent an overpressure situation.

Pressure relief valves are installed on, at least, one end of a pipeline system. These valves are sized to handle the full flow of the pipeline in the event the prime mover does not shut off. Their set point is to open and relieve into a tank or another device to prevent the pipeline from being overpressured and possibly rupturing.

14.3 CATHODIC PROTECTION

Cathodic protection systems are installed on pipelines and related facilities to prevent corrosion, which reduces pipe and fitting wall thickness that can eventually develop into a leak. This subject is covered in detail in Chapter 13.

14.4 CORROSION COUPONS

In cases where the product to be transported by pipeline is known or thought to have corrosive properties to the materials of the pipe, valves, or other components of the pipeline system, corrosion coupons are inserted into the line to monitor the corrosive properties. Corrosion coupons are inserted into the flow stream of a pipeline, usually in a facility at a location where a pipeline "pig" will not be passing by the coupon.

The coupons are removed and analyzed at predetermined time intervals. The material composition of the coupon is the same as the material in the pipe. By comparing the weight of a coupon over a period of time, a determination can be made whether internal corrosion of the pipe is occurring. If unchecked, internal corrosion of the pipe will develop, which will cause pits or thinning of the pipe wall thickness and the eventual development of a leak.

Corrosion inhibitors can be injected into the flow stream to eliminate or greatly reduce the amount of internal corrosion of the pipe. The pipe wall thickness can be increased above what would normally be required for pressure requirements during installation to add a corrosion allowance to the pipe.

14.5 PIPELINE MARKERS

There are generally three types of pipeline markers used:

- Markers placed along the pipeline route, indicating that a pipeline is present
- Aerial markers
- Water crossing markers

14.5.1 Markers

Pipeline markers are installed at locations along the pipeline, to mark the location of the line and identify that a pipe is located there to anyone in the area. The intent is that if a contractor or anyone else is planning to do any work in the area, they will see the pipeline markers and call the number on the marker to obtain more detailed information as to the precise location and depth of the line. This is to prevent a piece of construction equipment from hitting the pipe and causing a leak or weakening the pipe at a point where a leak may develop at a later time. In urban areas, the local "One Call" office can be contacted with a request to locate any underground utilities and pipelines in the area of concern.

Figure 14.3 shows what a typical pipeline marker looks like.

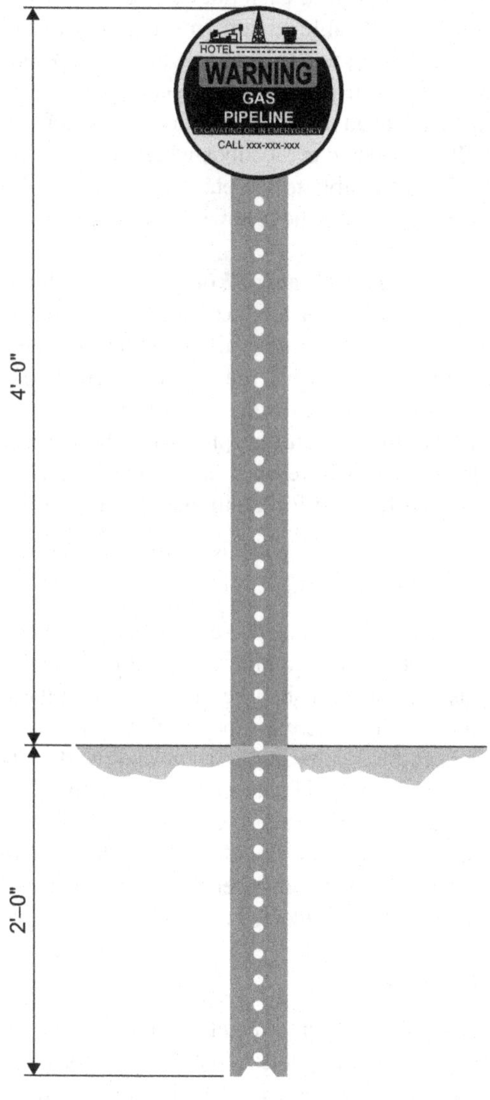

FIGURE 14.3 Typical pipeline marker.

ASME B31.4, Paragraph 434.18, establishes the location for pipeline markers for hazardous liquid pipelines.

1. The code states that adequate pipeline location markers for the protection of the pipeline, the public, and persons performing work in the area shall be placed over each buried pipeline in accordance with the following:
   a. Markers shall be located at each public road crossing, at each railroad crossing, at each navigable stream crossing, and in sufficient numbers along the remainder of the buried line so that the pipeline location including direction of flow the pipeline is adequately known. It is recommended that markers be installed on each side of each crossing whenever possible.
   b. Markers shall be installed at locations where the line is above the ground in areas that are accessible to the public.
2. The marker shall state at least the following on a background of sharply contrasting colors:
   a. The word "Warning" or "Caution" or "Danger" followed by the words "Petroleum [or the name of the hazardous liquid transported] Pipeline" or "Carbon Dioxide Pipeline," all of which, except for markers in heavily developed urban areas, must be in letters at least one inch high with an approximate stroke of one-quarter inch.
   b. The name of the operator and telephone numbers (including area code) by which the operator can reached at all times.
3. API RP 1109 should be used for additional guidance.

ASME B31.8, Paragraph 851.7, establishes the locations for markers for gas transmission and distribution lines.

1. Signs or markers shall be installed where it is considered necessary to indicate the presence of a pipeline at road highways, railroads, and stream crossings. Additional signs and markers shall be installed along the remainder of the pipeline at locations where there is a possibility of damage or interference.
2. Signs or markers and the surrounding right-of-way shall be maintained so that markers can be easily read and are not obscured.
3. The signs or markers shall include the words "Gas [or name of gas transported] Pipeline." The markers should also include the name of the operating company and the telephone numbers (including area code) by which the operating company can be contacted.

14.5.2 Aerial Markers

Aerial markers are installed at approximately 1-mile intervals. These are used as a reference point by aerial patrol pilots so that if they spot a leak, a construction or encroachment activity near or over the pipeline, or any other activity near the pipeline, they will call the pipeline operations center to report the activity, providing the milepost where the problem is located as a reference point.

Figure 14.4 shows what a typical aerial pipeline marker looks like.

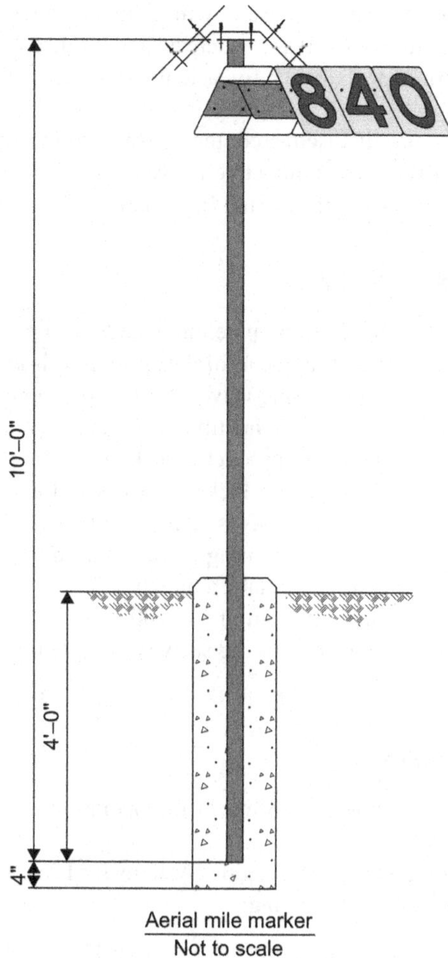

FIGURE 14.4 Shows what a typical aerial pipeline marker looks like.

14.5.3 Water Crossing Markers

Large signs are installed on the shore of both sides of pipeline water crossings to identify that a pipeline is buried under the river. This is extremely important in navigable waters, where a boat or ship may be dragging an anchor or a dredging operation may be taking place, to prevent damage from occurring to the line.

14.6 SMART PIGGING

Inserting and running a smart pig in a pipeline segment is used to determine if there are any abnormality issues present or if any may be developing with the pipe. These issues may include internal and external corrosion, abnormalities such as indentations in the pipe caused by mechanical damage to the pipe, and potential problems with welds. If these defects are not discovered and repaired, then there is a strong probability for a leak to develop in the area of the abnormality.

When an abnormality is discovered, the line will be excavated at that point, the area will be analyzed, and protective measures or repairs will be made.

Pipeline pigging is covered in detail in Chapter 15.

14.7 PIPELINE SECURITY

Security measures must be taken by pipeline operators to prevent unauthorized entry, operation, and misuse of any part of the system. This unauthorized operation might include opening or closing valves, starting or stopping pumps or other equipment, or turning off power to a facility. Such actions may result in an upset in the system operation and cause product to leak out of the pipeline or facility.

Security measures include chain locking valves in their normal operating position, installing fences around valves sites and other facilities with locking gates, and installing intrusion monitoring systems in unmanned facilities with alarms that will sound both locally and remotely.

Pipe that is installed above the ground outside of fenced facilities is sometimes protected from sabotage by placing sleeves over the pipe to protect it from intentional damage.

14.8 REGULATIONS

Pipeline regulations require monitoring of the operation of pipelines to check for and identify leaks.

ASME B31.4, Pipeline Transportation Systems for Liquid Hydrocarbon and Other Liquids, requires the following:

> Patrolling the route is to be completed on 2-week intervals, except for LPG or ammonia pipelines, which are to be patrolled at 1-week intervals.
> Selection and implementation of the leak-detection system should take into account the risk of both the likelihood and consequence of a leak.

Factors that could reduce the risk when determining the type and frequency of monitoring include the following:

- Service
- Location
- Construction
- Operating at low stress levels of the pipeline system
- Leak history

Response time should also be taken into consideration. The longer the response time, the more it supports the need for a faster detection system.

A leak-detection system can include the following:

- Right-of-way patrol
- Analysis of blocked in pressure
- Monitoring of changes in flow or pressure
- Volumetric line balance
- Other methods that can detect a leak in a timely manner

If a computerized monitoring system (CPM) is used, API RP 1130 should be followed. An abstract from RP 1130 is as follows:

14.9 PURPOSE

This recommended practice focuses on the design, implementation, testing and operation of CPM systems that use an algorithmic approach to detect hydraulic anomalies in pipeline operating parameters. The primary purpose of these systems is to provide tools that assist Pipeline Controllers in detecting commodity releases that are within the sensitivity of the algorithm. It is intended that the CPM system would provide an alarm and display other related data to the Pipeline Controllers to aid in decision-making. The Pipeline Controllers would undertake an immediate investigation, confirm the reason for the alarm and initiate an operational response to the hydraulic anomaly when it represents an irregular operating condition or abnormal operating condition or a commodity release.

The purpose of this recommended practice is to assist the pipeline operator in identifying issues relevant to the selection, implementation, testing, and operation of a CPM system. It is intended that this document be used in conjunction with other API standards and applicable regulations.

ASME B31.8, Gas Transportation and Distribution Piping Systems, requires the following:

> Each operating company of a transmission line shall provide for periodic leakage surveys of the line in its operating and maintenance plan. The types of surveys selected shall be effective for determining if potentially hazardous leakage exists. The extent and frequency of the leakage surveys shall be determined by the operating pressure, piping age, class location, and whether the transmission line transports gas without an odorant.

Each operating company having a gas distribution system shall set up in its operating and maintenance plan a provision for making periodic leakage surveys on the system.

The types of surveys selected shall be effective for determining if potentially hazardous leakage exists. The following are some procedures that may be employed:

- Surface gas detection surveys
- Subsurface gas detector surveys

- Vegetation surveys
- Pressure drop tests
- Ultrasonic leakage tests

B31.8, Appendix M, Gas Leakage Control Criteria, provides criteria for detection, grading, and control of gas leakage.

The DOT regulations Part 192, Section 706, *Transmission Lines: Leakage Surveys* state the following:

> Leakage survey of a transmission line must be conducted at intervals not exceeding 15 months, but at least once per calendar year.
>
> If the pipeline is being operated without odorant, then in
>
> Class 3 locations, the intervals shall not exceed 7.5 months, but at least two times per calendar year;
> Class 4 locations, the intervals shall not exceed 4.5 months, but at least four times per calendar year.
>
> The DOT regulations Part 195, Section 402 states the following:
>
> Operator must take necessary action, such as emergency shut down or pressure reduction to minimize the volume of hazardous liquid or CO_2 that is released from any section of a pipeline system in the event of a failure.

14.10 INTERMEDIATE BLOCK VALVES

During the construction of a pipeline, depending upon the length of the line, intermediate block valves are installed along the line. The minimum spacing of the valves is governed by regulations in some instances. See Chapter 10 for more details of valve installation. Due to terrain, location of access roads, and other factors, the operator may elect to install valves at intervals less than that required by regulations.

If a leak is detected and as soon as the general location of the leak is determined, the block valves on either side of the leak location are shut to confine the leak to the pipe distance between the block valves.

The pipeline operator may decide to install automatic or remotely operated actuators on some of the valves. If a leak is detected, the pipeline controller can shut the operation of the pipeline down and remotely close the remote actuated valves. This will prevent the leak from continuing to be fed by the line.

If the pipeline operator has installed a sophisticated computerized pipeline control and monitoring system to operate the pipeline system, the computer system may be programmed to shut the system down and close the valves once a leak is detected.

14.11 CHECK VALVES

Check valves are installed along pipelines at strategic locations. These locations may be at the downstream side of major water bodies and at low elevation points in the line. If a leak were to develop at a water crossing, once the pipeline is shut down or the pressure is reduced in the area, the check valve will close and prevent product in the line, downstream from the check valve, from flowing backwards and fueling the leak from the downstream direction.

14.12 PATROLLING

Patrolling pipeline routes will detect if there is a construction activity occurring in the vicinity of the pipeline, which could lead to the pipeline being damaged by construction equipment.

14.13 DETECTION

There are many different types of leak-detection systems that are used. The type is normally based on factors such as the pipeline system length, pipe size, operating pressure, throughput of the pipeline, and whether the pipeline traverses heavily populated areas or not.

Types of leak detection include the following:

- Patrolling the pipeline on foot or by air with a helicopter or fixed-wing aircraft. Patrolling is used to look for leaks such as liquid product on the surface of the ground, listening for noise, such as listening for a high-pitched sound when product or gas is escaping from the pipe under high pressure, erosion caused by high-pressure liquid, or gas escaping from the pipe to the surface of the ground, and sniffing for odor from gas leaking out of a pipeline that has had odorant injected into the gas stream. The injecting of odorant into a pipeline normally applies to distribution lines. Odorant is not commonly used in gas transmission lines.
- Measuring the product flow as it enters the pipeline and then again when it leaves the pipeline and comparing the volumes over a specific time period. A significant variation may indicate the possibility of a leak and further investigation should be made.
- Monitoring operating pressure at locations along the line for a significant drop in pressure, which may indicate a leak.
- A sophisticated computer-based modeling and control system that is used to remotely operate and monitor a pipeline can also have leak detection incorporated into it. Systems are monitored 24 hours a day, 365 days a year. These systems take into consideration measurement, operating pressure, flow rates, line temperature, and other operating dynamics of the system.

Provisions may include the following:

- Maintaining the right of way of the pipeline so that it is not overgrown with trees and brush and ensuring that structures have not been constructed over the pipeline right of way so that the right of way of the pipeline is clearly identified and if there is a leak it may be more easily spotted.
- Installation of measurement and pressure-monitoring equipment at the beginning and end points and any injection or delivery points along the pipeline and, perhaps, intermediate locations along the line.
- Installation of remote-operated block valves.

14.14 MEASUREMENT

In most cases, the amount of product or gas being injected into a pipeline is measured by some means. This measurement may be the custody transfer point from the shipper of the product to the pipeline operator as well as to monitor the amount of product injected into the line. For the same reasons, the product is measured again when it is delivered out of the pipeline, both for custody transfer and to monitor the amount of product delivered out of the line. The measurement readings are also used for leak-detection purposes. The volumes of product injected into the line are compared to the volumes delivered out of the line during a prescribed time period. In addition to the volume, pipeline operating pressures and temperatures, and perhaps the specific gravity of the product, are measured and monitored at the measurement locations. The volumes are compensated for temperature, pressure, and specific gravity so that "net" volumes are compared for a given length of time. If the measurement readings indicate that the volume being delivered out of the pipeline is less than what is being injected into the line over a period of time, then a system leak may have developed. If there is a large discrepancy in a relatively short amount of time, then a large leak may be indicated and the line may be shut down and operations' crews notified. If the discrepancy is small, then there may be other issues such as those with the measurement equipment and further investigation should be taken.

There are many different types of measurement systems that are used in the operation of pipelines. The descriptions of the different types of measurement are not the intent of this chapter. The type or system of measurement must be of a type that provides data that are accurate enough for quality measurement for leak-detection purposes.

14.15 SUPERVISORY CONTROL AND DATA ACQUISITION SYSTEM

Supervisory Control and Data Acquisition Systems (SCADA) are used to remotely monitor and control many functions of pipeline operations. Remote Terminal Units are installed at each location where data will be sent back to a control center and where control functions are received from the control center. A control center is a location where one or more pipelines can be monitored or controlled. This typically is at the headquarters of the pipeline operating company for larger pipeline operators.

The control center and the remote locations communicate through a communication system such as satellite, hard wire, radio, and the Internet.

Depending upon the size of operation, the control center may be manned 24 hours a day, 365 days a week. The control center can consist of an elaborate computer-based system that monitors, displays, provides some automatic controls, and gives the center operator the functionality of remotely controlling many aspects of the pipeline operation. Trends can be followed and reports generated.

The degree of automatic control is determined by the pipeline operating company. The control center operator will normally send the signals to remote locations to control valves, pumps, compressors, etc. Human interpretation of the data being monitored and taking the appropriate action are generally better than allowing a computerized system to take those actions.

When the operation changes beyond normal operation set points, the system will sound an alarm indicating a potential problem. The control center operator will then analyze and take corrective actions.

This is the case for leak detection. If the system detects a change in operating pressures, or a deviation of the volume of product entering the pipeline compared to the product exiting the line, an alarm is sounded to alert the operator of a potential problem. Depending upon the situation, the operator may determine that the line should be shut down and take those actions, which may include closing mainline block valves. The operator will also notify the pipeline operations department so that personnel can be dispatched to analyze the situation in the field.

Small pipeline systems may not have a central control center. In those cases, the relevant field data may be transmitted to a field office where it can be monitored. If a system deviation is detected, a local alarm will sound. During times when the office is not manned, the system can automatically call a "call out" phone number to notify operations personnel on call that a potential problem exists.

14.16 HYDROSTATIC TESTING

If a leak is suspected but it is too small to enable the detection of the location of the leak, the pipeline can be hydrostatically tested. If the pipeline is a gas pipeline, the gas in the pipeline or a segment of the line must be displaced with water prior to testing the line. For a hazardous liquid pipeline, in some cases the pipeline can be tested with the product that is in the pipeline, i.e., crude oil or gasoline. In other cases, the product in the line will be displaced with water before the testing can begin.

Depending upon the length and the elevation profile of the pipeline, the entire line will either be tested all at once or in segments, such as between two existing block valves. After the line has been filled with water, pressure will be applied. The test pressure will normally be 125% of the specified minimum yield strength of the pipe or 110% of the ANSI rating of the valves or other components in the pipeline. If the static test pressure cannot be maintained between the two valves, the pressure will be maintained by pumping additional water into the line and the route of the pipeline will be patrolled for the test segment to see if the product or water does come to the surface of the ground over a period of time.

If the pressure continues to bleed off and there is no noticeable liquid coming to the surface of the ground, the pressure will be bled off the line and the segment will be divided in two by closing a valve between the endpoints of the previous test segment. If there are no valves in the line segment to further divide the segment, then the line may need to be cut and capped to divide the line.

Once this has been completed, each of the remaining two segments will be hydrostatically tested. One of the segments should hold the test pressure and the pressure in the other should bleed off. The line segment for which the pressure bleeds off will be patrolled with the expectation that the cause for the loss of pressure will be found.

This process will continue until the leak is discovered. Once the location of the leak is determined, the pressure will be bled off the pipe and it will be repaired.

BIBLIOGRAPHY

ASME B31.4-2006, Pipeline Transportation Systems for Liquid Hydrocarbons and Other Liquids, American Society of Mechanical Engineers, New York, NY, 2006.

ASME B31.8-2007, Gas Transportation and Distribution Piping Systems, American Society of Mechanical Engineers, New York, NY, 2007.

CFR 49, Part 195, Transportation of Hazardous Liquids by Pipeline: Minimum Federal Safety Standards, U.S. Government Printing Office, Washington, DC.

CFR 49, Part 192, Transportation of Natural or Other Gas by Pipeline: Minimum Federal Safety Standards, U.S. Government Printing Office, Washington, DC.

Chapter 15

Pipeline Pigging and Inspection

Barry G. Bubar, P.E.

Chapter Outline

INTRODUCTION

Pigs have been used in the pipeline industry for about 80 years. The first pigs were made out of straw or wood and wrapped tightly with wire or barbed wire and propelled down the pipeline with the crude oil or product to do a cleaning job inside the pipe. This scraping tool made a squealing sound, like a pig, as it went by in the pipeline and that is where the name "pig" came about.

The next generation of pigs was known as the Go-Devils. They had leather driving disks and spring steel scraper blades, and some had spurs to chip off harder material and scale. They worked fine until during World War II (WWII) when gas pipeline diameters got much bigger. The Go-Devils were not able to withstand the higher forces resulting from the larger diameter pipelines; hence, new designs were necessary.

The leather disks were replaced with rubber cups, and the spring steel scrapers were replaced with wire brushes arranged in a circular configuration, which resulted in a very robust design. Modern pigs have come a long way from those first primitive ones.

15.1 PIG USE

Pipeline operation and maintenance becomes more effective and efficient when essential operation and maintenance jobs are performed using pigs. Many procedures that are needed for increased pipeline longevity, improved pipeline efficiency, and risk reduction, such as reducing the effect of internal corrosion, cannot be done without having pigging capability. Operators need to run cleaning pigs to dewax and descale the inside surface of the pipe and remove debris, which help improve pipeline performance, and they can do all this without significantly interrupting product flow to the customer.

Other pigging tasks include dewatering or swabbing, drying of gas pipelines after hydrostatic testing, batching products to minimize mixing of two different products at the liquid interface, and internal inspection of the entire pipeline system. Pipeline internal inspection can be done by using gauging plates, caliper pigs, or smart pigs. Smart pigs, also known as intelligent pigs or inline inspection (ILI) devices, will be discussed and analyzed in great detail later in this chapter.

Of the many reasons for pigging, the most important one is to maintain the original pipeline design performance and extend the useful life of the pipeline. Some experts are now stating that a properly designed, operated, and maintained pipeline has no lifetime limit and should be able to operate indefinitely.

15.2 PIPELINE PIGGING

Pipeline pigging is accomplished by inserting a pig into a launcher, see Fig. 15.1. The launcher hatch is then closed, and the valves are operated to allow pipeline pressure to push the pig out of the launcher and along the pipeline until it reaches the receiver trap, which catches it. The valves at the receiver are set by the pipeline operators so that the pig will go into the receiver barrel, and the pipeline flow will then bypass the barrel and go back into the main line without significantly interfering with the pipeline flow rate. The pipeline will continue in that manner until operating personnel arrive and open mainline valve A and then close the trap valve B and kicker/bypass valve C.

Figure 15.1 also shows some of the required vent and drain piping for a launcher/receiver system along with pressure relief valve piping and scraper trap instrumentation.

Note that code requires each launcher and receiver to be equipped with a pressure-relief device capable of safely relieving pressure in the barrel before inserting or removing scrapers (pigs). Also, a suitable pressure-indicating gauge must be operational so that the operator can be sure that barrel pressure has been relieved for safety before opening the hatch.

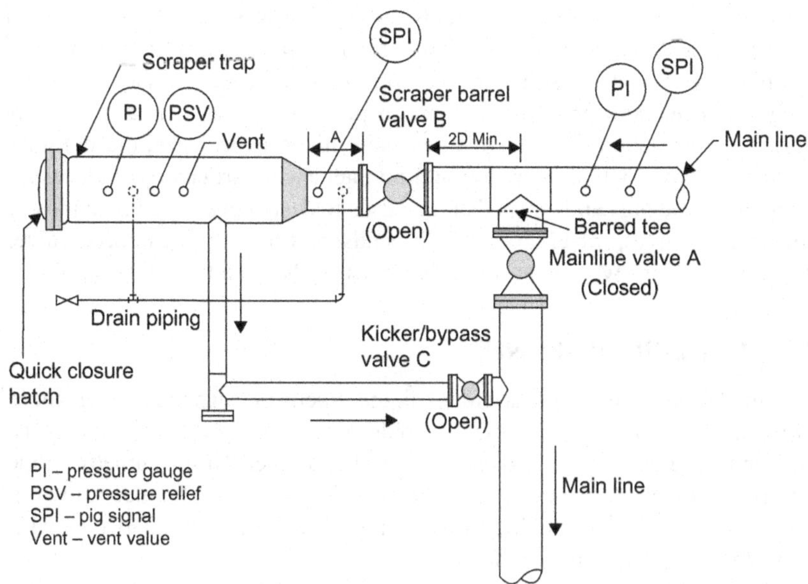

FIGURE 15.1 General piping for pig launcher/receiver station – plan view.

15.3 PROBLEM PIPELINES

The problem is that most of our aging pipeline systems were not equipped or designed to accommodate pipeline pigging. This is especially true in the smaller diameter, hazardous liquid pipelines located in urban areas. Because of the high demand for pipeline systems capable of transporting crude oil from production to refineries and products from refineries to distribution terminals and marine terminals, during and after WWII, pipeline designer's made mistakes, and did not consider long-term maintenance as part of their pipeline design. As a result, many liquid pipelines were not designed for pigging of any kind.

These older pipelines evolved over time by connecting different pipelines together and by constructing new pipeline extensions or laterals between existing pipelines, to supply product and crude oil as necessary to meet market demand. These pipelines had changes in diameter size, and many were constructed with short (or tight) radius bends and unprotected branch connections (meaning pipe tee connections without bars which are needed to keep pigs from catching and blocking the flow). Some pipelines were operated with reduced port valves or square port valves, and of course, all were without any pig launcher/receiver capability. Even worse, some of the old pipelines had mitered elbows and reduced port check valves. When corrosion leaks occurred, they were repaired by inserting a screw in the leaking corrosion pit and wrapping the pipe at that location with a steel band and rubber gasket.

The discovery of natural gas fields, in the United States, led to a boom in pipelining, and there was a rapid transition from small diameter, low-pressure gas pipeline systems, which were the norm before the war, to large diameter, long distance, high-pressure systems in just a few years. Like liquid pipe designers, the natural gas pipeline designer, due to expedience, did not focus on subsequent pipeline operation and maintenance, particularly with respect to pigging. The pig suppliers at that time tried to design pigs that would traverse square-port valves, travel around tight bends, and pass through check valves. These pig designs worked at times, but most of the time they did not.

15.4 PIGGABLE PIPELINES

Any modern natural gas or hazardous liquid pipeline designed and constructed in the last 20 years or so is now or should be equipped to launch and receive conventional pigs, which as mentioned earlier, are needed for numerous pipeline purposes. In most cases, these launcher/receiver barrels, also known as scraper traps, must be modified to accommodate the use of smart pigs when an internal inspection is warranted.

Figure 15.1 shows the general piping of a typical bidirectional pig launcher/receiver station. For smart pig use, which will be discussed later in this chapter, the pipe length between the trap valve and the barrel, needs to be much larger. This increase in length, which we call dimension A, is needed to support the heavy instrumented pig when it enters the trap. For a 16-in.-diameter pipeline, dimension A needs to be 13 ft long, and for a 24-in.-diameter pipeline trap, dimension A can be shorter at approximately 10 ft in length. This difference in length is because the smaller diameter instrumented pig must be made longer to package all the required smart pig equipment. Once modified for smart pig use, the typical pig trap station will work fine for conventional pigs as well. The receiver trap will be able to catch and hold a number of service pigs or conventional pigs before operators arrive to drain the barrel and remove the trapped pigs. They then clean up the scraper barrel and make ready the receiver station for catching the next incoming pig when necessary.

Piggable pipelines have their launchers and receivers spaced at reasonable distances, some spaced at 50 or 70 miles apart. The pipeline should be designed with long radius shop bends. Bared tees should be used at all pipeline branch connections to eliminate the problem of pigs catching at a pipeline lateral. All mainline valves must be full-port design, in accordance with API-6D, and check valves if required in the mainline must be of the pig passing design. With pipelines designed in this manner, operations will have little problem in launching and passing pigs of any kind.

Government code CFR Title 49, Part 195.120 has addressed this pipeline design issue by stating in paragraph (a) "Except as provided in paragraphs (b) and (c) of this section, each new pipeline and each line section of a pipeline where the line pipe, valves, fittings or other line component is replaced, must be designed and constructed to accommodate the passage of internal inspection devices." Paragraph (c) states, in effect, that if it is not practical due to construction problems, schedules, etc., and the operator determines and documents why it is not practical to comply with paragraph (a), he must petition within 30 days after discovery, for approval, that the design and construction would be impractical.[1]

15.5 PIG PROPULSION

When launched, a pig is propelled down the pipeline by differential pressure. You can think of it as a piston in a tube being moved by differential pressure, and the amount of differential pressure required is determined by the friction force exerted on it by the pipe, see Fig. 15.2.

The friction changes when the internal diameter or the pig direction changes. When a pig is stopped at a reduction in diameter or when it enters a tight radius bend, the friction on the pig increases. Then, the pig will be held at that restriction until sufficient differential pressure builds up to overcome the friction and lets the pig move again. Once the pig starts moving, it will accelerate to a much higher velocity, which will cause it to loose its seal with the inside surface of the pipe, making the pig inefficient by leaving some amount of liquid, gas, or debris behind in the pipe.

This problem is much more dynamic in a low-pressure gas pipeline due to the low viscosity of natural gas. Once stopped at a sharp inside diameter reduction, the pressure can build to, say, double the operating pressure, causing high acceleration. If this occurs close to the end of the pipeline, where there is little resistance in front of the pig, the speed of the pig could reach or exceed 100 mph (161 km/h).

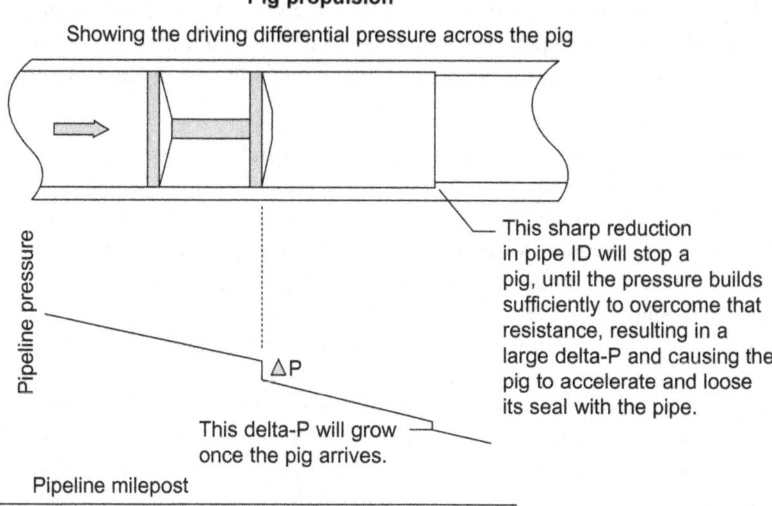

FIGURE 15.2 Pig propulsion.

When designing for excellent pigging performance in both gas and liquid pipelines, it is essential to keep the "hole" in the pipeline constant. So, designers should specify a constant internal diameter for all pipe sections and components that make up the piggable pipeline. Quite often, designers use thicker wall pipe when the code calls for more conservative design factors, such as $F = 0.60$ or even less for offshore, but the disadvantage is thicker wall pipe will have a smaller inside circumference than the mainline pipe. This difference in inside circumference can be significant. This is very important because the pig cup/disc design and satisfactory sealing are based on the inside circumference of the pipe. To have a difference in wall thickness, make sure that a transition piece is used to eliminate the sudden step, see Fig. 15.2.

15.6 UTILITY PIGS

The four main reasons for conventional or utility pigging of pipelines are as follows:

1. To provide maximum pipeline throughput by keeping the pipeline clean and pipe wall friction as low as possible.
2. To reduce contamination of pipeline product by preventing dissimilar products being batched in the pipeline from mixing at the batch interface. Often, when mixing of products occurs, a large volume of product must be downgraded or discarded.
3. To reduce internal corrosion in a liquid pipeline by removing entrained and trapped water and corrosives in the water, and in the case of natural gas pipelines by removing trapped condensate and water.
4. To reduce energy cost per unit of product delivered by keeping throughput high by sweeping the pipelines free of water slugs, condensate, and entrained air and vapor. Those slugs of water, condensate, and pockets of air and vapor in the pipeline reduce the pipeline's flow area requiring greater horsepower demand.

In past years, utility pigs were sold for special uses, such as for water removal, or for pipe wall cleaning, and for product batching, which is not the case today. Today, most conventional pig manufacturers make a multiuse pig that can be configured as a water removal tool, a pipe wall cleaning or scraping tool, a pipeline size gauging tool, or a batching pig to separate two different products being shipped in the pipeline. Looking at the literature, I find that most of the pipeline pig competitors make a rugged steel mandrel design with a large bolt circle for rigidly attaching replaceable cups and discs that can be configured in eight or more ways to meet major pipeline needs, see Fig. 15.3. Some of these needs are hydrostatic test water displacement, paraffin removal, pipe wall brushing and cleaning, product batching, pipeline drying, and pipe wall thickness gauging. These pigs can be equipped with optional accessories including magnets for welding rod retrieval and transmitters to locate pipeline

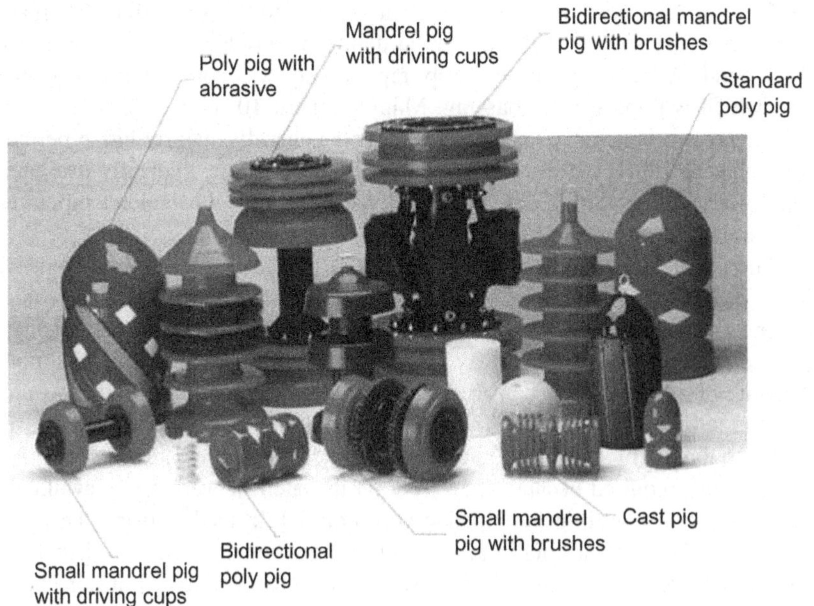

FIGURE 15.3 Pipeline cleaning and utility pigs. *Source: Courtesy of Girard Industries.*

blockage if it occurs. A short, but incomplete, list of manufacturers producing these utility pigs is TDW, Girard, Enduro, and Rosen.

The pigs can be configured for unidirectional or bidirectional operation by choosing replaceable conical cups or discs or a combination of cups and discs made from polyurethane, neoprene, or nitrile rubber. This popular mandrel design, with large bolt circle flanges for ridged attachment of elastomers, is targeted for the big bore pipelines, with diameters of 16 in. up to 48 in. (406 up to 1220 mm) and even larger. You can get these pigs with a bypass hole option, which produces a jetting action in front of the pig, to improve cleaning efficiency. This is similar to the Rosen's spider nose jetting system, which helps to keep removed wax and scale in suspension, as the pig pushes the debris down the pipeline and into the receiver trap.

Of course, the above-mentioned pig manufacturers and others make a variety of pigs for smaller pipeline sizes, 3 in. (76 mm) up to 14 in. (355 mm), with some of these pipelines having shorter radius bends to deal with, changes in pipe diameter and other pig passing restrictions such as reduced port valves. Most of these problems are associated with smaller pipelines, 3–6 in. (76–152 mm) and sometimes 8 in. (203 mm), because as stated earlier many small diameter lines were not designed for pigging. On the other hand, most of the 10-14 in. (254–355 mm)-diameter pipelines were designed properly with pipeline pigging as a goal and were constructed with long radius shop bends. For example, a 10-in. (254-mm)-diameter shop bend would have a bend radius of approximately 10 ft

(3048 mm). The design of 12- and 14-in. (305- and 355-mm)-diameter pipelines would use shop bends with similar diameter-to-radius geometry. Because these bends were formed in the shop, pipe ovality was controlled to less than 3%, which is good for pig passing. Many of these 10–14 in. (254–355 mm) lines were constructed with full-port valves; generally natural gas pipelines use ball-type block valves, and hazardous liquid pipelines typically use gate-type block valves. So as constructed, those larger, small diameter pipelines were much better for pigging.

The problem I found, while working as a district engineer, was some of those piggable pipelines had been "downgraded" by Operations and Project Engineers over the years, doing expedient pipeline O&M work while conducting required pipeline right-of-way relocations, done quickly to meet shipping demand, or a third-party's construction schedule. This pipeline "downgrading" that I'm talking about, which affects pipeline pigging more than anything else, relates to the use of shorter 3D radius bends for convenience when rerouting the pipeline was required because they were on the shelf or were easily available, and used when rerouting the pipeline was required, instead of using the long-radius shop bends as originally designed. Downgrading also relates to installing valves with reduced ports, just because they were available, putting check valves in the mainline for some operational reason, which seemed like a good idea at the time, or even replacing sections of pipe with thicker wall pipe because it was available and pretested. So often, "downgrading" the pipeline can make the implementation of a pigging program extremely difficult, requiring the removal of pipeline "bottlenecks" to allow pigging and often requires the selection of less-effective pigging tools, done because of the high-cost of retrofitting the entire pipeline for ideal pigging.

15.7 SELECTING PIGS FOR SMALL BORE AND DOUBLE DIAMETER PIPELINES

So, getting back to the problem of selecting the proper service pig for smaller size pipelines, say size 3–8 in. diameter, it is best to talk with the pig vendors to determine what they recommend for your particular pipeline configuration and needs. The smaller size utility pigs are designed with slender steel mandrels made for center-hole attachment of cups and discs. As with the larger size mandrel pigs, the center-hole type discs and cups can be customized or configured for a specific duty. These small mandrel pigs also have the flexibility to attach brushes, magnets, gauging plates, and transmitters. Probably the best feature of a mandrel pig, both large and small, is that they are rebuildable. You can buy replaceable conical cups, scraper cups, and scraper discs for your pigs once they are worn or damaged, allowing pig refurbishment for the life of the pipeline.

Figure 15.3 shows a variety of utility pigs in pipeline service today. The mandrel pigs are featured top center, with and without brushes. The slender steel mandrel with center-hole attachments are shown bottom center and bottom far left.

The popular poly pigs are shown, some with abrasive brushes on the side walls, and most are shaped for unidirectional flow. Also, Fig. 15.3 shows cast pigs of various size and combination shapes, which are molded in one piece, using polyurethane material, and can be used in problem pipelines with short radius bends because of their flexibility.

For cleaning and dewatering pipelines that were constructed with two-diameter pipe (DD), you can use the TDW (DD) WCK type pig, which is a good choice for a pipeline with changes in diameter, because of its excellent cleaning capability. The DD pig uses tempered steel springs that force the steel brushes up against the pipe wall but still have flexibility inward to avoid pipe obstructions. This pig design looks like two conical cup pigs with attached brushes connected to each other in the middle by a spring joint to allow articulation through tight bends. You should call T.D. Williamson or other pig vendors who make this type of DD pig to determine bend radius limitations, especially in the smaller size pipelines, so that you have confidence that the pig will meet your pipeline cleaning needs.

In the case where pig passing is concerned, small diameter pipe size requires relatively larger bend radius elbows; this is fundamental because the smaller pig design tends to have a larger length to diameter ratio than their bigger counterparts. Originally, the 3-, 4-, and even 6-in. (76-, 101-, and even 152-mm)-diameter pipelines should have been constructed with 6–10D radius bends, as a minimum. If your small bore pipelines actually have tighter bends than that, I would recommend launching poly pigs first to evaluate the pipelines pig passing ability. In smaller size pipelines, it is important to know the locations and radius of all the tighter bend elbows.

15.8 POLY PIGS

Many of the major pig manufacturers produce and supply poly pigs. Poly pigs are made with open-cell polyurethane foam of various densities and have various external coatings. The bullet shape design is to aid the pig in traversing fittings and valves but makes the pig unidirectional. You can also buy bidirectional poly pigs; Fig. 15.3 shows a small bidirectional poly pig located bottom left of center. The length of the pig is approximately twice the diameter to reduce the possibility of tumbling in the pipe. The diameter of the poly pig is larger than the inside diameter of the pipeline to insure a frictional drag between the foam pig and the inside surface of the pipe. They typically have a concave base plate with 90A durometer polyurethane coating. This provides a maximum rear sealing surface for the propelling force of the liquid or gas being used.

Swabs or bare poly pigs that only have coating on the base are used for pipe drying and batching purposes. External coatings on foam pig bodies consist of crisscross type spirals of high-durometer polyurethane. The spirals give strength and greater scraping action than using bare foam. Some of the pigs have imbedded silicon carbide abrasive in their reinforced spirals to add scraping power. These pigs are used quite often during pipeline construction and hydrostatic testing and can be used effectively for pipeline O&M. They are also popular for use in a pipeline emergency situations because of their flexible nature.

15.9 MY FIRST EXPERIENCE WITH POLY PIGS

In the early 1980s, I was the project engineer responsible for electrifying 10 in. (254 mm) Line-1 from Newhall Station to Jefferson and Hauser in Los Angeles city and then on to the Los Angeles refinery, near Long Beach, California. Along with this pump station upgrade, I also installed a 10-in. launcher at Newhall and a receiver/launcher station at Jefferson and Hauser some 30 miles (48 km) downstream, and a 10 in. (254 mm) receiver at the end of the 55-mile (88 km) pipeline, located inside the Los Angeles refinery.

Line-1 had been constructed during WWII, and since commissioning, it had been shipping sweet crude from Bakersfield area production to the Los Angeles refinery, on a 24-hour basis, for some 40 years. During that entire time, the pipeline had no pig launching facilities, and there was no record of it ever being pigged.

The first pig that was launched from Newhall to the receiver at Jefferson and Hauser came into the receiver barrel pushing a 3″ × 4″ pipe skid, which was probably used during pipeline construction and somehow got into the pipeline. In addition, a bunch of welding rod and a lunch pail were also found. More interesting was that the poly pig looked like it had been through a shark attack. Apparently, the pig had passed through more than one location where corrosion leaks on the 10 in. (254 mm) line had been repaired using screws, as explained earlier. These screws were left penetrating into the pipe flow area. A more ridged pig would probably have gotten stuck. I believe that after seeing the damage to the poly pig, operations tracked down the screw locations and replaced those corroded sections with new pretested pipe. Once the launcher/receiver stations were operational on Line-1, the pipeline operators set up a routine pigging schedule for the pipeline.

One of the most important aspects of having pipeline pigging capability is to develop a pigging plan and routinely follow it, which will keep pipeline performance at its highest level. Should an internal inspection be planned, using a smart pig of some kind to verify pipeline integrity, the pipeline would not need a great deal of preparation work because the inside surface of the pipe has been kept clean, and the pipeline is known to be piggable.

15.10 PIG TRAINS

Before we discuss "smart pigging" and internal inspection of pipelines, we need to look at some examples of pig trains. Pig trains can be used for many pipeline operation and maintenance endeavors. The first example presented here is to clean a pipeline that was used for pumping salt water.

The pig train requires four batching pigs. The front pig is pushing a slug of fresh water, which in turn is pushing out the salt water. The second pig is pushing a second slug of fresh water and, of course, the first pig as well. The fresh water desalinates the previous salt water flooded pipe. The third pig separates two slugs of glycol, which are pushing the train, and the glycol is dehydrating the pipeline. The fourth, and last, pig is being driven by nitrogen and pushes the last glycol slug. So, this pig train cleans and dries the pipeline and prepares it for the introduction of new product. Portable compressors are needed to fill and push the pig train using nitrogen.

This next example shows how a pig train can be used to clean up a crude oil pipeline, which is to be hydrostatically tested, and then converted to natural gas service. We will start by shutting down the pump stations and leaving the entire pipeline laid down in crude oil. But first, before shutdown, we need to repeatedly launch cleaning and scraping pigs during normal crude oil shipments necessary to skewer inside pipe walls and remove the major paraffin and scale. Success of this internal pipe cleaning program will be determined by the decreasing amount of wax and scale being pushed into the receiver with each pig. Once sufficiently cleaned, the in-service pigging will be stopped, and the pipeline will be shutdown.

The pig train configuration will be accomplished in the following manner: the first pig launched will be a bidirectional spider nose displacement pig with transmitter, which starts pushing the crude oil out of the pipeline, followed by a 1250 barrel water slug with surfactant cleaning solution at 100 ppm for contamination removal and pipe flushing. This cleaning solution is a biodegradable grease cutter that cleans the pipe wall to bare metal and leaves a water-wet condition. This slug is followed by another bidirectional pig and 1250 more barrels of water using an even higher concentration of surfactant, followed by two more spaced bidirectional pigs, followed by the test water filling the pipeline as the crude oil is being pushed out of the pipeline and into tankage. After the completion of hydrostatic test, the remainder of the surfactant cleaning solutions and test water will be pushed into available storage tanks for water treating and disposal. The pipeline will then be dried to 0°F dew point, by using portable air compressors, pushing poly pigs and swabs. Once the desired pipe wall dryness is reached, the pipeline will be protected with a 5 psi (35 kPa) blanket of nitrogen gas. The nitrogen will protect the pipeline from corrosion until it is put into natural gas service.

15.11 SMART PIGS

Smart pigs, also known as intelligent pigs, internal inspection devices, and in-line inspection (ILI) tools are now being used extensively for inspecting in-service pipelines. There use has been increasing rapidly due to their proven benefits and expanding capabilities. These sophisticated internal inspection devices travel through the pipe and measure and record irregularities in the pipe wall that may represent corrosion, gouges, and other deformations known as pipe anomalies. Because these inspection devices run inside the pipe, in a manner similar to conventional pigs, they are referred to as "smart pigs."

California now, in some cases, allows smart pig inspections of regulated intrastate pipelines as an alternative to their scheduled hydrostatic testing program.

Fig. 15.4 shows a magnetic flux leakage (MFL) smart pig being inserted into a pipeline launcher, and one being removed from a pipeline receiver, as well as the robust magnetic sensors used on the MFL pig, and the odometer wheels located on the back end of the pig for determining exactly where the MFL pig is in the pipeline. Also shown at the bottom of Fig. 15.4 is an ultrasonic testing pig supported on its shipping pallet. This smart pig is designed to detect pipe wall metal loss and is in competition with the MFL pig.

15.12 SMART PIG TYPES

There are basically four types of smart pigs that we want to introduce, which are as follows:

First is the popular magnetic flux leakage pig, which is an electronic pig that identifies and measures metal loss (caused by corrosion, gouges, etc.) in the pipe wall using a temporarily applied magnetic field. As the pig passes through the pipeline, this tool induces a magnetic flux in the pipe wall between the north and the south magnetic poles onboard. A homogeneous steel pipe wall, one without defects, creates a homogeneous distribution of flux. Anomalies in the wall, such as metal loss caused by, say, corrosion inside or outside of the pipe, result in a change in flux distribution. This flux distribution leaks out of the wall. Sensors onboard the pig detect and measure the amount and distribution of the flux leakage. This flux leakage signal is processed, and the resulting data is stored onboard the MFL pig for later analysis and reporting.

Smart pigs that use this principal of magnetic flux leakage to detect anomalies in pipelines are in widespread use today. The primary purpose of using MFL pigs is to detect corrosion defects and give an estimate of the size of the defect found. The latest pigs are able to detect defects of less than 5% of wall thickness. Defects that are so small that they are difficult to see visually. One limitation of the MFL pig is that they have very limited crack detection capability. The pig may detect cracks in girth welds but is not able to locate and detect cracks in the pipe's long axis or longitudinal weld seam.

Second is the ultrasonic testing (UT) pig that measures pipe wall thickness and metal loss. These pigs are equipped with transducers that emit ultrasonic signals perpendicular to the surface of the pipe. An echo is received from both the internal and the external surfaces of the pipe, and by timing these return signals and comparing them to the speed of sound in steel pipe, the wall thickness can be determined. So, cleanliness of internal surfaces and removal of wax

Magnetic sensors are used to examine the pipeline. The sensors are radically positioned and staggered to cover the entire inside surface of the pipeline.

Smart pig being inserted into the pipeline launcher. The pig was transported on a special pallet.

Odometer wheels measure exactly where the pig is in the pipeline. The wheels are always located on the end of the pig.

MFL smart pig being removed from the pipeline receiver. The front section of the pig contains the magnetizing brushes. On this pig, the sensors are in the second and third sections. Note the U-joint between the first and second sections.

Ultrasonic testing pig delivered to a pipeline for internal inspection. These pigs are equipped with transducers that emit ultrasonic signals perpendicular to the pipe surface. An echo is received from both internal and external surfaces. By timing these return signals, the wall thickness can be determined.

FIGURE 15.4 Smart pigs.

and scale buildup in the pipe is extremely important for assuring a successful UT test. This is especially important for crude and heavy oil pipelines where paraffin buildup inside the pipeline is common. Ultrasonic inspection pigs intended to detect metal loss will tend to report similar patterns of internal and external corrosions to those reported by MFL pigs.

15.13 CRACK DETECTION

A modified version of the UT pig is called the shear wave ultrasonic pig and is designed to reliably detect longitudinal cracks in suspected pipeline, longitudinal ERW weld defects, and crack-like defects (such as stress corrosion cracking). This type of smart pig is important because the most dangerous pipe cracks or crack-like defects are found in the long axis of the pipeline. The pipe's hoop stress works directly on longitudinal cracks and if not detected will ultimately result in rupture.

This shear wave UT pig is categorized as a liquid-coupled tool. It uses shear waves generated in the pipe wall by the angular transmission of UT pulses through a liquid coupling medium (oil, water, etc). The angle of incidence is adjusted such that a propagation angle of 45° is obtained in the pipeline's steel. This technique is appropriate for longitudinal crack inspection. These crack detection pigs have been in development for years, and their utility and accuracy are improving.

The fourth type of smart pig is the geometry pig (GP), also called a caliper pig, that uses a mechanical arm or electromechanical means to measure the bore of the pipe. In doing so, it identifies dents, deformations, and ovality changes. It can also sense changes in girth weld size and pipe wall thickness. In some cases, these pigs can also detect bends in the pipe and even buildup of debris at the bottom of the pipe. This geometry pig used to detect deformation anomalies such as dents should be of the type that provides orientation, location, and depth measurement of each dent or other deformation anomaly. This type of pig can be used to inspect both hazardous liquid and natural gas pipelines.

15.14 PREPARATION FOR SMART PIG INSPECTION

Now that we know a little more about smart pigs, and how they work, and have selected the smart pigs we intend to use, we are ready to start preparing the pipeline for internal inspection. Note that this pipeline preparation, which is necessary for a successful smart pig inspection, provides a good example in the use of utility cleaning pigs and gauging pigs. Please read the following pipeline cleaning logic:

1. An extensive cleaning operation is necessary before "smart pigs" are launched, especially when the pipeline was in crude or heavy oil service. Poly pigs or scraper pigs should be run until the amount of wax, dirt, and scale decrease to less than 50 lbs (23 kg) per run.

2. A detailed pigging log should be maintained: the log should list launching times and receiving times; condition of the pig removed from the receiver; volumes of dirt, wax, and debris received; and pipeline pressure and flow rate during pigging. This information will help determine the effectiveness of the pipeline cleaning program. Solvents can be added to the crude oil being shipped to help remove paraffin from the pipe wall.
3. Steel mandrel pigs with conical cups should be used to start the second cleaning phase. A minimum of five pig runs should be planned, but pigging should continue until the listed objectives of Phase 1 are achieved.
4. The next cleaning phase should utilize similar conical cup configured pigs with steel brushes and magnets, if necessary, and the brushes will gradually break up deposits on the pipeline wall. Launching should be continued until the objectives of Phase 1 are achieved.
5. The final cleaning should use bidirectional pigs with urethane discs and an added spider nose assembly. This disc type pig scrapes the remaining wax off the pipe walls, and the spider nose creates turbulence to mix and dilute the paraffin and keep it suspended in front of the pig, until the debris enters the receiver trap.
6. Next, check for pipe out of roundness greater than 10%, which can damage a smart pig, by launching a bidirectional pig equipped with an aluminum gauging plate of 90% of minimum ID. If this gauging plate is undamaged, the pipeline is ready to launch the first smart pig.

Figure 15.5 presents logic for pipeline cleaning and size inspection, which as explained earlier is necessary for running a successful "smart pig" inspection.

After cleaning and gauging is complete, the geometry pig or caliper pig can be run in the pipeline to exactly locate dents, pipe ovality, and other geometric defects. So next, let's launch the geometry pig to locate and measure pipe deformation. Refer above for the description of how this geometry pig (GP) works. If dents or bore reductions are greater than 8% of pipe diameter, they must be located and repaired before launching the corrosion detection MFL pig.

15.15 MFL SMART PIG

We have selected the MFL pig over the UT pig because the MFL pig is not as sensitive to pipe wall contamination, such as wax and scale. For inspection of a natural gas pipeline or clean products pipeline, the UT tool may be the smart pig of choice. Also, it was chosen, because "pipeliners" now have more experience using the MFL pig, with its widespread use for corrosion detection, and now there is a wealth of experience in reading and evaluating metal loss data. Before running this MFL pig, bench marker positions must be selected and placed along the right-of-way. These above ground markers will correlate pig measurements to actual ground locations, and thus calibrate the data.

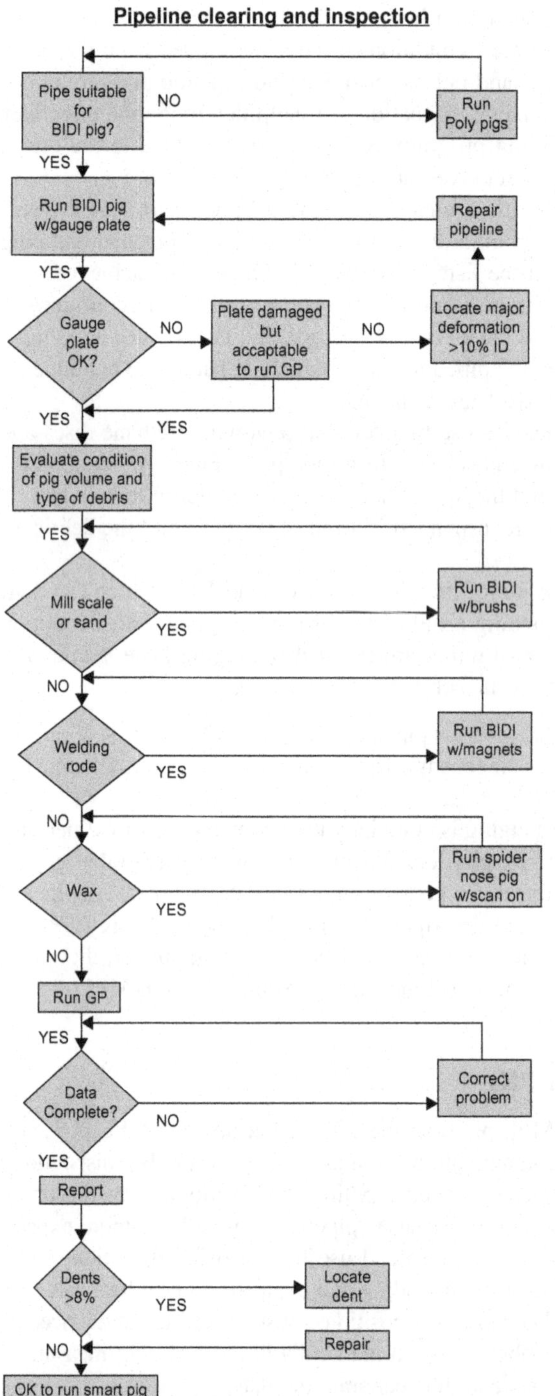

FIGURE 15.5 Pipeline cleaning and inspection for smart pig run.

Also, pig supporting and lifting equipment must be placed at launchers and receivers to handle, insert, and remove these complex internal inspection pigs. Of course, the pipeline launcher/receiver stations must be modified to accommodate the longer MFL pig, as discussed earlier, see Fig. 15.1.

Sometimes, a dummy smart pig is launched before running the instrumented smart pig when there is no previous pigging history on the pipeline. This assures that the smart pig will not be damaged during launch, travel in the pipeline, and entering the receiver trap and that the chances for a successful internal inspection are much greater.

15.16 POST SMART PIG INSPECTION

A preliminary assessment should be made, after the smart pig run is completed, and onboard data is printed out. At this point, the pipeline operator and smart pig vendor should evaluate the internal inspection data together and do a "broad brush" analysis of all the data.[3]

This pipeline data can fall into three categories:

1. Showing a bad pipeline condition. The pipeline has severe anomalies requiring immediate expert assessment, which is needed to determine requirements for:
   a. an immediate pipeline shutdown or
   b. operating pressure reduction.

   Confirm test data showing anomaly size and location with the smart pig vendor. Do inspection digs at the locations of the most severe anomaly to confirm actual pipeline condition and verify pig data. Determine what repair work is needed and the amount of rehabilitation necessary to return the pipeline to design condition.
2. Showing the pipeline in good to fair condition. The pipeline has some or no major anomalies. Consider the implications for future pipeline integrity:
   a. Assign a Level-1 Integrity Assessment condition for the pipeline that is satisfactory and set a future inspection schedule.
   b. Assign a Level-2 Integrity Assessment, which states that you fully understand the problem and have considered whether it is safe to operate the pipeline in a partially damaged condition.
3. Unsatisfactory smart pig performance. This requires making the necessary pipeline repairs or smart pig repairs and reinspecting the pipeline.

15.17 EXPERT DATA EVALUATION

In the earlier case, where pig run data looks good, but the pipeline is found to have severe anomalies, it is important to correlate those anomalies with real internal and external pipeline corrosion experience or third-party damage experience to help validate the pig data analysis.

Figure 15.6 shows gouges in the top quadrant of the pipeline caused by third-party damage. Gouges will be reported as metal loss and appear long and

FIGURE 15.6 Gouges on top of pipe.

narrow, usually on the top side of the pipe. There may be some evidence of denting. This anomaly may be located at a place where excavations are known to have happened, such as at a road crossing or nearby construction project.

Figure 15.7 shows a pipeline with multiple deep corrosion pits probably caused by disbonded coating and problems with pipeline cathodic protection (CP). This active external corrosion may be isolated due to a combination of small coating defects, or it may be extensive due to widespread coating degradation and poor CP. External corrosion is relatively common in pipelines as soil conditions and any local infrastructure can combine to damage the coating and make CP difficult.

15.18 EXTERNAL CORROSION

Different patterns or types of external corrosion that may be reported by a smart pig inspection are as follows:

- Isolated pits at any orientation on a pipeline. These may be active corrosion sites with relatively high corrosion growth rates.
- General external corrosion toward the bottom of the pipe is typical, where the pipe coating has become degraded. This corrosion tends to be on the bottom of the pipe because of differential aeration and at low points along the pipeline route, where ground water tends to accumulate more than at high ground.

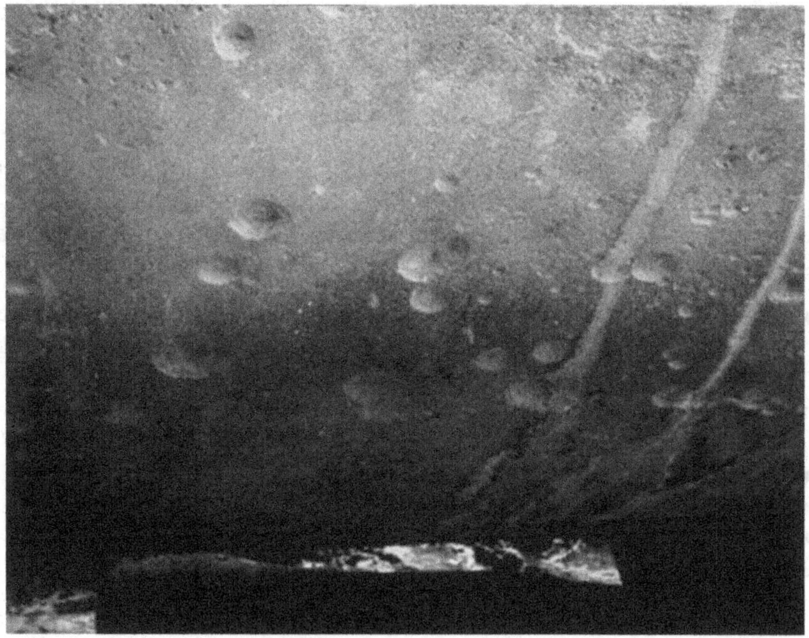

FIGURE 15.7 Multiple deep corrosion pits.

- Multiple deep corrosion pits may be active and may be related to microbially assisted corrosion. The growth rate is likely to be high. These multiple deep pits are critical when they line up closely in the longitudinal direction.
- Corrosion along the seam weld may be active and could be related to tenting of the tape wrap-type coating.

15.19 INTERNAL CORROSION

Active internal pipe corrosion is likely to be extensive and will occur in more than one location. It also tends to occur at the bottom of the pipe, where water and debris can collect.

Different patterns of internal corrosion that may be seen in the smart pig data are as follows:

- Shallow general internal corrosion is found at the low points along the pipeline route and at the bottom of the pipe. This could be an operational corrosion. The possibility that this is an active corrosion should be considered. For example, if the pipeline transports waxy crude oil, but cleaning pigs are run weekly, then it is unlikely that this corrosion is active.
- Numerous deep internal corrosion pits at the bottom of the pipe and at the low points along the pipeline route on a crude oil pipeline that is not used

continually and are rarely pigged. In this case, the corrosion is likely to be active microbially assisted corrosion.

15.20 POSTINSPECTION CRITERIA

It should be kept in mind when evaluating internal inspection data that a smart pig cannot detect all defects all the time and that pig measurements have associated errors. Below is a list of defects that after evaluation require immediate scrutiny and action:[2]

- Localized corrosion pitting where the deepest pit is 70% or more through the pipe wall, and when the pits are closely connected and aligned in the long axis of the pipeline.
- Dents (or bore restriction) greater than 7% of pipe diameter. This requirement has more to do with smart pig passing than pipe strength.
- Dents with associated gouging or cracking.
- Dents on pipe seam or girth welds.
- Buckles.
- Cracks.

Note, as mentioned earlier, longitudinal cracks in the pipe cannot be detected by using a magnetic flux leakage (MFL) pig or, of course, when using a geometry (GP) pig. One of the limitations of a MFL pig is that its crack detection ability is limited to finding major cracks in the pipes' circumferential direction and possibly to finding cracks in girth welds.

If pipeline cracking is suspected, a shear wave UT pig must be used to locate and evaluate longitudinal cracks, stress corrosion cracking, cracks in the ERW seam, and possible fatigue cracking in pipelines subjected to severe cyclic loading (refer to Section 15.13, "Crack Detection").

SUMMARY

As you can see, just running a smart pig for internal inspection of a pipeline with data evaluation is a significant project in itself, with potential operational implications! The good news is that smart pigs have evolved into sophisticated pipeline wall metal loss inspection tools. This is especially true with MFL type pigs, which provide realistic analysis of pipeline condition, both internally and externally. The bad news is that when you run a smart pig for the first time, in your pipeline, which has been operating as designed and without incident for years, you really do not know what to expect. If the condition of the pipe fails the "fitness-for-service" criteria, such as ASME B31G or other recognized pipeline codes using traditional fitness methods, the pipeline must be shut down or the operating pressure must be reduced for safety.

In the case of BP Alaska's Prudhoe Bay field, where a gathering line failure was detected in July 2006, they said they were totally caught off guard by the smart pig inspection results. They reported that the smart pig detected and measured 16 anomalies in 12 locations, which exceeded their operating criteria, and they were obligated to shut down the 400,000 bbl/day field, until repairs and pipe replacements were made. That was a very costly shutdown. When asked why they had not run a "smart pig" before, they said that the lines were totally above ground and were inspected visually for leakage.

So, surprise can be your biggest problem when doing smart pig inspections on pipelines that have been working normally. If you have been routinely running utility pigs in your pipeline and the line has been kept clean, based on the amount of wax and debris that is pushed into the receivers, an internal inspection should indicate minor or no internal corrosion.

Still, external corrosion and third-party damage to your pipeline can be your worst enemy. A pipeline operator must keep in mind that his pipeline has two sides, and of course, since it is usually buried, the outside can only be inspected by doing exploratory digs or running a smart pig. So, here is a list of concerns to keep in mind that can cause external metal loss and damage to your pipeline's OD, and if present, will be detected when an internal inspection is performed:

- Cathodic protection (CP) problems due to soil conditions or CP interference at pipeline crossings, pipe casings, and other substructure, which reduce the effect of CP and allow external corrosion to occur.
- General pipe coating problems such as disbonding of outside coating, which exposes the pipe wall to corrosion.
- Tape coat wrinkling that can occur, due to high soil shear stress, and allows pipe wall exposure to external corrosion (see results, Fig. 15.7).
- Rock damage to the pipe coating, usually found at the bottom of the pipeline, which damages the coating and exposes the pipe to localized corrosion.
- Third-party damage to the pipeline, such as gouges and dents, usually found at the top of the pipe and at locations near road crossings and adjacent construction sites (see Fig. 15.6).

REFERENCES

[1] Code of Federal Regulations, Title 49, Part 195, Transportation of Hazardous Liquids by Pipeline, 2010.

[2] ASME Code for Pressure Piping, B31.4, Addenda, 1991, 451.6.2 Disposition of Defects.

[3] Understanding the Results of an Intelligent Pig Inspection, a paper by, Roland Palmer-Jones, Prof. Phil Hopkins, Penspen Integrity, Newcastle upon Tyne, UK and Dr. David Eyre, Penspen Integrity, Richmond, UK, 2004.

BIBLIOGRAPHY

Code of Federal Regulations, Title 49, Part 192, Transportation of Natural and Other Gas by Pipeline, 2010.

Chapter 16

Pipeline Construction

Glenn A. Wininger

Chapter Outline

INTRODUCTION

Pipeline construction must act in accordance with all federal, state, district, and local regulations and mandated specific construction methods in some instances. Although federal regulations are predominately unaffected in most instances by the geographic region whereupon a pipeline is constructed, the regulations of states and the districts within those states, as well as local regulations, vary somewhat between each location. Although a pipeline contractor must always adhere to the specific regulations, each energy company maintains specifications relative to construction of a pipeline unique to that company. Many times, some specification differences are diminutive between companies, with other specification variations from company to company being extensive.

This chapter will delve into generic pipeline construction, and while providing some level of detail, the material covered herein does not imply any all-inclusive construction technique as regulations, site construction methods, and company specifications should be adhered to as indicated above.

16.1 PIPELINE CONSTRUCTION SEQUENCE

Construction of a pipeline, in most cases, must follow a rigid progression of events. As in an assembly line of a factory, each portion of the operation must rely and be dependent on the preceding portion that should be complete to maintain continuity in the progression. If there is a delay or problem in one portion of any operation, it delays the project and possibly ultimately delays completion of the overall project. As an example, in a factory setting such as an automobile assembly line, should one portion of the assembly break down and come to a halt, the following segments will respectfully cease operations. Similarly, during pipeline construction if a preceding operation is in delay or even comes to a standstill, the subsequent operations will also experience delays and possibly standstills. For example, grading of the terrain cannot begin or continue along a right-of-way until the designated right-of-way is cleared along the construction corridor. Each portion of the pipeline construction sequence maintains the appropriate combined crew personnel relative to the right-of-way width, terrain, schedule, size of line pipe to be installed, and weather conditions to name a few. Figure 16.1 shows a pipeline construction sequence with each portion of the sequence detailed. The accompanying descriptions associated with this figure represent a general summary of each crew and its role in getting the pipeline constructed within regulatory guidelines while maintaining a safe and productive working environment with the ultimate goal of completing the construction of the pipeline on schedule.

Pipe joints (individual pipe segments) are preferably delivered from their point of manufacture (usually indicative of the pipeline construction project and length)

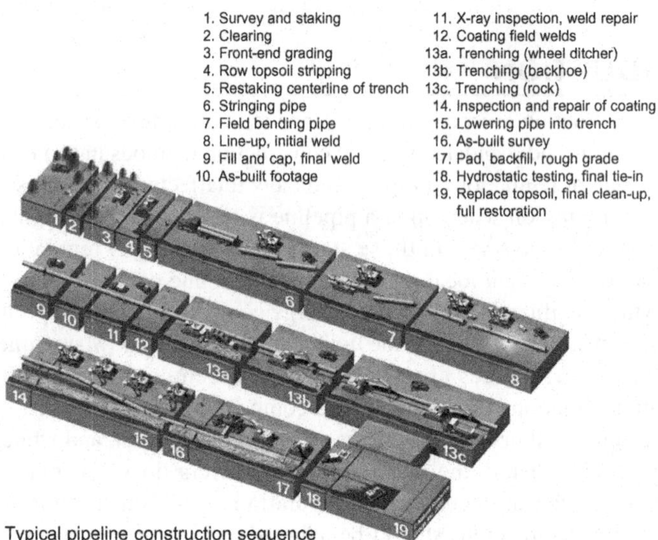

FIGURE 16.1 Pipeline Construction Sequence (Sequence may vary i.e. – 13a, 13b and 13c may preceed 6 – 12).

by rail to a rail off-loading yard conveniently located to the construction sites and then trucked to pipe predetermined storage yards convenient to the construction corridor, which are approved and permitted through the governing agency. Numerous storage yards/construction yards may be implemented to support individual pipeline construction spreads. Depending on the rail delivery schedule, trucks will make continuous trips between rail off-loading areas and pipe storage areas until the entire pipe assigned to the storage areas has been delivered. This delivery process could conceivably go well into the construction timeline of a project. Although their primary purpose is temporary storage of pipe, these storage yards could be used for "double jointing" two pipe segments before their delivery to the construction corridor. Double jointing is only applicable in locations where the process is conceivable and permissible.

In addition to the pipe storage areas close to the construction corridor, other construction yards, not used for pipe storage are utilized to store equipment. In addition, material relative to the project and key essential items necessary for the construction of the pipeline along with a location for the congregation of workers could occupy several acres depending on project size.

16.1.1 Clearing and Grading Crew

This particular crew clears the construction corridor of brush, trees, large rocks, and other obstructions from the pipeline's right-of-way. As with all the remaining succeeding operations within the construction progression of the pipeline, this crew must stay within a predescribed boundary or limit of construction along the corridor and access the corridor at approved access points.

In past years, the construction limits were not definitively defined along a pipeline construction corridor in many cases, which gave flexibility to a pipeline contractor for accessing the right-of-way along with clearing and grading more than was essentially necessary. On recognizing the impacts to shared, private, and communal lands, as well as the environmental impacts, regulations were initiated to administer requisite directions that would reduce the impact to the aforementioned. The personnel involved in the clearing and grading include operating engineers, laborers, and truck drivers. The operators run the heavy equipment such as dozers, backhoes, graders, and tree-cutting machines to name a few, along with a helper or apprentice assigned to each. The mechanics and service people provide the necessary service to maintain the equipment. Laborers aid in clearing the trees and do any manual digging and moving required. The truck drivers operate the buses for the crew to and from the construction corridor and the necessary trucks to haul any material from the corridor. The staked right-of-way corridor would be cleared and graded to provide as smooth and level a work area as is possible to facilitate the safe continual movement of equipment and personnel along the corridor. A grader with a blade would be used to knock down vegetation within as much of the construction corridor as is required and necessary to provide a safe and productive working area.

To provide an even and level surface as is practical, grading usually requires cutting and filling to achieve a more uniform grade for the pipeline construction progression. This grading may include ripping rock close to the surface when the terrain dictates. The grading also may be required to provide suitable working areas in instances of excessively steep slopes and side slopes or at approaches to water body crossings, and at established additional temporary extra work areas such as stream and river crossings, wetland crossings, road crossings, before residential areas, etc.

In the occurrence of top soil segregation (stripping), where the top soil is bladed to a depth and width as defined by the governing regulatory agency, the top predetermined obligatory amount of soil would be salvaged and stockpiled on the side of the construction corridor, and then spread back over the area after final grading. The location or site of the topsoil storage area along the construction corridor is predicated on diverse factors that may include, but not be limited to, adjacent or paralleling pipelines or other infrastructure, residential or commercial areas, adjacent noncontacted landowners, and some communal and state/federal lands.

Figures 16.2 through 16.6 show an indication of acceptable segregated topsoil storage areas. Although the ideal situation for segregated topsoil storage is indicated in Fig. 16.2, where the segregated topsoil is stored adjacent to the

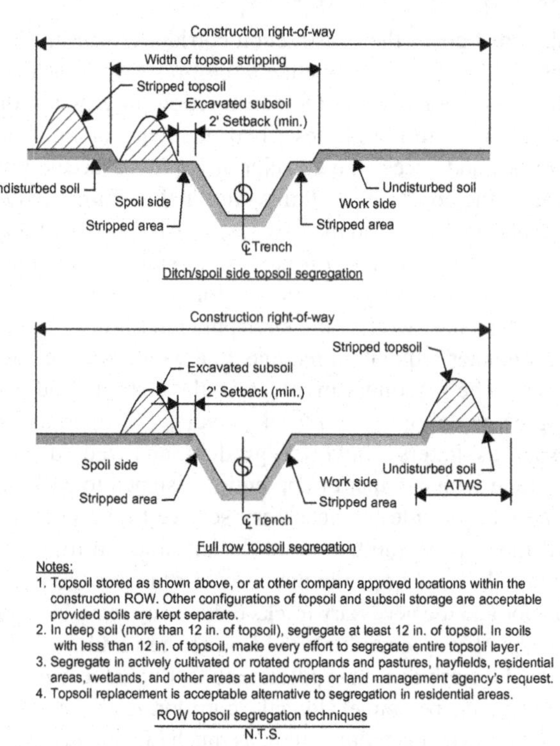

FIGURE 16.2 Segregation of Topsoil Within Construction Boundaries.

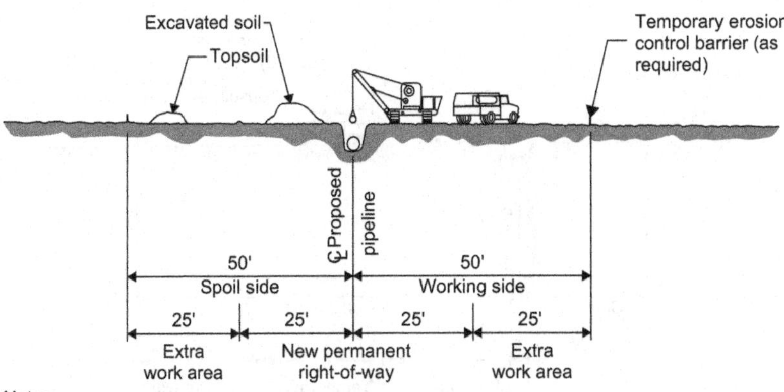

Notes:
1. Configuration does not include additional temporary work space at crossings.
2. Proposed pipeline to maintain a minimum of one foot clearance between subsurface utility crossing.

Proposed pipeline typical construction right-of-way
N.T.S.

FIGURE 16.3 Construction R.O.W. 50' Either Side of Pipeline with Topsoil Segregation on Spoil Side.

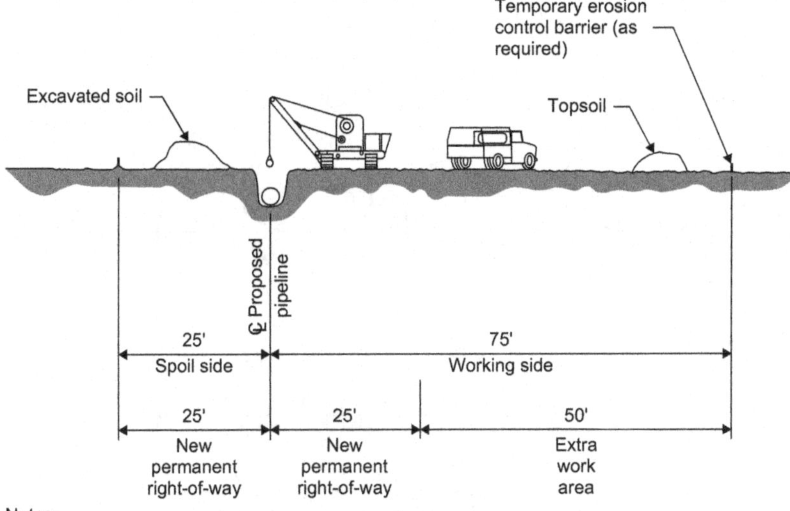

Notes:
1. Configuration does not include additional temporary work space at crossings.
2. Proposed pipeline to maintain a minimum of one foot clearance between subsurface utility crossing.

Proposed pipeline typical construction right-of-way
N.T.S.

FIGURE 16.4 Construction R.O.W. 25' × 75' When Additional Workspace is Required – example shows Topsoil Segregation on Working Side.

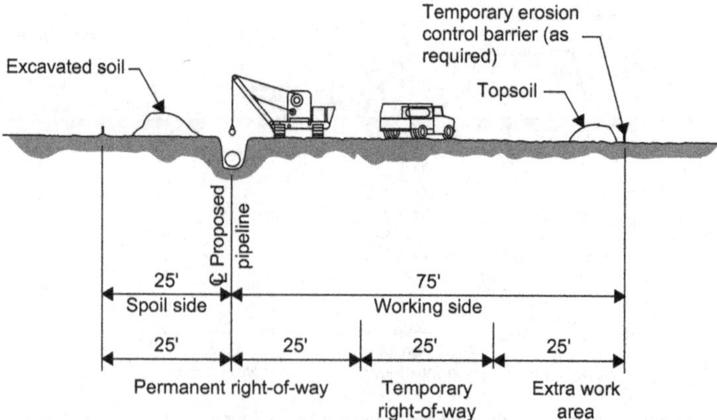

FIGURE 16.5 Construction R.O.W. 25' × 75' with Extra Workspace Defined for Topsoil Segregation on Working Side.

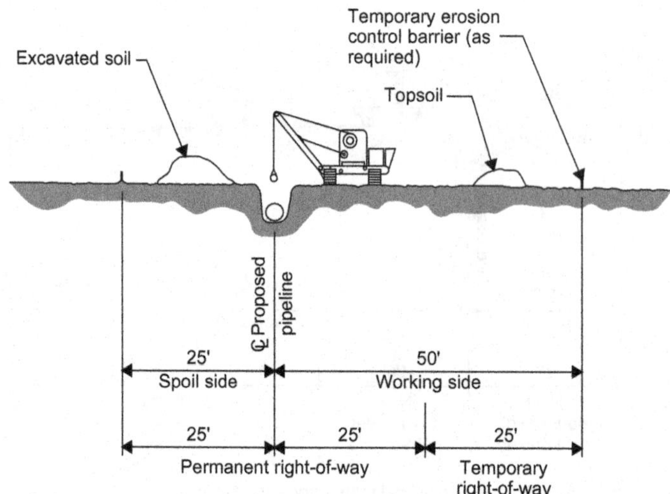

FIGURE 16.6 Construction R.O.W. 25' × 50' with Topsoil Segregation on Working Side.

excavated trench material, it is apparent that there are other circumstances that require variations to where topsoil is stored. In most cases, with the exception of minimal working area, the construction corridor required should be at a minimum of 25 ft wider to accommodate the segregated topsoil storage during construction operations.

16.1.2 Soil Classifications and Considerations

Six geologic groups of soil deposits (in alphabetical order) are aeolian, alluvial, colluvial, glacial, marine, and residual. The physical characteristics that are of particular relevance to soils include relative water content, dispersion or saturation, level, void ratio (volume of pore space to the total volume of soil), and porosity (the volume of space between soil particles). The Unified Soil Classification System classifies all soils as coarse, fine grained, or predominantly organic. Catalog constraints for soils include shrinkage limit, plastic limit, liquid limit, shear strength (amount of force required to separate soil particles), deformability (the relative changes of shape in response to the external forces), and the relative density. The characteristics of the soil allow for engineers to determine load analysis along with structural relation to the long-term effects after disturbance.

The following illustrates various soils encountered along a particular pipeline route:

- Othello-Elkton-Mattapex: This association consists of moderately well-drained and poorly drained, fine silty mixed soils that are nearly level or gently sloping. These soils are deep and have moderately slow to slow permeability. Soil acidity is strongly acidic to extremely acidic unless limed.
- Sunnyside-Christiana-Muirkirk: This association consists of very deep, well drained to somewhat excessively drained, moderately slow to slowly permeable soils on uplands. These soils have slopes from 0% to 50%. Most of this association lies in wooded or idle fields.
- Beltsville-Croom-Leonardtown: This association consists of deep to very deep soils that range in drainage from poor to well drained and have variable permeability. These soils are found in coastal plains and uplands. Slopes range from nearly level to moderately sloping. Most of this association is not used as cropland, and there is high erosion potential for much of this association.
- Neshaminy-Lehigh-Glenelg: Soils in this association are deep to very deep and well drained to somewhat poorly drained fine loamy. Permeability is moderately slow to slow. This association ranges from forested and very stony to cultivated soils. Soils are nearly level to steep. Most of this association has high erosion potential.
- Manor-Glenelg-Chester: This association consists of deep to very deep, steep to gently sloping, somewhat excessively drained and well-drained soils. This association occurs on hilly uplands. Most of this association has high erosion potential. Most of the soils are used for farming and a limited amount for pasture.

- Chrome-Conowingo-Neshaminy: Soils in this association are moderately deep to very deep, somewhat poorly drained to well-drained soils. Much of this association has high erosion potential. This association consists of nearly level to moderately sloping soils in well-dissected uplands. Much of this association is used for farming and pasture. Other areas are wooded or used for urban and suburban communities.
- Chester-Glenelg-Manor: This association consists of deep to very deep and gently sloping to steep soils. These soils are well drained to excessively drained. Most of the land area is used as cropland and to a limited extent, pasture. There is a high potential of erosion for approximately 50% of the association.
- Hagerstown-Duffield-Clarksburg: This association consists of deep and very deep, well-drained soils with moderate permeability. Some areas are less well drained with slow to moderately slow permeability. These soils weathered mostly from limestone. Most of this association is used as cropland or pasture.
- Edgemont-Highfield-Buchanan: This association consists of deep and very deep, well-drained soils. These soils formed from light-colored rocks, notably quartzite. They have moderate to moderately rapid permeability. This association is located on sloping to steep hills, ridges, and valleys and can be stony. Land use is a mixture of wooded areas and cleared areas for crops and orchards.

16.1.3 Trenching Crew

This crew excavates trenches on the right-of-way in which to lay pipe. The equipment operators run the equipment to excavate the material for the trench. The laborers perform any necessary hand digging and necessary general field labor activities. The truck drivers drive the buses to mobilize the crew to and from the site and the trucks to haul any unusable backfill material and rocks.

In flatter more forgiving terrain, a wheel trencher would be used to dig an approximately 60-inch-wide, 84-inch-deep trench, stacking the dirt alongside the ditch. The wheel rotates continuously as the machine moves along the pipeline route, and excavated material is continuously deposited alongside the ditch. In hilly terrain and in rocky areas along with areas where the pipeline changes direction when a wheeled trencher cannot be used, an excavator would perform the required excavation. The ditch would be excavated to a minimum depth adequate to allow for predetermined and legal depth of cover over the pipeline. The minimum cover over the pipeline varies according to requirements of regulatory agencies, landowner agreements, specific geographical locations, and features along the pipeline route. Spoil and topsoil would be windrowed and stockpiled separately, preferably along the nonworking side of the trench. Some circumstances, when topsoil segregation is required, mandate that the topsoil be segregated along the working side of the corridor as indicated in the figures 16.4, 16.5 and 16.6.

In geographical locations that necessitate the movement of livestock, ditch plugs should be left in place at agreed-upon intervals. Furthermore, these locations should be left open when stringing the pipe alongside the ditch. At these locations, one joint of pipe should be set back at the ditch plug location.

The trench is usually established to one side of the center of the working corridor rather than in the center. This allows adequate room for pipe to be strung along with construction equipment and operations. At some locations, blasting or special rock-cutting equipment such as a hoe ram on backhoe equipment or rock saw may be required when the ditch must pass through solid rock.

The width of the pipeline ditch varies according to the size of the pipeline along with regulated and agreed-upon requirements. Figure 16.7 illustrates a typical pipeline ditch and the references on determining the width of the ditch.

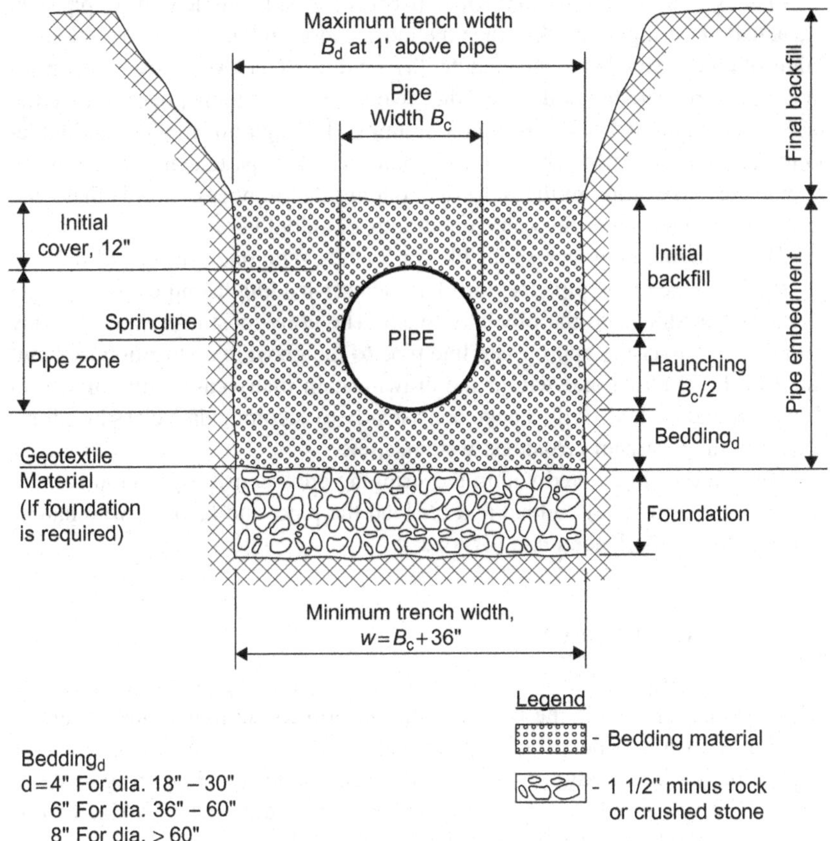

FIGURE 16.7 Pipeline Trench Detail.

16.1.4 Stringing Crew

Normally, pipe segments are delivered to the pipe storage yards at locations close to the pipeline corridor where the stringing trucks will subsequently deploy the joints along that corridor. This operation, by having the pipe joints in yards close to the corridor, ensures that each joint needs only to be moved from the storage yard to the corridor after the trench is completed when it is ready to be bent and subsequently welded into the pipeline string alongside the trench. By handling the pipe joints at a minimum to save cost and time, it also reduces the prospective for damage to the pipe before installation.

In geographic locations, and where permits allow, longer lengths of pipe can be hauled via stringing truck to the pipeline corridor. In this case, two sections or joints of pipe are hauled to the job after the welding of the two sections transpires in a pipe yard. The method of welding two sections, or joints, of pipe together prior to stringing is known as double jointing. Through this method, a double-joint yard is established with a double-joint rack. This particular method, when applicable, saves time in the field by reducing the number of welds that need to be made in less-than-favorable conditions of a double-jointing yard.

This stringing crew is responsible for loading pipe onto trucks and delivering the line pipe to the pipeline corridor, then off-loading the pipe and stringing it onto skids alongside the pipeline trench. The crew must take care on many projects to deliver the appropriate line pipe to the geographic location as stipulated in the contract and within the drawings. In many cases, line pipe wall thickness and/or grade will change consistently to accommodate class locations, road crossings, stream crossings, wetlands, etc.

The personnel involved within the stringing crew consist of truck drivers to haul the pipe, operators to load and off-load the pipe, and laborers who handle the line pipe to be loaded and off-loaded.

16.1.5 Bending Crew

This crew is responsible for bending the joints of line pipe, so the pipeline, when welded alongside the trench, will conform to the trench and terrain of the right-of-way. A bending engineer calculates the amount of bend in the pipe with the knowledge of the specific provisions within piping systems codes. The codes indicate the minimum bending radius for each size of line pipe. The bending is typically performed by a bending machine that is a track-mounted hydraulic machine, similar to a bulldozer track, which bends a pipe to a precise angle specified by the bending engineer. Bending machines are used to perform cold bending on a job site. Even large-diameter pipe can be accommodated in today's modern bending machines, but it may also be necessary to make some bends in a shop on a special machine. A bending machine is suitable for a large range of diameters, while a separate bending set is used to adapt the machine to the correct pipe diameter.

Hydraulic controls give the operator complete command of all bending machine operations. A hydraulically-driven winch moves the pipe through the machine. A calibrated indicator rod allows the operator to make consistently uniform bends. A pin-up shoe automatically grips the pipe to prevent distortion.

The personnel comprising the bending crew consist of the bending engineer(s), equipment operators, and laborers: the bending engineers operate the machine, the operators run the winch and the tractor, and the laborers manage the line pipe into the bending machine and then back into position alongside the trench.

16.1.6 Pipe Gang and Firing Line Welders

Front-End Welding Crew (Pipe Gang)

The pipe gang follows the pipe stringing and bending crews. The pipe gang positions the pipe, aligns it, and makes the welds. The laborers set the skids and prepare the pipe for the weld, except the final buff that is done by the welder helper. The pipe gang uses one or two side booms to set the pipe up on skids to support the pipe off the ground and line up the pipe with the contour of the trench. Alternatively, an automatic welding machine may be utilized to support and align the pipe. The pipeline welding crews align the pipe for welding and complete the welding of the pipeline above the trench. Pipe ends are aligned and clamped in place. Once the pipe is in the line-up clamps and proper alignment has been achieved, two welders perform the first pass called the root pass. On completion of the root pass, the line-up clamp is removed and the pipe gang moves on to the next joint of pipe. Immediately after the pipe gang has moved ahead, the next group of welders applies the second bead or hot pass.

Firing Line Welding Crew

The firing line crew can consist of many welders using welding machines mounted on pickup trucks. The welding crews follow and place the remainder of the weld material into the joint, including the final "cap" weld. When completed, there are a total of four or five layers of welds on the pipe joint. All welders who work on a joint have unique identifying codes. The codes are marked on the area adjacent to the pipe, so complete records of the welding are maintained.

16.1.7 Coating Crew

Although the pipe would arrive at the right-of-way with a corrosion-resistant coating, crews apply additional coating to the weld areas and repair any damage to the factory-applied coating to prevent corrosion. The pipe coating is a special material that coats pipes for pipelines. It prevents water from contacting the steel of the pipe and causing corrosion. The pipe is cleaned and primed first with a self-propelled machine that removes any loose material from the pipe surface with a rotating set of brushes or buffers.

The pipe surface around the girth welds must be free of mill scale and corrosion products and need to meet the Near White standard as stipulated by the coating manufacturer's specifications.

The pipe surface must be free of extraneous matter and contaminants, which are as follows:

1. Welding residues and spatter, dirt, abrasive residues, masking tape leftovers
2. Organic contaminants such as grease and oils
3. Soluble salts, both unreacted and those having reacted with iron during the corrosion process

The pipe surface must have a suitable anchor pattern.

16.1.8 Lowering-In Crew

This crew lays pipe in a ditch. The pipe is lowered as part of the coating operation or separately by a lowering-in crew. The personnel involved are operating engineers, laborers, and truck drivers. The operators run the equipment. The laborers move the skids and do any other manual labor required. The truck drivers drive the trucks that haul the skids and the fuel truck.

When the welding and coating are complete, the pipe is suspended over the ditch by side-boom tractors, which are crawler tractors with a special hoisting frame attached to one side.

Then, the pipeline is gradually lowered to the bottom of the ditch. In rocky soil or solid rock, it is sometimes necessary to put a bed of fine soil in the bottom of the ditch before lowering the pipeline. The fine fill material protects the pipe coating from damage.

16.1.9 Backfill Crew

This crew covers the pipeline, so adequate material is underneath and above the pipe. This includes replacing the topsoil and returning the right-of-way to its original or better condition. This involves the operating engineers and laborers. The operators run the equipment and the laborers do the manual work (e.g., removing rock, hand fill).

With new pipelines, or when conducting a maintenance activity for existing pipelines, backfilling and bedding must be provided in a manner that will offer firm support for the pipeline and not damage either the pipe or the pipe coating by the type of backfill material used or subsequent surface activities.

If the backfill material contains rocks or hard lumps that could damage the coating, care must be taken to protect the pipe and pipe coating from damage by such means as the use of mechanical shield material; backfilling procedures must not cause a distortion of the pipe cross section that would be detrimental to the operation of the piping and the passage of cleaning or internal inspection devices.

All pumping of water should comply with existing drainage laws and local ordinances relating to such activities.

Prior to backfilling, all drain tile should be permanently repaired, inspected, and the repair documented. Prior to backfilling, trench breakers should be installed on slopes where necessary to minimize the potential for water movement down the ditch and potential subsequent erosion. During backfill, the stockpiled subsoil should be placed back into the trench before replacing the topsoil. Topsoil should not be utilized for padding the pipe. Backfill should be compacted to a minimum of 90% of preexisting conditions in which the trench line crosses tracks of wheel irrigation systems (pivots). To reduce the potential for ditch line subsidence, spoil should be replaced and compacted by backhoe bucket or by wheels or tracks of equipment traversing down the trench.

The topsoil cover of actual depth or less than 4 ft should not be backfilled with soil containing rocks of any greater concentration or size than existed prior to pipeline construction in the pipeline trench, bore pits, or other excavations.

16.1.10 Tie-In Crew

This crew completes those construction tasks bypassed by regular crews. These include welding road and river crossings, valves, portions of the pipeline left disconnected for hydrostatic testing, and other fabrication assemblies. The tie-in crew follows behind the lowering-in crew to make the final cuts on the pipe to connect the entire pipeline. They also include coating the final tie-in welds. The personnel involved are welders, operating engineers, laborers, and truck drivers. The welders align the pipe and complete the welds. The operators move the pipe and piping accessories such as valves, components, etc. The laborers do any hand digging that might be required and the coating of the welds. The truck drivers haul the equipment and related material.

16.1.11 Testing Crew

This crew tests the line to prove its structural soundness and ability to fulfill its design function. In hydrostatic testing, the line is filled with water and then pressurized in accordance with the required contractual agreements.

Hydrostatic testing is sometimes done by the general contractor and other times subcontracted to a specialty contractor. The trades involved are laborers and operators. The laborers do the required manual labor. The operators run any equipment needed. All newly installed pipelines, including pipe segments that have been replaced in existing pipelines, undergo hydrostatic testing before being put into service. Hydrostatic testing involves isolating that portion of the pipeline undergoing testing, filling it with water, and then pressurizing the line to a specified pressure to check for leaks. A batching pig driven ahead of the water is used to remove any air and forms an efficient seal to isolate that portion undergoing testing. Without a pig in downhill portions of the line, the water will run down underneath the air, trapping pockets at the highest points within the pipe. Long pipelines will normally be tested in sections; short pipelines may be tested as single units. Temporary connections for filling and draining the pipeline are used, and a pump is used to "pressure up" the line. Once the specified pressure is attained, the pump is shut off and the "static" leak test commences. A leak is indicated if the pressure falls over the period of the test.

Once hydrostatic testing is completed, the water is removed. Although the majority of the water should be removed simply by draining the water at appropriate and approved locations along the segment undergoing a test, some water will still remain. Typically, a pig is used that is designed specifically to capture water and deliver it to a point where it can be removed. This dewatering pig serves a dual purpose, removing water and also removing construction debris that may still remain in the pipeline.

16.1.12 Clean-Up Crew

This crew completes the final cleanup on the construction site. This involves operating engineers, laborers, and truck drivers. The operators complete the backfilling. The laborers do the manual cleanup. Truck drivers haul material and equipment. Cleanup should happen immediately following backfilling operations when weather or seasonal conditions allow.

All garbage and construction debris (e.g., lathing, ribbon, welding rods, pipe bevel shavings, pipe spacer ropes, end caps, pipe skids) should be collected and disposed of at approved disposal sites.

The right-of-way should be recontoured with spoil material to approximate preconstruction contours and as necessary to limit erosion and subsidence. Loading of slopes with unconsolidated spoil material should be avoided during slope recontouring. Topsoil should be replaced after recontouring of the grade with subsoil. The topsoil should be replaced on the subsoil storage area and over the trench so that after settling occurs, the topsoil's approximate original depth and contour (with an allowance for settling) should be achieved. Subsoil should not be placed on top of topsoil.

Surface drainage should be restored and recontoured to conform to the adjacent land drainage system. Erosion control structures such as permanent slope breakers and cross ditches should be installed on steep slopes where necessary to control erosion by diverting surface runoff from the right-of-way to stable and vegetated off right-of-way areas.

During cleanup, temporary sediment barriers such as silt fence and hay bale diversions should be removed, accumulated sediment recontoured with the rest of the right-of-way, and permanent erosion controls installed as necessary.

After construction, all temporary access should be returned to prior-to-construction conditions unless specifically agreed with the landowner or otherwise specified by the company.

Warning signs, aerial markers, and cathodic protection test leads should be installed in locations that do not impair farming operations and are acceptable to the landowner. All bridges, fences, and culverts existing prior to construction should be restored to meet or exceed approximate preconstruction conditions.

Caution should be utilized when reestablishing culverts to ensure that drainage is not improved to a point that would be detrimental to existing water bodies and wetlands. All temporary gates installed during construction should be replaced with permanent fence unless otherwise requested by the landowner.

16.2 RESTORATION OF DISTURBED CONSTRUCTION R.O.W.

The objectives of reclamation and revegetation are to return the disturbed areas to approximately preconstruction use and capability. This involves the treatment of soil as necessary to preserve approximate preconstruction capability and the stabilization of the work surface in a manner consistent with the initial land use.

The following mitigative measures should be utilized unless otherwise approved or directed by the company based on site-specific conditions or circumstances. However, all work should be conducted in accordance with applicable permits.

Compacted cropland should be ripped a minimum of three passes, at least 18 in. deep, and all pasture should be ripped or chiseled a minimum of three passes, at least 12 in. deep, before replacing topsoil.

Areas of the construction right-of-way that were stripped for topsoil salvage should be ripped a minimum of three passes (in cross patterns, as practical) prior to topsoil replacement. The approximate depth of ripping should be 18 in. (or a lesser depth if damage may occur to existing drain tile systems).

The decompacted construction right-of-way should be tested by the contractor at regular intervals for compaction in agricultural and residential areas. Tests should be conducted on the same soil type under similar moisture conditions in undisturbed areas immediately adjacent to the right-of-way to approximate preconstruction conditions. Penetrometers or other appropriate devices should be used to conduct tests. Topsoil should be replaced to preexisting depths once ripping and disking of subsoil is complete up to a maximum of 12 in. Topsoil compaction on cultivated fields should be alleviated by cultivation.

Chapter 17

Welding and NDT

Barry G. Bubar, P.E.

Chapter Outline

INTRODUCTION

Welding plays a major role in the construction, modification, and repair of hazardous liquid pipelines and natural gas pipelines. Even though there has been a large amount of research and development (R&D) related to the advancement of semiautomatic and automatic pipe welding machines, which should significantly increase weld deposition rate and improve pipeline production, most pipeline welding today is still being done manually, using SMAW (stick) electrodes.

There has also been a large amount of R&D work in the field of nondestructive testing (NDT) and especially for pipeline weld inspection using ultrasonic methods. But radiographic NDT (X-ray) is still preferred to evaluate pipe weld integrity, in both large and small pipeline projects. This is partly because weld defect indications are easily seen on X-ray film, and the associated criteria necessary to evaluate a defect and make a pass or fail decision is well established, trusted, and documented in API 1104.[1]

So today in the pipeline industry, we are still "stick welding" on most pipeline projects and still using X-ray inspection to qualify pipeline girth welds made during construction and in-service repair: but change is in the wind!

Automatic (or mechanized) welding is starting to be used on large diameter gas pipeline construction to improve production rate and reduce cost. Big diameter pipeline projects, such as the 36-in. Cheyenne Plains Natural Gas Pipeline Project and the 36-in. Alliance Natural Gas Pipeline system, used automatic welding, and they both selected a highly accurate and automated ultrasonic testing system (AUT) to ensure girth weld integrity. AUT was necessary to provide quality inspection with speed sufficient to keep up with the automatic weld production. Even so, X-ray inspection was required on all pipeline tie-ins. We will talk more about automatic welding and nondestructive testing using ultrasonic methods later in the chapter.

17.1 PIPELINE WELDING PROCEDURES

Pipeline welding must be performed by a qualified welder, in accordance with a written qualified welding procedure specification[2] and the requirements of section 5 of API 1104 or section IX of the ASME Boiler and Pressure Vessel Code Qualifications. In this chapter, we discuss the welding requirements and testing specified by API 1104, Edition 20, which is the latest edition at this time. API 1104 is the pipeline welding standard for weld quality and innovation and is recognized as such around the world.

This standard is extremely flexible by design, allowing most welding processes available today to be used in production pipeline welding and pipeline repair: from shielded metal arc (stick electrode), gas metal arc welding, flux-cored arc welding, plasma arc welding, oxyacetylene welding, flash butt welding, and others. The one ridged requirement is, before using any of these allowed welding processes to weld pipelines, a detailed procedure specification must be written and qualified to demonstrate that pipeline welds made, using that particular welding process, have suitable mechanical properties such as strength, ductility, hardness, and soundness. To qualify a suitable welding procedure, the quality of the welds made, by a qualified welder, can only be determined by using distractive testing, in accordance with section 5.6 of API 1104. This distractive testing of pipe weld specimens include tensile strength tests, nick-break tests, and jig guided-bend tests.

API 1104 specifies exactly where to cut out these test specimens from the qualifying butt-joint weld, for each pipe diameter range. Pipe size is categorized by diameter and wall thickness, ranging from 2.375 to 4.50 in. (60.3–114.3 mm), 4.50 to 12.75 in. (114.3–324 mm), and greater than 12.75 in. (324 mm), each with wall thickness less than or equal to 0.500 in. (12.7 mm) and with wall thickness greater than 0.500 in. The welding standard also specifies the number of test specimens required for each diameter range with all details necessary to prepare the weld specimens for testing. The test method for each required test is clearly stated, and the criteria for passing each test are given in detail.

For example, testing a welding procedure specification for butt-joint welding of 30 in. (762 mm) diameter, 0.450 in. (10.2 mm) wall thickness, X-65 pipe must be in strict compliance with a written procedure, following the guidelines, as shown in the sample procedure specification form in Fig. 17.1. This welding procedure form lists all critical parameters and essential variables to be followed during welding and must be strictly complied with parameters such as joint design, weld position, direction of weld, filler metal (the type of electrode), number of beads (or weld passes), shielding gas and flow rate, number of welders required, type of lineup clamp.

To qualify the 30-in. (762 mm) diameter pipe, butt joint requires cutting and testing 16 specimens from the weld, four specimens from each quadrant. This is because API 1104 specifies that for pipe size greater than 12.75 in. (324 mm) diameter and with wall thickness less than or equal to 0.500 in. (12.7 mm), the following tests are required: four tensile strength, four nick-break, four root-bend, and four face-bend tests.

17.2 SPECIMEN PREPARATION

Let's use the tensile strength test as an example: the four tensile strength specimens must be approximately 9 in. (230 mm) long and 1 in. (25 mm) wide with the welded joint approximately in the center of the specimen. They may be machine cut or oxygen cut, but the sides should be smooth and parallel. Also the weld reinforcement should be left as is, and not removed from either side of the specimen.

17.3 TESTING

The tensile strength test specimens shall be broken under tensile load using equipment capable of measuring the load at which failure occurs. The tensile strength (in pounds per square inch) will be computed by dividing the maximum load at failure by the smallest cross-sectional area of the specimen, as measured before the load was applied.

17.4 CRITERIA FOR WELD ACCEPTANCE

The tensile strength of the weld including the fusion zone in each specimen will be greater than or equal to the specified minimum tensile strength of the pipe material. If the specimen breaks outside the weld and its fusion zone (in the parent pipe material) and meets the minimum tensile strength requirements of the pipe specification, the weld will be accepted. If the specimen breaks below the specified minimum tensile strength of the pipe material, the specimen will be set aside and a new test weld will be made.

Refer to API 1104 for additional requirements regarding specimen preparation, testing, and acceptance criteria for nick-break, root-bend, and face-bend

Reference: API Standard 1104

PROCEDURE SPECIFICATION NO. ________________

For ____________________ Welding of ____________________ Pipe and fittings
Process ____________________
Material ____________________
Pipe outside diameter and wall thickness ____________________
Joint design ____________________
Filler metal and no. of beads ____________________
Electrical or flame characteristics ____________________
Position ____________________
Direction of welding ____________________
No. of welders ____________________
Time lapse between passes ____________________
Type and removal of lineup clamp ____________________
Cleaning and/or grinding ____________________
Preheat/stress relief ____________________
Shielding gas and flow rate ____________________
Shielding flux ____________________
Speed of travel ____________________ Plasma gas flow rate ____________________
Plasma gas composition ____________________
Plasma gas orifice size ____________________
Sketches and tabulations attached ____________________

Tested ____________________ Welder ____________________
Approved ____________________ Welding supervisor ____________________
Adopted ____________________ Chief engineer ____________________

Standard V-bevel butt joint

Sequence of beads

Note: Dimensions are for example only.

Electrode Size and Number of Beads

Bead Number	Electrode Size and Type	Voltage	Amperage and Polarity	Speed

FIGURE 17.1 Procedure specification form, showing standard V-bevel butt joint and the weld bead sequence.

specimens. The following is only an introduction to the required additional destructive tests. The nick-break test is designed to determine the soundness of the weld: it should show complete penetration and fusion. Each specimen shall be broken at the nick and visually examined. The greatest dimension of a gas pocket should not exceed 1/16 in. (1.6 mm) and the combined area of all gas pockets should not exceed 2% of the exposed surface area. Slag inclusions should not be more than 1/32 in. (0.8 mm) in depth and should not be more than 1/8 in. (3 mm) in length. There should be at least 1/2 in. (13 mm) separation between adjacent slag inclusions.

The root-bend and face-bend tests are designed to ensure that the weld specimen has adequate ductility and toughness. Both root and face-bend test specimens should be approximately 9 in. (230 mm) long and approximately 1 in. (25 mm) wide, and their long sides should be rounded. The weld-cap and root-bead reinforcement must be removed flush with the surfaces of the specimen. The surfaces must be smooth, and any scratches that exist must be light and transverse to the weld. The test method requires that both face-bend and root-bend specimens are bent in a guided-bend jig with plunger, specified by API 1104. The plunger, placed over the specimen, is forced into the guide gap until the specimen is U shaped.

Both bend tests should be considered acceptable, if no cracks or other imperfections exceeding 1/8 in. (3 mm) occur in the specimen after bending. All 16 tested specimens are needed to qualify a welding procedure specification. Once qualified by destructive testing, these specifications of welding procedures must be strictly adhered to by the welder, when welding a pipeline using that welding procedure.

17.5 CLASSIC PIPELINE WELDING

Pipeline welding is a relatively simple procedure, in concept, requiring a machine-cut bevel or oxygen-cut bevel on each end of each pipe section and butting the two pipes together and holding them by using a pipe clamp. For the standard V-bevel, each pipe after grounding at 30 degrees must have a 1/16 in. (1.6 mm) high-flat face, called a land and the mating lands must be spaced approximately 1/16 in. (1.6 mm) apart. Once the required pipe fit-up is obtained, and the pipe clamp is tightened to hold the pipes in place, tack welding of the joint is done to ensure that the fit-up does not slip, see Fig. 17.1 showing the standard V-bevel butt joint. But, in reality, the fit-up with 1/16-in. pipe gap is not as easy as it looks because each pipe is not perfectly round, and careful fitting of the pipe is of paramount importance.

Figure 17.1 shows the standard pipe butt-joint fit-up, it also shows the weld bead sequence, and it has a list of essential variables making up the written weld procedure specification. For the wall thickness shown schematically, there are five weld beads or weld passes needed to complete the full-penetration weld. Notice that the first bead, called the root pass, fuses the two land ends, picking up both inside edges of the pipes. So, this root pass, also called the "stringer pass" by welding crews for years, is the most important bead (or pass) of the entire butt weld. Because pipe joint fit-up, tacking the two pipes in place and welding the root pass, requires the most welding expertise and artistry in production welding, so the welders assigned to this task usually have the most experience. Once the root bead is completed, helpers will grind out the roughness and remove all slag and sometimes wire brush it if necessary, prior to starting the second pass. When pipeline size is 12.750 in. (305 mm) in diameter and larger, two welders, one on each side, will work as a team, putting in the root pass and in production, they will also weld the next pass, called the hot pass, before moving on to their next pipe joint.

In new pipeline construction projects (or production welding) where pipe is strung along the trenched right-of-way, welders work in teams, almost like an assembly line, but where the welders move, not the pipe. There are maybe two root-pass or hot-pass teams fitting and welding their butt joint and once completed, they will "leapfrog" to the next pipe joint to be fitted, clamped, and welded. Behind the root-pass teams, there may be two or more fill-pass teams and sometimes one or two fill-pass/weld-cap teams, completing each full-penetration butt weld, as shown in Fig. 17.1. The fill-pass teams and weld-cap teams do not need as much expertise as the root-pass welders, but still must be precise using their stick electrode or in some cases doing semiautomatic wire-feed welding, so that they don't accidently hit the pipe outside the V-shaped weld joint, causing an arc burn. The arc burn must be removed according to code requirements by either grinding it out and leaving the pipe wall smooth, but with sufficient thickness left to be in pipe wall tolerance, or if that cannot be accomplished, they must totally cut out the weld joint and grind new bevels, fit up the pipe, and reweld the joint.

Regarding welders expertise, for production type pipeline welding, where maximum pipe footage per day is the goal, some qualified welders just specialize in welding the fill passes and doing the cap weld, as required to complete the butt welds, as a team. Once 800 ft or more of production welded pipeline, using single- or double-jointed sections, is complete, the pipe is lifted up, off the skids where it was supported during welding, weld inspection, and weld-joint coating, and lowered into the trench to be tied-in to the previously lowered in pipeline.

17.6 DOUBLE JOINTS

Often pipeline is double jointed, meaning two pipe sections have been welded in a field shop or brought out to the pipeline right-of-way welded together, to reduce the number of production welds required. Usually, the shop-welded double joints are rolled, as compared to the position butt welds, which are done in the field by field crews.

Figure 17.2 shows a section of a 16-in.-diameter pipeline during a relocation project being lowered into the trench. The section is ready for tie-in, using an external clamp. Welders, their helpers, and a side-boom operator will maneuver

Lowering in a section of field bent and welded 16" pipe, ready for tie-in welding using an external clamp

FIGURE 17.2 Lowering in a section of field bent and welded 16 in. pipe.

the pipe into position, fit up the pipe joint, clamp it into position, and tack-weld it to prevent slippage. Figure 17.3 shows two welders welding a 16-in.-diameter butt weld, using SMAW (stick) electrodes. Note that the external clamp has been removed to allow more freedom for the welder to achieve a quality weld and complete all required weld passes.

The two welders are welding vertical down, each starting from the 12 o'clock position. This pipeline relocation project was done more than 25 years ago, but the stick welding technique today has not changed much. There has been improvement in truck-mounted diesel engine–driven welding generators and in filler metals such as wire electrodes and special stick electrodes, tailored for high-strength pipe, where only very low levels of diffusible hydrogen is allowed in the weld.

Also the so-called semiautomatic wire feeders and wire guns are now being used for pipeline production welding, along with internal pipe clamps, which greatly improve the pipe fit-up and increase productivity.

Truck-mounted welding machines have gotten better and are more automatic and as a result, make pipe welding easer. One of the new welding machines called PipePro 300 uses a small diesel engine to drive an inverter type power source and is based on Miller Electric technology. It has good dig control (or arc force

FIGURE 17.3 Butt welding a 16 in. diameter pipe, with 0.25 in. wall thickness.

control) and allows a selection of settings. This arc force is the amount of "drive" created at the end of the electrode that pushes the molten metal into the joint. Setting the inverter at lower voltages automatically boosts the amount of amperage proportionally, when needed, keeping the overall welding power at the end of the electrode sufficient to sustain the proper arc.

Welders say that they like the smoothness of the arc and the ability to run both stick and wire feed at much lower fuel consumption than with their old DC generator. Fuel consumption is one of the reasons to select it, but ease of use and versatility are the deciding factors.

17.7 USING HIGHER X-GRADE PIPE

To reduce pipeline cost at construction, both natural gas pipeline and liquid pipeline companies want to reduce the amount of tonnage used in their pipeline design, because pipe is sold by the pound. Using large diameter, higher-strength/thinner wall pipes can amount to substantial saving in construction costs. The companies not only save money by buying thinner pipes but also save because they use less filler metal (welding rod) to weld the thinner wall pipe and with less welding time needed. They will also enjoy a significant saving in shipping cost because of lower tonnage. Since higher-strength pipe will allow shipping gas and liquid products at higher pipeline pressures, the companies will be transporting more cubic feet of gas per hour or more barrels of liquid products per hour. The result will mean better pipeline efficiency and profit for the company.

The disadvantage is that there are some welding issues related to using higher-grade pipe, such as X-70 and X-80. One of the main concerns is hydrogen cracking. Metallurgists and engineers are working on that problem, along with welding equipment companies like Lincoln Electric, Miller Electric or CRC-Evans and filler metal suppliers like Lincoln Electric or Hobart Brothers and others. Those developers and suppliers have also been working with pipeline welding contractors in accordance with API 1104 recommendations to develop and qualify welding procedures for 24-in.-diameter and larger, high-X-grade pipe.

On a recent 36-in.-diameter natural gas pipeline construction project, most of the girth welding was done using CRC-Evens automatic welding equipment, welding the pipe both internally and externally. Refer to Section 17.10 in this chapter for the details. But even though the bulk of the pipe welding was done automatically, all of the tie-in welds had to be done manually. For example, each time they lowered 800 or 1000 ft of production welded pipe into the trench, the automatically welded pipe was tied-in manually to the previously lowered piping. Each pipeline tie-in needed to be aligned using a heavy external pipe clamp and all the welding was done manually from outside the pipe. Other required manual tie-in welding on the project included butt welding pipe of different wall thickness, or welding on flanges for pipeline block valves or for other pipeline equipment.

The most experienced pipeline welders are needed to do these tie-in welds, working with their helpers and crane operators. This 36-in.-diameter pipe had to be fitted, aligned, and clamped for welding. Typically, for large diameter, high-grade pipe tie-ins, both the root pass and the hot pass are welded using special stick electrodes in the downhill direction and the two fill passes and cap weld are put in using semiautomatic wire-feed guns, requiring an uphill welding procedure.

This mixed welding process works best on higher-grade pipes where hydrogen cracking is a problem and high production is important. Welders use and qualify a special E-6010 rod for the root pass and use special E-8010 or E-9010 rod from Hobart Brothers or others for the hot pass, thus completing the first two passes. Then, they switch to low-hydrogen flux-cored wire with 25/75 argon/CO_2 blanking gas to put in the two fill passes and the cap weld. The flux-cored wire allows the highest welding speed possible. The properties of these electrodes were selected to be compatible with properties of the base metal, helping to address hydrogen cracking concerns, yield strength, and toughness, and were verified during welding procedure development. Of course, this mixed welding process tie-in procedure just described was qualified by destructive testing in accordance with API requirements, prior to making any production welds.

The semiautomatic welding with flux-cored wire feed is about two times faster than stick, even though the welder must still position his wire-feed gun manually. This speed advantage with wire is because you don't need to change rods, one after another, but uphill welding using low-hydrogen flux-cored wire is necessary to eliminate slag inclusion in the weld bead. Also semiautomatic welding with wire feed requires blanketing gas, which must be protected from the wind to keep the weld bead from developing porosity, which could disqualify the weld. So, welders must use some form of wind protection at the pipe for semiautomatic tie-in welding.

17.8 WELDERS' QUALIFICATION

The purpose of welder qualification is to determine the welder's ability to make sound butt and fillet welds using qualified procedure specifications discussed earlier, before any production welding is done.

Welders must be qualified according to the applicable requirements of API 1104, sections 6.2 through 6.8. A welder who satisfactorily completes the qualification test procedure is then a qualified welder, with that procedure, provided the number of test specimens, required by section 6.5, have been removed, have been tested, and have met the acceptance criteria of section 5.6.

Prior to starting a qualification test, the welder will be allowed reasonable time to adjust the welding equipment to be used. The welder must use the same welding technique and proceed with the same speed as he will use; if he passes the test, and he is permitted to do production welding. This qualification of welders must be conducted in the presence of a representative, acceptable to the company. Section 6.6, "Radiographic Butt Welds Only," is a "loop hole" to required destructive testing of specimens, which must be cut from each butt-weld test. At the company's option, the welder's qualification of butt weld may be examined by radiography in lieu of the destructive tests specified in section 6.5. This option will be more expedient, when qualified welders are in short supply and are needed quickly.

According to API 1104, section 6.22, a welder with a single welding qualification is qualified within the limits of the listed essential variables, as described in the written welding procedure specification. If any of the essential variables are changed, such as welding direction (from downhill to uphill), filler metal (electrode type), pipe diameter from 2.375 in. (60.3 mm) through 12.750 in. (324 mm), and nominal wall thickness from 0.188 in. (4.8 mm) through 0.750 in. (19.1 mm), the welder to be using the changed procedure variables, must be requalified.

To eliminate this problem, it is best for a welder to have a multiple qualification. For a multiple qualification, a welder must successfully complete two tests described here, using previously qualified procedures. For the first test, the welder should make a butt weld in the fixed position, with the axis of the pipe either in the horizontal plane or inclined from the horizontal plane at an angle of not more than 45 degrees. This butt weld should be made on a pipe, with an outside diameter greater than or equal to 12.75 in. (324 mm) and with a wall thickness equal to or greater than 0.250 in. (6.4 mm), without using a backing strip. Specimens should be removed from the test weld at locations shown in figure 12 of API 1104.

For the second test, the welder should lay out, cut, fit, and weld a full-sized branch-on-pipe connection. This test should be performed with a pipe diameter of 12.75 in. (324 mm) or larger and with a nominal wall thickness of at least 0.250 in. (6.4 mm). A full-size hole should be cut in the run. The weld should be made with the run-pipe axis in the horizontal position and the branch axis extending vertically downward from the run. The finished weld should exhibit a neat, uniform workman-like appearance and complete penetration around the entire circumference. The completed root bead should not contain any burn-through of more than 1/4 in. (6 mm). The sum of the maximum dimensions of separate unrepaired burn-throughs in any continuous 12 in. (300 mm) length of weld should not exceed 1/2 in. (13 mm). Four nick-break specimens should be removed from the branch connection weld at the quarter points and should be prepared and tested in accordance with API 1104 sections 5.8.1 through 5.8.3.

According to API 1104 section 6.3.2, a welder who has successfully completed this multiple qualification with the large diameter butt weld and large diameter branch connection shall be qualified to weld in all positions: on all wall thickness, joint designs and fittings, and all pipe diameters. Still, this *qualified welder* will need to be trained and checked out before using semiautomatic and automatic welding equipment to make production welds on pipelines.

17.9 WELDERS' RESPONSIBILITY

One of the reasons that stick electrodes (SMAW) are still preferred for welding the root pass is that they allow the welder more flexibility and control, as he travels vertical down from the 12 o'clock position to 6 o'clock position, watching the puddle and fusing both inside edges of the fitted pipes. Pipe butt welds or girth welds as they are called in the industry, shown in Fig. 17.3, are quite often rejected by X-ray or other nondestructive testing (NDT) methods for inadequate penetration, or inadequate penetration, due to high-low. The short form for inadequate penetration is IP, and for inadequate penetration due to high-low is IPD, see Fig. 17.6b showing high-low due to a pipe fit up problem. This condition occurs when the welder's electrode depositing filler metal during the root pass fails to fuse both inside edges of the fit-up pipe joint. To minimize this problem, welders, their helpers, and also the side-boom crane operator must use care when fitting each butt joint and if needed, they may rotate one of the pipes to get the best fit. This common fit-up problem is aggravated by the fact that some pipes are not perfectly round as said earlier, even though they may be in pipe specification tolerance. To ensure that both inside edges are fused during the root pass, welders need to watch the molten puddle and angle their electrodes as necessary, as they travel in fixed position, vertical down, being careful not to burn-through or blow out the root bead, as it could be a cause for weld rejection.

Today, in natural gas pipeline construction, with pipe size 24 in. (610 mm) in diameter and larger, the use of high-production automatic welding equipment is starting to gain some popularity. This is because joint fit-up and the root-bead welding is not as critical as explained above. This automatic welding equipment with special inside pipe clamps has been developed to increase pipeline production and is being used to align, clamp, and weld the root-pass bead from inside the pipe. This faster more accurate internal welding machine is called an IWM. It combines a lineup mechanism that clamps the pipe ends in alignment and has internal pipe welding equipment, which automatically deposits the root bead in less than 1.5 min. The machine then releases and quickly travels down the pipe either 40 or 80 ft, ready to clamp and weld the next pipe. The limitations of this amazing internal clamp and welding machine combination are that each butt-welded pipe must have the same inside diameter, and as of now, the IWM machines are only available for pipe sizes 24 in. and larger.

A self-contained electrical source provides power for the shielding gas, the wire-feed motors, and other integrated systems in the IWM. A self-contained pneumatic system operates the travel motors, brakes, drive wheels, aligners, and clamping shoes. Manipulation of the IWM is controlled by an operator or a welder at the end of a 40/80 ft reach rod and control box. The lightweight tubular reach rod contains the control cable, arc welding leads, gas charge hose, and other required leads. The welding heads are located symmetrically around the rotating ring and perform the root-pass welding from the inside. The number of welding heads on the IWM varies between four and eight depending on the pipeline size. Because of the design, the internal root-pass technique is the most tolerable to mismatch (or high/low) than any other external root-pass technique.

For girth welds on pipelines with smaller diameter than 24 in. (610 mm) or with larger diameter pipe having different wall thickness, only conventional pipe fit-up using external clamps and manual welding can be done.

17.10 AUTOMATIC PIPELINE WELDING

It is important to note that while automatic (mechanized) welding requires fewer skilled welder/operators, it takes more support personnel than manual welding. People call it automatic welding but it is really not. It would be more accurate to call it mechanized welding. It does require a large number of welders/operators and other skilled support people, so it is a good thing for the pipeline industry.

The following is a rundown of what might be called typical for automatic pipe welding of a 36-in.-diameter pipeline to demonstrate the differences between automatic and manual welding:

- The start is with an automatic end-facing machine to change the 30-degree factory bevel on the pipe to a modified "J" bevel. This bevel, unique to the process, requires less filler metal and reduces the welding time.
- Next, the pipe is preheated to 200°F–400°F (93°C–204°C) using a special electric heater. This is required for higher X-grade pipe to minimize hydrogen cracking.
- An internal lineup clamp, with welding heads called an IWM, traveling inside the pipe aligns the two pipe sections and pneumatically clamps them in position, as explained earlier.
- After the welder/operator checks fit-up, six weld heads mounted on the internal clamp weld the root pass. Each head welds 1/3 of one half of the pipe. From clamping to weld completion takes less than 2 min.
- The next weld is the external hot pass. The first welding rig, like the nine that follow, carries two truck type welding machines each using a small diesel engine to drive an inverter type power sources and a side crane that hoists the welding shack. The shack prevents wind from affecting the shielding gas and provides a platform for two welders/operators and two welder helpers. The helpers affix the welding track to the pipe and clamp the automatic external welding head (also called a bug) to the track and connect the wires as necessary. The first welder/operator then pushes the contactor button, which preflows the shielding gas. The welder then pushes the arc start button and the track-mounted bug starts traveling down from the 12 o'clock position, powered by a constant speed DC motor. Since the track band may not be perfectly parallel to the joint, the welder/operator must carefully watch the bead, using a horizontal adjustment knob to recenter the welding head in the bevel when necessary. After the first bug clears the top of the pipe, the welder/operator on the opposite side of the pipe repeats the process, with his weld bead being completed at 6 o'clock position as well.
- Shacks 2 and 3 each put in the first fill pass; shacks 4 and 5 each put in the second fill pass; shacks 6 and 7 each put in the third fill pass; and shacks 8 and 9 each put in the cap. Each of these shacks welds every other joint. The hot-pass bug moves quickly (44–50 in./min) compared to the fill passes and the cap. These later passes require slower travel speeds because they deposit more filler metal and the torch oscillates to fill the joint and to let the weld pool cool slightly. During these passes, the welder/operator can also control torch oscillation width and arc length, making adjustments as necessary to maintain a good bead.

Overall, it is estimated that automatic pipeline welding now has a production advantage of about 20% over pipelines welded manually with SMAW stick electrodes, using experienced crews and both using about the same number of welders and helpers. This less-than-expected advantage is partly because not all of the pipeline can be welded automatically. After 800 or 1000 ft of production welded pipe using CRC-Evans internal and external automatic welding machines, with all the automatic welds ultrasonically inspected for weld defects, and with girth welds coated with FBE coating, the production welded pipe is lowered into the trench for tie-in welding. These tie-in welds must be done manually. So at most, this production pipeline is only 90% automatically welded.

Of course, this automatic welding performance as outlined above is being improved by advancements in external track-mounted welding machines (bugs). The future welding bugs will have through-the-arc tracking to adjust for welding head alignment problems, smart card improvements, making the welding more automatic, and the welding heads will use increased amounts of pulsed GMAW with a dual-head and dual-wire bug design to significantly increase welding rates. More R&D work needs to be done with external clamping of pipe and pipeline tie-in welding. At this time, it seems possible to achieve an additional 40 percent overall increase in weld productivity compared with manual welding, with the help of future automatic welding advancements.

Of course, the automatic welding procedures that we have been discussing here were qualified in the lab and field using qualified welders and were tested by destructive testing to ensure that required tensile strength, ductility, and soundness were achieved, before using the candidate procedure for welding pipelines.

17.11 VERIFYING AUTOMATIC WELD INTEGRITY

To ensure quality, all automatic welds were quickly ultrasonically tested using AUT. Where X-ray may take up to 10 minutes per weld joint, AUT takes about 90 seconds to inspect and interpret the entire girth weld, quick enough to keep up with automatic production welding. AUT provides the ability to look at a weld in detail by sending a sound wave into the pipe and having the computer map the echo. We will talk more about ultrasonic testing as an alternative to X-ray testing.

17.12 SEMIAUTOMATIC WELDING

With high-strength pipe, welders need to be concerned about hydrogen cracking when making tie-ins. This requires preheating the pipe using propane torches or electric blankets as the heat source and heat sticks to verify the desired preheat was achieved before butt-welding the tie-in. Also, the use of low-hydrogen electrodes is of utmost importance due to the chemistry of high-strength pipe. It has been found that root pass integrity with X-70 and X-80 pipes is critical, so welders working in teams, along with their helpers, are fitting up, clamping, and then completing the root-pass weld, usually using a vertical downhill method, as mentioned earlier, using E6010 stick electrodes for better fluidity and ductility. The E6010 electrode flows easily and its ductility helps prevents cracking. Then after grinding as necessary and wire brushing the root bead, the hot pass is welded using a E8010 or E9010 low-hydrogen stick electrode, also vertically downhill, which consumes most of the low yield strength root bead and brings the weld metal strength up to that of the pipe base metal. Instead of welding out the entire butt joint with stick rod, the tie-in welding can be sped up by changing the welding process to semiautomatic wire-feed welding, to complete the two or more fill passes and the cap weld. It was found during laboratory testing with various welding processes and filler metals and during qualification tests that semiautomatic low-hydrogen wire welding was at least two times faster than using low-hydrogen stick rod. Both of the low-hydrogen welding processes required welding uphill to prevent slag inclusion in the weld bead. This speed advantage and the ease of using wire-feed guns to fill and cap the tie-in welds convinced most welding contractors on the Cheyenne Planes Project to use the semiautomatic (wire-feed) welding system to speed up tie-ins (see Fig. 17.4).

Most welders had no experience with wire-feed, flux-cored welding, but with a little training, they became proficient quickly. First, they needed to learn how to set up there welding trucks and welding machines for wire feed, and next they needed to learn the proper welding technique.

Wire-feed welding requires having a lot more equipment than is needed for SMAW stick welding, such as a wire feeder that holds 14 lb of wire electrode, and a wire-feed gun such as a Bernard flex neck gun, which allows the welder to bend the nozzle to get comfortable with his position. The welder needs to also have a tank holding the blanketing gas compatible with the wire used and all required hoses and welding cables, etc.

Once set up, the welder must practice maintaining his gun angle relative to the weld puddle as he keeps the wire electrode in the center of the joint. Once the welder gets the hang of it, it is not a problem. The flex neck gun makes the process more forgiving by allowing a welder access to hard to hit areas of the pipe.

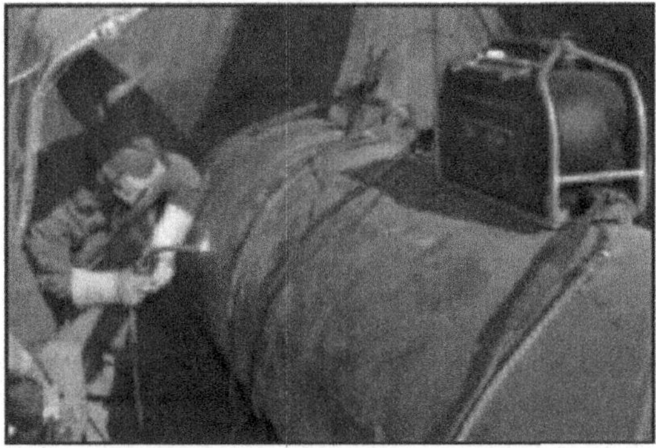

FIGURE 17.4 Welder using flex neck gun with wire-feed case on top pipe.

17.13 STRENGTH OF WELDED PIPELINES

Let me go back in time to compare the modern 30-degree bevel butt joint shown in Fig. 17.1, with an old style oxyacetylene butt weld. I'm doing this to show how adding a simple bevel made a world of difference in pipe strength and significantly advanced pipeline welding.

While working as district engineer on our Bakersfield gathering to LA Refinery crude oil system during the January 1994 Northridge, California earthquake, our 10-in.-diameter crude oil pipeline broke and separated in about 19 different locations, in the San Fernando Valley area, causing crude oil to spill out of each rupture. The pipeline was immediately shut down for evaluation.

This 6.7-magnitude earthquake caused severe ground movement and earth strain, which took out freeway overpasses and bridge supports and caused major structural damage to commercial buildings, homes, etc. The shock or high acceleration imparted large bending and shear loads on buried pipelines in that area.

The 10-in. pipeline, called Line-1, was constructed before World War II and was oxyacetylene welded. In those days, the pipe ends were spaced about 1/16 in. apart and butt welded without beveling. The torch fused the 5/16 in. wall pipe on the outside surface, penetrating into the joint approximately 1/8 in. or more. See the sketches of oxyacetylene butt weld in comparison to the 30-degree V-bevel joint in Fig. 17.5. This type of weld joint was definitely not a full-penetration pipe weld, which is achieved when complying with Fig. 17.1 requirements. This oxyacetylene weld joint actually was a good example of a notch-sensitive stress riser, something you don't want in a highly seismic location.

Even though the weld had a large cap, only about half the pipe wall was welded. Today, that weld would be rated as 100% inadequate penetration (IP) and rejected without question. The current NDT criteria for that weld defect is when an indication of only 1 in. or more of IP is found in 12 in. of continuous welding.

This pipeline had been hydrostatically tested to the maximum pipe flange test rating numerous times without pipe failure. So, even though the weld joint could take full pipeline test pressure, the weld was not strong enough to endure

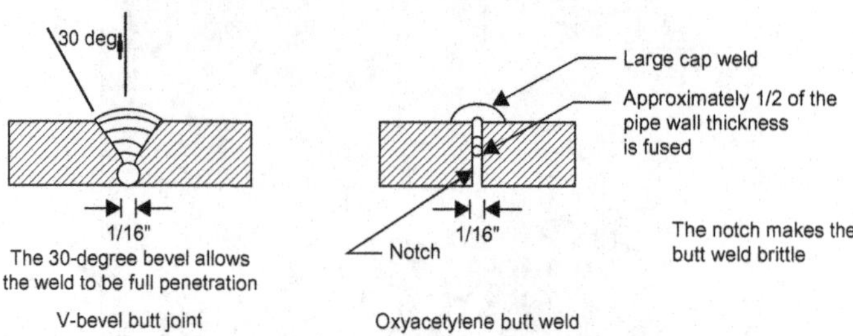

FIGURE 17.5 Showing pre-WWII pipe weld joint in comparison with modern V-bevel joint.

bending and shear loads resulting from the earthquake. Refer to Appendix 9, which develops both the pipe hoop stress formula and pipe longitudinal stress formula, showing that for a given pipeline pressure, the longitudinal stress is only one half the magnitude of the pipe hoop stress. It is this lower longitudinal stress that acts on each butt weld. So, pipe pressure testing alone is not sufficient to verify pipe butt-weld integrity. This speaks to the fact that a butt weld must be as strong and flexible as the pipe itself. So this is one reason that API 1104 requires each welding procedure specification used to weld pipeline to be qualified by distractive testing to ensure that the welds have adequate strength, toughness, and ductility.

Other more modern and recently constructed pipelines, having full-penetration welds, were located in the same pipeline right-of-way as Line-1 or nearby, and survived the forces of the earthquake without problem. The 10-in. Line-1 was repaired to allow removal of crude oil in the line, but was not put back into pipeline service for safety reasons.

17.14 NONDESTRUCTIVE TESTING OF PIPE GIRTH WELDS

Weld testing and acceptance standards, presented in section 9 of API 1104, apply to weld imperfections located by NDT methods such as radiographic (X-ray), magnetic-particle, liquid-penetrate, ultrasonic testing, and visual inspection. In this chapter, we will only concern ourselves with radiographic inspection, ultrasonic testing, and visual inspection methods, since they are used most frequently for testing pipeline weld integrity. If interested in the other two methods, refer to sections 9.4 and 9.5 of API 1104 for weld acceptance standards and defect indications when using both magnetic-particle and liquid-penetrate test methods.

As mentioned earlier, AUT has been used to inspect automatic pipe girth welds recently on at least two large natural gas pipeline construction projects. Ultrasonic testing was selected to keep up with the automatic welding rates and was used to test all production girth welds on the Cheyenne Planes Project and also on the Alliance Pipeline Project. But because AUT is relatively new, and without a large track record, all pipeline tie-in welds and welds at road crossing in urban locations were 100% X-ray inspected to comply with code requirements.

AUT has several advantages over X-ray, and I'm sure that its use and popularity will grow with time. First, it's faster, as a girth weld AUT inspection is about five times faster than X-ray, it's safer because there is no radiation to deal with, and ultrasonic NDT inspection provides a 3D defect indication compared with the classic 2D defect indication given when reading X-ray film. This more extensive defect information could lower repair costs, because knowing defect height and location in the pipe wall may let inspection technicians tell repair crews which weld pass contains the defect. That could save time by going directly to the defect to make the repair.

17.15 RADIOGRAPHIC NDT

Let's go directly to radiographic (X-ray) inspection. It is the most important, because almost all NDT of pipeline butt welds are done by X-ray. Weld inspection for both small and large diameter pipelines during construction, maintenance, and repair is done using the X-ray method. Of course, for liquid pipelines, the product must be emptied out of the pipe before weld inspection.

To X-ray a pipe butt weld, the film is placed on one side of the pipe and centered on the weld bead and a radiographic source positioned at the correct distance and angle is used to expose the film. The film is quickly taken into a dark room, developed, given a serial number, and is read.

Table 17.1 lists possible girth weld defects that an X-ray technician looks for when he reads the film. I thought it would be handy to have this abbreviated defect table ready, which lists the symbols for weld defects, names the defect, defines the defect, such as "IP – incomplete filling of the weld root," and gives most of the criteria to determine if the weld is defective according to section 9, API 1104, or acceptable. Of course, for the complete pipeline weld acceptance criteria and defect definition, please refer to section 9.

17.16 REPAIR OF DEFECT

The code specifies that pipeline weld repair and defect removal is required once an "indicated defect" is found unacceptable by NDT inspection methods. Both federal codes CFR Title 49, Part 192.245 for Natural Gas Pipeline[2] and Title 49, Part 195.230 for Hazardous Liquid Pipeline[3] have the same wording and require the following:

1. Each unacceptable weld must be removed or repaired.
2. Each weld that is repaired must have the defect removed down to sound metal, and the segment to be repaired must be preheated if conditions exist that would adversely affect the quality of the weld repair. After repair, the segment with the weld repair must be inspected to ensure its acceptability.
3. Repair of a crack or of any defect in a previously repaired area must be done in accordance with written weld repair procedures that have been qualified. The repair procedure must provide that the minimum mechanical properties specified for the welding procedure used to make the original weld are met on completion of the final weld repair.

17.17 WELDING REJECTION CRITERIA

The questions that people have been asking for decades regarding API 1104 welding defect inspection criteria is how did they decide on the amount of weld bead imperfection that was allowable and was part of that based on strength tests? And how did they determine the length of weld imperfection necessary to be

TABLE 17.1 Most Butt-Weld Defects Found by NDT and the Pipe Weld Rejection Criteria

Symbol	Defect	Defined As	Criteria Rejects Weld If
IP	Inadequate penetration without high-low	Incomplete filling of the weld root, as shown in Figure 17.6a	Individual IP exceeds 1 in. or the total length of IP in any continuous 12 in. weld is >1 in.
IPD	Inadequate penetration due to high-low	One edge of the root is not bonded because adjacent pipe is misaligned, as shown in Figure 17.6b	Length of individual IPD exceeds 2 in. or the total length of IPD in any continuous 12 in. weld is >3 in. (76 mm)
ICP	Inadequate cross penetration	Subsurface imperfection between the first inside pass and the first outside pass, as shown in Figure 17.6c	Length of individual ICP exceeds 2 in. or the total length of ICP in any continuous 12 in. weld is >2 in. (51 mm)
IF	Inadequate fusion	Surface imperfection between the weld metal and the base metal that is open to the surface, as shown in Figure 17.6d	Length of individual IF exceeds 1 in. or the total length of IF in any continuous 12 in. weld is >1 in. (25 mm)
IFD	Incomplete fusion due to cold lap	Imperfection between two adjacent weld beads or between the weld metal and the base metal, Figure 17.6e	Length of individual IFD exceeds 2 in. or the total length of IFD in any continuous 12 in. weld is >2 in. (51 mm)
IC	Internal concavity	The root pass is slightly below the inside surface, shown in Figure 17.6f	Any internal concavity is acceptable, provided the thickness at the cavity is greater than the thinnest pipe wall thickness
BT	Burn-thorough	Excessive penetration in the root bead causing the weld puddle to blow into the pipe	When the maximum dimension exceeds 1/4 in. (6 mm)
ESI & ISI	Slag inclusion: ESI – elongated slag ISI – irregular shape slag	Nonmetallic solid entrapped in the weld metal	Length of ESI (wagon tracks) >2 in. (51 mm). Total length of ISI in any continuous 12 in. weld is >1/2 in. (13 mm)

(*Continued*)

TABLE 17.1 Most Butt-Weld Defects Found by NDT and the Pipe Weld Rejection Criteria—cont'd

Symbol	Defect	Defined As	Criteria Rejects Weld If
P	Porosity	Gas trapped by the solidifying weld metal before the gas can rise to the surface	Size of individual pore >1/8 in. (3 mm). Size of individual pore >25% of thinnest wall. Distribution of scattered > permitted by API
C	Cracks	Crack in the weld	A crack of any size located in the weld, but is not a star crack width/length 5/32 in. (4 mm). Star cracks are found at weld bead stop points.
EU & IU	Undercutting: cover pass or root pass	Grove melted into parent metal adjacent to toe or root of the weld and not filled	Total length of EU/IU in any continuous 12 in. weld is >2 in. (51 mm)

considered a defect, requiring repair or removed from the weld? Many have said that they thought small defects even larger than unacceptable to API would have little or no effect on pipe strength, except for cracks. Cracks can grow in depth and length over time, when subjected to cyclic stresses, caused by ground movement or high-pressure surges, and must be removed from the weld.

The API 1104 committee's answer to the questions was that their job was to foster welding excellence. Their weld quality acceptance criteria, used as a pass or rejection standard when inspecting pipe welds by NDT, came from a large welder database and represent the average welder's workmanship and welding ability. This level of welder excellence has been demonstrated over time by qualified and experienced pipeline welders.

As a result of the many pipeline weld acceptance criteria questions received over the years, API has now added Appendix A, "Alternative Acceptance Standard for Girth Welds," to their pipeline welding standard API 1104.

The appendix explains that the welding acceptance standard given in section 9 is based on empirical criteria for workmanship and has placed primary importance on imperfection length. This criteria has provided an excellent record of reliability in pipeline service for many years. The use of fracture mechanics and fitness-for-purpose for determining acceptance criteria is an alternative method and incorporates the evaluation of both imperfection height and imperfection

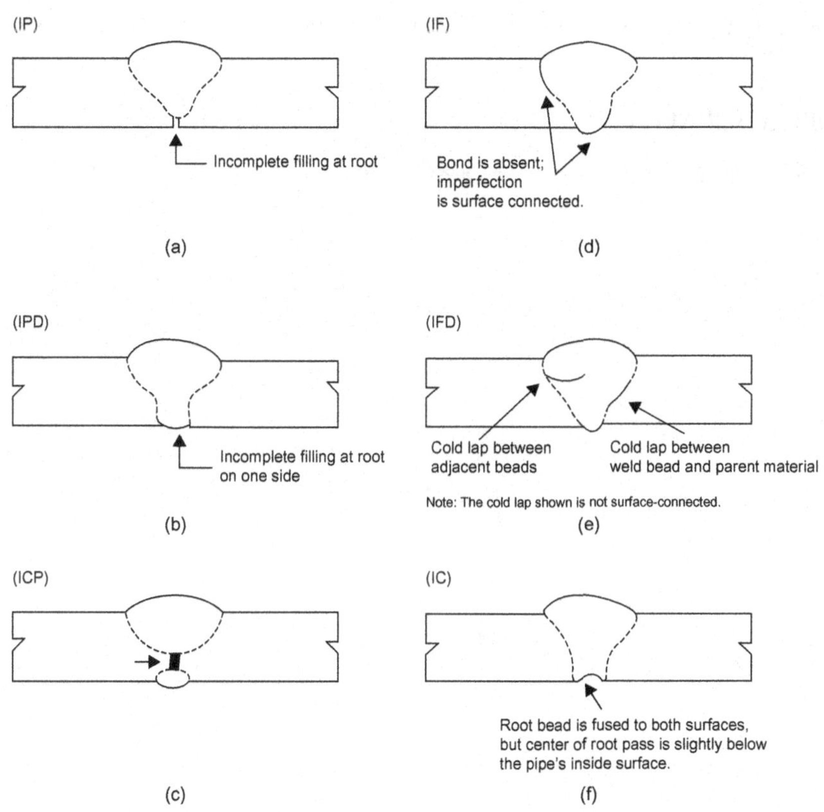

FIGURE 17.6 Sketches showing some pipe weld imperfections located by NDT. (a) Inadequate penetration without high-low. (b) Inadequate penetration due to high-low. (c) Inadequate cross penetration. (d) Incomplete fusion at root of bead or top of joint. (e) Incomplete fusion due to cold lap. (f) Internal concavity.

length. Typically, but not always, this fitness-for-purpose criteria provides a more generous allowable imperfection length. Qualification tests, stress analysis, and inspection are required to use this alternative fitness-for-purpose method of weld acceptance. We will not elaborate more on this alternative criteria for weld defect analysis, but to say that Appendix A provides very interesting technical reading and is available for use, if necessary.

REFERENCES

[1] Welding of Pipelines and Related Facilities, API Standard 1104, twentieth ed., American Petroleum Institute, 2005.

[2] Code of Federal Regulations, Title 49, Part 192, Transportation of Natural and Other Gas by Pipeline, 2010.

[3] Code of Federal Regulations, Title 49, Part 195, Transportation of Hazardous Liquids by Pipeline, 2010.

BIBLIOGRAPHY

ASME Code for Pressure Piping, B31.4, Addenda, 1991.

Chapter 18

Hydrostatic Testing

Barry G. Bubar, P.E.

Chapter Outline

INTRODUCTION, INCLUDING RISK-BASED ALTERNATIVES TO TESTING

No pipeline operator may operate a pipeline, or a new segment of pipeline, unless it has been pressure tested in accordance with CFR Title 49, Part 195 for Hazardous Liquid Pipelines and CFR Title 49, Part 192 for Gas Pipelines, without leakage. Also, no person may return to service a segment of pipe that has been relocated or replaced until it has been tested as mentioned above and each hazardous leak has been located and eliminated.

In this chapter, we will be talking primarily about liquid lines, but we will present the differences in testing requirements and procedures for natural gas pipelines. In general, there is not much difference in planning and hydrostatic testing a gas transmission pipeline, than with a hazardous liquid pipeline, except that after test and test water removal, the gas pipeline must be cleaned and dried to 0°F dew point. In some cases, repair sections, pipe replacement, and short relocations are tested with air, inert gas, or natural gas prior to start-up so that water removal and internal drying of the inside of new pipe is not required.

Figure 18.1 shows a large diameter natural gas pipeline under construction and being lowered in for tie-in. Once the tie-in welds are complete, X-rayed and coated, the line will be backfilled and ready for testing. A large amount of filtered test water will be pumped into the pipeline from a nearby water source, such as a river or lake. The pipe will be hydrostatically tested in sections between block valves, most likely equipped with body bleeds to assure zero leakage at the valve, and also between test heads designed for pipeline testing. The length of each hydrostatic test section will depend on pipeline elevation differences. After testing, the water will be pushed down the pipeline and reused for testing the next series of sections. We will discuss much more about testing cross-country pipelines, later in this chapter.

FIGURE 18.1 Lowering in for tie-in and hydrostatic testing. *Source: Courtesy of Pipeline and Gas Technology magazine.*

Pipeline hydrostatic testing is one of the most important and challenging procedures an engineer or pipeline operator is required to do with his or her pipeline system. It is extremely important that the personnel responsible for conducting the pipeline test have proper training and a good understanding of the dangers of subjecting line pipes, valves, and fittings to excess pressure. Overpressuring the pipeline can cause yielding of the pipe wall, which may plastically deform the pipe, damage flange gaskets, valves and fittings, or even worse, cause the pipeline itself to rupture.

The test pressure for each test conducted in accordance with Part 195[1] must be maintained throughout the piping system being tested, for a minimum of 4 h continuously at a pressure equal to at least 125% of the maximum allowable operating pressure (MAOP), when the test pipe is above ground or is in an open trench, and can be visually inspected for leaks. If the pipeline is buried, as usually is the case, the pipeline must be tested for an additional 4 h at a pressure of 110% or more of the MAOP. In some cases, the pipeline MAOP is equal to the internal design pressure, but it can never exceed the internal design pressure. The internal design pressure is calculated by using the pipe hoop stress formula:

$$P = 2 \times t \times (\text{SMYS}/D) \times E \times F \tag{18.1}$$

where

P – Pipe internal design pressure (psi)
t – Pipe wall thickness (inches)
SMYS – Specified minimum yield strength of the pipe being used (psi)
D – Pipe outside diameter (inches)
E – Pipe seam joint factor, which is equal to 1.0 for seamless or post-1970 ERW pipe
F – Pipe design factor equal to 0.72

The design factor for offshore liquid lines and high-risk locations is 0.60, and in some cases even lower. Please refer to Appendix 9 for the derivation of hoop stress and pipeline longitudinal stress formulas.

Operators must not operate the pipeline at pressure greater than the MAOP to insure adequate pipeline safety margin and to comply with federal code. The pipeline must also be protected from overpressurization caused by pressure surges in the line resulting from upset or non–steady state condition. Pressure safety devices such as pressure relief valves and high-pressure switches must be installed and set to limit pipeline pressure surges from exceeding 110% of the MAOP.

There are some alternatives to the pipeline hydrostatic testing requirements, as outlined above, but they are used only in limited situations and for special cases. Some relate to pipeline risk-based alternatives to mandatory pressure testing for older hazardous liquid and carbon dioxide pipelines, and others relate to very low stress liquid pipelines. Also, in California, the state board allows running smart pig inspection, in some cases, on piggable pipelines, as an alternative to performing the required hydrostatic test. We discussed more about using smart pigs as an alternative to hydrostatic testing in Chapter 15.

This risk-based criteria approach, as an alternative to testing aging pipelines as mentioned, requires determining the risk classification of each segment of each pipeline, with risk classification A being lowest risk and risk classification C being the highest risk. Each risk level is determined by using four pipe segment indicators: the location indicator (high if not rural), product indicator (high if shipping highly toxic, highly volatile or flammable product), volume indicator (high if the pipeline is 18 in. (450 mm) in diameter or greater), and the probability of failure indicator (high if the segment has experienced more than three failures in the last 10 years due to time-dependant defects such as corrosion). Refer to Part 195 Appendix B, entitled "Risk-Based Alternatives to Pressure Testing" for more about how to evaluate pipeline risk using the four-indicator-risk-based approach. The appendix also specifies what must be done with each pipe segment once the risk classification is determined. For example, with Classification A, the code states no further testing is required, but for pipeline segments determined to be classified as C, the code requires that pipe testing must be done using water as a test media and specifies the time deadline for completion. This is all we will mention here about this alternative approach, except that even with this risk-based integrity approach, the higher risk pipe still requires hydrostatic testing to insure integrity. For a better understanding of this alternative approach, visit Part 195, Appendix B as stated above.

18.1 TESTING PIPE

The simplest, most straightforward method of hydrostatic testing is when pipe testing can be done above ground using water as the test medium. The pipe segment can be filled with water while pushing a pig or sphere through the pipe, which displaces entrained air from the pipe section, while the entire section is being filled for test. The test section will be pressurized and the test pressure, pipe wall temperature, and ambient air temperature must be monitored for the 4-h test duration. Because the pipe can be visually inspected for leakage, the validity of the test can be quickly determined, verifying that the pipe was successfully tested for strength without leakage. It is probable that some test water will need to be bleed off during the test to control and/or limit rising pressure caused by the sun heating or ambient temperature change to prevent the maximum test pressure from being exceeded. If ambient air cooling occurs during the duration of the test, water injection may be needed to bring the test pressure back to the desired test level. Measuring the amount of water injected or bleed is not critical in this case because leakage can be determined visually. The details for setting up the necessary hydrostatic test equipment, instrumentation, and methodology will be presented later in this chapter.

When hydrostatic testing a pipeline or even just a pipe section that is buried, it is much more difficult to verify that the pipe strength test was successful and that it passed without leakage. In the case where a small pipeline leak is suspected, it is often difficult to find the leak location. It is common for a test crew to add green dye to the test water, which will help them with finding the leakage as they dig in suspected locations. If during the hydrotest, the pipe ruptures, the results are straight forward: the pressure drops off instantly and quite often the discharging liquid blows out the cover above the pipeline, allowing the test water to cascade off the pipeline right-of-way. We will discuss more about this type of rupture later in this chapter.

So, hydrostatic testing buried pipelines, or pipe sections, requires sophisticated instrumentation and methodology to recognize small leaks, which could be caused by small cracks, corrosion pits, or third-party damage. Often, the leakage can be quite small, and it has often been misdiagnosed as air or product trapped in the test water. Posttest leak analysis will be explained in detail, later in this chapter.

Some examples of buried pipeline tests conducted by the industry are pipe replacements (common), pipeline relocations required due to planned residential or commercial construction that conflict with the pipeline right-of-way (common), upgrading an existing liquid pipeline to higher operating pressure to increase pipeline throughput (somewhat common), and finally, conversion of a pipeline to different use such as changing it from hazardous liquid service to natural gas service (less common).

18.2 CLASSIFYING IN SERVICE PIPELINES

Hazardous liquid pipelines are classified as interstate and intrastate. Interstate pipelines transport liquid or natural gas between states, across the nation, and include foreign commerce. An intrastate pipeline, on the other hand, is a pipeline that operates in a particular state or local area and is not interstate. Interstate pipelines, once tested after construction, can operate indefinitely, unless the pipeline is altered, relocated, or damaged. Only the new pipe installed or replaced on an interstate pipeline must be hydrostatically tested to substantiate its MAOP, in accordance with Part 195. The original pipe need not be retested unless the pipeline and/or pump stations are to be upgraded to increase pressure and throughput. Another reason to retest all or part of an interstate pipeline is after a smart pig inspection with test data showing significant pipe wall metal loss, or other critical damage, and where the integrity of the pipeline is questionable. Also as part of a pipeline risk-management program, some operators may launch smart pigs during a normal shipping operation to evaluate pipe integrity, including corrosion, possible pipe cracking, or damage to the outside surface of the pipe. It is recommended that Chapter 15 be visited for much more discussion on the use, advantages, and disadvantages of smart pigs.

18.3 INTRASTATE PIPELINES

Intrastate pipelines are typically regulated by a state agency and in some cases also regulated by city, harbor, or airport authorities, at locations where the lines are. Many of these pipelines are old, with extensive leak histories. They are located in city streets and near residential neighborhoods. To obtain permits to continue operating these intrastate lines, the operator must agree to a testing frequency based on their leak history. The worst lines may require testing annually, others on a 3-year schedule, and the rest, even without leak history, on a 5-year schedule. Testing any of these pipelines requires significant planning and coordination with the pipeline control center, field operators, and the permitting agency.

For example, for a typical pipeline hydrostatic test, a large amount of test water must be available at the pump station or testing location, which is normally not available at that location. So, arrangements must be made to obtain the water. This test water must push product or crude oil, etc., out of the pipeline and into tankage for storage. Reliable pipeline block valves must be operated in accordance with the test plan, and blinds must be set and pipeline vents installed as required. Test equipment and instrumentation must be brought to the test location and setup. The pipeline pressure relief valve, which protects the pipe from overpressure during normal operation, must be taken out of service before test pressurization. The pressure relief system's isolation valve on the pipeline must be closed and locked during test, but put back into service for safety, before pipeline test water flushing commences. The pipeline is then pressurized to test requirements and the data evaluated to assure successful test completion. After the test, valves and blinds are operated and aligned to allow the test water to be pushed out of the pipeline, using pipeline product, crude oil, etc., and pumped into available tankage, where the test water can be treated for disposal, in a manner that will minimize contamination of the environment.

Now, the pipeline is refilled with product or crude and can be returned to the pipeline control center (CC) for shipping to customers as scheduled.

You can see that just doing one pipeline hydrotest, maybe in a network of pipelines scheduled to be tested that year, requires a well–thought out test plan with experienced field personal. Working closely with the CC and field operations, and in strict accordance with a hydrostatic test plan, is critical to test success. Because most of these hydrostatic tests are in urban areas and are located where it is difficult to dig up the pipe to verify leakage, good data recording during the test is essential to determine that the test was successful and that the pipe passed without leakage, as required by code.

18.4 PRETEST PLANNING FOR AN INTRASTATE PIPELINE

Prior to any hydrotest, the test engineer or pipeline operation manager should determine the minimum and maximum test pressure range for the entire pipeline. This test pressures should take into account the current and future status of pipeline operations, the maximum operating pressure, and federal and state testing regulations. The test pressure during this strength test must be maintained to at least 125% of the maximum allowable operating pressure (MAOP) for the pipeline.[3] In general, the lowest pipeline elevation and the weakest section of the pipeline will be factors that could limit the MAOP of the system. Pipe pressure ratings and the location of valves, flanges, and other pipeline components must be identified and recorded by milepost.

The best approach to study the test pressure influence over the length of the pipeline is to develop a spreadsheet for the entire pipeline, with pipe data, elevation gradient, pipeline operating pressure, test pressure (in psi and ft), and pipe yield strength, all vs. milepost. This requires that pipe diameter, wall thickness, pipe grade, and internal design pressure are known for each section of pipe to be tested. The best source of this information can be obtained from pipeline profile charts, pipeline alignment drawings, and pipe plot plans.

Table 18.1 presents a good example of pretest planning and analysis, done by spreadsheet, and presents the pipeline design and performance by milepost. This analysis can be extremely useful before hydrostatic testing to insure that no section of the pipeline will be subjected to excess pressure, and it also provides a baseline for future test planning. I recommend spending a little time examining this spreadsheet, until you are comfortable with the methodology.

The pipeline pressure (shown in column 7), over the entire length of the line was obtained by running a pipeline hydraulic program and comparing the results with actual test data. Table 18.1 also presents a method for evaluating pipe strength by determining local operating pressure (Local) as percent of pipe yield strength, hydro pressure (Hydrostatic Pressure) as percent of pipeline operating pressure (column 13), and hydro pressure as percent of pipe yield strength (column 14), for the entire pipeline.

Note that the 900 psig MAOP at milepost 0, (column 7), determined by test, is not close to the internal pipe design strength, with only 30.4% yield, and the weakest section of pipeline is located at milepost 12.5 with hydro pressure at 62% of yield strength. Also, the amount of pipeline product to be removed and test water required to fill the pipeline in barrels is calculated. Analysis of this type can thoroughly evaluate candidate pipelines before testing, so that overpressurization of pipe does not occur.

The pipe data and calculated results can be plotted (elevation feet vs. milepost) and presented graphically as shown in Fig. 18.2. This graph shows pump station pressure at MAOP and the minimum test pressures as required by code, over the entire pipeline, as well as the actual hydrostatic test pressure, starting at 1149 psig (7928 kPa) but not constant across the 19.51 mile (31.5 km) pipeline, due to pipe elevation change. This graph clearly shows the operating margin of safety over the entire length of pipeline and can be used as a baseline for future pipeline expansion.

Looking at the pipe data given in Table 18.1, it is clear that the pipeline is not piggable because of the wide range of pipe diameter: from 6 in. to 8 in. to 10 in. and back to 6 in. (150 to 200 to 250 and back to 150 mm). Pig suppliers now make double-diameter (DD) pigs that can do an acceptable job in pipeline with two diameters with cleaning and batching, but this wide range in diameters exceeds their capability. So, a batching pig or scraper cannot be used successfully to separate product from the test water as test water pushes product out of the pipeline and into storage tankage. Since pigs cannot be used, the best approach is to push the

TABLE 18.1 Pretest Analysis of Hypothetical Pipeline with Multiple Diameters*

Analysis of a Hypothetical Liquid Pipeline with Multiple Pipe Diameters and Wall Thickness, Located in an Urban Area†													
1	2	3	4	5	6	7	8	9	10	11	12	13	14
Pipe Mile–Post	Elevation (ft)	Pipe O.D. (in.)	Pipe W.T. (in.)	Segment Volume (bbls)	Pipe Pressure (ft)	Pipe Pressure (psig)	Test Pressure (ft)	Test Pressure (psig)	Local % of Yield	Hydrostatic Pressure (ft)	Hydrostatic Pressure (psig)	Hydro % of Line	Hydro % of Yield
0.00	15	6.625	0.280		2076	900	2595	1125	30.4	2653	1149	128	39
0.44	15	8.625	0.322	83	2059	891	2574	1114	34.1	2653	1149	129	44
0.65	15	10.75	0.365	69	2055	890	2569	1112	37.4	2653	1149	129	48
1.06	400	8.625	0.322	211	1649	714	2061	892	27.3	2268	983	138	38
1.08	780	10.75	0.365	7	1268	549	1585	686	23.1	1888	818	149	34
1.74	860	8.625	0.322	340	1154	500	1443	624	19.1	1808	783	157	30
3.03	960	8.625	0.322	421	974	422	1218	527	16.1	1708	740	175	28
6.34	36	10.75	0.365	1081	1815	786	2269	982	33.1	2632	1140	145	48
12.5	58	10.75	0.280	3172	1648	713	2060	982	39.1	2610	1131	159	62
13.1	40	6.625	0.280	298	1559	675	1949	844	22.8	2628	1139	169	38
14.2	30	6.625	0.280	219	1312	568	1640	710	19.2	2638	1143	201	39
18.6	20	6.625	0.280	834	494	214	618	267	7.2	2648	1147	536	39
19.5	10	6.625	0.280	164	341	148	426	185	5.0	2658	1152	778	39
			Total = 6899										

Model information: pipe strength, A53 Grade B, SMYS = 35,000 psi; product, oil with SG = 0.97; viscosity = 18 cSt; MAOP = 900 psi; pipeline discharge pressure = 148 psig; flow rate = 590 GPM; required pump station horsepower = 400 hp; column 10: local (pressure)% of yield = 100(7)*(3)/(2*35000*(4)); column 13: hydro % of line (pressure) = 100*(12)/(7); column 14: hydro % of yield = 100*(12)*(3)/(2*35000*(4)); number in () indicates column value.*

†*This is an example of an intrastate pipeline only 19.51 miles long.*

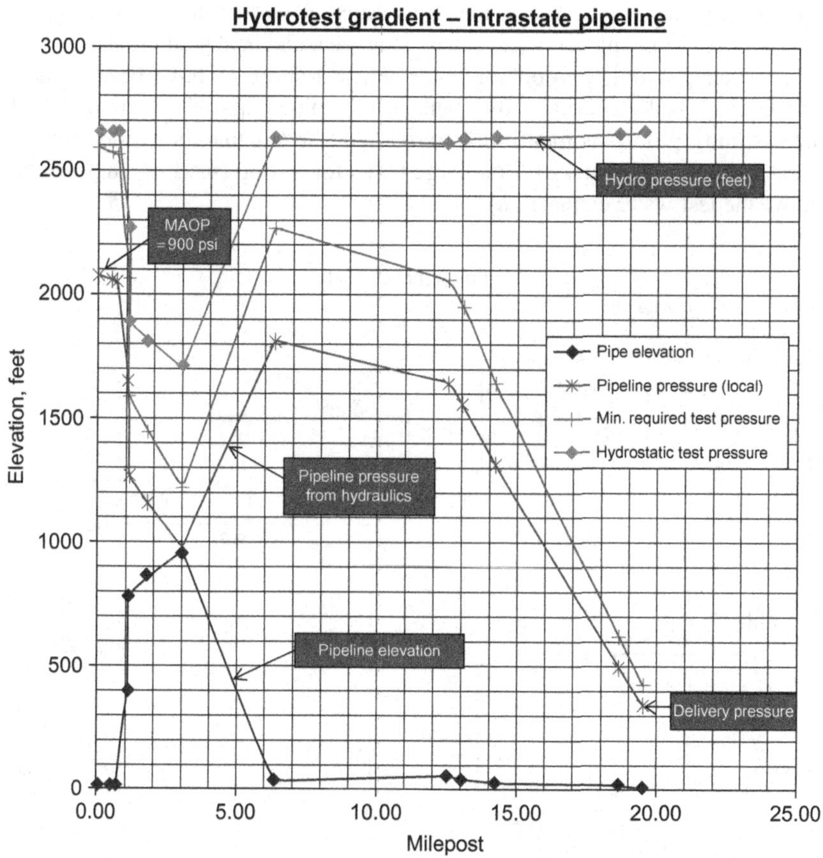

FIGURE 18.2 Hydrostatic test gradient for hypothetical intrastate pipeline.

water at a turbulent rate, along with the 150 barrel overflush. The overflush volume was determined by using 2000 ft (613 m) of 8 in. (200 mm) pipeline as average.

It is believed, as the pipeline is filled and overflushed with the test water, that most of the product residue, air, and vapor are pushed into tankage along with the entire pipeline product.

The amount of test water including the flush water is about 7100 barrels (1,128,900 liters). Since probably there will not be water at the test site, arrangements must be made to obtain it. Water can be piped to the test site from a nearby fire hydrant or some other water source by using the services of companies, such as Rain-for-Rent. They bring in temporary, portable 6 in. pipe, or larger, and string it above ground from the fire hydrant to a connection on the suction side of the shipping pump when available or to a portable line fill pump as needed. They can arrange for permits with fire departments to connect to hydrants using backflow preventers to protect the water supply from contamination, water filters when needed, and flow meters to account for water usage. They deliver the portable pipe and equipment to the test site and do the setup. After the hydrostatic test is complete, they remove the piping and water safety equipment and haul it all off-site.

Another possibility to obtain water at a test site is to have Baker Tanks deliver clean tanks to the site. The tanks can be piped up to the suction side of the shipping pump, as mentioned above, and arrangements would then be made to have water trucks fill the large tanks for testing (see Fig. 18.3, which shows the test water connection).

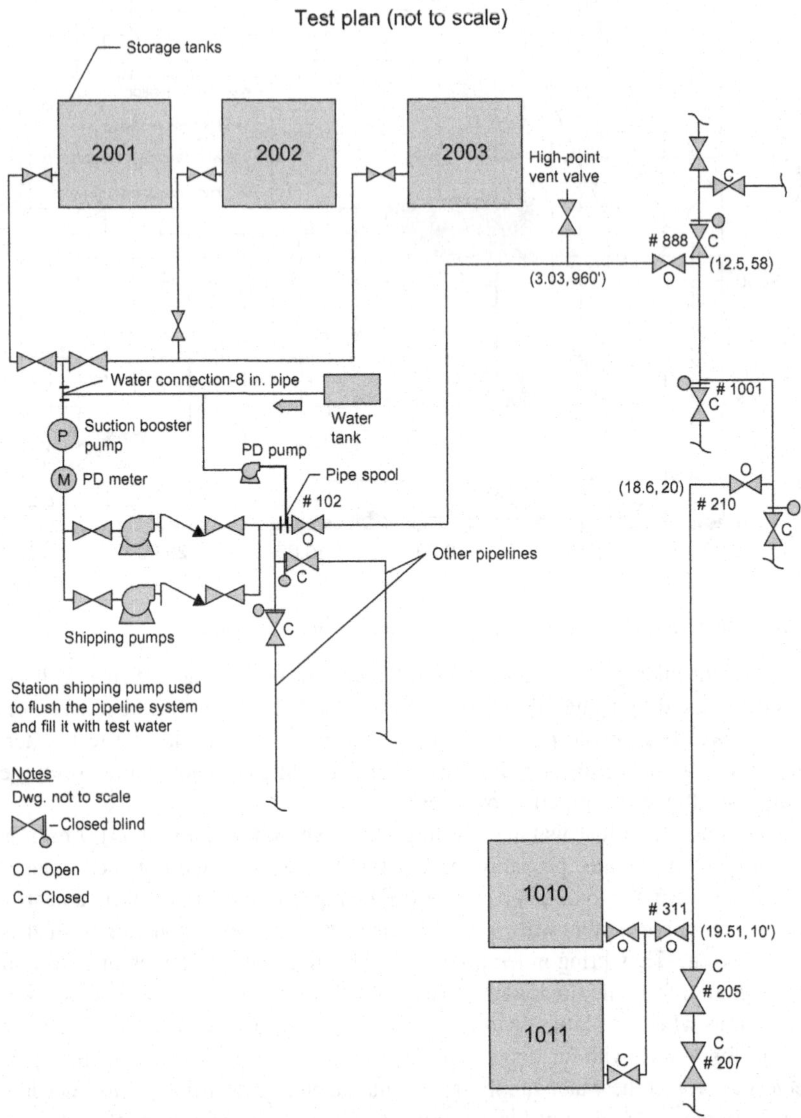

FIGURE 18.3 Test plan – Intrastate pipeline, 19.51 miles long.

A third option is to temporally use one of the oil storage tanks at the pumping facility to hold test water. It may make good economic sense to reuse the test water when a number of pipelines in that location are scheduled for testing, and the existing pipeline system is sufficient to allow pushing the test water, after hydrostatic test, back into the designated storage tank.

The pipeline shipping pumps or portable pumps must be able to flush the pipeline and safely fill the line with test water. Although the test water system is being connected to the suction side of the pumps, arrangements should be made to install a high-point vent valve at or close to the highest point of the pipeline if one does not exist. For this hypothetical pipeline (which is presented schematically in Fig. 18.3), the high point is at milepost 3.03. Figure 18.3 shows a graphical presentation of the pipeline system being tested. We will discuss more about the importance of venting air and vapor from the line after the water fill is complete.

Now that the test pressure level for the line has been determined and test water will be available at the site, this pretest plan can be completed.

18.5 TEST WATER DISPOSAL

A plan for test water disposal must be developed, which will minimize contamination or damage to the environment. One method that has been used for years is after the test is completed and considered successful, the test water is pushed out of the line using pipeline product or crude, maintaining a turbulent rate in the pipeline, and into receiving storage, including an overflush to help insure the line is packed with product and free of test water.

Then, let the water/oil settle out in the storage tank by difference in densities. Drain the tank water into an oil/water separator and treat the waste water to a safe surface disposal condition. Consult with waste water experts and local disposal companies to insure that waste water quality requirements are met.

18.6 SAFETY AND EQUIPMENT PROCEDURES DURING TEST

Precautions and safety procedures are an important part of the pre test plan, such as training field personnel in the proper operation of facility valves and reviewing with them the proper lockout and tagout policies and procedures.

18.7 TURNING AND OPERATING VALVES

No pipeline valve or facility valve shall be operated by field operators, test crew members, and test contractors without permission from the Pipeline CC or unless the facility has been released by an approved Facility Release.

18.8 TRAINING AND JUDGMENT

One of the most dangerous procedures connected with pipeline testing, occurs during the process of pushing product or crude out of the pipeline and into storage, using test water as the pushing fluid. This is partly because the test director must rely on the expertise of the field crew properly operating pipeline valves and other equipment. There have been large and costly crude oil and product spills because one or more valves were not operated correctly but were reported to be positioned in accordance with the test plan. Field personnel need to have experience and training with the valves they are assigned to operate. First, they must be comfortable with the way they turn the valve to close it or to open it. A good training rule of thumb is "right-is-tight." Some pipeline valves are different from the norm, for example, to open a General Twin Seal valve from closed position to fully open takes more than just a complete turn of the hand wheel to the left. It takes several complete turns of the hand wheel just to release the valve seats and more turns to slip the wedges apart, and finally more turning to rotate the internal plug to full open. When in the full open position, the hand wheel will not rotate any further to the left.

In one major pipe spill, which I experienced as a district engineer, crude oil ran out of the ruptured pipe and onto a heavily traveled highway, covering the roadway curb-to-curb. The rupture was caused by a valve reported to be open but accidently left closed. The field operator unlocked the hand wheel and rotated it to the left several turns, thinking that it was open "wide enough." Then, he locked it in position and called the Control Center, stating the valve was open and locked. In reality, the inner plug of the valve had not yet rotated, and the valve was still in the closed position. Valve operating error was not the only major problem that occurred during this critical pipeline hydrostatic test.

Another error during that test was made by the test director himself in having the pipeline pressure relief valve system closed and locked off from the pipeline, before the line was completely laid down in test water and ready for testing. It was necessary to disable the relief system before pressure test but not before crude oil removal and completion of test water filling. If that relief system had not been disabled, the pipeline would have been protected during this critical phase, and the oil spill would have been avoided.

What actually occurred was the pipeline shipping pumps were started and were pumping test water into the line, pressurizing the crude oil, but without flow into tankage. This was because the inline valve was mistakenly left closed. The pressure increased quickly resulting in pipeline rupture, and spilling crude oil, not environmentally safer test water.

So, field operator training with equipment that they are expected to operate is necessary. Also, test directors and engineers must use good judgment when developing test plans that have built-in safeguards by design, such as making sure that the line is protected during unusual pipeline operation, as explained above.

18.9 BACK TO TEST PROCEDURE

The test procedure must specify how to operate pipeline valves, facility valves, and setting blinds in accordance with the Facility Release, as necessary to fill the 19.51 mile (31.4 km) pipeline with water, using the shipping pumps shown on the Test Plan, Fig. 18.3. As part of this work, a temporary test water connection is to be installed on the suction side of the station pumps. 7100 barrels of water will be needed to fill and flush the line. Once line filling is complete, the storage tank valves must be closed and blinded, and all entrained air and vapor collecting at the high point of the pipeline must be vented, near milepost 3. The pipe spool next to valve #102 must be removed, after allowing the existing shipping pumps to push product with test water to downstream storage tanks and leaving the pipeline in water, see Fig. 18.3.

Now with the 19.51 mile pipeline totally filled with test water, close valve #102 and assemble the test head and the high-pressure PD pump and piping and instrumentation for data collection, in accordance with Fig. 18.4. Also, refer to Section 18.11, the hydrostatic test equipment specification list, entitled Equipment for Hydrostatic Test to verify that the selected equipment and instruments meet test requirements.

Backfill and compact around the exposed test head, pipeline valve, and the wrapped temperature sensor to insulate test piping from ambient temperature changes. Pressurization of the pipeline should not begin until the water temperature of the filled pipeline has stabilized.

18.10 PRESSURIZATION

The test director should maintain continuous surveillance over the operation and ensure that the test is carefully controlled. Open valve #102 and start the PD pump pressurizing the pipeline. The pipeline should be pressurized at a moderate

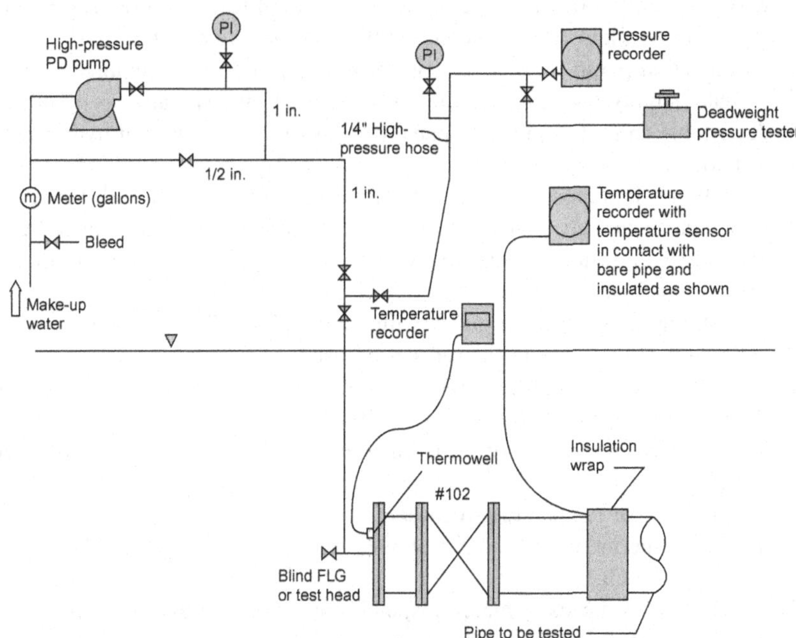

FIGURE 18.4 Pressure test and data collection equipment.

but constant rate, and when 80% of the specified hydro pressure is reached, the pumping rate should be controlled to minimize pressure variations by reducing the pump stroke but continuing to pressurize until the specified test pressure is reached. Care should be taken as the pressure reaches target value not to over-pressurize. Monitor test pressure using the pressure gauge and pressure recorder connected in parallel with the dead weight tester.

When the hydro pressure is achieved, pumping should be stopped and all valves and connections to the pipeline should be closed and inspected for leakage. A period of observation should follow where test personnel verify that the specified hydro pressure is being maintained and that the water temperature has stabilized. At this time, a pipeline line rider should be at the high-point vent valve and vent all remaining trapped air at that location. On completion of this work and test pressure level verification, the PD pump discharge valve should be closed and the pump and instrument piping should be checked again for leaks. Next, the pressure and temperature recorder charts should be set to real time and chart recording of test pressure and temperature should be started.

18.11 LIST OF EQUIPMENT FOR HYDROSTATIC TEST[5]

Refer to Fig. 18.4 for setup of instrument and test equipment.

1. High-volume pump capable of filling the pipeline at a minimum velocity of 2 mph or use facility pipeline shipping pumps when available.
2. A variable speed, positive displacement pump capable of pressurizing the line to at least 100 psi in excess of the specified test pressure. The pump should have a known volume per stroke, with stroke counter.
3. An injection pump to inject corrosion inhibitor and dye if required.
4. A water supply line filter capable of insuring clean test water, as required.
5. A portable tank, if required, capable of providing the source of hydrostatic test water.
6. Two large diameter Bourdon tube–type pressure gauges with pressure range and increment divisions necessary to indicate anticipated pressure.
7. A deadweight tester certified for accuracy and capable of measuring increments of 1.5 psi (10 kPa).
8. A 24-h recording pressure gauge with chart and ink. This gauge should be calibrated and deadweight tested before use.
9. A 24-h recording thermometer with chart and ink capable of recording temperatures from 32°F (0°C) to 122°F (50°C). A recording thermometer with thermowell capable of measuring temperatures with output resolution of 0.1°F.
10. An ambient air temperature recorder.
11. Two glass thermometers capable of recording temperatures from 32°F (0°C) to 122°F (50°C).
12. Temporary test heads or manifolds, and piping with fittings as necessary to connect pressure and temperature instruments as shown in Fig. 18.4.

18.12 TEST ON

The line rider should continue driving the pipeline right-of-way with special concern in highly populated areas and staying in communication with the test director.

Test pressure, fluid temperature, and pipe wall temperature should be monitored and recorded continuously during the entire test duration.[5]

Deadweight tester readings should be taken and recorded at the start and at termination of the test, and at least every hour during the test duration. If injection or bleeding of test water is required to maintain the test pressure between predetermined minimum and maximum values, the added or subtracted volumes must be measured as accurately as possible and recorded at the time taken on the Hydrostatic Test Results and Pipeline Data form, shown in Fig. 18.5. Deadweight tester readings, test water temperature, pipe wall temperature, and ambient air temperature readings must all be recorded hourly as a minimum on the back of the hydrostatic test results form for record, including the exact time and volume of the injections and bleeds.

This completed and signed hydrostatic test form, along with the pressure and temperature recording charts, and pipeline profile and Test Plan as shown in Fig. 18.3 must be kept for the life of the pipeline.

18.13 POSTTEST RESULTS

If the specified hydro pressure during the test stayed within the acceptable range, the hydrostatic test was successful and passed without leakage. The pipeline then can be put back into service, in accordance with the Facility Release. The pressure test head and temporary test water connections must be removed, the disassembled shipping pump piping must be reinstalled, and then valve blinds must be opened as required. Valves also must be operated to allow the pipeline to fill with product and push test water out of the pipeline and into tankage for waste water treatment and disposal. Now, the pipeline can be turned over to the Control Center and is ready to go back into service.

If the hydrostatic test pressure does not hold during test and injection is required to keep it in the acceptable range, further investigation and analysis must be done by the test engineer/director to validate test success. For example, if pressure data does not trend with temperature changes in the fluid, then data must be carefully analyzed.[4] Data that does not trend correctly may indicate the existence of a leaking pipe at a defect or be due to a corrosion pit. If the pressure had been holding and suddenly drops off quickly, the pipe probably ruptured and the line riders must find the location and take responsibility until emergency personnel arrive. In that case, the pipeline failed and must be repaired, and put back on test as quickly as possible.

In the next section, we will look into posttest analysis to determine pass or fail using $\Delta V/\Delta P$ and other diagnostic tools. $\Delta V/\Delta P$ is defined as a change in volume for an associated change in pressure, for a known pipeline volume, under pressure.

LIQUID PIPELINE SAFETY
HYDROSTATIC TEST RESULTS

PIPELINE DATA		Test Date
Pipeline Operator		Company conducting test if other than operator
Kind of Test [ ] New [ ] Replacement [ ] Annual [ ] 3 Year [ ] 5 Year [ ] Other		
Pipeline Identification (line number, name, etc.)		
Pipeline Location (milepost, street, station, etc.) From: ____ To: ____		
Normal Product Transported	Normal Operating Pressure P.S.I. at (location)	
Maximum Operating Pressure P.S.I. at (location)		

PIPE DATA

Pipe O.D.	Wall Thickness	Specification & Grade (SMYS)	Length of Pipe Being Tested	Volume (Barrels)

TEST DATA

Test Medium [ ] Water [ ] Petroleum*		*Has waiver been granted?	
Location of Pressure Recording Equipment			Elevation
Other Elevations	Pipeline – High Point	Pipeline – Low Point	
Test Equipment	Make & Model of Deadweight Tester	Serial #	Date Last Calibrated
	Make & Model of Chart Recorder	Serial #	Date Last Calibrated
	Make & Model of Temperature Recorder	Serial #	Date Last Calibrated

FIGURE 18.5 Hydrostatic test results and pipeline data form.

18.14 POSTTEST LEAK ANALYSIS

During initial pipeline pressurization, we can stop when the pressure reaches 50 psig (344.8 kPa) or above and take data to evaluate the field value of $\Delta V/\Delta P$. We will compare this field value with the theoretical value of $\Delta V/\Delta P$ to quantify the presence of entrapped air or other gases in the test section.

First, take an accurate pressure reading using the deadweight tester. Then, bleed approximately 10 psi (68.9 kPa) and allow the pipeline pressure to stabilize. Take another accurate pressure reading. Carefully measure the volume that was bled and convert it to gallons. Now, to get the field value of $\Delta V/\Delta P$, we simply take the volume bled and divide it by the difference in the pressure readings.[4] Record this calculation of $\Delta V/\Delta P$ and keep it for posttest analysis. After a 4-h or 8-h test, it is a good idea to calculate a second value of $\Delta V/\Delta P$ before the test section is totally depressurizing the pipeline, using the 10 psi ΔP method, as done above.

The theoretical $\Delta V/\Delta P$ for buried pipe is calculated as follows:

$$\Delta V/\Delta P = V[(D/(E\times t))\times(1-\mu^2)+C] \tag{18.2}$$

where

V – Fill volume of the segment for the individual pipe diameter, D (gal)
D – Outside diameter of pipe (in.)
E – Elastic modulus of steel pipe = 30,000 psi
t – Wall thickness of pipe (in.)
μ – Poisson's ratio of steel pipe = 0.3
C – Compressibility of water = 3.20×10^{-6} in.^{3}/in.^{3}/psi

Note: For pipeline with multiple diameters, you must sum the individual $\Delta V/\Delta P$s of each pipe diameter to obtain the total $\Delta V/\Delta P$ for the entire pipeline.

This $\Delta V/\Delta P$ constant can provide useful data with which to predict how much fluid will be required to bring the pipeline to the specified pressure. For example, for a fully packed line at 0 psig, with a desired hydro pressure of 1000 psi and a $\Delta V/\Delta P$ constant measured and calculated at 0.15 gal/psi, the amount of fluid needed to bring the test section to pressure will be 1000 psi × 0.15 gal/psi = 150 gal. This will be the minimum amount, but if trapped air is present, the required amount of fluid to reach hydro pressure will be much more.

$\Delta V/\Delta P$ is defined as a change in volume for an associated change in pressure, for a known pipeline volume, under pressure. The theoretical $\Delta V/\Delta P$ will give us the expected volume for the associated change in pressure assuming that the volume under pressure is free of entrained gas and air. The field value of $\Delta V/\Delta P$ is the actual volume for an associated change in pressure for the specified volume under test. If air exists in the line under pressure, the field value will be larger than the theoretical value. So, air or vapor should be vented one more time at the high-point vent valve, before starting the test.

The three things that cause the most problems during posttest analysis are as follows:

1. Entrained air, vapor, or product in the test section
2. Leaking isolation valves and pipe fittings
3. Changes in test fluid temperature during the test

18.15 ENTRAINED AIR AND VAPOR

First, we will discuss entrained air. Entrained air in the pipeline may prevent pressure stabilization of the test water over the test duration or mask the presence of an actual leak. The first indication of a large amount of air left in the pipeline will usually occur during pressurization. The pressure will not build at first as expected because water injected in the pipeline during pressurization must displace this large volume of highly compressible air. This expanding air may prevent detection of an actual pipe leak. If a small leak is present, the expected pressure drop due to that leak may not be apparent because expanding air in the pipeline will tend to hold pressure constant. This is the primary reason for using $\Delta V/\Delta P$ to determine the percentage of air that is present and for eliminating that air in the test section before testing.

To obtain accurate results, it is important that the entire piping system be reasonably free of air, entrapped product, and vapor from the product. Eliminating trapped air, gases, and product increases the accuracy of the test results, and as mentioned, too much air or gas can hide small pipe leaks during the test. The usual method for filling a piggable line requires using a batching pig or sphere to displace the pipeline product with test water. If this is not feasible, the water flush must be at a rate high enough to ensure a completely turbulent interface between the test water and the displaced product. In the case where the pipeline has changes in diameter, ensure that turbulent flow is maintained in the largest diameter.

As mentioned earlier, it is important to install a high-point vent valve on the pipeline being tested. After high-point vent installation, bleed all trapped air and gas completely from the line.

18.16 LEAKING ISOLATION VALVES AND FITTINGS

Leaking valves and pipe fittings are the next concern when pressure is dropping during a test. Test sections should be isolated at pipeline block valves by using slip blinds to insure no leakage. If the test section cannot be blinded but the valves are double blocked instead, the operator must measure pressure increase in the adjacent section between the double-blocked valves to insure a tight seal exists. You need to be careful when using a thin "fire blind" at an isolation valve because under pressure the thin blind will deform and the blind cannot be removed without removing the entire valve. This often requires calling in vacuum trucks to remove product on the opposite side of the test valve being removed.

So, leakage through valves and fittings jeopardizes the chances for a successful test and may lead to data that cannot be correlated, and in that situation, the pipeline must be retested.

18.17 CHANGING TEST WATER TEMPERATURE

Changing test water temperature during the 8-h test can have a large effect on the test pressure and must be accounted for. As mentioned earlier, the test director did wait for the test fluid temperature to stabilize as much as possible. But test fluid temperature is affected by a variety of factors including changes in ambient air temperature, weather conditions such as rain, pipe location, buried or above ground, and test media temperature. For buried pipe, the expected temperature fluctuation due to changes in ambient air temperature will be small because of the insulation provided by the soil. But heat transfer from the soil to the test fluid or from the test fluid to the soil will have an effect on test fluid temperature, which will cause a relative change in test pressure. For example, for a 12-in.-diameter pipe with an initial water fill temperature of 60°F and soil temperature of 70°F, the test water temperature will rise approximately 2.5°F over the 8-h test. Larger diameter pipe would have a smaller temperature increase. A rough estimate for the increase in test pressure due to the 2.5°F increase in the temperature of the fluid is

$$\Delta P_t = (\Delta P/\Delta T) \times 2.5 = 47.5 \text{ psi} \tag{18.3}$$

where $\Delta P/\Delta T$ is approximately 19 for steel pipe tested with water.

So, when pressure does not trend with fluid temperature changes, the data must be carefully analyzed and accounted for.

18.18 POSTTEST REPORT

Posttest calculations must be done to correlate pressure, temperature, and volume bled or injection. A good way to validate test success or failure is to do a volume balance, where the analysis must account for volume lost or gained.

This requires evaluating the change in pressure with respect to temperature, $\Delta P/\Delta T$, and the volume change due to temperature ΔV_t is:

$\Delta P/\Delta T = (B - (2 \times A))/[(D/(E \times t) \times (1 - \mu^2)) + C]$ equals approximately 19 as stated above.

Using $B = 9.60 \times 10^{-5}$ in./in./°F, the coefficient of expansion for water and $A = 8.40 \times 10^{-6}$ in./in./°F, the coefficient of linear expansion for steel,

$$\Delta V_t = (\Delta P/\Delta T) \times \Delta T \times (\Delta V/\Delta P) \tag{18.4}$$

where $\Delta V/\Delta P$ is the measured field value, say = 0.15.

18.19 VOLUME ANALYSIS

For illustration purposes, the recorded data during a 4-h pressure test on a hypothetical intrastate pipeline, with state required 5-year testing scheduled, are as follows:

Start pressure of 900 psi and end pressure of 915.5 giving $\Delta P = 15.5$ psi
Start temperature of 60°F and end temperature of 62.6°F giving $\Delta T = 2.6$°F

Note that at 1.5 h after the test started, the pressure was dropping, so 4 gal was injected into the pipeline. We will use the pretest field value of $\Delta V/\Delta P$, which was determined to be = 0.15 gal/psi.

You must be careful with the algebraic signs when evaluating volume lost or gained. So, for volume lost due to pressure ΔV_p, the sign must be negative because you would need to remove 2.32 gal to bring the balance back to starting pressure, see calculation below.

$\pm\Delta V_p$, volume accounted for by pressure change

$$\Delta V_p = \Delta P \times (\Delta V/\Delta P) = -15.5\,\text{psi} \times 0.15 = -2.32\,\text{gal}$$

$\pm\Delta V_t$, volume accounted for by temperature change

$$\Delta V_t = (\Delta P/\Delta T) \times \Delta T \times (\Delta V/\Delta P) = 19 \times (-2.6) \times 0.15 = -7.41\,\text{gal}$$

Volume lost plus injected

$$\Delta V_p = -7.41$$
$$\Delta V_t = -2.31$$
$$\Delta V_i = 4.0\,\text{injected}$$

Total: −5.72 gal or −1.43 gal/h

California Government Code allows the calculated leakage for a pipeline of this volume to be 1.45 gal/h plus a 1 gal/h tolerance or a total of 2.45 gal/h.

Results: So, with a calculated 1.43 gal/h leakage, this pipeline has successfully passed its hydrostatic test, PASSED.

18.20 TESTING INTERSTATE LIQUID AND NATURAL GAS TRANSMISSIONS LINES

Let us turn now to testing interstate pipelines. These cross-country pipeline routs typically have terrain with large elevation differences, which require breaking the pipeline into separate sections for testing. This is necessary to insure that no part of the pipe is subjected to excess pressure during the test. Since each test section is a separate hydrostatic test, minimizing the number of sections will significantly reduce the overall project cost. The cost of each test section depends partly on the cost of test manifold fabrication, and handling, transportation to the field, installation in the pipeline, and removal. Also, there is the cost required to fill each section, test, dewater and in the case of natural gas pipelines, to dry each section. So, where a pipeline could require 80 or more test sections at a cost of $50,000 or more per test section, it becomes clear that planning to control cost is very important. One way of significantly reducing the cost of hydrostatic testing cross-country pipelines is to increase the test pressure range or design the pipeline using higher X-grade pipe and flanges. But before we discuss saving money on hydrostatic testing, we need to understand how to determine the test section length and number of pipeline sections required.

Refer to Table 18.2 to understand the method of analysis necessary to evaluate pipeline elevation profile and determine test section details. In this case, we are using a test pressure range from 90% SMYS to 100% SMYS. You will need to refer to Fig. 18.6, showing the pipeline elevation vs. milepost for approximately the first 30 miles of a typical cross-country pipeline. The figure also shows the variation in length of the first four test sections and test levels due to elevation change. This information is also tabulated in Table 18.2.

The procedure used in this spreadsheet (Table 18.2) is as follows:

1. Select a test pressure range for the entire pipeline. The minimum pressure at the high point of the section should be 90% SMYS. The highest pressure of the section will be at the low point and for analysis can range from 98% SMYS to 105% SMYS. For this example, we have used 100% SMYS for the high-pressure, low elevation point.
2. Plot the pipeline elevation profile as accurately as possible, see Fig. 18.6. It is necessary to have civil survey data providing pipeline elevation vs. milepost because the elevation data needs to be as accurate as possible.
3. Evaluate the high and low pressure (psig) for the pipe to be tested using the hoop stress formula, based on SMYS. When designing a new pipeline, use a higher X-grade pipe, for example, X-70 instead of X-52 or X-60.
4. Starting with the low-point elevation, calculate the test section elevation at the high point. This is easily done using the Bernoulli energy equation for water being the test fluid: $P1 \times 2.308 + Z1 = P2 \times 2.308 + Z2$ and solving for $Z2$ elevation.
5. The test section length and location of test section end can be determined graphically by the intersection of the high-point elevation $Z2$ and the plotted pipeline elevation profile. Drop vertically down from the intersection point and read the milepost at section end. Then, subtract the starting milepost from it to get section length. Of course, a computer program can be written to determine the test section length by curve fitting the pipeline elevation data and iterating on $Z2$ for an exact milepost solution.

Studying Table 18.2 provides the clear approach for determining test section length and the number of test sections required for any pipeline hydrostatic test. For analysis purposes, there are four places to choose inputting the starting elevation and proper pressure level (low point = 1625 psi and the high point = 1463 psi) based on our pressure range. They are the beginning section, section low point, section high point, and section end point. Note that in test section 1, the low-point elevation of 461 ft controls, with pressure of 1625 psi, and results in the high-point elevation of 834 ft at the end of the 9.55 mile section. In test section 2, the beginning point is the same as the low point with elevation of 834 ft and pressure of 1625 psi, and the resulting high point based on the energy equation is the same as the end point with elevation of 1208 ft at milepost 13.1. The resulting section length is 3.5 miles. Next looking at section 10, the beginning point is the same as the high point

TABLE 18.2 Determine Length of Each Test Section for Hydrostatic Test*

Testing Cross-Country Pipelines with Large Elevation Changes†

T.S. No.	T.S. Length	Beginning			Low Point			High Point			End Point		
		MP	Elevation	psi	MP	Elevation	psi	MP	Elevation	psi	MP	Elevation	psi
1	9.55	0.0	812	1463	4.5	461	1625	9.6	812	1463	9.6	812	1463
2	3.5	9.6	812	1625	9.6	812	1625	13.1	1186	1463	13.1	1185.9	1463
3	1.6	13.1	1185.9	1625	13.1	1186	1625	14.6	1560	1463	14.6	1559.79	1463
4	2.5	14.6	1559.79	1625	14.6	1560	1625	17.1	1934	1463	17.1	1933.69	1463
5	1.5	17.1	1933.69	1625	17.1	1934	1625	18.6	2308	1463	18.6	2307.58	1463
6	1.8	18.6	2307.58	1625	18.6	2308	1625	20.3	2681	1463	20.3	2681.48	1463
7	3.0	20.3	2681	1625	20.3	2681	1625	23.3	3055	1463	23.0	3055	1463
8	0.5	23.3	3055	1625	23.3	3055	1625	23.8	3429	1463	23.8	3429	1463
9	0.3	23.8	3429	1625	23.8	3429	1625	24.1	3600	1463	24.0	3600	1463
10	1.6	24.1	3600	1463	25.7	3226	1625	24.1	3600	1463	25.7	3226	1625
11	2.0	25.7	3226	1463	27.7	2852	1625	25.7	3226	1463	27.7	2852	1625
12	2.3	27.7	2852	1463	30.0	2452	1625	27.7	2852	1463	30.0	2478	1625

Criteria: Using 100% SMYS to determine the pressure at lowest elevation of each test section; using 90% SMYS to determine the test pressure at the highest elevation of each test section. Pipe specification, 16 in. diameter, API 5L-X52; wall thickness 0.25 in.; 100% pressure = 2.25*52000/16 = 1625 psi, low point; 90% pressure = .9*2*.25*52000/16 = 1463 psi, high point.*

†*This analysis determines the length in miles of each test section for the specified pressure range and the number of test sections necessary to test the entire pipeline.*

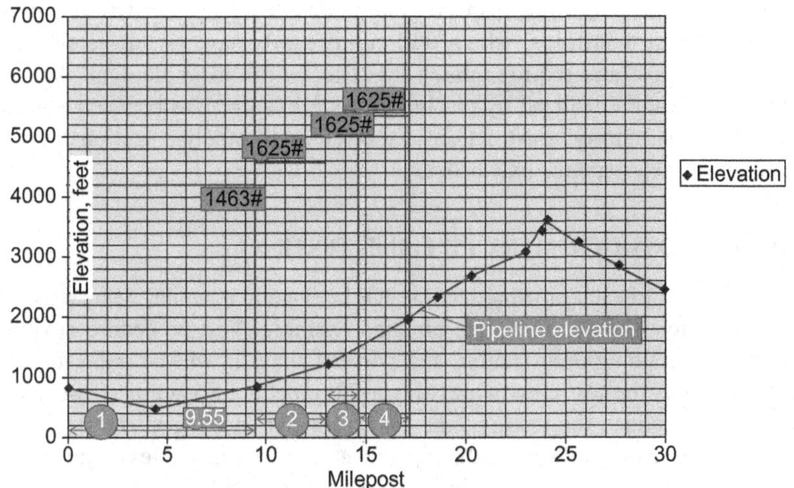

FIGURE 18.6 Elevation profile for cross-country pipeline, showing test sections.

with elevation of 3600 ft and pressure set at 1463 psi. The gradient is negative starting at the high point at milepost 24.1 and resulting in an elevation at the end point of the section (low point is the same as the end point) of 3226 ft. This occurs at milepost 25.7 and results in a section length of 1.6 miles. To continue with this analysis, see below.

18.21 TEST SECTION 12

Test section 12 beginning at milepost 27.7 is the last one shown in the table, with test section length equal to 2.3 miles. But by following this method of analysis to the end of a pipeline, all test sections can be determined, including the total number required.

It is estimated that a 300-mile cross-country pipeline could require as many as 80 or more test sections, depending upon terrain, calculated in this manner.

Now that we know how to evaluate pipeline hydrostatic test sections using pressure range, if we raise the maximum test pressure to 105% SMYS, we know by experience that we could reduce the number of test sections by approximately 20%. So, instead of 80 sections, we would need to test approximately 64 sections. We can do this responsibly now by using continuous yield determination-computerized instrumentation that makes it possible to test to higher pressures, with potential yielding being a known factor before it occurs. Studies have shown conclusively a traditional overextension of specified minimum yield by all pipe miles. Remember that most of the pipe rated API 5L X-52 actually tested much higher than 52,000 psi but not high enough to be rated for the next X grade.

Other reduction in cost can be obtained by testing against freeze plugs instead of test manifolds, which will significantly save on fabrication and installation cost and requires much less excavation to implement. Another cost saving can be realized by placing pipeline block valves between test section locations, during construction of the relocated pipeline sections or construction of new pipeline extensions.

18.22 CROSS-COUNTRY PIPELINE TESTING

Actual testing of each pipeline section will be done in a similar manner as presented earlier with the intrastate pipeline example. Test procedure development, test section pressurization, and posttest analysis will be done for each test section as the test crew works down the pipeline. Sometimes, two crews are used to reduce the testing time involved: one starting at the beginning and the other crew midway on the pipeline at a place where a test water source is available. For cross-country pipelines, the process will be easier because the test water in the pipeline after test completion will be pushed downstream to be reused.

While the first sections are on test, the test crews are working downstream, installing test heads and equipment to facilitate testing the next sections. There is less risk of rupturing pipe and dealing with resulting spills because the pipe is new, or if an older pipeline is being tested, the pipe will have been internally inspected prior to product removal, using an internal inspection device or smart pig.

Test equipment and instrumentation required at each test section will be the same as specified in the list entitled "Equipment for Hydrostatic Test," except that portable compressors and batching pigs are now needed to move the test water down the pipeline for use in the next test section.

When hydrostatically testing gas pipelines, the test water must be removed after test by launching numerous bidirectional pigs designed for water removal, as well as using pigs with spring loaded brushes to scrub the inside of the pipe. Portable air compressors should be sized to provide initial pig velocity of 5 ft/s after most of the water, dirt, and debris are removed from inside the pipeline. The final drying will be done by using polyswab pigs to wipe and remove the remaining dirt and moisture from the inside surface. Once a dew point of 0°F is reached, the pipeline will be given a 5 psi nitrogen blanket to limit corrosion until it can be put into gas service.[2]

18.23 PIPELINE RUPTURE

Part of my responsibility as the district engineer was to conduct pipeline hydrostatic tests on company lines that were permitted to operate in the Los Angeles city limits. I was responsible for testing a 14-in.-diameter crude oil pipeline on a

3-year testing schedule. The testing was mandatory, even though the pipeline was rated as an interstate line, bringing crude oil from Bakersfield production to Los Angeles refineries and then connecting with pipelines that shipped oil out of state. The reason for the 3-year schedule was because the line went up over the Hollywood Hills, with a route using highly traveled roadways, and also because its safety record up to that point was somewhat tarnished. We tried to be extremely careful in removing the crude oil from the pipeline and with pipe testing because of the test record and the sensitive location of the pipeline. As part of this scheduled hydrostatic test, it was believed that the test section from Newhall Pump Station to Jefferson and Hauser in Los Angeles would test high enough to allow us to increase the MAOP by 80 psi at our Newhall Station. With this higher pump station pressure, pipeline performance would have been significantly increased, and that increased throughput was needed to meet shipping commitments in the Los Angeles area.

As the pipeline test engineer, I did not know all that stuff at the time. My job was to test the pipeline at the pressure levels given me by management.

So, the crude in the line was successfully pushed out of the pipeline and into Long Beach tankage by test water obtained at Newhall. The pipeline test section was overflushed to insure that most of the residual oil was out of the line, and the pipeline was packed with green test water. Two or more test sections were pressurized using the Newhall Station pumps, which took the pressure to approximately 80% of test pressure. The subject test section was isolated for test by closing appropriate block valves. These WKM valves had a grease ring seat design to positively seal the gate for hydrostatic testing. The test section from Newhall to Jefferson and Hauser was then taken up to test pressure, which was approximately 90% SMYS. The pressure was monitored at both ends of the test section. I had two pipeline line riders out driving the right-of-way, once it was on test pressure. The pressure looked good and was holding for the first 15 min of the required 8-h test. Then, all of a sudden the pressure chart pen dropped to almost zero instantly, and it was clear that the pipeline had ruptured. I called the line riders and told them to start searching for the location of failure. The pipe rupture occurred in the longitudinal seam, and the pipe split wide open like a "watermelon." The test water blew out pipeline cover and street asphalt located directly above. Water cascaded out from the pipeline covering the canyon road curb to curb. A 70-year-old lady was driving over the location just as the pipeline ruptured and the rushing water on the street surface, caused her car to hydroplane and she lost control, spun around, and hit the curb. She was not injured but was in a state of shock as the line rider arrived on the scene.

After the incident, the pipeline was repaired and back on test in less than 20 h. The hydrostatic test pressure, by the way, was reduced to the previous test value. A 20 ft (6.1 m) section of the failed pipe was removed from the roadway and replaced with a 20 ft section of seamless API 5L A53 Grade B pipe.

The error in this case was that the strength of WWII welded seam pipe was not well known. The old electricfusion-welded seam pipe, manufactured during the war, should have been treated as having a seam factor $E = 0.80$. In hindsight, using that seam factor would not have allowed testing the pipeline to 90% SMYS, and this pipeline rupture would never have happened. But it did happen and we learned from it!

REFERENCES

[1] Code of Federal Regulations, Title 49, Part 195, Transportation of Hazardous Liquids by Pipeline, 2010.

[2] Code of Federal Regulations, Title 49, Part 192, Transportation of Natural and Other Gas by Pipeline, 2010.

[3] ASME Code for Pressure Piping, B31.4, Addenda, American Society of Mechanical Engineers, 1991.

[4] California Code of Regulations, Title 2, Division 3, Chapter 1, Article 5.5 (2 CCR 2560-2571) Marine Oil Terminal Pipelines, 1997.

[5] Recommended Practice for the Pressure Testing of Liquid Petroleum Pipelines, API 1110, second ed., American Petroleum Institute, 1981.

Chapter 19

Commissioning

Hal S. Ozanne

Chapter Outline

INTRODUCTION

In preparing for the commissioning and start-up of the operation of a new, upgraded, or downgraded pipeline system, a commissioning plan must be developed that takes into consideration all aspects of the preparation, start-up, and operation of the system. The more complex the system is, the more detailed the plan must be. A good starting point in developing a plan is to review and understand the system process flow diagram (PFD) and the facilities piping and instrument diagrams (P&IDs).

These diagrams show the major components, the pipe sizes of the system, and the instrumentation that is used to monitor and control the system. A review and understanding of these drawings will provide a good knowledge about the major components of the system to be used and the operation of it.

See Figs 19.1 and 19.2 for an example of a system PFD and a facility P&ID.

If the project consists of something as small as replacing a section of pipe, then the commissioning plan can be relatively minor. On the other hand, if a project consists of the installation of a new pipeline and facilities like pump or compressor stations and measurement facilities, then the plan will be very detailed, lengthy, and require a significant amount of time to develop, usually with a team of personnel.

The team preparing the commissioning and start-up plan may include the operating company project manager, the engineering project manager,

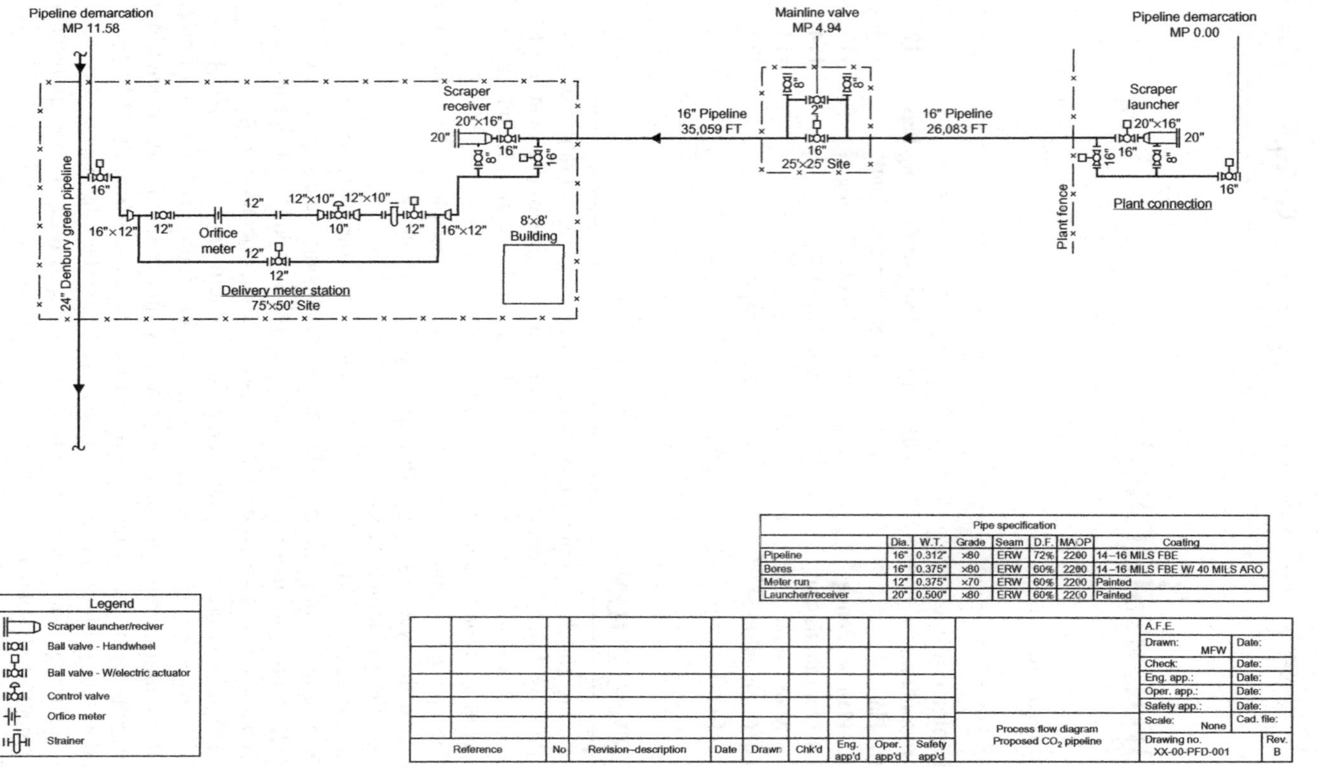

Pipe specification							
	Dia.	W.T.	Grade	Seam	D.F.	MAOP	Coating
Pipeline	16"	0.312"	×80	ERW	72%	2200	14–16 MILS FBE
Bores	16"	0.375"	×80	ERW	60%	2200	14–16 MILS FBE W/ 40 MILS ARO
Meter run	12"	0.375"	×70	ERW	60%	2200	Painted
Launcher/receiver	20"	0.500"	×80	ERW	60%	2200	Painted

FIGURE 19.1 System PFD.

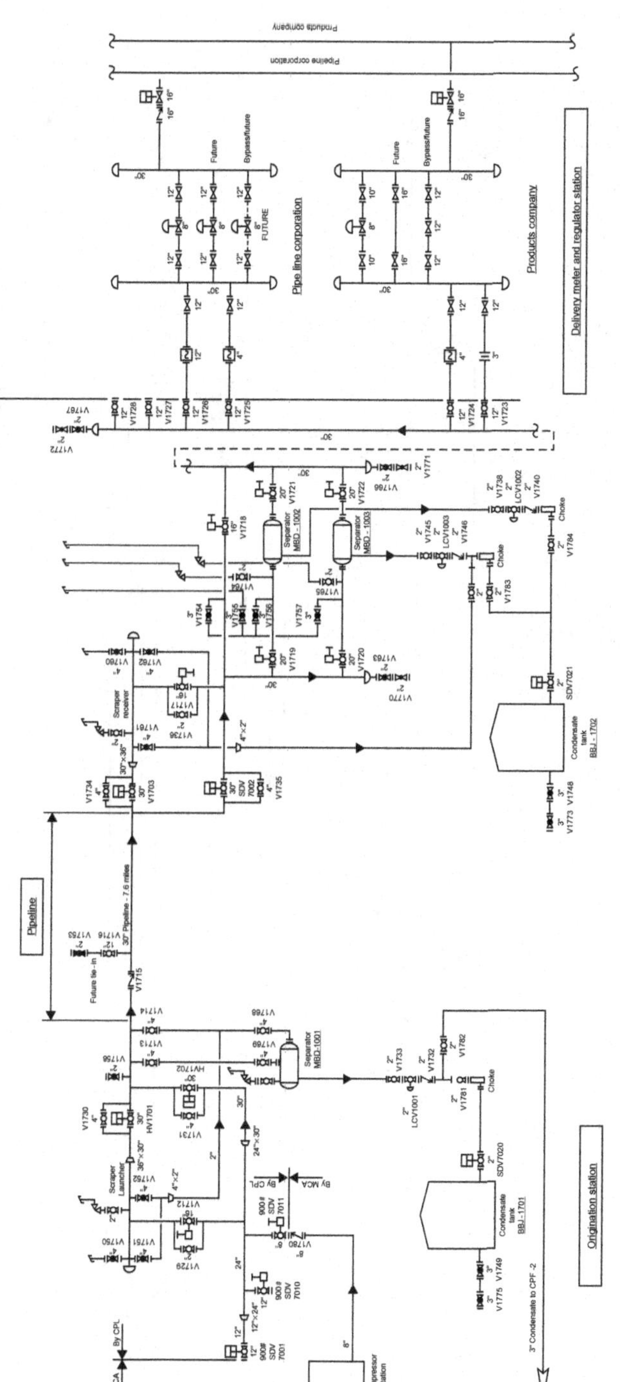

FIGURE 19.2 Meter facility P&ID.

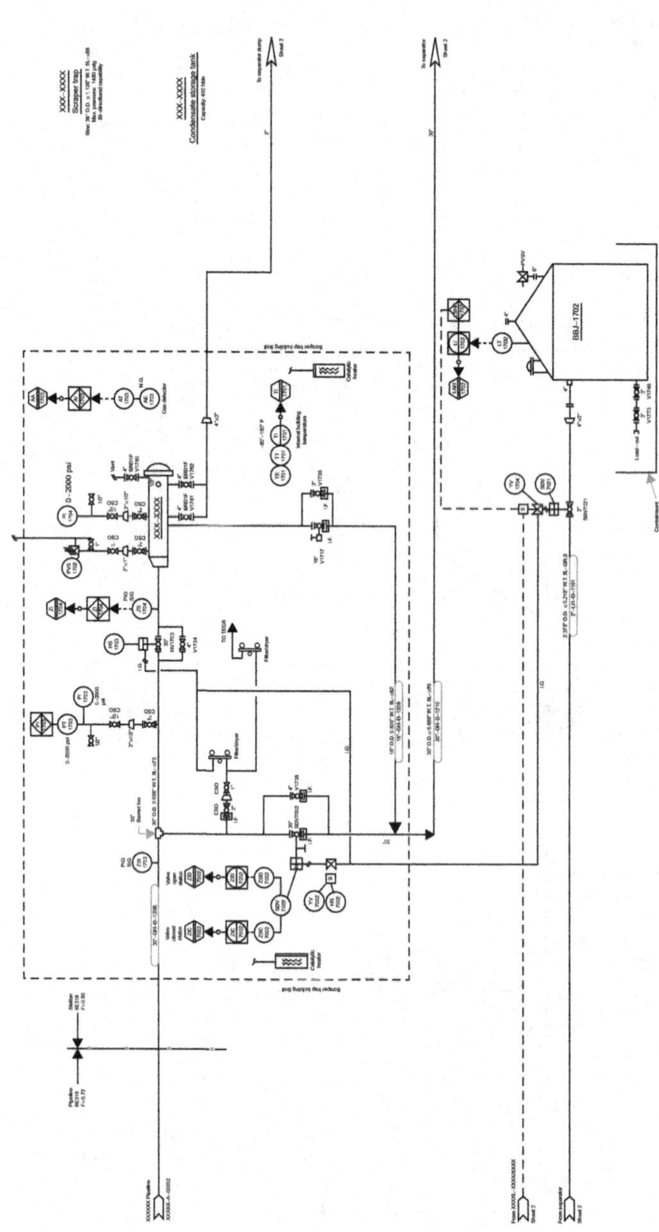

FIGURE 19.2 cont'd.

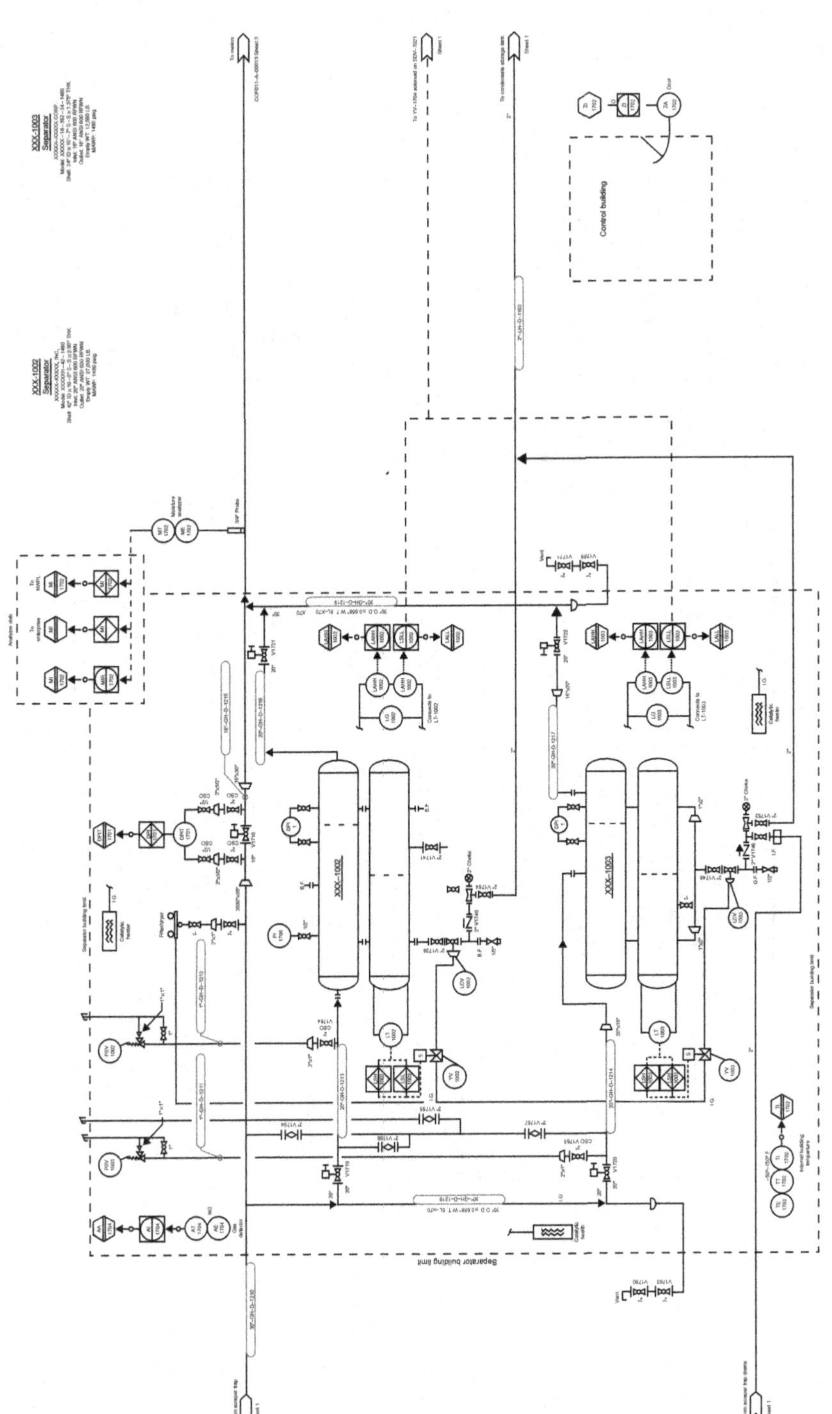

FIGURE 19.2 cont'd.

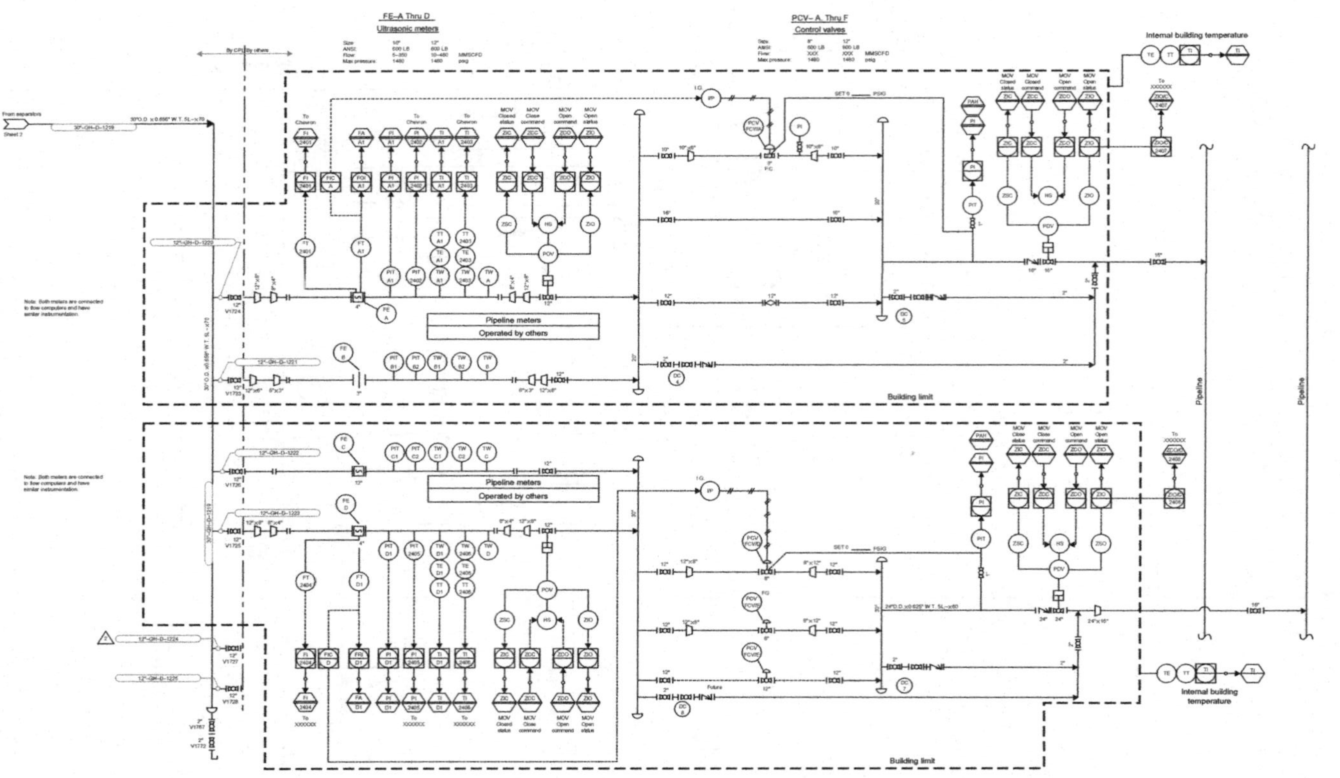

FIGURE 19.2 cont'd.

operations manager, control center manager, and other personnel who have critical knowledge of the components of the system and the intended operation of the system.

19.1 PLAN

The development of a commissioning plan will begin long before the construction and installation of the pipeline and/or facilities have been completed. The plan should be very detailed and include step-by-step instructions as to how the system should be commissioned. By doing this, the risk of a problem developing will be minimized, and there will be less chance of any safety problems developing.

A typical plan will include the following:

- A detailed sequence of events that lead to operations that must be completed
- Verification that the completion of operating and maintenance manuals is complete
- Verification that all construction and installation activities for the project are complete
- Verification that all piping components of the system have been hydrostatically tested
- Verification that the pipeline has been dried if required
- Filling the pipeline with product (crude oil, products, or gas)
- Operator training
- Running a sizing pig
- A check that communications to all facilities, where applicable, are operational
- Verification that power is connected and energized for all facilities/locations where required
- Coordination with receipt and delivery companies
- Personnel are trained and prepared for the plan

19.2 PLAN SEQUENCE

A detailed plan sequence will be prepared prior to commissioning the system. The plan will incorporate a step-by-step instruction for the operations team to follow from the time the construction contractor turns the system over to the pipeline operator until that system is ready for normal operation.

In some cases, the entire system may not be placed into operation at the same time. For instance, one pump station on a multistation liquid pipeline might be placed into operation first, with the others to follow at a later time. The same may be the case at a tank farm where not all of the tanks will be completed and placed into operation at the time the line is placed into initial operation. In situations like these, the commissioning plan will be developed and implemented in phases as each phase of the project is ready to be put into operation.

19.3 OPERATIONS AND MAINTENANCE MANUALS

As the plan is being prepared, a verification of the status of the operations and maintenance manuals should be made. Pertinent data can be obtained from the operations manual, which can be used in the preparation of the plan. The commissioning plan is a prelude to the full operation of the system. See Chapter 21 for details of operations and maintenance manuals.

19.4 COMPLETION OF CONSTRUCTION

During the construction of a pipeline system, there will be a team of construction inspectors and managers monitoring the construction to verify that the various aspects of the construction are completed according to the construction specifications, drawings, and other construction documents. Part of the contractor's responsibility is to verify that the components of the system all operate correctly. For instance, valves can be opened and closed, a motor can be energized, lights can be turned on or off, etc.

Before the contractor leaves the project, an operations team should visit each location to verify that all components will operate as intended. They may include making sure that valves are installed in the right direction, meters are installed in the right direction, all on/off switches work correctly, all tie-down anchors and bolts have been installed, etc. This is a double check to ensure that everything has been installed correctly and nothing has been overlooked.

In most cases, there is a warranty clause in the construction contract, where the contractor must take care of any issues of their work product. During the system checkout if it is discovered that items have been not been completed or things have not been installed correctly, the contractor or vendors will be required to make the necessary corrections. When major pieces of equipment, such as pumps, motors, compressors, and turbines, are purchased, part of the order will include having a technician from the vendor on-site during checkout and start-up of the piece of equipment.

19.5 SIZING OR GAUGING PIGS

When the contractor has finished installing the pipe before the line is hydrostatically tested, a gauging or sizing pig is run through the line and transported by air. The diameter of the pig is sized to be slightly less than the inside diameter of the pipe. It may consist of thin plates between the cups of the pig. If the pig encounters any indentations, obstructions, or narrowing in the line, the plates will bend to indicate there is a problem in the pipe. It will be up to the construction contractor to determine the problem that occurred, the location of the problem, and to repair it before the line is put into operation.

19.6 SYSTEM CHECKOUT

A team will be dispatched to each facility, including scraper trap locations, intermediate valve stations, metering facilities, storage terminals, and compressor or pump stations, to thoroughly check out each one. Each component will be checked for correct operation, instruments will be calibrated, and pumps and motors will be checked for alignment and correct rotation. Communication systems at each location will be checked to ensure that communications are functioning back to the control center. All instrumentation including pressure gages and transmitters, temperature transmitters, and pressure switches will be calibrated.

The list can be quite extensive, and that is why it is important that the commissioning plan be very detailed. The system PFD and P&IDs should include all of the components for each location, including all instrumentation and communication paths, so that a comprehensive commissioning plan can be prepared using these drawings as a basis.

Dependent on the size and number of facilities in the system, the checkout may require from a few days to several weeks to complete. The process can begin while the contractor is still working on cleanup, fencing, and other site work without interfering with the checkout team.

If the pipeline facilities include some sophisticated equipment, it is generally preferable to have representatives from the equipment manufacturer present for the checkout and system start-up.

19.7 PIPELINE DRYING

After the pipeline has been successfully hydrostatically tested, the water must be removed from the line. This process is normally completed by inserting pipeline pigs or scrapers into the line, and air is used to push the pigs or scrapers through the line, pushing the water out of the line. This process will be repeated until no more water comes out of the line.

Prior to hydrostatically testing the pipeline, arrangements will need to be made ahead of time for the proper disposal of the water. This arrangement is usually handled during the permitting phase of the project. The permitting agencies require the pipeline company to indicate where the water for hydrostatic testing is going to be acquired and where it is going to be disposed. Testing of the water will normally be required before it is discharged to verify there are no hazardous materials in it.

Depending on the service for the pipeline, additional steps may be required to further dry the pipeline. The delivery points for pipelines in natural gas, nitrogen, and chemical service will have specifications of very low moisture content for the product that is delivered into their system. In such cases, additional pipeline drying steps will be required.

This may require that the air used to push the pigs through the pipeline is dried before it is compressed into the line. In some situations, nitrogen is used instead of air. Swabbing pigs are pushed through the line until they are completely dry when they are removed from the pipeline. There are companies that provide pipeline drying services if the construction contractor does not have the necessary equipment.

19.8 LINE FILL

A pipeline is filled with the product that it will transport immediately after it has been dried so that moisture will not be reintroduced into it. The beginning of the line will be connected to the source of its product. There may be a connection to another pipeline, a terminal (for a liquid line), oil or gas wells, a gas plant, or a refinery.

The initial volume of product for the line fill of the pipeline is normally owned by the pipeline operating company. Because the fill or inventory stays in the line, it does not belong to the companies that ship products through the line. The product is measured as it is injected into the line so that the line fill volume can be accounted for.

As the line is being filled for the first time, air is bled out of the line at the end of the line. During construction, vent valves are installed along the line. The vent valves are normally located at the same locations as intermediate pipeline block valves and at locations where the elevation is at high points on the line. The line is not backfilled at these high-elevation locations until after the line is filled and placed into operation.

The first intermediate block valve from the beginning point of the lines is closed, the vent valves are opened, and the fill is started. When the product reaches the vent valve(s), it is closed and the intermediate block valve is opened. The procedure continues at each block valve until the line has been completely filled with product.

19.8.1 Example

The following is an example of a commissioning plan for a 30 inch natural gas pipeline. Due to the amount of large elevation difference along this particular pipeline, a portion of the lines was tested with nitrogen instead of water to prevent overpressuring the lower section of pipe due to the elevation head of water.

30 inch Pipeline System Commissioning Procedure

30 inch Hydrotest: Segment 1

1. After valves 1716 and 1753 are opened halfway, Segment 1 of the 30 inch pipeline from the top of the hill to the delivery facility will be hydrotested in five sections because of the elevation differences and the effects of head pressures. The profile of the test sections will be such that the maximum

pressures at the lowest point will not exceed 95% of yield. The test pressure to maintain will be 1.25 × 1480(required MAOP) = 1850 psig. The test water will be trucked to the first section. After the first test, the test water will be transferred to the next section to be tested and the process will be repeated until the line comprising Segment 1 is tested.

2. The pipeline will be physically tied into the delivery facility piping.

Hydrotest: Facilities

1. The delivery facility will be hydrotested at the prefab facilities and will be assembled already tested on-site.
2. The origin facility will be hydrotested at the prefab facility and will again be tested after assembled on-site along with the pipeline.

Cleaning and Drying: Top of the Hill to Delivery

1. After Segment 1 of the 30 inch line has been hydrotested and welded out, the cleaning and drying will begin.
2. A temporary pig launcher will be installed at the beginning of Segment 1.
3. Close valves 7002, 1717, 1734, 1735, and 1736 at the far end of Segment 1.
4. Install hose on pig receiver vent valve 1760 and attach to a portable frac tank suitable to receive remaining water from the hydrotest.
5. Open valve 1703 to receive pigs.
6. Install swab pigs in temporary launcher at beginning of Segment 1.
7. Run as many swab pigs with dry air until the dew point at the far end of Segment 1 is below 0°F. (Note: Valves 1716 and 1753 must be drained and purged to remove any water trapped in valve bodies.)
8. Install one more swab, but do not run. This pig will maintain the separation between the nitrogen and hydrotest water.

Cleaning: Segment 2

1. Remove the temporary pig launcher from Segment 1 and install on the beginning of Segment 2 pipe.
2. Open the clapper on check valve 1715 and lock it in place.
3. Close valves 1714, 1713, 1702, 1731, 1730, 1712, 1729, 1751, and 1752. Open pig trap closure and use side boom and deflector shield to block pig trap opening and knock any construction debris down on the ground at the end of the trap.
4. Open main receiver trap valve 1701 and vent valve 1750 to receive pigs.
5. Install a test head with swab pigs at the high point of Segment 2.
6. Launch swab pigs until pipeline is determined to be clean, using dry air.

Pneumatic (Nitrogen) Test: Segment 2

1. The fuel and instrument gas feed valve will be disconnected, and the valves are plugged to reduce the potential for high pressure to damage any equipment.

2. Valves 1714, 1713, 1702, 1731, 1701, 1730, 1712, 1729, 7011, 7010, 7011, 1768, 1769, 1751, 1752, and 1750 all at the origination facility will be half opened to purge the valve bodies and reduce the impact to the seats.
3. Valves 1733, 1737, and 7020 will be locked and tagged out by operations personnel to prevent high pressure from getting introduced into the storage tank.
4. The Segment 2 pipeline and origination facility will be tested to 1.1 × 1480 (required MAOP) = 1650 psig (max 1700 psig or 80% of SMYS) using nitrogen as the test medium. The test will be maintained for a period of 8 h. The main field road will be blocked and the area will be cleared by non-essential personnel to reduce the risk to personnel and traffic during the test. Vehicles will be allowed to pass one at a time and be cleared on the other side of the isolation area.
5. Nitrogen will be introduced through the vent valve downstream of pig trap 1758. During the initial pressuring up of the origination facility with nitrogen, valves 1750 and the test valve on the far end of the segment will be used to purge the air from the pipeline and origination facility. After nitrogen reaches the vent valve and is vented, these valves will be closed.
6. The test pressure will be raised to 200 psig and then stopped while the flanges are checked for leaks. Close valves 7001, 7010, 7011, 1768, and 1769.
7. Then, the test pressure will be increased in increments of 200 psig. Pressure will be held steady at each interval for 10 min until the test pressure of 1200 psig is reached. Then, close valves 1714, 1713, 1702, 1731, 1701, and 1730 to isolate the origination facility piping from the test section.
8. Continue to raise the pressure until 1650 psig is achieved. Once the nitrogen test is complete, the nitrogen inlet valve 1758 on the pipeline will be closed, and the nitrogen hose is depressured and disconnected from the valve. (During the pressuring up process, all flanges at the origination facility will be tested continuously for leaks. If any leaks are identified, the pressuring up process will be suspended until the leaks are repaired. Note: Once the 8-h test is complete and as the pressure is bled off the test section and reaches 1200 psig, valves 1714, 1713, 1702, 1731, 1701, and 1730 will be opened so as to not trap pressure in the origin facility.)

Purging 30 inch: Segment 1

1. A test head will be installed on the beginning of Segment 1 and a temporary welded 4 inch jumper with a throttling regulating type valve will then be installed between the nitrogen section of tested pipe and Segment 2, and the far end of the segment.
2. Close valves 7002, 1717, 1734, 1735 1761, 1762, and 1736 and open 1703 and 1760 at the delivery facility. Open the pig trap building doors and roof vents. Institute traffic monitoring and controls to stop and alert traffic to the nitrogen danger. Alert all personnel working on-site that nitrogen venting will occur at the site and the danger posed.

3. Open valves 1716 and 1753 to vent the air out of the system.
4. Nitrogen will be introduced from the section of nitrogen-tested pipe to the segment to purge the pipeline of air from the segment and vent through the vent valve 1760 on the pig receiver. The previously installed swab will be used to keep a positive interface between the nitrogen and the air being removed from the segment.
5. During the transfer of nitrogen to Segment 2, the pressure on the segment will be monitored at that point so that the pressure does not exceed 150 psig.
6. Monitor the pressure, and when the pig passes the side-tap valve, close valves 1716 and 1753.
7. After receiving pig at delivery facility and venting the air from the system with valve 1760, the pipeline is now free of air.
8. Isolate and bleed down trap by closing valves 1703 and 1734 and verify 1760 is open. Once the trap is bled down, open the trap and remove the pig. Equalize the nitrogen-tested pipe with the pipe from the segment (pressure on the pipeline is expected to be at 100–120 psig).

Purging Delivery Facility

1. Open valves 1718, 1719, 1720, 1721, 1722, 1754, 1756, and 1757.
2. Close valves 1772, 1767, 1723, 1724, 1725, 1726, 1727, 1728, 1755, 1761, 1762, 1717, and 1736.
3. Slowly open valve 1735, pressure the delivery facility to 25 psig of nitrogen, and close valve 1735.
4. Bleed down station through valves 1767 and 1772 until just above atmosphere and close valve 1772.
5. Slowly open valve 1735 and pressure the delivery facility to 25 psig of nitrogen and close valve 1735.
6. Bleed down station through valve 1772 until just above atmosphere and close valve 1772 in that order. Repeat steps 3 through 5 above, one more time (the delivery facility will be pressure and vented a total of three times).
7. Using valve 1735, pressure up station to pipeline pressure and close valve 1735.
8. Partially open valves 7002, 1717, 1718, 1719, 1720, 1721, and 1722, then crack the body bleed valves on each valve to purge the air. Also, crack open the separator vent valves 1764 and 1765 to bleed the air. Open and close valves 1734 and 1736.
9. Crack open valves 1727 and 1728, and open the body bleeds to purge the air. Note: Valves 1723, 1724, 1725, and 1726 will be commissioned with nitrogen with the meter stations.
10. After equalizing delivery facility, close valves 1735, 7002, 1717, 1736, and 1734.
11. Note: If possible coordinate purging of both delivery facility meter stations at this time.

Top of the Hill Tie-In

1. Bleed nitrogen pressure on pipeline to just above atmosphere by opening the pig trap vent valve 1760 at the delivery facilities and pig trap vent valve 1750 at the origination facilities and the side-tap valves 1716 and 1753.
2. Cut off temporary fittings between the nitrogen-tested pipe and the Segment 2 section and insert swab pigs at each opening to contain an atmosphere of nitrogen in both sections of pipeline.
3. Using the 10–20 ft pipe spool pup, complete final pipeline tie-in work on the top of the hill.
4. X-ray all welds and coat.
5. Unlock clapper in check valve 1715 from the open position and fix to original (normal) position.

Pipeline Disposition

1. Origination facility and 30 inch pipeline are now idle with a blanket atmosphere of nitrogen except for the small areas with air due to the final tie-in. The delivery facility is idle with approximately 100 psig of nitrogen, which will be used to purge the two delivery meter stations at a later date. During the commissioning of the system with gas, the two swab pigs that were inserted and left in the pipeline will be retrieved from the pig receiver at the delivery facility.

Origin Facility and Pipeline Commissioning with Natural Gas

1. At the origin facility, close valves 1701, 1730, 1712, 1729, 1751, 1752, 1750, 1702, 1731, and 7011 and open valves 7001 and 7010 halfway to isolate the origin facility and prepare for commissioning with natural gas.
2. Swap the spectacle blind and open valve(s) on 8 inch pipeline slowly to allow natural gas to flow to the origin facility in the 8 inch input pipeline up to the 8 inch inlet valve 7011 until pressure is 25 psig, then block 8 inch pipeline valve and then vent upstream of valve 7011 until an LEL reading is achieved. Then, bleed 7011 body bleed to vent nitrogen.
3. Open the 8 inch feed valves and pressure up to feed pipeline pressure.
4. Open valve 7011 and pressure up the origin piping up to 8 inch pipeline pressure. Bleed the pressure off the bodies of valves 1712, 7011, 7010, 7001, and 1702 until an LEL reading is achieved. Crack valve 1701 to purge valve body.
5. Open valves 1769 and 1768 (separator block valves) and close the valve 1733 and the separator vent. Open valves 1751 and 1752 and pressure up to 8 inch pipeline pressure. Crack open valves 1713 and 1714 to purge nitrogen into 30 inch pipeline. Open the vent on the separator until an LEL reading is achieved and close the vent on the separator.
6. Close valves 7001, 7010, 1701, 1730, 1712, 1729, 1751, and 1752 and vent pig trap. Then, open pig trap and insert one Criss-cross Poly Pig (#5 foam pig with polyurethane coating) in the pig trap as far as can be pushed into

the reducer. (The pig will be launched to maintain the separation between the nitrogen and natural gas when we purge nitrogen from the system.)

7. Notify personnel at the delivery facility and side-tap valve to monitor the pressure since preparing to launch a pig with natural gas. (Note: Operations must start raising pressure on the 8 inch pipeline by opening wells.) At the delivery facility, close valves 7002, 1735, 1734, 1717, 1736, 1761, and 1762.
8. Close the pig trap and open valve 1730 and 1701 to prepare to launch the pig.
9. Open valve 1731 partially and then slowly open valves 1729 and 1712 to launch the pig in the 30 inch pipeline.
10. Once the pig sigs trip downstream of the pig trap, open valves 1730, 1731, 1713, and 1714, successively.
11. Notify delivery facility and side-tap valve personnel that pig was successfully launched.
12. Delivery facility personnel monitor the pressure at the delivery facility until pressure reaches 50 psig, and then open valve 1760 (pig trap vent) to allow nitrogen to be vented to atmosphere.
13. Side-tap valve personnel, open valve 1716 (slide tap valve) and 1753 (vent at the side tap valve) and monitor for LEL and close both valves 1753 and 1716 (in that order) once a sufficient reading is achieved.
14. At the delivery facility, monitor vent and watch for pig sig trips to alert to pig arrival.
15. Once pig sigs trip at delivery facility, notify origin facility personnel and close valves 1703 and 1734 and bleed trap down with valve 1760 open. Verify trap is depressured and open pig trap to verify pig has been received. If received, remove pig and close pig trap. Pressure pig trap to pipeline pressure.

Delivery Facility Commissioning with Natural Gas

1. Open valves 1718, 1719, 1720, 1721, and 1722 and open valves 1727 and 1728 halfway. Open valves 1735 and 7002 to raise pressure of station piping to 25 psig and close 1735 and 7002.
2. Open valves 1767 and 1772 (ball and gate) and vent the nitrogen until a LEL reading is achieved or pressure is just above atmospheric, and close valves 1772 and 1767 in that order.
3. Open body bleeds on valve 1727 and 1728 and close right before the atmospheric pressure is reached.
4. Open valve 1735 to raise pressure of station piping to 25 psig and close 1735.
5. Open valves 1763, 1770, 1766, and 1771 to vent nitrogen until pressure drops 5 psig and close valves 1770 and 1771. Open valves 1754, 1756, and 1757 (equalization valves) and then valve 1755 (separator vent) to vent the nitrogen until pressure is just above atmospheric, and close valves 1755, 1754, 1756, and 1757 in that order.

6. Open valves 1735 to raise pressure of station piping to 25 psig and close 1735.
7. Open valves 1767 and 1772 (ball and gate) and vent the nitrogen until a LEL reading is achieved or pressure is just above atmospheric, and close valves 1772 and 1767 in that order.
8. Open valve 1735 to raise pressure of station piping to pipeline pressure.
9. The delivery facility is ready for start-up. Once power is established, the facility is ready for instrumentation commissioning and communication validation.

BIBLIOGRAPHY

ASME B31.4-2006, Pipeline Transportation Systems for Liquid Hydrocarbons and Other Liquids, American Society of Mechanical Engineers, New York, NY, 2006.

ASME B31.8-2007, Gas Transportation and Distribution Piping Systems, American Society of Mechanical Engineers, New York, NY, 2007.

CFR 49, Part 195, Transportation of Hazardous Liquids by Pipeline: Minimum Federal Safety Standards, U.S. Government Printing Office, Washington, DC.

CFR 49, Part 192, Transportation of Natural or Other Gas by Pipeline: Minimum Federal Safety Standards, U.S. Government Printing Office, Washington, DC.

Chapter 20

Specification Writing, Data Sheet Production, Requisition Development, and Bid Analysis

Glenn A. Wininger

Chapter Outline

INTRODUCTION

In this chapter various aspects of specifications for pipe, material, and certain equipment used in pipeline and facilities construction will be addressed along with an overview of construction specifications. In addition, material requisition development will be discussed, and a general guideline for bid analysis will be dealt with as well.

20.1 SPECIFICATION WRITING

The single most important reason for writing clear and precise specifications is to ensure that the vendors or contractors, engineers, and bidders are well informed enough to provide the final product required that results in quality without additional work or modifications resulting from poorly executed specifications. One other reason why the specifications must be clear is to provide fairness across the board to all potential suppliers of material/goods and services when requesting quotations from a diverse group of vendors or contractors. The specifications must be written clearly and contain the appropriate amount of detail so that more than one supplier can satisfy the obligation related to what is being supplied whether it be material or services. When a bidder or supplier fully understands the specifications for bidding purposes, this allows for increased competition among the bidders and ultimately achieves better value for your money.

The specifications should be authoritative not restraining. The objective of writing specifications is to explain to the suppliers of material or services what is required to achieve the final product. If there is insufficient data and detail to support the final goal the bidder may be confused. This results in a product or service that will not satisfy the objective. Simultaneously, the specifications should not be too detailed in manners that restrict the bidders needlessly. It is important that as many bids as possible are received to improve competition and increase the chances of purchasing equipment or services that meets the requirements at the best possible price. In summary, specifications must be detailed enough to leave no question in the bidders mind as to what is required, but should be generic enough to allow multiple manufacturers' equipment to be offered.

Specifications should be written in some type of bullet point format. The beginning of the specification should indicate the most important distinctive account of what is to be provided. Each subsequent bullet point should address only one feature without referencing any additional attribute or information. The bullet points should be sequentially numbered starting from 1.

Only key attributes that are relevant to the expected final product should be included within the specifications under each point. For example, in specifications related to pipeline construction, there is no need to specifically indicate the size of bulldozer required for grading a pipeline corridor as this information is not a factor in the awarding of any bid. The key attribute in this case is to provide relevant information related to the grading requirements while concentrating on specifics that would include a safe and productive corridor for ensuing construction activities.

Keep sentences short and simple. It doesn't matter if specifications read like a grade-school textbook. Many times, sentences can become so long and complex that the writer is even confused. This results in conveying something very different from what was intended, or in some cases, the sentence conveys nothing.

20.2 MATERIAL SPECIFICATIONS

20.2.1 Pipe

Pipe that is used to transport petroleum products and natural gas is subject to ANSI/ASME guidelines. For petroleum products, the particular guideline is ANSI/ASME B31.4 and for natural gas, the guideline is ANSI/ASME B31.8. The pipeline incorporates a design safety factor, prescribed by the

US Department of Transportation (DOT) regulations, that ensures continual safe operation of a pipeline. In addition to the specific guidelines outlined in ANSI/ASME, should compliance be required under the Federal Pipeline Safety Regulations, additional guidelines are required under the DOT Part 192 for gas pipelines and DOT Part 195 for liquid pipelines.

Regarding gas pipeline standards that fall under the DOT Part 192, the criteria for specification relates to the following:

§ 192.112 Additional design requirements for steel pipe using alternative maximum allowable operating pressure.

For a new or an existing pipeline segment to be eligible for operation at the alternative maximum allowable operating pressure (MAOP) calculated under § 192.620, a segment must meet the following additional design requirements. Records for alternative MAOP must be maintained, for the useful life of the pipeline, demonstrating compliance with these requirements:

To address this design issue:	The pipeline segment must meet these additional requirements:
(a) General standards for the steel pipe:	(1) The plate, skelp, or coil used for the pipe must be micro-alloyed, fine grain, fully killed, continuously cast steel with calcium treatment. (2) The carbon equivalents of the steel used for pipe must not exceed 0.25 percent by weight, as calculated by the Ito-Bessyo formula (Pcm formula) or 0.43 percent by weight, as calculated by the International Institute of Welding (IIW) formula. (3) The ratio of the specified outside diameter of the pipe to the specified wall thickness must be less than 100. The wall thickness or other mitigative measures must prevent denting and ovality anomalies during construction, strength testing and anticipated operational stresses. (4) The pipe must be manufactured using API Specification 5L, product specification level 2 (incorporated by reference, see § 192.7) for maximum operating pressures and minimum and maximum operating temperatures and other requirements under this section.

The wall thickness (W.T.) of a pipeline is determined by the maximum allowable operating pressure (MAOP) and is related to the type of construction and population density (class location) along the pipeline route. Additionally, federal regulations and published industry standards play a key role in the design specification of a pipeline and components.

Line pipe is milled to the American Petroleum Institute (API) engineering and metallurgical specifications. Line pipe is manufactured and then milled from high-strength carbon steel.

The API Specification 5L, as noted in the above regulation, defines the specific requirements for pipe manufactured to transport natural gas, oil, and water. This particular API specification for manufacturing line pipe includes stringent standards of the carbon steel with regards to physical and chemical characteristics.

All line pipe is tested at the source or pipe mill to ensure that it meets all requirements of steel chemistry, strength and toughness, and dimensional characteristics. There are numerous pipe mills around the world that manufacture line pipe under API 5L specifications for the pipeline industry. These mills produce two types of line pipe. The first is seamless pipe that is formed from a cylindrical bar of steel. This bar is heated to a very high temperature and then a probe is inserted to create a hole through the cylinder. The cylinder is then transferred to rollers, which size the cylinder to the specified diameter and wall thickness. Seamless pipe is used from diameters as small as 0.5" to approximately 16" generally. Some mills can produce seamless pipe up to 24" in diameter.

Interstate natural gas pipelines along with localized pipeline systems use the welded variety of line pipe. In cases in which larger diameter pipe is required the welded line pipe is necessary. This is the second type of line pipe made within API 5L specifications. These mills manufacture line pipe using a welding process to close the seam after forming a steel plate or coil into a cylindrical shape. As the purchaser of the welded pipe, the operating pressure of the proposed line pipe would be provided to the mill. The mill uses ultrasonic and/or radiological inspection methods to ensure the quality of the weld seam and then initiates pressure tests on each joint of pipe to levels that exceed the proposed operating pressure of the pipeline.

Each company that utilizes the various mills for the production of the line pipe maintains the manufacturing and test records of the pipe while the pipeline remains in service.

As a general summary, line pipe specifications to accommodate the required volume that the proposed pipeline will deliver and ultimately submitted for bidding purposes to pipe mills should include the following parameters:

P – Design pressure, in lb/in^2
t – Nominal wall thickness (W.T.) that will be required
Grade of Steel – to determine SMYS
S – Specified minimum yield strength (SMYS) according to the grade of the line pipe
D – Nominal outside diameter (O.D.) of the line pipe required
F – Design factor – this could be diverse dependent on the class location changes found throughout any given project. Some projects may see only one or two class location changes while others recognize all four class locations throughout a project length.
E – Longitudinal joint factor
T – Temperature derating factor

The maximum allowable operating pressure (MAOP) that will be the eventual maximum operating pressure of the proposed pipeline is calculated with the above information and is noted in the following equation:

$$P = (2St/D) \times F \times E \times T \qquad (20.1)$$

20.2.2 External Coating of Line Pipe

When it became necessary, years back, to establish a way to transport oil and natural gas from one region to another, cross-country pipelines were built. In many locations, bare pipe was laid with overcoat protection. Throughout this period, knowledge about corrosion protection was very limited; as such those forerunners built the best pipelines possible with the available technology.

Eventually, after pipeline failures occurred at locations along a pipeline that were not the result of dents in the pipe from large rocks on the pipe in the ditch or weak pipeline welds, it became understood that an electrical field created by the earth caused corrosion on underground steel pipe.

After considerable research, it became apparent that some sort of outer coating of a nonconductive material installed on the pipe joints would contest this electric field and thwart the effects of the earth's electrical field from affecting the pipe. In those early days, before coatings could be applied at the mills or coating plants under favorable conditions, a system was used to apply coating to the bare pipe at the construction site. One of the first preferable field applied coatings used was coal tar enamel. The coal tar was brought in solid form to the construction site and then heated until it became a liquid. After it reached the liquid stage, the applicators applied it using hoses to the pipe. The coal tar enamel would harden, thereby providing a nonconductive coating to the pipe. Although this system worked to some degree, it was laborious and time-consuming not to mention costly.

With the advent of technology, a variety of coating systems have been undertaken through the years, which have advanced with ingenuity and the understanding and use of new materials. After years of research, and with the ability to coat line pipe in a controlled facility, five major coating systems are commonly used for pipelines today. The different systems are accepted and utilized by pipeline companies based on various factors including, but not limited to, the following:

- Cost
- Management on handling
- Regional availability of the coating material
- Transportation from the source to the pipeline installation site
- Methodological and procedural reasons

The following list includes the five major coating systems used today in pipelines:

1. Three-layer PE (3LPE): 3LPE systems consist of an epoxy primer, and a grafted copolymer medium-density (MDPE) adhesive to bond the epoxy primer with a high-density (HDPE) topcoat. The large acceptance of 3LPE is represented by its ability to withstand very rough transportation and field treatment during construction and installation without excessive

damage to the coating. 3LPE also maintains a broad operating temperature range (from −45°F to +185°F). 3LPE coating is dominant worldwide for onshore pipelines, especially in Europe. The only exception of the application of 3LPE is in North America. Projects in Asia, and the Middle East are noticing a predominant increase in the use of 3LPE as well.

2. Three-layer PP (3LPP): 3LPP system consists of an epoxy primer, and a grafted copolymer PP adhesive to bond the epoxy primer with a PP topcoat. 3LPP systems are predominantly used for offshore projects as an elevated operating temperature (32°F to +284°F) of the system is recognized as a major advantage. In addition, 3LPP provides an excellent combatant under the extreme mechanical stress on the pipes. Projects in many regions throughout the world have utilized 3LPP coatings, which provide access to deeper gas and oil fields.
3. Fusion-bonded epoxy (FBE or Dual FBE): FBE is dominant in North America, the United Kingdom, and a few other countries. Although FBE is popular, the tendency to use the system is declining in favor of 3LPE and PP systems. FBE is a powder-based coating and is a thermoset polymer coating. FBE possesses the same properties as traditional epoxy and is used to coat and protect steel pipe, pipe fittings, and valves. Fusion bonding comes from the process by which it adheres to a substrate – when the hardener and epoxy react and the coating takes solid form, the chemicals involved are cross-linked – the process is irreversible, even with severe heat applications. The liquid FBE film wets and flows onto the steel surface on which it is applied and soon becomes a solid coating by chemical cross-linking, assisted by heat.
4. Coal tar enamel (CTE): CTE and asphalt enamel are both still used in some countries. However, both systems are declining in use and suffer from health and environmental concerns. When used with synthetic primer and inner and outer wraps, it forms a protective coating that offers the following benefits: high resistance to mechanical damage, good thermal stability, resistance to root growth penetration, complete resistance to soil bacteria and marine organisms, and a high resistance to petroleum products.
5. Asphalt enamel and polyurethane (PUR): PUR systems are most commonly used for pipeline rehabilitation projects or girth weld coating. As with CTE, PUR systems also suffer from health concerns.

In addition to the five most commonly used and accepted corrosion resistance pipe coatings, an extra precautionary pipe overlaying system or measure may be recognized for use in road and stream crossings, horizontal directional drills, and in areas where boulders or rock are encountered. Abrasion-resistant overcoat (ARO) is an epoxy-based polymer concrete that is applied over FBE, 3LPE, 3LPP, or any other chosen coating system. ARO is usually applied at a mill over the corrosion prevention coating to act as a protective measure for the coating system. The specification for mil thickness is generally 40 mils minimum.

The following specification for external application procedures for plant applied fusion-bonded epoxy (FBE) coatings and an abrasion-resistant overlay (also referred to as overcoat) (ARO) coating to steel pipe is courtesy of the National Association of Pipe Coating Applicators (NAPCA; Fig. 20.1). The specification

NAPCA Bulletin 12-78-04

EXTERNAL APPLICATION PROCEDURES
FOR PLANT APPLIED FUSION BONDED EPOXY (FBE)
COATINGS AND ABRASION RESISTANT OVERLAY (ARO)
COATINGS TO STEEL PIPE

1. **General**
   a. These specifications may be used in whole or in part by anyone without prejudice, if recognition of the source is included. The National Association of Pipe Coating Applicators (NAPCA) assumes no responsibility for the interpretation or use of these specifications.
   b. The intended use of these coatings is to provide corrosion protection for buried pipelines. Above ground storage of coated pipe in excess of 6 months without additional Ultraviolet protection is not recommended.
   c. The following definitions apply:
      i. Applicator - The contractor who applies the coating to the pipe.
      ii. Company - The purchaser of the coated pipe or the entity for whom the Applicator coats the pipe.
      iii. SSPC - The Steel Structures Painting Council.
      iv. NACE - NACE International.
      v. Manufacturer - The company that makes the coating materials which are applied to the pipe.
      vi. FBE—Fusion bonded epoxy pipe coating
      vii. ARO—Abrasion resistant overlay coating

2. **Scope**
   a. The Applicator shall furnish all labor, equipment and material required, shall prepare all surfaces to be coated and shall apply the coating to all surfaces to be coated.
   b. Corrosion protection, as provided under this specification, is furnished by the application of fusion bonded epoxy to the exterior of pipe to be placed underground.

3. **Pipe Conditions**
   a. Pipe delivered to the Applicator for coating shall be free of protective oils, lacquers, mill primer, dirt or any other deleterious surface contamination which may affect the application of the coating. The pipe surface shall be as free as possible from scabs, slivers and laminations. Removal of such contaminants shall be as agreed between the Applicator and the Company.

FIGURE 20.1 External application procedures for plant applied fusion-bonded epoxy (FBE) coatings and an abrasion-resistant overlay (also referred as overcoat) (ARO) coating to steel pipe. *(National Association of Pipe Coating Applicators (NAPCA)).*

b. Any paint markings or stenciling of the pipe surface shall be of the type and thickness that can be removed easily during normal surface preparation.

4. **Handling of Bare Pipe**

a. Proper equipment for unloading, handling, and temporary storage of bare pipe shall be used to avoid any damage to the pipe or pipe ends.

b. If internally coated pipe is received at the Applicator's plant, care shall be taken to avoid damage to the internal coating or the obliteration of the internal pipe markings during any phases of work covered by this specification. Internal coatings must be capable of withstanding the processing conditions necessary for the application of the external coating.

c. The Applicator shall visibly inspect the pipe upon receipt for damage such as dents, flat ends, and bevel damage. Any damage observed at this point shall be noted on the inbound tally, and the Company shall be informed within 24 hours of receipt of the pipe. Any non-visible defects such as slivers, scabs, laminations, burrs, dents, etc. will be observed after the pipe is blast cleaned and at the Company's request, removed as an extra work item.

d. When the roughness of the pipe surface is such that the normal coating rate will be impaired or will result in coated pipe that will be in non-conformance with these specifications, at the Company's option, the Applicator shall take whatever steps necessary to rework the pipe surface and/or add additional coating thickness to the pipe so as to provide coated pipe which can be processed at a normal rate and which will comply with these specifications. The cost of reworking the pipe surface and/or applying additional coating thickness will be agreed upon in advance by the Applicator and Company and paid for by the Company.

5. **Materials and Workmanship**

All material furnished by the Applicator shall be of the specified quality. All work shall be done in a thorough workmanlike manner. The entire operation of pipe receiving, stockpiling, surface preparation, coating application, storage and loadout shall be performed under the supervision of and by experienced personnel skilled in the application of protective coating.

6. **Equipment**

The Applicator's equipment shall be in such condition as to permit the Applicator to follow the procedure and obtain results prescribed in these specifications.

7. **Coating Material**

a. All coating materials, including repair or patch materials, purchased or used under these specifications, shall be packaged in suitable and approved containers. The containers shall be plainly marked with the name of the Manufacturer, type of material and batch or lot number where applicable. Bulk shipments shall be allowed provided the above information is included in the bill of lading.

b. The coating material shall be packaged in containers suitable to keep the contents clean and dry during handling, shipping and storage. Storage and handling conditions shall be in accordance with the Manufacturer's recommendations.

FIGURE 20.1 cont'd.

c. Precautions shall be taken during the handling, shipping and storage of all materials to prevent damage to the containers that would result in contamination of the coating materials. All contaminated, or otherwise damaged materials shall be discarded.

d. Time/temperature limitations for FBE Powders shall be in accordance with the Manufacturer's recommendations.

8. Surface Preparation

a. Before blasting, all oil, grease, mill lacquer and other deleterious material on the surfaces of the metal to be coated shall be removed by suitable means.

b. In cold weather or any time when moisture tends to collect on the steel, the pipe shall be uniformly warmed for sufficient time to dry the pipe prior to cleaning. The pipe temperatures shall be maintained at least 5 degrees F above the dew point during the cleaning and coating operations. Pipe temperature shall not exceed 160 degrees F as a result of preheat.

c. Pipe surfaces shall be blast cleaned to a Near-White metal finish in accordance with SSPC-SP-10 or NACE #2 requirements.

d. NACE, Swedish Pictorial, SSPC or other mutually agreed upon standards shall be used to judge the degree of cleaning.

e. A consistent abrasive working mix shall be maintained by frequent additions of small quantities of new abrasive commensurate with consumption. Infrequent large quantity additions of abrasive shall be avoided.

f. Following cleaning and prior to coating the pipe, abrasive remaining on the outside of the pipe shall be removed by air blast, vacuum or other suitable methods. If air is used, the air should be dry and free of contaminants, and all particles removed from the surface shall be collected in such a manner as not to contaminate clean pipe.

g. Following cleaning and prior to coating, the pipe surface shall be inspected for adequate cleaning and surface condition. Pipe not properly cleaned shall be rejected and recleaned.

h. Blast cleaned pipe surfaces shall be protected from conditions that would allow the pipe to flash rust before coating. If flash rusting occurs, affected pipe shall be recleaned.

i. Surface imperfections such as slivers, scabs, laminations, burrs and weld spatter shall be removed by hand filing or light grinding as long as such removal does not:
   i. Adversely affect the quality of pipe cleaning.
   ii. Adversely affect the normal production rate of the plant.

9. Coating Application (FBE)

a. The pipe shall be heated to a temperature within the tolerances recommended by the Manufacturer of the coating material to be applied. The pipe shall not be heated to a temperature in excess of 600 degrees F at any time during the process. Blue oxide formation shall not be used as an indicator that maximum temperature has been exceeded or that damage to steel properties has occurred.

b. The pipe shall be monitored for proper temperature prior to coating by use of a temperature sensitive crayon.

FIGURE 20.1 cont'd.

c. Coating material shall be applied uniformly utilizing the best commercial practices and in such a manner so that the minimum cured film thickness is that specified by the Company.

d. The coating shall be applied to the full length of each pipe except for a "cutback" of not less than 0.5 inches (1.27 cm).

e. Use of recycled coating material is permitted if adequate recovery and screening equipment is used and maintained. An adequate recycle system must properly blend recycled and virgin coating material into the delivery system.

f. The Applicator shall supply samples of the coating material to the Company, at any time requested, for such test as the Company may wish to run, to assure that the quality of the coating material is being maintained. Company inspection and acceptance of coating materials must be completed prior to application of said materials. Subsequent testing may be conducted to confirm that the coating material has not suffered from improper handling or storage.

g. During coating and curing periods, the coated pipe shall be handled so as to avoid any damage to the coating.

h. After the coating is cured, the pipe may be force cooled to facilitate coating inspection and repair, provided care is taken to avoid any damage from thermal shock to the newly applied coating. The maximum temperature for inspection and repair is 250 degrees F.

10 Coating Application (ARO)

a. The corrosion barrier and the abrasion resistant overlay shall utilize a separate fluidized bed.

b. Two separate spray booths are preferred. Reclaimed FBE and ARO powder may be used if they are applied in two separate spray booths, but may not be used if they are applied in the same booth. Powder from the topcoat shall not be recycled into the base coat powder except as directed by the ARO powder manufacturer.

c. The ARO topcoat shall be applied in accordance with the coating manufacturer's recommended procedure and prior to gelling of the FBE corrosion base coat.

11 Inspection and Testing

a. The entire procedure of applying the protective coating material as herein specified will be rigidly inspected from the time the bare pipe is received until the coated pipe is loaded on the carrier for shipment.

b. If the Company designates an Inspector, the Inspector shall be provided free access to the Applicator's plant at any time during any operation involving the pipe, with the right to inspect and to accept or reject work performed.

c. The Applicator's Quality Control Inspector shall be responsible for stopping operations when conditions develop which could adversely affect the quality of the completed work.

d. Although the principal purpose of the coating inspection by the Company and Applicator is to insure compliance of the coating with these specifications, such inspection shall also include examination for previously undetected defects in the pipe, pipe surface or on the pipe ends. Pipe having such defects shall be set aside for subsequent repair or replacement by the pipe supplier and for any necessary coating repair. Recoating or coating repair that may be

FIGURE 20.1 cont'd.

necessary by reason of these defects in the pipe which do not involve fault on the part of the Applicator shall be done at the Company's expense.

e. When Company's Representative exercises Company's right of approval at the Applicator's plant, the Company's Representative shall conduct final inspection on the Applicator's out-bound rack. Accepted pipe shall be presumed to be produced as specified unless test results indicate a discrepancy.

f. Coating Thickness Measurements

   i. An appropriate film thickness gauge, calibrated to the National Bureau of Standards' Certified Coating Thickness Calibration Standards shall be used to perform coating thickness measurements on FBE and ARO coatings. A tooke gauge or grinding/filing down the ARO to the base FBE shall be used to perform measurements of base coat and the top coat or a piece of masking tape can be applied to the FBE before the ARO is applied and the measurements can be taken with an appropriate dry film thickness gauge.
   ii. The agreed upon coating specification shall state the absolute minimum thickness to be applied. All thickness readings measured at or above the absolute minimum shall be accepted for both FBE and ARO coatings.
   iii. The range of coating thickness measurements shall be recorded for each joint of pipe.
   iv. NAPCA recommends that no milage be specified below 10 mils minimum.

g. Electrical Inspection

   i. Holiday inspection of the entire coated surface shall be performed with an approved high voltage Holiday Detector to indicate any flaws, holes, breaks or conductive particles in the protective coating.
   ii. The Holiday Detector shall have sufficient D.C. voltage and be equipped with a positive signaling device. The search electrode shall be made of conductive rubber, or other applicable material. The Holiday Detector shall be operated in such a way as to audibly and/or visually detect the presence of all holidays.
   iii. The voltage to be used shall be per the guidelines of NACE RPO490 or the ARO powder manufacturer.

h. Adhesion of the cured coating shall be checked by pushing a sharp knife blade through the cured coating to the pipe surface. The coating will not strip or peel when the knife is moved in a "whittling" motion against the steel surface, if proper adhesion of the cured coating to the pipe surface has been attained.

i. A minimum of one bend test shall be performed for each day of production. The acceptance criteria shall be 1.5 degrees per pipe diameter permanent strain with no tears, cracks, or disbonding of the coating as outlined in the following procedure. Either the mandrel or four point bend methods are acceptable for producing the permanent strain. Bend testing is not applicable for ARO coatings.

   i. A ring sample at least 12 inches long shall be cut from production coated pipe and strips measuring 1 inch by 8 inches

FIGURE 20.1 cont'd.

shall be prepared. The 8 inch dimension shall be parallel to the longitudinal axis of the pipe.

ii. Condition the straps to 32 degrees F in a freezer or with dry ice.

iii. If the mandrel method is to be used, the required mandrel radius shall be calculated by the following equation.
R=57.3t/s - t/2
Where t = effective strap thickness
R = bend radius of outer curve of strap
When a mandrel of the calculated radius is not available, the mandrel of the next smaller size shall be used.

iv. For the four point method, estimate the required bend using the equation explained in the Mandrel method.

v. The straps shall be bent so that the uncoated side is in contact with the mandrel/support pins. The bend shall take at least 10 seconds and no more than 30 seconds to complete.

vi. Remove the strap from the bend apparatus and measure the permanent radius by matching the strap against a chart of known radius curves.

vii. Calculate the permanent strain in degrees per pipe diameter by the following equation:
deg/pd = 57.3t/R - (t/2)
Where t = effective strap thickness
R = bend radius of outer curve of strap

viii. Allow the straps to warm to room temperature and visually inspect them for cracks, tears in the coating, and disbonding of the coating. Disregard any defects within 0.1 inches of the strap edge or within 0.5 inches of the support pins. The presence of strain marks alone does not constitute a failure.

j. A minimum of one 24-hour cathodic disbondment test shall be performed for each day of production. The acceptance criteria shall be a maximum of 12 mm radius from the edge of the intentional holiday as outlined in the following procedure. Disbondment greater than 8 mm radius generally indicates that problems exist in either the process or powder and shall be investigated.

i. The sample shall be a four inch square segment cut from the test ring.

ii. Drill a 1/8-inch diameter holiday in the coating at the center of the sample.

iii. Glue a plastic cylinder onto the specimen with the holiday at the center of the cylinder.

iv. Pour approximately 350 ml of electrolyte into the cylinder. The electrolyte shall be composed of 3% by weight sodium chloride (NaCl) in distilled water.

v. Place the test cell on a hot plate controlled heat transfer medium (steel shot or grid in a metal pan), insert a thermometer so that it is immersed and resting on the sample and adjust the temperature to 150 +/-5 degrees F.

vi. Connect the negative lead from a variable voltage DC power supply to the specimen. Attach the positive lead to a platinum or platinum-coated wire immersed at the center of the cell.

vii. Adjust the voltage to 3.5 volts DC using a calomel reference electrode.

FIGURE 20.1 cont'd.

viii. After 24-hours, remove the test cell and immediately drain the electrolyte. Rinse with tap water, dismantle the cell, and air cool to room temperature.

ix. Within one hour of removal from the hot plate, make radial cuts from the edge of the holiday outward using a utility knife. The cuts shall be at least 0.8 inches in length and through the coating to the substrate.

x. Immediately insert the blade of the utility knife under the coating and, using a prying action, chip off the coating. Continue until the coating demonstrates a definite resistance to the prying action.

xi. Measure the radius of the disbonded area from the holiday edge along each radial cut and average the measured results.

12 **Repair Procedures**

a. All defects disclosed by the Holiday Detector and other obvious defects shall be repaired by the Applicator.

b. Holidays which are the result of slivers, scabs, laminations, or other steel conditions beyond the control of the Applicator shall be repaired at the Company's expense.

c. Areas of repair to the coating shall be holiday inspected by the Applicator on a 100 percent basis.

d. Pinhole type holidays may be patched using the hot melt patch stick method. Abrade the adjacent coating surface with either a hand file or coarse sand paper. The abrasion should be on the surface only and should not remove a significant thickness of the coating. The surface to be touched up will be heated with a small torch until the stick starts to melt, then the stick should be rubbed over the heated surface, building up a small puddle of patching compound over the entire area being patched. Such holidays may also be patched by use of a two part, 100% solids, liquid epoxy compound specified by the Manufacturer using the method set forth in the following paragraph. Liquid epoxy repairs of pinhole type holidays shall be considered an extra work item to be performed at a price agreed upon between the Company and Applicator.

e. Where larger areas of damaged coating are to be repaired and the use of patch sticks is not practical, a two part, 100% solids, liquid epoxy compound specified by the Manufacturer shall be used. The damaged area shall be abraded by hand filing or use of carborundum cloth. Application shall be made to a minimum thickness of 25 mils (0.64mm) and shall overlap the undamaged area a minimum of 0.5 inches (1.27cm).

f. The liquid patch compounds shall not be applied when the pipe temperature is below 50 F unless provisions are made for heat curing the patch material using methods and temperatures in accordance with procedures recommended by the coating Manufacturer.

g. All repairs performed on ARO coatings shall be made with an accepted 100% solids liquid epoxy compound.

13 **Coated Pipe Handling, Storage and Loading Requirements**

a. Pipe shall be stored, handled and transported in a manner to prevent damage to the pipe walls, beveled ends and the coating.

FIGURE 20.1 cont'd.

b. Storage racks shall be so designed as to protect the coated pipe from standing water, direct soil contact, and sharp or hard objects that might damage the coating.

c. All individual coated pipe, that is handled, stored, or shipped, shall be protected by padding, separators or dividers. These separators shall be affixed to the exterior surface of the pipe. Suggested separators consist of 1) polypropylene rope rings for larger and smaller diameters of pipe joints, with rope thickness varying from 3/8 to 5/8-inches in diameter depending on pipe wall thickness; 2) spiral cardboard rings with waxed outside surface for 4 inches nominal and smaller diameter pipe joints, with ring being 2 inches wide and 0.25 inch thickness; 3) 0.25 inch minimum thickness rubber; or 4) other material approved as separators and acceptable to the Company.
   i. Each 40 foot joint of coated pipe shall have affixed to its exterior surface a minimum of three (3) separators, randomly spaced, but spaced relatively close to the ends of each joint and near the center of the pipe. The minimum number of separators for nominal 60 foot pipe shall be four (4); and for nominal 80 foot pipe, the minimum number of separators shall be five (5).
   ii. The type and thickness of separators shall be chosen to prevent joints of pipe from coming in contact with each other.

d. The coated pipe shall be shipped using sufficient and proper dunnage to adequately protect the pipe and coating.

e. All pipe shipped by rail shall be loaded in accordance with API Specifications RP 5L1, Latest Edition.

14 Supplementary Details Supplied by the Company

When possible, the Company shall supply the following supplemental information:

a. Length and diameter of pipe.
b. Grade, wall thickness and/or weight per foot of pipe.
c. Source and approximate shipping date from the pipe mill.
d. Method of shipment from the mill.
e. Approximate shipping date to the destination.
f. If pipe is to be stored, the approximate length of time it is to be stored.
g. Length, style and post preparation of cutback.
h. Minimum weight per car or truck required to protect lowest outbound rate.
i. Name and type of carrier.
j. Stacking and/or loading instructions.
k. Pipe manifest, preferably electronic.

DISCLAIMER: These specifications may be used in whole or in part by anyone without prejudice, if recognition of the source is included. The National Association of Pipe Coating Applicators (NAPCA) assumes no responsibility for the interpretation or use of these specifications. The intended use of the coatings identified herein is to provide corrosion protection for buried pipelines. Above ground storage of coated pipe in excess of six months without additional Ultraviolet protections is not recommended.

FIGURE 20.1 cont'd.

should be used only as a guideline and not construed as an established specification as all companies differ in specification development and related procedures for implementing and distributing specifications.

20.2.3 Fittings, Valves, and Components

The specifications for fittings, valves, and various other components that are specific to any pipeline project fall under the same guidelines as line pipe. As with pipe there are certain aspects and criteria that are relative to preparing proper definitive specifications where a vendor or supplier can provide an accurate quotation for the required material that will suit the needs and fulfill the project parameters.

When preparing the specifications for fittings, valves, and miscellaneous components, the design pressure and deliverable volumes need to be known as well as class location to compute the design factor, SMYS, temperature, rating, etc.

Below is a design specification worksheet that would be an example reflecting specific items regarding a pipeline project (Fig. 20.2).

20.2.4 Induction Bends

Induction bends are factory manufactured bends formed from straight pipe using a pipe bending machine and the electric induction heating process. Induction bends are referred to by pipeline construction field personnel as "hot bends." Pipe bends produced from this process find significant application in the pipeline industries, especially in areas such as suburbs and metropolitan areas where cold bending pipe requires much more area than is feasibly obtainable in the field.

Induction bending is simultaneously a shaping of the bend and electric heat induction process. Induction bending is an active process that requires close monitoring to ensure that dimensional properties and conveyed material meet specified standards.

For pipeline personnel, induction bending allows the installation of piggable bends directly from bare uncoated pipe. The bends should be produced from the pipe that is allocated directly for the project.

When producing specifications for the production of induction bends for a project, it is important to specify whether the induction bend will be segmentable or nonsegmentable. When a specified amount of degree is required from the field surveys, a nonsegmentable induction bend can be specified. Many companies order extra 90-degree segmentable induction bends for those locations in the field that requires segmenting or cutting the induction bend to fit a required alignment.

REQUEST FOR QUOTATION			
BOM No.			
REV. No.	1/0/1900		
Project:	xxxxxxxxxxxxxxxxxx		
Project No.:			
Item	***Total Quantity***	***Unit of Measure***	***Description***
			Fabrication Pipe
			Pipe must be manufactured by one of the following: American Steel Pipe, California Steel Industries, Iiva America, JFE/Kawasaki, JFE Nippon, Lone Star Steel, Stelpipe, Stupp, Sumitomo, Tenaris or Tex Tube. Coating must be applied by one of the following: A&A Coating, Bayou Pipe Coaters, Dura Bond Coating, L.B. Foster, LaBarge Pipe & Steel, Midwest Pipe Coating or Shaw
1	120	FT	16" Pipe, 16" o.d. x 0.375" w.t., API-5LX-60 or better, ERW, Plain ends, DRL, w/12 - 14 mils DFT, Fusion Bond Epoxy.
2	20	FT	10" Pipe, 10 3/4" o.d. x 0.365" w.t., API-5LX-42 or better, ERW, Plain ends, SRL or DRL, w/12 - 14 mils DFT, Fusion Bond Epoxy.
3	10	FT	2" Pipe, 2 3/8" o.d. x 0.218" w.t., ASTM A53 or API-5L Gr. B/X-42, ERW, Plain ends, SRL or DRL, w/12 - 14 mils DFT, Fusion Bond Epoxy.
4	20	FT	12" Pipe, 12 3/4" o.d. x 0.375" w.t., API-5LX-52 or better, ERW, Plain ends, SRL or DRL, Bare.
5	20	FT	10" Pipe, 10 3/4" o.d. x 0.500" w.t., ASTM A53 or API-5L Gr. B/X-42, ERW, Plain ends, SRL or DRL, Bare.
6	20	FT	6" Pipe, 6 5/8" o.d. x 0.432" w.t., ASTM A53 or API-5L Gr. B/X-42, ERW, Plain ends, SRL or DRL, Bare.
7	10	FT	1" Pipe, 1.315" o.d. x 0.179" w.t., Sch. 80, ASTM A106, A53 or API-5L Gr. B, SMLS, Plain ends, SRL, Bare.
			Valves
			Flanged ball valves shall be manufactured by one of the following: Cooper Cameron, WKM, Delta, Grove or PBV Threaded ball valves shall be manufactured by one of the following: Worchester or Jamesbury
8	2	EA	Ball Valve, 10", ANSI 600, RF. FLG., API Spec. 6D, Bolted Body Construction, Gear Operated, Full-Open Port Ball.
9	16	EA	Ball Valve, 1", Valve Body Rating 2500 - 3000# MWP @ 100 deg. F., 1000# Min. Differential Pressure Rating @ 100 Deg. F., Carbon Steel Body, FNPT Ends, Bolt-through Body Design, Handle Operator with locking device, Marpac E325-12-JL-L or equal.
			Forged Steel Weldneck Flanges
			Forged steel flanges must be manufactured by one of the following: Boltex, Coffer, Galperti, Hackney, Ladish or National Flange.
10	1	EA	Flange, 12", ANSI 600, WNRF, 12.000" bore, STD. WT., MSS-SP-44, F52.
11	2	EA	Flange 10", ANSI 600, WNRF, 10.020" bore, STD. WT., MSS-SP-44, F42.
12	4	EA	Flange 10", ANSI 600, WNRF, 9.750" bore, XS, ASTM A105.
13	4	EA	Flange, 6", ANSI 600, WNRF, 5.761" bore, XS, ASTM A105.
14	1	EA	Flange, 2", ANSI 600, WNRF, 1.939" bore, XS, ASTM A105.
			Forged Steel Blind Flanges
			Forged steel blind flanges must be manufactured by one of the following: Boltex, Coffer, Galperti, Hackney, Ladish or National Flange.
15	1	EA	Flange, 2" x 1", ANSI 600, Slip-on Reducing Flange, ASTM A105.

FIGURE 20.2 Design specification worksheet.

REQUEST FOR QUOTATION

BOM No.			
REV. No.	1/0/1900		
Project:			
Project No.:			

Item	*Total Quantity*	*Unit of Measure*	*Description*
			Stud Bolts (Standard Flanged Connections)
16	60	EA	Stud Bolts 1 1/4" dia. x 9 1/2" lg., ASTM A193 Gr. B7, c/w 2 Hex Nuts each, SFOQ, ASTM A194 Gr. 2H & 1 Flat Washer ASTM A325, Stud, Nuts & Washser w/IMF-3W coating, All Assembled & Tag 12", ANSI 600, R.F. Fig.
17	80	EA	Stud Bolts 1 1/4" dia. x 9 1/4" lg., ASTM A193 Gr. B7, c/w 2 Hex Nuts each, SFOQ, ASTM A194 Gr. 2H & 1 Flat Washer ASTM A325, Stud, Nuts & Washser w/IMF-3W coating, All Assembled & Tag 10", ANSI 600, R.F. Fig.
18	60	EA	Stud Bolts 1" dia. x 7 1/2" lg., ASTM A193 Gr. B7, c/w 2 Hex Nuts each, SFOQ, ASTM A194 Gr. 2H & 1 Flat Washer ASTM A325, Stud, Nuts & Washser w/IMF-3W coating, All Assembled & Tag 6", ANSI 600, R.F. Fig.
19	16	EA	Stud Bolts 5/8" dia. x 4 3/4" lg., ASTM A193 Gr. B7, c/w 2 Hex Nuts each, SFOQ, ASTM A194 Gr. 2H & 1 Flat Washer ASTM A325, Stud, Nuts & Washser w/IMF-3W coating, All Assembled & Tag 2", ANSI 600, R.F. Fig.
			Stud Bolts (Insulated Connections)
20	60	EA	Stud Bolts 1 1/4" dia. x 9 3/4" lg., ASTM A193 Gr. B7, c/w 2 Hex Nuts each, SFOQ, ASTM A194 Gr. 2H & 1 Flat Washer ASTM A325, Stud, Nuts & Washser w/IMF-3W coating, All Assembled & Tag 12", ANSI 600, R.F. Fig Insulation Flange
21	48	EA	Stud Bolts 1 1/4" dia. x 9 1/2" lg., ASTM A193 Gr. B7, c/w 2 Hex Nuts each, SFOQ, ASTM A194 Gr. 2H & 1 Flat Washer ASTM A325, Stud, Nuts & Washser w/IMF-3W coating, All Assembled & Tag 10", ANSI 600, R.F. Fig Insulation Flange
			Gaskets (No substitutions)
22	4	EA	Gasket, 12", ANSI 600, Flexitallic Style CGI, 304SS with Flexite (TM) Super Non-Asbestos Filler, Inner Ring I.D. 12 1/16".
23	12	EA	Gasket, 10", ANSI 600, Flexitallic Style CGI, 204SS with Flexite (TM) Super Non-Asbestos Filler, Inner Ring I.D. 8 1/8".
24	8	EA	Gasket, 6", ANSI 600, Flexitallic Style CGI, 304SS with Flexite (TM) Super Non-Asbestos Filler, Inner Ring I.D. 6 3/16".
25	2	EA	Gasket, 2", ANSI 600, Flexitallic Style CGI, 304SS with Flexite (TM) Super Non-Asbestos Filler, Inner Ring I.D. 2 3/16".
			Flange Insulation Kits (No substitutions)
26	3	EA	Flange Insulation Kit for 12", ANSI 600, R.F. Fig., PSI Type F, Phenolic Gasket with Nitrile Seal, Double G-10 & Steel Washer Set and Single Piece G-10 Sleeve.
27	3	EA	Flange Insulation Kit for 10", ANSI 600, R.F. Fig., PSI Type F, Phenolic Gasket with Nitrile Seal, Double G-10 & Steel Washer Set and Single Piece G-10 Sleeve.
			Butt-Weld Tee Fittings
			All butt weld fitting (tees, ells, reducers, etc.) must be manufactured by one of the following: Custom Alloy, EZEFlow, Flo-Bend, Hackney, Ladish, Mills Iron, Steel Forgings, Taylor Forge, Tube Forgings or Weldfit.
28	2	EA	Tee, 16" x 16" x 10", Reducing, 0.375" w.t. x 0.365" w.t., XS, MSS-SP-75, WPHY-60, Weld.
29	2	EA	Tee, 12", 0.375", STD. WT., MSS-SP-75, WPHY-52 or better, Weld.

FIGURE 20.2 cont'd.

REQUEST FOR QUOTATION			
BOM No.			
REV. No.	1/0/1900		
Project:			
Project No.:			
Item	***Total Quantity***	***Unit of Measure***	***Description***
			Butt-Weld Ell Fittings (1.5 D - Long Radius)
30	3	EA	Ell, 16", 90 deg., L.R., 0.375" w.t., STD. WT., MSS-SP-75, WPHY-60, Weld.
31	2	EA	Ell, 10", 90 deg., L.R., 0.365" w.t., STD. WT., MSS-SP-75, WPHY-42 or better, Weld.
			Butt-Weld Reducers
32	1	EA	Reducer, 16" x 12", Conc., 0.375" w.t. x 0.375" w.t., STD. WT., MSS-SP-75, WPHY-60 or better, Weld.
33	2	EA	Reducer, 10" x 6", Conc., 0.500" w.t. x 0.432" w.t., XS, ASTM A234, WPB, Weld.
			Caps
34	1	EA	Cap, 16" 0.375" w.t., XS, MSS-SP-75, WPHY-60, Weld.
35	2	EA	Cap, 12", 0.375" w.t., STD. WT., MSS-SP-75, WPHY-52, Weld.
			Forged Fittings (Socket Weld)
36	1	EA	Ell, 1", 90 deg., 3000#, ASTM A105, Socket Weld.
			Forged Fittings (Olets)
37	1	EA	Weldolet, 2" on 16", Std. WT. Header x 2" Extra Strong, ASTM A105.
38	15	EA	Thredolet, 1" on 36" - 2", 3000#, ASTM A105.
			Forged Fittings (Misc.)
39	16	EA	Plug, 1", Pipe, Hex Head, 3000#, ASTM A105, Threaded.
			Pipe Nipples
40	15	EA	Pipe Nipple, 1" dia. x 3" lg., ASTM A106 Gr. B, SMLS, Sch. 80, TBE.
41	1	EA	Pipe Nipple, 1" dia. x 6" lg., ASTM A106 Gr. B, SMLS, Sch. 80, POE x TOE.

FIGURE 20.2 cont'd.

20.3 CONSTRUCTION SPECIFICATIONS

The general outline in developing specifications was covered earlier in this chapter where clarity and definitiveness were emphasized. A large pipeline project construction specification can contain many chapters and relevant pages that should encompass all aspects of construction throughout a project. The construction specification representation contained in this section deals with a plant or facility that was small in comparison to a cross-country pipeline project. Nonetheless, the comparative material contained within the specifications identifies the relevant necessary work while identifying procedures and reporting protocols that remain clear and decisive.

As before, the following specification (Fig. 20.3) is for a guideline only as all companies have certain standards and acceptable procedures in generating company specific specifications.

	CONSTRUCTION SPECIFICATION	

SPECIFICATION FOR STANDARD CONSTRUCTION (PLANT/FACILITY)

THIS PAGE IS A RECORD OF ALL REVISIONS TO THE SPECIFICATION. EACH TIME THE SCOPE OF WORK IS CHANGED, ONLY THE NEW OR REVISED PAGES ARE ISSUED.

FOR CONVENIENCE, THE NATURE OF THE REVISION IS BRIEFLY NOTED UNDER REMARKS, BUT THESE REMARKS ARE NOT A PART OF THE SCOPE OF WORK. THE REVISED PAGES ARE A PART OF THE SCOPE OF WORK AND SHALL BE COMPLIED WITHIN THEIR ENTIRETY. COVER SHEET AND SCOPE OF WORK PAGES ISSUED PREVIOUSLY ARE TO BE DESTROYED AS REVISED COPIES ARE ISSUED.

REV.	DATE	BY	APPROVED BY (INITIALS)			PAGES	REMARKS

	Revision Date:	Spec. Number: 10-XXXXXX-CONST-XXX	Page 1 of 22

FIGURE 20.3 Construction specifications.

	CONSTRUCTION SPECIFICATION	

TABLE OF CONTENTS

SPECIFICATION FOR STANDARD CONSTRUCTION (PLANT/FACILITY)

	Revision Date:	Spec. Number: 10-XXXXXX-CONST-XXX	Page 2 of 22

FIGURE 20.3 cont'd.

	CONSTRUCTION SPECIFICATION	

1.0 DEFINITIONS

1.1 COMPANY: Entity responsible for the contract award and owner of asset.

1.2 CONTRACTOR: Successful Bidder, general contractor responsible for the installation scope of work.

2.0 SITE PREPARATION AND EXCAVATION

2.1 General

2.1.1 Company will have all of COMPANY'S pipe and equipment on the construction site located for the CONTRACTOR by means of visible markers. Use of equipment and/or making excavations around said pipe and equipment shall be performed only in accordance with and in the presence of the COMPANY Representative. Extreme caution shall be exercised by CONTRACTOR while excavating near a high pressure pipe and equipment and may not be performed without prior approval by COMPANY Representative.

2.1.2 All benchmarks of the construction site shall be visibly marked by a surveyor provided by COMPANY unless otherwise specified in the Contract Documents. CONTRACTOR shall be responsible for establishing all necessary bench marks used during the construction period. Such bench marks shall be referenced on the construction drawings.

2.1.3 Before opening any excavation, One Call and utility companies shall be contacted to determine if there are underground utility installations in the area. Utility lines or cables shall be located and flagged prior to excavation operations. CONTRACTOR will provide spotters to hand excavate as needed to locate lines or cables and maintain visual contact with these lines or cables until excavation is completed. CONTRACTOR will provide temporary supporting for lines or cables as needed.

2.2 Clearing of Site and Grading

2.2.1 The construction site shall be cleared of all obstructions as shown on the detail drawings and as required by COMPANY Representative and all debris shall be disposed of as directed by the COMPANY Representative.

2.2.2 The CONTRACTOR shall, as directed by COMPANY Representative, provide adequate temporary or permanent protection to all significant above or below ground items that may be damaged or destroyed during construction.

2.2.3 Prior to grading and leveling, the topsoil shall be removed and stored in an area as directed by the COMPANY Representative. Said topsoil may then be used as a final ground cover after the leveling has been completed. Soil to be used as fill shall be free of rocks and other debris, and shall be approved by COMPANY Representative before being used. Soil which is removed shall be disposed of as directed by COMPANY Representative.

2.3 Compaction

2.3.1 All fill material for subgrades and base materials shall be thoroughly compacted in approximately six to eight-inch layers to the density

	Revision Date:	Spec. Number: 10-XXXXXX-CONST-XXX	Page 3 of 22

FIGURE 20.3 cont'd.

	CONSTRUCTION SPECIFICATION	

specified in the Scope of Work or as required by COMPANY Representative. Each layer shall be wetted, if required by COMPANY Representative, and thoroughly compacted until such fill is completed to the desired elevation. CONTRACTOR shall supply all water required for compaction and ground wetting. The COMPANY Representative shall have the right to reject any fill material considered unsuitable for proper compaction or base material.

2.3.2 Clods or lumps of material should be broken and mixed by blading or similar methods so that a mixture of uniform density is attained in each layer.

2.3.3 Wetting of the construction area to prevent blowing and shifting soil shall be performed by CONTRACTOR as directed by the COMPANY Representative. CONTRACTOR shall comply with all environmental codes for fugitive dust emissions.

2.4 Roadways and Parking Areas

2.4.1 The location of roads and parking areas will be designated on the drawing in the Scope of Work or determined by the COMPANY Representative. Generally, the subgrade and base course will consist of natural ground. In areas where fill material is not required, natural ground shall be cleaned, grubbed and leveled to the elevation and grade required. Whenever subgrade fill material is required, it shall be placed and compacted in accordance with Section 2.3.

2.4.2 The base course for roadways and parking areas shall consist of either natural ground or a minimum of six inches (6") of crushed stone or coarse gravel. The crushed stone or gravel shall consist of material which is predominant in the local area, and shall be free of soil, clay and other objectionable debris. The COMPANY Representative shall have the right to reject any base course material which is unsuitable. Natural ground used as the base course shall be cleaned of all brush and debris by blading and shaped to the elevation and grade required. CONTRACTOR shall maintain and repair base course or natural ground roadway until the construction is complete and accepted by the COMPANY Representative.

2.4.3 The finishing of roadways and parking areas shall be performed as specified in Section 16.0. Generally, roadways and parking areas which shall not bear heavy traffic may be finished at any time designated by the COMPANY Representative. Those areas of heavy traffic shall not be finished until the construction of facilities has been completed.

2.4.4 Culverts shall be installed as required for proper drainage of water from the surrounding terrain. The CONTRACTOR shall install the culverts as shown on COMPANY drawings or as directed by the COMPANY Representative. Culverts shall be installed at the intersection of all COMPANY roads and/or access roads and public roads. Such installations shall be in accordance with the appropriate state, county, or local specifications.

2.5 Fencing

2.5.1 The CONTRACTOR shall install fencing around the installation as listed in BID FORM or the COMPANY drawings. All fencing shall be installed inside the defined property lines. Unless the installation of fencing is detrimental to

	Revision Date:	Spec. Number: 10-XXXXXX-CONST-XXX	Page 4 of 22

FIGURE 20.3 cont'd.

	CONSTRUCTION SPECIFICATION	

the construction of facilities, the fencing shall be installed after grading or leveling and prior to the construction of the various facilities.

2.5.2 All fenced areas shall have a minimum of two gates or accesses. At least one of these gates shall be a minimum 12 ft. wide drive gate. Each gate shall have mounted on it secure locking hardware acceptable to COMPANY Representative.

2.6 Excavations

2.6.1 All excavations shall be performed in a neat and workmanlike manner and in accordance with established good construction practices, all local construction codes and in accordance with the COMPANY Representative. Excavations shall be properly barricaded at all times to prevent personal injury. All excavations shall be in compliance with OSHA 1926 Subpart P. CONTRACTOR employees shall not be allowed to enter an excavation until it has been inspected and found safe by a competent person at the beginning of each shift. The competent person shall be a CONTRACTOR'S Representative as defined in OSHA 1926.650 to 652. Additional inspections shall be made during each shift as required by conditions. Sloping, shoring or trench shields, if utilized, shall meet the requirements of OSHA 1926.652. If inspections reveal unsafe conditions, all work shall cease and personnel shall be removed from excavation until it meets compliance and is approved by the competent person.

2.6.2 Should excavation expose rock before the required elevation is reached, the COMPANY Representative will have the option to accept the top of rock elevation or to require CONTRACTOR to remove the rock to the required elevation. Rock excavation shall be paid for at the unit price specified in the Bid Form or, as agreed upon by COMPANY Representative and CONTRACTOR in writing prior to the start of excavation.

2.6.3 The finished ditch in excavated rock shall be deep enough so that the minimum cover required by the COMPANY Representative is obtained over the installed piping. The finished ditch excavated in solid rock shall have a six-inch (6") earth or sand padding for the pipe to rest on and shall be backfilled with the same material to six (6) inches above the top and along both sides of the pipe.

2.7 Footings and Foundations

2.7.1 Footings and foundations shall be set on undisturbed firm soil, in accordance with drawings and shall <u>not</u> be placed on fill soil. If acceptable to the COMPANY Representative, the CONTRACTOR may use the soil as the form for concrete rather than a fabricated form for concrete in excess of 6" below finish grade. The soil must be firm and undisturbed before soil forming may be used. All concrete above grade and extending 6" below grade will be formed in accordance with Section 4.1.

2.8 Backfill

2.8.1 The fill shall be free from organic and other compressible material. The COMPANY Representative shall have the right to reject any fill material considered unsuitable. All backfill or other fill located under floors, walks and/or driveways, shall be layered, with wetting and tamping of each layer

	Revision Date:	Spec. Number: 10-XXXXXX-CONST-XXX	Page 5 of 22

FIGURE 20.3 cont'd.

	CONSTRUCTION SPECIFICATION	

until such elevation as required is reached in accordance with Section 2.3. Sand or approved equal fill under slabs shall be tamped and leveled to the satisfaction of the COMPANY Representative and no slabs shall be poured until authorized by COMPANY Representative. Fill containing frozen material shall not be used. Pipe trenches shall be backfilled with soil compacted in layers to an elevation slightly higher than ground level and then repacked or tamped. If further settlement of the backfill occurs, said procedure shall be repeated as necessary and as directed by the COMPANY Representative. Excess soil from excavations which is not required for grading or backfilling shall be stockpiled or disposed of as directed by the COMPANY Representative.

3.0 DELIVERY, HAULING AND STORAGE OF MATERIAL

3.1 Material Handling and Storage

3.1.1 COMPANY will have component parts and miscellaneous material required for construction available at the UnionTown warehouse, as designated in the Scope of Work of the Bid Documents. CONTRACTOR will provide adequate equipment and personnel to transport, unload and store all materials for the work and to load and haul material from storage sites to construction site. CONTRACTOR shall be responsible for selection of the unloading and storage site.

3.1.2 CONTRACTOR must visually inspect all equipment, instruments, joints of pipe, elbows, fittings, valves, flanges, bolts and nuts and other COMPANY-supplied material for any apparent damage which may impair the serviceability of the item. CONTRACTOR shall set aside any item with apparent damage and advise the COMPANY Representative immediately. CONTRACTOR remains responsible for condition and security of the item until COMPANY provides for removal of the item. Contractor shall be responsible for safe unloading and proper storage of materials to avoid damage, loss and theft.

4.0 CONCRETE WORK

4.1 General Requirements

4.1.1 Forms and shoring required for concrete work shall be supplied, installed and removed by the CONTRACTOR. Forming shall be true to line, rigid and sufficiently braced to hold the established dimensions as indicated on the drawings, during and after placing concrete.

4.1.2 All reinforcing steel and/or wire mesh not supplied by CONTRACTOR shall be as specified on drawings by the COMPANY Representative. Reinforcing steel shall not be installed with any portion of or supports for the steel imbedded in contact with the natural earth. All splices of reinforcing steel will have a minimum of twenty-four (24) inches overlap between adjacent bars. Splices should be staggered whenever possible to maintain maximum strength of reinforced concrete. All formed rebar shall be cold bent. Flame cuts are not allowed. All rebar unless otherwise specified or approved by COMPANY Representative shall conform to ASTM A-615/A615A including supplementary requirement S1 and shall be Grade 60. Welded wire fabric shall be plain wire fabric and in accordance with ASTM A185/A185A.

	Revision Date:	Spec. Number: 10-XXXXXX-CONST-XXX	Page 6 of 22

FIGURE 20.3 cont'd.

	CONSTRUCTION SPECIFICATION	

4.1.3 All concrete, whether job mixed or ready mix, shall be mixed in the proper proportions as specified on drawings and/or by the COMPANY Representative. Aggregate type and concrete mix shall be approved by the COMPANY Representative, and unless otherwise specified by drawings, all concrete shall be at least a 3000 psi design mix. Ready mixed concrete shall be mixed and transported in accordance with "Specifications for Ready Mixed Concrete", ASTM C-94.

4.1.4 Concrete shall be vibrated as required by COMPANY Representative to insure that the concrete fills all sections of forms and that subsequent batch pourings form a homogeneous mixture.

4.1.5 Concrete shall be allowed a curing time of at least three (3) days before removing forms. While curing, concrete shall be protected as necessary from adverse weather conditions. CONTRACTOR will only remove concrete forms after receiving approval from COMPANY Representative. COMPANY reserves the right to require CONTRACTOR to remove and replace any concrete, at CONTRACTOR'S expense, the COMPANY Representative determines is unacceptable.

4.1.6 Concrete structures, such as building foundations which will be receiving heavy loading shall be allowed a minimum of ten (10) days curing time before being loaded. Actual cure time shall be determined by the COMPANY Representative based upon concrete test results.

4.1.7 Concrete blocks requiring epoxy grout installation for mounting hardware shall have at least 12 days cure time before the grout is installed.

4.1.8 The CONTRACTOR shall install building foundations as specified on drawings and as directed by COMPANY Representative. The building CONTRACTOR shall furnish to COMPANY all necessary foundation anchor bolts as specified on drawings prior to the start of construction unless otherwise specified by the Companying the Scope of Work.

5.0 DITCHING

5.1 General Requirements

5.1.1 If ditching is required for the project, CONTRACTOR shall be responsible for making the "one call" notification to respective state hotline number prior to start of work. CONTRACTOR shall be responsible for locating and protecting all underground facilities and lines, and shall immediately notify COMPANY Representative and officials having jurisdiction in the event such facility or line is damaged. CONTRACTOR shall repair or have repaired all damage to an underground facility or line at its sole expense.

5.1.2 Excavations exceeding four feet (4') in depth will require sloping, benching (stair-stepping) or shoring in accordance with OSHA 1926.652. Trenches four feet (4') deep or more shall have an adequate means of exit such as ladders, steps or ramps located so as to require no more than 25 feet of lateral travel. In excavations which personnel are required to enter, excavated or other material will be stored and retained at least two feet (2') from the edge of the excavation. Open ditches and trenches within operating plants or facilities shall be barricaded in accordance with OSHA 1926.202 to protect personnel and traffic from excavations.

	Revision Date:	Spec. Number: 10-XXXXXX-CONST-XXX	Page 7 of 22

FIGURE 20.3 cont'd.

	CONSTRUCTION SPECIFICATION	

5.1.3 The ditch shall be excavated to a width not less than six (6) inches greater than the nominal outside diameter of the pipe to be placed therein.

5.1.4 The ditch shall be cut to a uniform grade and to a depth which will provide a minimum cover of thirty-six (36) inches measured from the top of the pipe to the average elevation of the original ground on the two sides of the ditch unless consolidated rock is encountered.

5.1.5 (NOT REQUIRED FOR THIS BID) Where other pipelines, drainpipes, telephone conduits or other foreign lines are to be crossed, the new pipeline is to be installed under the existing line unless the CONTRACTOR is specifically instructed otherwise by COMPANY. Lines to be undercrossed within the limits of the pipe ditch shall be completely exposed by hand excavation only. The ditch shall be excavated deep enough to provide a minimum vertical clearance of twelve (12) inches between the pipeline to the installed and the existing line crossed.

5.1.6 The CONTRACTOR shall provide and maintain adequate pumps and other equipment for the removal and disposal of surface and ground water entering the trench. The trench shall be kept relatively dry until the pipeline is complete to eliminate damage from floatation or otherwise.

5.1.7 All ditches and trenches inside station fences shall be compacted during backfill operations. Compaction types shall range from mechanical compaction at 6" lift intervals while maintaining optimum moisture content for soils, to water packing soils during backfill operation. Type of compaction will be designated by the COMPANY Representative based on variables including expected vehicle or equipment traffic, personnel access, etc.

6.0 GROUTING

6.1 Concrete surfaces to be grouted shall be roughened by chipping to remove all surface contaminants and expose aggregates. Surfaces prepared for epoxy grout or dry pack cement grout must remain dry and clean. Surfaces prepared for conventional cement grouting are to be thoroughly cleaned and wetted, with no free standing water, before placing grout.

6.2 All areas not be grouted, such as pipe supports shall be sealed to prevent grout leakage. Other items not to be grouted, such as anchor bolts and leveling screws, shall be covered with thread lubricant, glazing putty, modeling clay, or other material as required by COMPANY Representative to keep the grout from adhering. All jack pockets shall be filled with cement grout after removing jacks and before any other grouting takes place. When grouting with epoxy grout, all forming material shall be thoroughly waxed or covered with polyethylene film before final assembly to facilitate removal after curing. Small openings in the grout forms shall be sealed with cement grout, putty, or other suitable material to prevent leakage of grout.

6.3 All equipment surfaces to be grouted shall be cleaned of all paint, oil, rust, scale, and dirt with acceptable cleaning material. Surfaces not to be grouted and exposed surfaces adjoining grouting area shall be protected from splashing grout with suitable covering.

6.4 Grouting material, especially epoxy grout, shall be installed in the shade and within the temperature limits specified by the manufacturer unless specifically directed

FIGURE 20.3 cont'd.

	CONSTRUCTION SPECIFICATION	

otherwise by COMPANY Representative. Direct sunlight shall be avoided during the pouring and curing of the grout. The CONTRACTOR shall, if necessary, erect a sunshade as directed by the COMPANY Representative.

6.5 Expansion joints shall be considered for pours with dimensions exceeding seven (7) feet. Joints will be designed and installed in accordance with grout manufacturers recommendations.

6.6 Epoxy grout shall be poured immediately after the CONTRACTOR mixes the component liquids. While pouring grouting materials, CONTRACTOR shall check for leaks frequently along forms and small openings in the forms. Leaks will not self-seal, and if not stopped, could cause voids under equipment. Epoxy grout material shall be non-shrinking type and shall be mixed and installed as directed by the grout vendor's specifications and as directed by the COMPANY Representative. All excess mixed grouting material will be disposed of as directed by COMPANY Representative.

6.7 CONTRACTOR shall remove grout forms after receiving approval from COMPANY Representative. COMPANY reserves the right to require CONTRACTOR to remove and replace, at CONTRACTOR'S expense, any grout that COMPANY Representative determines is unacceptable.

7.0 STRUCTURAL STEEL AND BUILDING

7.1 Structural Steel

7.1.1 All pipe clamps, guides, anchor bolts, and trench angles shall be fabricated as shown and specified on COMPANY drawing or as otherwise specified by the COMPANY for any particular application. The CONTRACTOR shall furnish labor, materials, and equipment to fabricate such items if such items are not furnished prefabricated by the COMPANY as noted in the Scope of Work.

7.1.2 The CONTRACTOR shall fabricate any structural steel support, bracket, platform, stairs or other related items not supplied by COMPANY as shown on drawings or as requested by the COMPANY Representative. All COMPANY provided prefabricated structural steel shall be erected and/or installed by the CONTRACTOR as provided for in the Scope of Work.

7.2 Buildings

7.2.1 The Building CONTRACTOR shall provide all labor, materials and equipment necessary to completely fabricate the building steel and install all hoists, siding, louvers, doors, sashes, glazing, hardware, and insulation at their fabrication facility unless otherwise arranged as detailed in the Scope of Work or as shown on drawings.

7.2.2 The Building CONTRACTOR shall secure written approval from COMPANY before subcontracting any portion of the work or material.

7.2.3 All buildings and related material shall be designed and fabricated by a competent and authorized Building CONTRACTOR to comply with COMPANY specifications. All building Contractors shall forward two copies of all fabrication, assembly detail drawings, and a complete material list to COMPANY prior to installation.

	Revision Date:	Spec. Number: 10-XXXXXX-CONST-XXX	Page 9 of 22

FIGURE 20.3 cont'd.

	CONSTRUCTION SPECIFICATION	

7.2.4 (NOT REQUIRED FOR THIS BID) Where building insulation is required, the insulation will be contained by chicken wire or other suitable material to prevent sagging.

7.2.5 The Building CONTRACTOR shall be responsible for the complete fabrication of all phases of the building. COMPANY reserves the right to inspect the building at any time and to require building vendor to correct any design or assembly deficiencies at vendor's expense. Building CONTRACTOR or erector must provide fall protection equipment for all workers in accordance with OSHA 1926 Subpart E.

7.2.6 In cases where piping projections through building walls are needed, the Building CONTRACTOR shall cut and install the insulation and/or siding to fit and shall furnish and install flashing around said projections as necessary, unless otherwise specified the COMPANY Representative.

7.2.7 In cases where building erection precedes installation of piping and/or other projections through the building wall, CONTRACTOR shall cut smooth openings where such projections extend through siding and insulation and furnish and install flashing around openings as required by COMPANY Representative. Whenever such cuts are made, adequate bracing of siding material on large openings shall be provided by the CONTRACTOR responsible for the work as required by COMPANY Representative.

8.0 ELECTRICAL WORK

8.1 General

8.1.1 The work shall be performed by skilled craftsmen under competent supervision. All equipment or material furnished by CONTRACTOR or approved subcontractor shall be COMPANY approved prior to installation.

8.1.2 The prescribed work shall be designed, installed and tested in accordance with the appropriate requirements of the latest editions of the following standards.

8.1.2.1 NEMA – National Electrical Manufacturers Association

8.1.2.2 NEC – National Electric Code

8.1.2.3 ANSI – American National Standards Institute

8.1.2.4 IEEE – Institute of Electrical & Electronic Engineers

8.1.3 The material used and the work performed shall meet all requirements of local, state and national authorities having jurisdiction over electrical equipment.

8.1.4 It is the intent of these specifications to outline minimum requirements. In instances where capacities, sizes, etc., of electrical equipment, devices or materials are in excess of these minimum requirements, such designated capacities shall prevail.

8.2 Conduit and Fittings

8.2.1 All rigid conduit shall be hot dipped galvanized steel. All rigid conduit installed underground shall be PVC coated rigid steel. No conduit shall be buried until approved by COMPANY Representative. COMPANY reserves the right to require CONTRACTOR to expose all conduit not approved before

	Revision Date:	Spec. Number: 10-XXXXXX-CONST-XXX	Page 10 of 22

FIGURE 20.3 cont'd.

	CONSTRUCTION SPECIFICATION	

burying, at CONTRACTOR'S expense. Underground conduit shall have 24" minimum cover with exceptions allowed by NEC.

8.2.2 Rigid conduit runs shall be installed as shown on the construction drawings and shall be installed in a workmanlike manner complying with accepted construction practices. All bends in conduit shall be made with an acceptable hand or machine bender. All conduit after bending shall have a true round with full inside cross sectional area maintained along the full length of the bend.

8.2.3 All conduit cuts shall be made square with the longitudinal axis of the conduit and all burrs or sharp edges removed from the inside by file or reamer. Field made threads on conduit shall be full and continuous and the thread die shall be completely engaged on the conduit along the full length before being removed.

8.2.4 Conduit couplings shall be made watertight and a minimum of five (5) full threads shall be fully engaged. An approved pipe thread lubricant shall be applied on the male threads only prior to assembly.

8.2.5 Conduit fittings, pull boxes, junction boxes and other similar devices shall be selected and installed so that openings in such devices will allow full and easy accessibility for replacing wiring, etc., without disassembly of rigid conduit. Conduit shall not be used to support any type of junction box or fixture.

8.2.6 All conduit runs shall be installed complete prior to pulling in wires. All conduit opening shall be plugged after installation until wire is pulled. Immediately prior to pulling in wires, CONTRACTOR will run a swab through the conduit run to remove debris and moisture. COMPANY Representative may require additional swabbing of the conduit if deemed necessary. When required by COMPANY Representative, all seal fittings shall be carefully sealed with Chico X Fiber and an adequate amount of sealing compound.

8.2.7 Rigid conduit shall be supported on not more than 8-foot centers. Additional conduit supports shall be installed as required by COMPANY Representative to support critical points in the conduit runs, such as bends, fittings, etc.

8.2.8 Minimum acceptable conduit size is 3/4 inch.

8.3 Wiring

8.3.1 Power and control wiring for 600 volts or less service shall be copper. The wire insulation used shall be suitable for 90° C dry and 75° C wet locations and moisture and chemical resistant. COMPANY Representative has the right to reject any wire that is not deemed suitable for use in the anticipated operating conditions or not complying with local or national codes.

8.3.2 Wire splicing in conduit shall not be allowed. Splices shall be made mechanically tight by use of pressure connectors or soldered joints. Proper soldering procedures must be utilized with soldered joints in order to eliminate cold soldered joints and acid core solder shall not be used. All soldered splices will be wrapped with plastic, UL approved electrical tape that exceeds the minimum dielectric strength and environmental requirements of the insulation on the spliced wire.

	Revision Date:	Spec. Number: 10-XXXXXX-CONST-XXX	Page 11 of 22

FIGURE 20.3 cont'd.

	CONSTRUCTION SPECIFICATION	

8.3.3 When pulling wire, CONTRACTOR shall feed the wire into the conduit so that damage to the cable insulation is avoided. Wire shall be fed into conduit in a neat group to prevent twisting or binding. An adequate application of wire lubricant shall be used to aid in pulling.

8.3.4 (NOT REQUIRED FOR THIS BID) Wire to be used for direct burial in the earth shall be designed for that use and approved by COMPANY Representative. When cable is laid in a trench, the trench need only be wide enough for convenient handling of the cables, and shall be deep enough to provide at least 24" of cover on the installed cables. No splices shall be installed underground unless specifically approved by COMPANY Representative. If approved, splices shall either be brought above ground into a pull box, if practical, or if made underground, the splice shall be coated with Scotchcast splicing kit or equal.

8.3.5 Conduit laid in a trench shall not be placed on or covered with rocks or material which may damage the conduit. Conduit emerging from the ground shall be rigid metal conduit from 24" below grade to a conduit or junction box at least six (6) inches above grade. Buried conduit may not be installed in the same ditch with a natural gas pipeline.

8.3.6 Grounding networks shall be installed in accordance with the requirements of the National Electrical Code. Ground wire shall be installed in conduit at grade and connected to a ground rod driven into moist earth. All ground circuits must be checked for continuity after installation and the circuit resistivity may not exceed one (1) ohm. CONTRACTOR will search and repair all deficiencies that cause a higher resistance reading at its expense. The electrical grounding system shall be tested using the 3-point test to assure that the resistance to ground is less than five ohms.

8.3.7 Circuit breakers and separable or adjustable trip units are to be tested to ensure they operate at the designed current trip points.

8.3.8 Low voltage power cable shall be megohmmeter tested.

9.0 INSTRUMENT AND CONTROL INSTALLATION

9.1 Instruments

9.1.1 All instruments will be mounted in a safe professional manner, free from vibration and adequately protected from mechanical, environmental, and physical damage.

9.1.2 All installed instruments shall conform to the latest Instrument Society of American standards.

9.1.3 All devices shall be marked or tagged with calibrations setting and date after completion of calibration and remarked when any changes have been made to calibration settings.

9.2 Tubing

All high pressure exposed tubing shall be 304 or 316 stainless steel tubing and shall meet ASTM-A-269 specifications. Bending of tubing shall be made by a tubing bender. Any tubing containing bends with flat spots or kinks shall be discarded. All tubing shall be cut with a tubing cutter and tubing ends will be reamed and deburred to the original ID.

	Revision Date:	Spec. Number: 10-XXXXXX-CONST-XXX	Page 12 of 22

FIGURE 20.3 cont'd.

	CONSTRUCTION SPECIFICATION	

Rigid tubing will be adequately strapped to prevent vibration and physical damage.

(NOT REQUIRED FOR THIS BID) Flame retardant polyethylene tubing may be used for low pressure tubing (50 psi or less) and must be approved by COMPANY Representative prior to installation. Polyethylene tubing shall be rated for a minimum burst pressure of 600 psi at room temperature and R-10 hardness. Tubing will be physically protected from excessive heat and physical damage by use of conduit, channeling, etc. as required by COMPANY Representative. Under no circumstances will polyethylene tubing be permitted to go through an opening in an enclosure or metallic material without a bulkhead union installed in the opening.

All tubing fittings shall be compression-type fittings and shall be installed as per manufacturers specifications. Tubing fitting types and material shall be approved by COMPANY Representative before being used. Tube fitting and ferrule material will match the tubing material used unless otherwise approved.

9.3 Pressure taps

9.3.1 All pressure taps shall be designed for process line pressure and must be able to isolate the instrument line. COMPANY Representative must be present or notified at least 24 hours prior to pressure tapping into operating lines. Only pressure taps and procedures approved by COMPANY Representative shall be used.

9.4 Testing for leaks

9.4.1 Prior to operation of instruments, controls, and shutdown devices, all connections shall be tested for leaks. All shutdown and control devices shall be tested for accurate operation by external supplied pressure, temperature, or other signals, prior to operation. Perform continuity check on all circuit loops prior to termination.

9.5 Instrument Calibration

9.5.1 Unless received sealed and certified by manufacturer, each transmitter, switch, receiver, recorder, controller and other device shall be calibrated throughout its range, complete with calibration sheets, to assure that it operates in correct proportion to its input signal. Perform functional check of all instruments.

9.6 Control Loop Testing

9.6.1 Test each control loop to assure that the entire loop operates as it was designed. This shall include stroking of all valves and operators to assure that they have full movement and operate in the correct direction when signaled by the controlling device.

10.0 WELDING SPECIFICATION

10.1 Work Required

10.1.1 The work required under this specification includes the furnishing of all supervision, labor, equipment, services and welding materials necessary for the joining of pipe, valves, fittings and other appurtenances by welding in a manner complying to COMPANY welding specifications. COMPANY requirements for welding will take precedence over these general welding specifications.

	Revision Date:	Spec. Number: 10-XXXXXX-CONST-XXX	Page 13 of 22

FIGURE 20.3 cont'd.

	CONSTRUCTION SPECIFICATION	

10.2 Welding – General

10.2.1 The facilities of COMPANY are designed and constructed in accordance with ASME B31.8. and all references to B31.3 have been changed to B31.8

10.2.2 Welding procedure, welding, welder qualification, destructive testing and non-destructive testing of welding shall be performed in accordance with ASME B31.8 which is incorporated herein by reference.

10.3 Materials

10.3.1 This specification shall apply to field welding of pipe manufactured conforming to ASTM Specifications A-53, A-83, A-106 or A-120; American Petroleum Institute Standards (API) 5L, Line Pipe, Grades A, B, X-42, X-46, X-52, X-60 and X-65; and shall also apply to field welding of pipe not manufactured in conformance to those specifications, provided the physical and chemical properties comply with such specifications. Unless otherwise stated by specific plans, pipe will be supplied in double random lengths. The average length, percentage of short pipe and the minimum length shall be as provided by the specification governing the manufacturer of the pipe.

10.4 Welding Procedure and Weld Qualification

10.4.1 The welding procedure for welding of steel pipeline facilities, including pipe, fittings, valves and other appurtenances requiring the joining of welding, shall be in accordance with the COMPANY weld procedures.

10.4.1.1 Type of Line-up Clamps - Line-up clamps shall be used for pipe welding and shall be used for fitting and tie-in welding, where practical. Either external or internal clamps shall be used for pipe diameter less than 16 inches. Internal clamps shall be used for pipe having a 16-inch O.D. or greater.

10.4.1.2 Removal of Line-up Clamps – For grade X-56 or less material and 8 5/8" or larger diameter pipe, line-up clamps shall be held in place until 75% of the root bead has been deposited. For pipe that is 6 5/8" and smaller in diameter, line-up clamps shall be held in place until the root pass is tacked enough so that no cracks can occur with normal pipe handling as judged by the COMPANY Representative.

10.4.1.3 Cleaning – All beads shall be thoroughly cleaned and all slag and scale removed after the weld material has been deposited. Cleaning may be accomplished with either hand tools or power tools. COMPANY reserves the right to require power brushing for cleaning of all beads and power grinding of the stringer bead and hot pass in order to eliminate slag or other injurious materials and to provide a surface contour which will facilitate the deposit of the next weld bead.

10.5 Welding and Weld Acceptability

10.5.1 Any weld shall, upon the request of COMPANY, be removed from the line and tested in accordance with ASME B31.8. Welds for routine testing shall, whenever possible, be selected at a time and location which will least interfere with efficient and orderly construction operations. Any weld so removed shall have coupons cut there from and tested in the manner prescribed in Section

	Revision Date:	Spec. Number: 10-XXXXXX-CONST-XXX	Page 14 of 22

FIGURE 20.3 cont'd.

	CONSTRUCTION SPECIFICATION	

10.085. The welder or welders may be disqualified, by COMPANY Representative, from further work if the weld fails to comply with the specified requirements. The number of welds selected for destructive testing shall not exceed three (3) percent of the average daily number of production welds or one production weld each day, whichever is the greater. CONTRACTOR will be reimbursed for the cost of removing the test weld and rejoining the pipe sections only in the event the destructive testing proves that the weld is acceptable.

10.5.2 CONTRACTOR will provide radiographic inspection of 100% of the finished welds. The CONTRACTOR shall arrange the work in a manner which will permit the maximum usage of weld inspection and assist in moving radiographic inspection equipment at no additional cost to COMPANY. The minimum number and the location of welds selected by COMPANY which are to be radiographically inspected shall be in accordance with the rules set forth in the ASME B31.8, and COMPANY requirements. Generally, welds on a pipeline to be operated at 20% or more of SMYS must be non-destructive tested. If the pipe is less than 6" OD or will operate under 40% of SMYS and the number of welds are so limited in number that NDT is impractical, then visual examination by a qualified welding inspector according to Section 10.082 above, is acceptable.

10.5.3 COMPANY shall apply the procedures and standards of acceptability set forth in ASME B31.8 for radiographic testing. Interpretation of the radiographic results will be made by COMPANY or by a third party acting for COMPANY.

10.5.4 COMPANY may require CONTRACTOR to soap test all pipeline welds before being coated. CONTRACTOR will provide labor, equipment and materials required to pressurize the pipeline section with air to 100 psi and soap test each weld. Any weld which indicates a leak or is defective will be marked and repaired or replaced, as required by COMPANY, at the CONTRACTOR'S expense prior to start of the final test.

10.5.5 CONTRACTOR shall furnish shielding material when epoxy coated pipe is being welded. The shielding material will cover the coating for at least six (6) inches on either side of the welding area. The shield material is to remain in place until the final weld cap is installed so as to protect the coating from weld splatter.

10.5.6 CONTRACTOR may not install by welding any miter joint greater than 3 degrees on steel pipe to be operated at 30% or more of SMYS. CONTRACTOR must bend pipe or install factory formed elbows to alter the direction of the pipe. On pipe operated at 10% of SMYS but less than 30% of SMYS, a miter joint can not deflect the pipe more than 12½ degrees and must be at least one pipe diameter from any other miter joint weld. On pipe to be operated at 10% of SMYS or less, a miter joint can not deflect the pipe more than 90 degrees.

10.6 Weld Repair and Repair of Steel Pipe

10.6.1 Should lamination, split ends or other defects in the pipe be discovered, the joint of pipe containing such defects shall be cropped, repaired, or removed from the line as directed by the COMPANY.

	Revision Date:	Spec. Number: 10-XXXXXX-CONST-XXX	Page 15 of 22

FIGURE 20.3 cont'd.

	CONSTRUCTION SPECIFICATION	

10.6.2 COMPANY may authorize repair of defects, except cracks, in the root, hot pass and filler beads. Any weld having a crack in the root, hot pass or filler bead shall be removed. Any weld that shows evidence of repair work having been done without authorization by COMPANY may be rejected. Repairs may be made to pin holes and undercuts in the final bead without authorization but the repaired weld must meet with the approval of COMPANY. CONTRACTOR will only be allowed to repair a weld one time. If a repaired weld is found to be unacceptable, the weld will be removed and the pipe rejoined at the CONTRACTOR'S expense.

10.6.3 Before repairs are made, injurious defects shall be entirely removed by chipping, grinding or flame gouging to clean metal. All slag and scale shall be removed by wire brushing. Preheating of such an area may be required by the COMPANY.

10.6.4 The CONTRACTOR shall remove all defects in the unwelded pipe or welded pipeline that meet or exceed COMPANY defect dimensions. The CONTRACTOR shall be paid as stipulated in the Bid Form for each defect not caused by CONTRACTOR which is removed and delivered to a COMPANY Representative. A defect shall include:

10.6.4.1 Any gross deformation in or near a beveled end that is to be welded which causes an unweldable joint.

10.6.4.2 Any dent occurring in the pipe that is greater than 0.250" deep, and one-half (1/2) the pipe diameter long in any direction or judged unacceptable by COMPANY Representative. The depth shall be determined by the gap between the lowest point of the defect and the original contour of the pipe.

10.6.4.3 Any dent containing a stress concentrator, such as a scratch, gouge or groove.

10.6.4.4 Any dent affecting a longitudinal or girth weld.

10.6.4.5 All arc burns.

10.6.5 CONTRACTOR shall remove the damaged portion by cutting it out as a cylinder. Any defect caused by the CONTRACTOR shall be removed by the CONTRACTOR at its expense and the unusable pipe may be billed to CONTRACTOR at COMPANY'S cost. CONTRACTOR may not be repaired by insert patching or pounding out.

10.6.6 If COMPANY required CONTRACTOR to grind a portion of the steel pipe, the remaining wall thickness must be equal to the minimum thickness allowed by the pipe manufacturing specification. Any area not meeting this requirement shall be removed by CONTRACTOR.

11.0 COATING – BELOW GROUND FACILITIES

11.1 CONTRACTOR shall furnish the field coating materials for pipe ends and bare fittings to be buried and the Contractor shall be responsible for their storage and application according to manufacturer's recommendations. Required coating system is Denso Protal 7200.

	Revision Date:	Spec. Number: 10-XXXXXX-CONST-XXX	Page 16 of 22

FIGURE 20.3 cont'd.

	CONSTRUCTION SPECIFICATION	

11.2 Pipe surface must be power brushed or mechanically cleaned prior to coating. Grease and oil shall be removed with a totally volatile solvent that leaves no oily residue. Welds shall be cleaned of all welding slag, splatter and scale. Sharp edges or burrs shall be removed by grinding or filing.

11.3 Shrink sleeves provided by COMPANY shall be used to coat all bare pipe and fittings delivered to the field, which will be buried. Pipe that is brought aboveground must have at least six (6) inches of coating above ground level.

11.4 Just prior to lowering in pipe in ditch, pipe to be electronically jeeped with calibrated holiday detector set at a minimum of 1200 volts and any holiday repaired.

11.5 Holidays will be repaired with epoxy sticks, or large areas will be primed and taped. Pipe surface must be power brushed or mechanically cleaned prior to coating repairs, taping or shrink sleeve applications.

12.0 COATING – ABOVE GROUND FACILITIES

12.1 Scope

12.1.1 All painting work shall include labor, tools, scaffolding, tarpaulins, drop cloths, and all equipment and services necessary for and incidental to the cleaning and preparation of the surfaces, application of paint as specified and cleanup to restore area and all unpainted surfaces to conditions present prior to start of work and acceptable to COMPANY Representative.

12.2 General

12.2.1 All oil, grease, rust, soil, dust, loose paint and other foreign material shall be removed from the surfaces to be painted prior to painting. Such foreign material shall removed by non-oily solvents, wire-brushing, sandblasting, power tools or other methods approved by the COMPANY Representative. All nameplates shall be covered with masking tape prior to painting.

12.2.2 Valve stems, glass, brick, glazed tile, flange faces and any equipment or other surfaces specified by the COMPANY Representative shall be adequately covered to protect the surface during painting operations. Such protection will be removed by the CONTRACTOR after the painting process is completed.

12.3 Paint Application

12.3.1 Paint is to be applied to surfaces prepared in accordance with the manufacturer's specifications. Paint will be applied in accordance with each specification. Drying times between coats and application rates will be adhered to.

12.3.2 Painting operations will not be permitted unless surface and surrounding temperatures will be in excess of 40F for at least 24 hours after painting operations for the day have been completed.

12.4 Drying

12.4.1 No paint shall be force dried under conditions which will cause checking, wrinkling, blistering or otherwise detrimentally affect the properties of the paint.

12.4.2 No drier additive shall be added to the paint on the job unless specifically required in the paint specifications furnished.

	Revision Date:	Spec. Number: 10-XXXXXX-CONST-XXX	Page 17 of 22

FIGURE 20.3 cont'd.

	CONSTRUCTION SPECIFICATION	

12.5 Exterior Coating

12.5.1 Surface Preparation.

Remove oil, dirt, grease, mill scale and other surface contaminants. Round off sharp edges and remove weld splatter. Abrasive blast clean to S.S.P.C. SP-10 or NACE #2 near white metal blast. Use 16-40 mesh abrasive to produce a maximum surface of 2.0 mils. Accustrip process may be requested and used in certain areas if paint manufacturer allows.

12.5.2 Prime Coat. Per COMPANY Paint Procedure.

12.5.3 Top Coat. Per COMPANY Paint Procedure.

12.6 High Temperature Coatings <1000 deg F

12.6.1 Surface Description.

Vessels, components, and piping operating at temperatures between 200 degrees and 1000 degrees F.

12.6.2 Surface Preparation. Same as below, if required.

12.6.3 Coating System.

12.6.3.1 One coat 859-06 Inorganic Zinc Primer unless previously primed with compatible primer.

12.6.3.2 One coat of Wilko 842-10 Straight Silicone ASA Gray applied to a dry film thickness of:

12.6.3.2.1 1.5 mils for 1000 degrees F service

12.6.3.2.2 2.0 mils for 500 degrees F service

12.6.4 Surface Description. Steel surfaces operating with skin temperatures from 450 to 1000F. Examples: mufflers, exhaust systems, stacks, etc.

12.6.5 Surface Preparation. N.A.C.E. #2 near white metal blast. Minimum surface profile depth of 1.0 mil. Maximum surface profile depth of 2.0 mils. Use 16 to 40 mesh abrasive. Prior to sandblasting, clean to remove oil, grease, dirt or contaminants.

12.6.6 Coating System

12.6.6.1 One coat 859-06 Inorganic Zinc Rich Primer to a dry film thickness of 2.0 to 3.0 mils.

12.6.6.2 One coat of 849-01 Straight Silicone Aluminum Spray - reduced 50% No. 1 thinner .5 mil wet seal only.

12.6.6.3 One coat 849-01 Straight Silicone Aluminum Spray - reduced 1/2 - 1 pint per gallon. Dry film thickness = 1.0 to 1.5 mils maximum.

13.0 TESTING AND REPAIRS

13.1 Gas Pressure Test

13.1.1 CONTRACTOR shall furnish all labor and equipment necessary to fabricate, install and remove test connections and appurtenances to piping.

13.1.2 COMPANY shall furnish gas required for gas test, unless otherwise specified in the Scope of Work. COMPANY may act as Testing Contractor and, if so, shall supply testing equipment and operator necessary to conduct test. This

	Revision Date:	Spec. Number: 10-XXXXXX-CONST-XXX	Page 18 of 22

FIGURE 20.3 cont'd.

	CONSTRUCTION SPECIFICATION	

will not relieve CONTRACTOR with assisting in the testing unless specifically stated by COMPANY Representative.

13.1.3 CONTRACTOR will furnish all labor and equipment required to locate and repair any leak as evidenced by such test performed by COMPANY or CONTRACTOR. CONTRACTOR shall bear the expense of locating and repairing leaks and retesting the line if the leak was due to faulty workmanship or defective material or equipment furnished by CONTRACTOR. COMPANY shall bear such expense if defect is due to defective material or equipment furnished by COMPANY.

13.1.4 Acceptance of CONTRACTOR'S work will be contingent upon line satisfactorily passing a minimum eight (8) hour shut-in pressure test at such pressure as stipulated in the Scope of Work or as designated by COMPANY Representative.

13.1.5 Prior to conducting a pressure test with gas, the CONTRACTOR shall disconnect, slipblind or otherwise isolate all tanks, equipment, vessels, or items other than pipe, valves, and fittings from any system to be tested. Such equipment shall not be tested with a gas.

13.2 Hydrostatic Pressure Testing

13.2.1 Spectra requirements for Hydrostatic Pressure Testing shall take precedence over these general testing requirements and a Spectra representative must witness the test.

13.2.2 CONTRACTOR shall be responsible for:

13.2.2.1 providing equipment and labor necessary to conduct, or help conduct if done by party other than CONTRACTOR, the hydrostatic test, including but not limited to, fabricating, installing and removing temporary manifolds, and all other work necessary in preparation for the test; and

13.2.2.2 Locating and repairing defects and removal of water, material and equipment after an acceptable test has been performed. CONTRACTOR shall bear the expense of locating and repairing leaks and retesting the line if the leak was due to faulty workmanship or defective material or equipment furnished by CONTRACTOR. COMPANY shall bear such expense if defect is due to defective material or equipment furnished by COMPANY.

13.2.3 CONTRACTOR shall furnish, or cause to be furnished, test gauges, and dead weight measurement devices, pressure recording charts, displacement devices and methanol for drying the pipeline, unless otherwise specified in the Scope of Work or approved in writing by COMPANY Representative. Dead weights and chart recorders shall have a calibration record not longer than six months from a certified calibration company, or per Spectra documentation requirements.

13.2.4 The test pressuring unit shall be located so that the pressure at the low point will not exceed the specified minimum yield strength of the pipeline, and all sections must be tested at the minimum test pressure for the specified time period.

	Revision Date:	Spec. Number: 10-XXXXXX-CONST-XXX	Page 19 of 22

FIGURE 20.3 cont'd.

	CONSTRUCTION SPECIFICATION	

13.2.5 The pressuring procedure shall not start until COMPANY Representative is present and has authorized pressurizing. Before the pressurizing can begin, the COMPANY Representative must determine that the required instruments are installed and are operating correctly. The required test instruments include: (a) Calibrated Pressure gauge with a maximum range of two times final test pressure and with a maximum of 20 psi divisions; (b) 24-hour pressure recorder with maximum range of four times final test pressure and maximum of 50 psi divisions; and (c) dead weight measurement device readable to 1.0 psi. The Testing Contractor shall keep a log of all pressure readings, time recorded and test fluid added throughout the test period. All instruments shall be operating during the entire test period.

13.2.6 If, during the testing operations, it becomes apparent that a leak has occurred on the line, the CONTRACTOR and COMPANY Representative shall immediately start keeping an exact record of labor, equipment and other charges necessary to locate and repair the leak so that proper charges can be made in the event the material furnished by COMPANY proves to be defective. This itemized breakdown of costs incurred will be subject to audit. If, during the location of a leak, it becomes apparent that the pipe itself has proven to be defective, the manufacturer's representative shall be promptly notified by COMPANY; however, this notice shall not interfere with the repair. After a leak or break repair has been made, the test shall be repeated until the section under test has been determined to be satisfactory.

13.2.7 The section shall remain on test at the pressure specified in the scope of Work or designated by COMPANY employee until COMPANY Representative has determined that the test is acceptable. No test section shall be accepted unless:

13.2.7.1 The test period is eight (8) consecutive hours or more and the test period may be extended, if COMPANY requests, to determine the nature of any pressure loss; and

13.2.7.2 With approval from the COMPANY Representative a test of four hours shall be acceptable if the test section meets the requirements of ASME B31.8.

13.2.7.3 The total volume of water added during the test period does not exceed three (3) gallons per hour.

13.2.8 The pressure lost may be adjusted for volume changes due to temperature fluctuations. However, such adjustment may not exceed the pressure lost per °F of ambient temperature change experienced during the stabilization period following the packing of the line after filling. If after applying this adjustment, the test is not acceptable and it is the opinion of the COMPANY Representative the cause is temperature oriented, then the test period shall be continued until the pressure stabilizes or a pressure reversal is experienced.

13.2.9 The COMPANY Representative shall indicate acceptance of the test by signing the test log and the pressure recorder.

13.2.10 The discharge of test fluid from the section shall be at locations designated or approved by COMPANY Representative. The discharge of test fluid shall

	Revision Date:	Spec. Number: 10-XXXXXX-CONST-XXX	Page 20 of 22

FIGURE 20.3 cont'd.

	CONSTRUCTION SPECIFICATION	

be done in such a manner as to not interfere with COMPANY existing facilities and their operations. The fluid shall be discharged so as not to cause flooding or erosion and must not be allowed to leave COMPANY property unless proper authorization has been received from state authorities. The COMPANY Representative shall prepare the required forms and obtain test kits for sampling of discharge fluid, and shall inform CONTRACTOR when discharge of test media may occur.

13.2.11 Testing Contractor shall furnish to COMPANY complete and current records on all phases of the testing program, including recording charts, dead weight log and all information on leaks or breaks.

13.3 Testing of Electrical Cable Insulation

13.4 Testing of Instrumentation Circuits

13.5 Testing of Ground Grid

14.0 INSTALLING MECHANICAL EQUIPMENT

14.1 General

14.1.1 All equipment shall be inspected by CONTRACTOR before beginning installation or upon receipt of equipment to insure that the equipment is clean, free of defects, and in good working condition. Any equipment found defective or unsatisfactory in any manner shall be reported to the COMPANY Representative immediately.

14.1.2 CONTRACTOR shall protect and secure equipment and material from the weather and other damaging natural and man-made hazards by a secured covering or other acceptable method of storage. All openings to equipment interiors shall be kept securely covered. CONTRACTOR shall repair any damages at its expense caused to equipment as a result of inadequate protection. CONTRACTOR may also be required to clean equipment because of inadequate protection. CONTRACTOR is also required to replace any equipment that has been damaged, lost or stolen while in their possession.

14.2 Installation

14.2.1 Installation and/or erection of equipment shall be performed by experienced personnel in accordance with manufacturer's drawings, specifications, recommendations and COMPANY Representative. All equipment shall be thoroughly tested in accordance with COMPANY Representative requirements to insure that is in proper mechanical working order prior to final acceptance of CONTRACTOR'S work.

14.2.2 All heat exchanger equipment will have thermowell and pressure tap connections installed on the inlet and outlet piping. All equipment normally operating with a measurable pressure differential will have pressure taps on the inlet and outlet connections as directed by COMPANY Representative.

14.2.3 Suction piping to compressors handling wet gas shall be installed according to COMPANY Representative so as to prevent any accumulation of condensate from being carried into the compressor. Check valves and relief

	Revision Date:	Spec. Number: 10-XXXXXX-CONST-XXX	Page 21 of 22

FIGURE 20.3 cont'd.

	CONSTRUCTION SPECIFICATION	

valves shall not be installed where COMPANY Representative determines condensate may accumulate and freeze up the valves.

14.2.4 All exposed hot piping, rotating or reciprocating machine parts, and other hazardous locations shall be suitably equipped with guards or marked as directed by COMPANY Representative to protect personnel.

15.0 RADIOGRAPHY

15.1 General

15.1.1 Each radiographer shall present a copy of his/her most recent radiography certificate to the COMPANY Representative. The radiographer shall follow the requirements of API 1104 Section 6 and 8 (latest edition) for pipeline, compressor station, and plant work, and ASME Section IX QW-191 for welding on pressurized piping systems covered by ASME B31.8 code.

15.1.2 100% of all welds within the fence or defined property boundary shall be radiographically inspected prior to backfill or start-up of facility.

15.1.3 Radiography shall be performed to COMPANY Requirements.

16.0 CLEANUP

16.1 General

16.1.1 CONTRACTOR shall collect and remove all paint barrels, skids, welding rod, defective material and all other construction debris within the plant, station or storage areas, etc. at the completion of construction and in areas of finished construction to the satisfaction of the property owners and COMPANY Representative. In no case shall the above-mentioned, or any other, material be buried and covered as a means of disposal.

16.1.2 All surplus material furnished by COMPANY shall be hauled and delivered by CONTRACTOR to points designated by COMPANY.

16.1.3 Large rocks, stumps, etc. shall be removed from the site by CONTRACTOR and hauled, at CONTRACTOR'S expense, to some location agreeable to the COMPANY Representative and the landowner or other authority having jurisdiction.

16.1.4 The CONTRACTOR will fill and recompact all ditches to minimize settling in accordance with Section 5.0. CONTRACTOR shall dispose of excess backfill to locations designated by the COMPANY Representative and meeting the requirements of such landowner or tenant.

16.1.5 Pipeline markers shall be furnished by the COMPANY and shall be installed by the CONTRACTOR at fence lines, road crossings and any other location designated by COMPANY Representative. The marker shall be placed approximately twelve (12) inches away from the pipeline and installed to a depth of approximately thirty six (36) inches into the backfill.

	Revision Date:	Spec. Number: 10-XXXXXX-CONST-XXX	Page 22 of 22

FIGURE 20.3 cont'd.

20.4 MATERIAL REQUISITION DEVELOPMENT

As material requirements are identified with regards to a specific need and relative to the engineering and design aspects of a project as outlined in the previous sections, they can be summarized into material requisitions. Requisitions are modified as required and can be used to request price and delivery from vendors. Once a vendor is selected, the requisition creates a purchase order (PO). The purchase order is then used to track the material through the purchasing process and to identify the material as it is received at a storage facility or the actual construction site. The PO identifies the material quantity and its price from a selected vendor. The status of POs or requisitions can be accessed at any time by either their identification number or the material item code.

Most companies can control orders and negotiate better deals with suppliers with computerized material requisition and PO status in an automated environment. Through automation, a company can automatically generate material requisitions and POs to increase order sizes. The companies can also schedule material shipments to meet order requirements, manage the vendors, and track supplier shipments and goods receiving. It remains imperative that any material requisition match the contract-specific specifications in order to allow for the internal network within a company to acquire the appropriate material for the given project.

20.5 BID QUOTATION AND BID ANALYSIS

20.5.1 Bid Quotation

Request for proposal (RFP) for a pipeline project must be in compliance with the construction specifications. As each line item addressed in the proposal is identified for bidding purposes, it cannot be in conflict with the construction specifications unless the line item for pricing indicates particular considerations due to mitigating circumstances such as agency permitting requirements. Each project is unique and as such there are instances within a project that require special pricing to accommodate certain requirements that are accepted by the owner in order to complete the project.

The proposal or bid documents are prepared by the owner or the company and then submitted to qualified bidders for their use in preparing bid proposals and pricing. The proposals generally contain relevant information including the following:

- Introduction information and abstract of the project: The introduction should contain the project title page and the table of contents. The abstract should not exceed 250 words and should provide a general understanding of the overall project. The abstract should not provide any detailed information; this would be addressed in the line items of the proposal. The abstract should be concise, and each portion should say something worthwhile.

- Bid solicitation: This portion should be brief; however, it should be detailed enough to allow the bidder to decide whether the project is within the realm of capabilities based on the complexity of the project or if their workforce is available and to determine whether the prospective bidder is interested in bidding.
- Instructions to bidders: This should specify the detailed terms and conditions of the bidding process for this particular project.
- Information available to bidders: This section should provide all project-related information as well as any conditions that would affect the bidders' pricing.
- Bid summary form: This form should be kept in a simplistic format and should require the prospective bidder to provide only information necessary to evaluate the bids.
- Contract agreements: In pipeline construction, there are usually four main types of contract agreements with regards to bidding. These include the following:
  - Fixed price portions or lump sum: This is utilized where the scope of work is clearly identified and a set price can be determined by the prospective bidder for this portion of the proposal.
  - Cost plus: This should be avoided as it relates to high costs due to higher spending and slower work. This is intended where the scope of work cannot be clearly defined resulting in an ambiguous estimation of the required equipment or man-hours to complete the task.
  - Unit pricing: This is intended to be used where the scope and quality of work are clearly defined yet the quantity is not as definitive. Although the quantities are estimated by the company, the bidder submits a fixed price for the estimated quantities and is paid based upon completed work for the actual quantities.
  - Time and material pricing: Although not as precarious as a cost plus scenario, time and material pricing can be costly if the companies' respective field representatives do not closely monitor the work being performed. In a T&M situation, the successful bidder submits rate sheets during the bidding process with respect to the labor and equipment that would be used for this type of work. Although the work may be performed, there may be equipment that is not being used at the site; however, a charge is submitted for that piece of equipment.
- Conditions of contract: The conditions of contract within the contract agreement section are defined using legal terminology to eliminate confusion in cases of dispute. Some items contained within the conditions of contract would include but not be limited to:
  - Object of the contract
  - Method of payment
  - Performance bond
  - Obligations of contractor
  - Obligations of company

- Commencement, delays, and liquidated damages
- Quality control of works
- Acceptance of works
- Retention
- Termination of contract
- Force majeure
- Settlement of dispute
- Effective date

20.5.2 Bid Analysis and Evaluation

First, the bids should be reviewed to ensure compliance with all aspects of the bid documents. Should there be a significant irregularity between the bid and the bid document, the company should reject all portions of that bid in its entirety. Should there be an insignificant or inconsequential irregularity between the bid and the bid document, the company may wish to waive the irregularity or permit the bidder to correct the irregularity.

Evaluation of the bids is most useful by using some sort of matrix comparison ranking against the various bidding line item criteria. In these various weightings, percentages are allocated toward the different line items toward the final bid selection. The highest percentage ranking should always be in the price of the overall bid with smaller percentage rankings for other criteria. Some selection criteria may include the following:

Bid price	55%
Bid quality	5%
Overall capacity	20%
Experience	10%
Scheduling	10%

To compare the different bid submissions in each criteria based on percentage of weighting, each company is given a percentage score out of 100 for each selection. The score is then multiplied by the selection weighting above and then added together for a total score. The company with the highest weighted score would have the best overall bid submission.

Chapter 21

Operations and Maintenance Manuals

Hal S. Ozanne

Chapter Outline

INTRODUCTION

Operations and maintenance manuals are required documents that are necessary for the safe and reliable operation of a pipeline system. They are used by pipeline operations and maintenance personnel for a step-by-step guide for operating and maintaining a pipeline system. The manuals are also used for training new operations and maintenance personnel in the systems operation.

Regulations also require that most pipeline operators have manuals in place for their system(s).

The content and amount of detail in the manuals is based on the complexity of the system.

21.1 OPERATING MANUALS

At a minimum, operating manuals should include the following documentation as applicable:

- Operating pressures
- Communications
- Line location and markers
- ROW maintenance
- Patrolling
- Integrity assessments and repair
- Pump station, terminal, and tank farm maintenance and operations
- Controls and protective equipment
- Storage vessels
- Fencing
- Signs
- Prevention of accidental ignition
- Corrosion control
- Emergency plan
- Records
- Training
- Modify plans when changes are made in the system.

21.2 REGULATIONS

ASME regulations prescribe the minimum requirements for maintaining operating and maintenance practices. Those regulations are summarized below.

ASME B31.8, Gas Transmission and Distribution Piping Systems – This regulation requires, at a minimum, that the pipeline operator have written operating and maintenance plans that will be adequate from the standpoint of public safety based on the provisions of the code, and the operator's knowledge of the facilities and the conditions in which they operate.

The basic requirements are as follows:

Each operating company having gas transmission or distribution facilities within the scope of the code shall

1. have a written plan covering operating and maintenance procedures in accordance with the scope and intent of the code;
2. have a written emergency plan covering facility failure or other emergencies;
3. operate and maintain its facilities in conformance with these plans;
4. modify the plans periodically as experience dictates and as exposure of the public to the facilities and changes in operating conditions require;
5. provide training for employees in procedures established for their operating and maintenance functions. The training shall be comprehensive and shall be designed to prepare employees for service in their area of responsibility;
6. keep records to administer the plans and train properly.

The essential features of the Operating and Maintenance Plan prescribed above shall include

1. detailed plans and instructions for employees covering operating and maintenance procedures for gas facilities during normal operations and repairs;
2. items recommended for inclusion in the plan for specific classes of facilities that are given in paragraphs 851.2, 851.3, 851.4, 851.5, 851.6, and 861(d);
3. plans to give particular attention to those portions of the facilities presenting the greatest hazard to the public in the event of an emergency or because of construction or extraordinary maintenance requirements;
4. provisions for periodic inspections along the route of existing steel pipelines or mains, operating at a hoop stress in excess of 40% of the specified minimum yield strength of the pipe material to consider the possibility of location class changes. It is not intended that these inspections include surveys of the number of buildings intended for human occupancy.

21.3 WRITTEN EMERGENCY PROCEDURES

Each operating company shall establish written procedures that will provide the basis for instructions to appropriate operating and maintenance personnel who will minimize the hazard resulting from a gas pipeline emergency. At a minimum, the procedures shall

1. provide a system for receiving, identifying, and classifying emergencies that require immediate response by the operating company;
2. indicate clearly the responsibility for instructing employees in the procedures listed in the emergency plans and for training employees in the execution of those procedures;
3. indicate clearly those responsible for updating the plan;

4. establish a plan for prompt and adequate handling of all calls that concern emergencies whether they are from customers, the public, company employees, or other sources;
5. establish a plan for the prompt and effective response to a notice of each type of emergency;
6. control emergency situations, including the action to be taken by the first employee arriving at the scene;
7. disseminate information to the public;
8. safely restore service to all facilities affected by the emergency after proper corrective measures have been taken reporting and documenting the emergency.

21.4 TRAINING PROGRAM

Each operating company shall have a program for informing, instructing, and training employees responsible for executing emergency procedures. The program shall acquaint the employee with the emergency procedures and how to promptly and effectively handle emergency situations. The program may be implemented by oral instruction, written instruction, and, in some instances, group instruction, followed by practice sessions. The program shall be established and maintained on a continuing basis with provision for updating as necessitated by revision of the written emergency procedures. Program records shall be maintained to establish what training each employee has received and the date of such training.

Detailed plans and instructions for employees covering operations and maintenance procedures for gas facilities during normal operations and repairs are as follows.

ASME 31.4, Liquid Hydrocarbon Pipeline Transportation Systems, requires

1. each operating company to develop operating and maintenance procedures that are adequate from the standpoint of public safety based on the provisions of the code, and the company's experience and knowledge of its facilities and the conditions under which they are operated;
2. the methods and procedures set forth herein serve as a general guide, but do not relieve the individual or operating company from the responsibility for prudent action that current particular circumstances make advisable;
3. its recognition that local conditions (such as the effects of temperature, characteristics of the line contents, and topography) will have considerable bearing on the approach to any particular maintenance and repair job.
4. the availability of suitable safety equipment for personnel use at all work areas and operating facilities where liquid anhydrous ammonia is transported. Such safety equipment shall include at least the following:
   a. full face gas mask with anhydrous ammonia refill canisters,
   b. independently supplied air mask,

   c. tight-fitting goggles or full face shield,
   d. protective gloves,
   e. protective boots,
   f. protective slicker and/or protective pants and jacket,
   g. easily accessible shower and/or at least 50 gal (190 L) of clean water in an open top container. Personnel shall be instructed in effective use of masks and limited shelf life of refill canisters. Protective clothing shall be of rubber fabric or other ammonia impervious material.

Code of Federal Regulations (CFR), Part 192, Transportation of Natural and Other Gas by Pipeline, requires a procedural manual for operations, maintenance, and emergencies in §192.605.

Each operator shall include the following in its operating and maintenance plan:

1. General: Each operator shall prepare and follow for each pipeline, a manual of written procedures for conducting operations and maintenance activities and for emergency response. For transmission lines, the manual must also include procedures for handling abnormal operations. This manual must be reviewed and updated by the operator at intervals not exceeding 15 months, but at least once each calendar year. This manual must be prepared before operations of a pipeline system commence. Appropriate parts of the manual must be kept at locations where operations and maintenance activities are conducted.
2. Maintenance and normal operations: The manual required by paragraph 1 of this section must include procedures for the following, if applicable, to provide safety during maintenance and operations.
   a. Operating, maintaining, and repairing the pipeline in accordance with each of the requirements of this subpart and Subpart M of this part.
   b. Controlling corrosion in accordance with the operations and maintenance requirements of Subpart I of this part.
   c. Making construction records, maps, and operating history available to appropriate operating personnel.
   d. Gathering of data needed for reporting incidents under Part 191 of this chapter in a timely and effective manner.
   e. Starting up and shutting down any part of the pipeline in a manner designed to assure operation within the MAOP limits prescribed by this part, plus the build-up allowed for operation of pressure-limiting and control devices.
   f. Maintaining compressor stations, including provisions for isolating units or sections of pipe and for purging before returning to service.
   g. Starting, operating, and shutting down gas compressor units.
   h. Periodically reviewing the work done by operator personnel to determine the effectiveness and adequacy of the procedures used in normal operation and maintenance and modifying the procedure when deficiencies are found.

i. Taking adequate precautions in excavated trenches to protect personnel from the hazards of unsafe accumulations of vapor or gas, and making available when needed at the excavation, emergency rescue equipment, including a breathing apparatus and a rescue harness and line.
j. Systematic and routine testing and inspection of pipe-type or bottle-type holders including
   i. provision for detecting external corrosion before the strength of the container has been impaired;
   ii. periodic sampling and testing of gas in storage to determine the dew point of vapors contained in the stored gas which, if condensed, might cause internal corrosion or interfere with the safe operation of the storage plant;
   iii. periodic inspection and testing of pressure limiting equipment to determine that it is in safe operating condition and has adequate capacity.

3. Abnormal operation: For transmission lines, the manual required by paragraph 1 of this section must include procedures for the following to provide safety when operating design limits have been exceeded:
   a. Responding to, investigating, and correcting the cause of
      i. unintended closure of valves or shutdowns;
      ii. increase or decrease in pressure or flow rate outside normal operating limits;
      iii. loss of communications;
      iv. operation of any safety device;
      v. any other foreseeable malfunction of a component, deviation from normal operation, or personnel error, which may result in a hazard to persons or property.
   b. Checking variations from normal operation after abnormal operation has ended at sufficient critical locations in the system to determine continued integrity and safe operation.
   c. Notifying responsible operator personnel when notice of an abnormal operation is received.
   d. Periodically reviewing the response of operator personnel to determine the effectiveness of the procedures controlling abnormal operation and taking corrective action where deficiencies are found.
   e. The requirements of this paragraph 3 do not apply to natural gas distribution operators that are operating transmission lines in connection with their distribution system.
4. Safety-related condition reports: The manual required by paragraph 1 of this section must include instructions enabling personnel who perform operation and maintenance activities to recognize conditions that potentially may be safety-related conditions that are subject to the reporting requirements of §191.23 of this subchapter.

5. Surveillance, emergency response, and accident investigation: The procedures required by §§192.613(a), 192.615, and 192.617 must be included in the manual required by paragraph 1 of this section.

CFR 49, Part 195, Transportation of Hazardous Liquids by Pipeline, requires a procedural manual for operations, maintenance, and emergencies in §195.402.

1. General. Each operator shall prepare and follow for each pipeline system a manual of written procedures for conducting normal operations and maintenance activities and handling abnormal operations and emergencies. This manual shall be reviewed at intervals not exceeding 15 months, but at least once each calendar year, and appropriate changes made as necessary to ensure that the manual is effective. This manual shall be prepared before initial operations of a pipeline system commence, and appropriate parts shall be kept at locations where operations and maintenance activities are conducted.
2. The administrator or the state agency that has submitted a current certification under the pipeline safety laws (49 U.S.C. 60101 *et seq.*) with respect to the pipeline facility governed by an operator's plans and procedures may, after notice and opportunity for hearing as provided in 49 CFR 190.237 or the relevant state procedures, require the operator to amend its plans and procedures as necessary to provide a reasonable level of safety.
3. Maintenance and normal operations: The manual required by paragraph 1 of this section must include the following procedures to provide safety during maintenance and normal operations:
   a. Making construction records, maps, and operating history available as necessary for safe operation and maintenance.
   b. Gathering of data needed for reporting accidents under Subpart B of this part in a timely and effective manner.
   c. Operating, maintaining, and repairing the pipeline system in accordance with each of the requirements of this subpart.
   d. Determining which pipeline facilities are located in areas that would require an immediate response by the operator to prevent hazards to the public if the facilities failed or malfunctioned.
   e. Analyzing pipeline accidents to determine their causes.
   f. Minimizing the potential for hazards identified under paragraph 3(d) of this section and the possibility of recurrence of accidents analyzed under paragraph 3(e) of this section.
   g. Starting up and shutting down any part of the pipeline system in a manner designed to assure operation within the limits prescribed by paragraph §195.406, consider the hazardous liquid or carbon dioxide in transportation, variations in altitude along the pipeline, and pressure monitoring and control devices.

h. In the case of pipeline that is not equipped to fail safe, monitoring from an attended location pipeline pressure during startup until steady state pressure and flow conditions are reached and during shut-in to assure operation within limits prescribed by §195.406.
i. In the case of facilities not equipped to fail safe that are identified under §195.402(c)(4) or that control receipt and delivery of the hazardous liquid or carbon dioxide, detecting abnormal operating conditions by monitoring pressure, temperature, flow or other appropriate operational data and transmitting this data to an attended location.
j. Abandoning pipeline facilities, including safe disconnection from an operating pipeline system, purging of combustibles, and sealing abandoned facilities left in place to minimize safety and environmental hazards.
k. Minimizing the likelihood of accidental ignition of vapors in areas near facilities identified under paragraph 3(d) of this section where the potential exists for the presence of flammable liquids or gases.
l. Establishing and maintaining liaison with fire, police, and other appropriate public officials to learn the responsibility and resources of each government organization that may respond to a hazardous liquid or carbon dioxide pipeline emergency and acquaint the officials with the operator's ability in responding to a hazardous liquid or carbon dioxide pipeline emergency and means of communication.
m. Periodically reviewing the work done by operator personnel to determine the effectiveness of the procedures used in normal operation and maintenance and taking corrective action where deficiencies are found.
n. Taking adequate precautions in excavated trenches to protect personnel from the hazards of unsafe accumulations of vapor or gas, and making available when needed at the excavation, emergency rescue equipment, including a breathing apparatus and a rescue harness and line.

4. Abnormal operation: The manual required by paragraph 1 of this section must include procedures for the following to provide safety when operating design limits have been exceeded:
   a. Responding to, investigating, and correcting the cause of
      i. unintended closure of valves or shutdowns;
      ii. increase or decrease in pressure or flow rate outside normal operating limits;
      iii. loss of communications;
      iv. operation of any safety device;
      v. any other malfunction of a component, deviation from normal operation, or personnel error which could cause a hazard to persons or property.
   b. Checking variations from normal operation after abnormal operation has ended at sufficient critical locations in the system to determine continued integrity and safe operation.

   c. Correcting variations from normal operation of pressure and flow equipment and controls.
   d. Notifying responsible operator personnel when notice of an abnormal operation is received.
   e. Periodically reviewing the response of operator personnel to determine the effectiveness of the procedures controlling abnormal operation and taking corrective action where deficiencies are found.
5. Emergencies: The manual required by paragraph 1 of this section must include procedures for the following to provide safety when an emergency condition occurs:
   a. Receiving, identifying, and classifying notices of events that need immediate response by the operator or notice to fire, police, or other appropriate public officials and communicating this information to appropriate operator personnel for corrective action.
   b. Prompt and effective response to a notice of each type emergency, including fire or explosion occurring near or directly involving a pipeline facility, accidental release of hazardous liquid or carbon dioxide from a pipeline facility, operational failure causing a hazardous condition, and natural disaster affecting pipeline facilities.
   c. Having personnel, equipment, instruments, tools, and material available as needed at the scene of an emergency.
   d. Taking necessary action, such as emergency shutdown or pressure reduction, to minimize the volume of hazardous liquid or carbon dioxide that is released from any section of a pipeline system in the event of a failure.
   e. Control of released hazardous liquid or carbon dioxide at an accident scene to minimize the hazards, including possible intentional ignition in the cases of flammable highly volatile liquid.
   f. Minimization of public exposure to injury and probability of accidental ignition by assisting with evacuation of residents and assisting with halting traffic on roads and railroads in the affected area, or taking other appropriate action.
   g. Notifying fire, police, and other appropriate public officials of hazardous liquid or carbon dioxide pipeline emergencies and coordinating with them preplanned and actual responses during an emergency, including additional precautions necessary for an emergency involving a pipeline system transporting a highly volatile liquid.
   h. In the case of failure of a pipeline system transporting a highly volatile liquid, use of appropriate instruments to assess the extent and coverage of the vapor cloud and determine the hazardous areas.
   i. Providing for a postaccident review of employee activities to determine whether the procedures were effective in each emergency and taking corrective action where deficiencies are found.
6. Safety-related condition reports: The manual required by paragraph 1 of this section must include instructions enabling personnel who perform operation and

maintenance activities to recognize conditions that potentially may be safety-related conditions that are subject to the reporting requirements of §195.55.

21.5 DETAILS

The following are subjects that must be included in an operating manual.

21.5.1 Operating Pressures

A pipeline or each segment of a pipeline system has an established maximum operating pressure or MAOP. The pressure is based on the MAOP of the system components as well as the pressure to which the system was hydrostatically tested to. Shutdown and pressure relief systems should be set to prevent the system from being over pressured. These pressures are required to be documented so that personnel involved with the operation are aware of the pressure limits so that they do not inadvertently operate the system at pressure exceeding those limits.

21.5.2 Communications

A communications system must be in place for the system operation. For simple systems, this may be a list of telephone numbers of personnel to call for various things. For larger more complex systems that have a central operations center, the communications system may be complex.

Communications will be required from each remote location to the operations center. The system may include hard wire telephone lines, satellite systems, and the use of the Internet and radio communications. In many cases, it may be a combination of all of the types above. The system may have some redundancy for some of the larger facilities that are monitored and operated from an operations center.

21.5.3 Line Location and Markers

As described in Chapter 14, pipelines must be clearly marked with markers to indicate that a pipeline is present with the name of the operating company and 24-h emergency telephone number. In the event of an emergency involving the pipeline, the operator or operations center can be notified so that any necessary actions can be taken. Aerial markers are generally installed at 1-mile increments. When an aerial patrol is being conducted, the pilot can use the aerial markers as a reference to report any encroachments or potential leaks that he observes.

21.5.4 ROW Maintenance

After a pipeline has been constructed and put into operation, the right of way is cleaned up and restored based on the conditions agreed to in the right-of-way easements with the land owners. The pipeline operator is required to keep the right of way cleared so that it is accessible and can be patrolled by foot and by air. This would include cutting down any trees or brush that grow in the right of way as well as mowing any heavy grass. If a problem develops with the pipeline, it will be easier to spot if the right of way is maintained.

21.5.5 Patrolling

As described in Chapter 14, pipelines must be patrolled on a regular basis to look for any sign of a leak or construction activity near or over the pipeline. The patrol is also a means to determine if there may be encroachment developing near the pipeline.

21.5.6 Integrity Assessments and Repair

A part of the operation of a pipeline system is to make integrity assessments on a regular basis. Federal regulations prescribe the frequency that assessments must be made. This usually includes running a smart pig through the pipeline to determine whether any irregularities are observed. These would include internal and external corrosion, mechanical damage, and other abnormalities that require attention. The plan will delineate the type of repair or actions that the operator will take when an abnormality is detected.

21.5.7 Pump Station, Terminal, and Tank Farm Maintenance and Operations

This section of the operating plan will be very detailed. The size of the facility or facilities will determine the amount of detail. This will include a step-by-step sequence of operating each component of the facility such as starting and stopping a pump, directing liquid flow to a tank, pulling suction from a tank to feed a pump and inserting a scrapper or pig into the pipeline leaving a facility, or starting and stopping a compressor at a compressor station on a gas pipeline.

An accurate, up to date, schematic drawing of the facility indicating all operable components is critical to accompany the written operating instructions. An operator can remotely visualize the facility he or she is operating. If the operator is operating a facility locally, they can see the actual components they are operating. In many instances, the piping between valves, pumps, tanks, and so on is underground and is difficult to trace. A schematic graphically indicates where the piping is between the components.

When a component at a facility such a pump, motor, or motor-operated valve is going to be worked on, the piece of equipment must be locked out and tagged so that it cannot be operated or started while it is being worked on. This will prevent injury to maintenance personnel.

21.5.8 Controls and Protective Equipment

Controls include on or off switches to open and close valves, start and stop pumps or compressors, and so on. In many cases, selecting one of these types of controls will cause a sequence of operations to take place.

Protective equipment monitors operating conditions such as pressure, level, temperature, and so on and will automatically stop an operation to prevent damage or an upset when the condition exceeds the set point of the device.

21.5.9 Storage Vessels

Storage vessels include liquid storage tanks, and natural gas liquid pressure storage vessels like spheres and bullets. Each vessel has a maximum amount of product that can be safely stored in it without the possibility of overflowing or over pressuring the vessel. Level gauges and transmitters are used to monitor the amount of product in the vessel. High-level and high-pressure switches are installed with set point alarms that will notify the operator when the vessel is close to reaching its high-level point so that the operator can switch the flow to another vessel or discontinue taking delivery of product into the vessel.

21.5.10 Fencing

A fence is normally installed around the facility to prevent unauthorized personnel from entering the facility. Larger facilities may have a key card access system for authorized personnel to gain access. In some cases, an intrusion system may be installed to notify the operator if someone is trying to enter the facility without the proper authorization. For instance, if someone was climbing over the fence, it would be detected and would be dealt with accordingly.

21.5.11 Signs

As with markers along the pipeline right of way, signs are normally posted at each facility identifying the company with an emergency telephone number to call in the event of an emergency.

21.5.12 Prevention of Accidental Ignition

The operator must take precautions to prevent accidental ignition for their system. This includes checking for any leaks when an operator visits a facility and is part of a leak detection system to monitor for leaks so that the proper actions can be taken including shutting the system down to minimize the product escaping the system.

Gas detectors are often installed on a gas pipeline system at meter facilities and compressor stations. In the event that gas is detected, an alarm is sounded and the facility is shut down.

Flame detectors are also installed at pipeline facilities so that if flames are detected, an alarm is sounded and the facility is shut down.

If work is going to be conducted at an operating facility that involves welding or another source of flame or flash, then the detection system may be disabled while the work is going on. An observer must monitor the facility during the time the system is disabled to take any corrective action in the event that an incident occurs.

21.5.13 Corrosion Control

Corrosion control is discussed in detail in Chapter 13.

The operator must have a cathodic protection system installed to prevent corrosion from occurring on their system. Cathodic protection test stations must be installed on the pipeline during the initial construction. A corrosion survey shall be conducted, at a minimum, on an annual basis. This involves taking pipe to soil conductivity readings. These readings are recorded and compared with the readings from earlier surveys. By comparing these readings, a technician can determine if an excessive amount of current is required in a section of the pipeline. If that is the case, it may indicate a problem with the pipe coating or the possibility of mechanical damage to the pipe that may cause corrosion to develop.

When this is discovered, the operator must excavate and uncover the pipe in the area of concern and make necessary repairs to correct the problem.

21.5.14 Emergency Plan

Pipeline operators must have emergency plans prepared for their system that delineate what actions will be taken in the event of an emergency. The plan must cover all possible emergencies that may occur that should include a leak, explosion, power outage, communications outage, spill, or injury to personnel, to name a few.

21.5.15 Records

Pipeline operators are required to keep accurate records and statistics of the operation of their system. These records include operating pressures, flow rates, operating temperatures, storage inventory, information, corrosion control records, and so on. Records of repairs and maintenance are also recorded. These records are subject to review by regulatory agencies including the Federal DOT

(Department of Transportation). The DOT wants to verify that the system is being operated within the mandated guidelines.

21.5.16 Training

The pipeline operator is required to prepare training manuals as part of the operating and maintenance manuals. The training manuals are used to train new personnel assigned to pipeline operations and maintenance, as well as to conduct refresher training to experienced personnel. The manuals are to be revised whenever there is a significant change to the system.

21.5.17 Modification to Plans

When any changes are made to the system that affect the operation, the manuals must be revised to reflect the change to the operation.

21.6 MAINTENANCE MANUALS

Maintenance manuals are prepared to prescribe the procedures to be used for preventive, as well as nonpreventive maintenance of components on the pipeline system or pipeline facility.

21.7 PREVENTATIVE MAINTENANCE

Preventative maintenance is important to the unplanned uninterrupted operation of a pipeline system. In many cases, the equipment manufacturers issue preventative maintenance recommendations for the equipment that they provide to a pipeline company. This will include the frequency of maintenance, as well as when certain wear parts should be replaced. These may include bearings on pumps and motors, valve seals, pump seals, pressure and temperature sensors, other instrumentation, and electronic components. Lubrication oil is checked and changed at certain time intervals.

Regulations require that devices, like mainline valves that are not normally operated, be completely closed and opened at least once a year to ensure that they will operate when needed.

21.8 PROJECT DATA BOOK

An important part of the operations and maintenance manual is a project data book(s). This book is developed at the end of a project to include the data for all of the components of the system. This information is used for maintenance purposes. All of the information is included in one location when it is needed so that a search is not required to find the information.

Following is an example of the table of contents of a project data book.

Project Data Book

Table of Contents – Volume 1

Equipment and Instrument Data

1. Valves

	Description	Manufacturer	Vendor	Tag No.
1.1	12-in. ESD valve	WKM/Morin	Wilson	SDV-7001
1.2	Ball valves, 16"+	Grove/Bettis	Cameron	
1.3	12", 8" Ball valves			
1.4	30" Check valve			
1.5	2" Dump valves	Fisher	Puffer-Swieven	LCV-1001, 1002, 1004
1.6	2" Tank Valves	WKM/Morin	Wilson	SDV-7020, 7021
1.7	Thermal relief valve	A-G	Brock Easley	PSV-1001, 1002, 1003, PSV-1701, 1702
1.8	2" Chokes	Cactus		

2. Vessels

	Description	Manufacturer	Vendor	Tag No.
2.1	Origin separator	Coastline	Coastline	MBD-1001
2.2	Delivery separator		Burgess-Manning	MBD-1002
2.3	Delivery separator		Perry Equip.	MBD-1003
2.4	Condensate tanks	Challenger	Challenger	BBJ-1701, BBJ-1702

3. Instrumentation

	Description	Manufacturer	Vendor	Tag No.
3.1	Filter/dryer	Welker	Welker	
3.2	Moisture/CO_2 analyzer	Spectra Sensors	Davis & Davis	MA/T-1701, ME-1701
3.3	Moisture analyzer	Spectra Sensors	Davis & Davis	MT-1702, ME-1702
3.4	Level transmitter	Orion	Davis & Davis	LIT-1001
3.5	Level transmitter	Orion	Davis & Davis	LIT-1002
3.6	Level transmitter	Orion	Davis & Davis	LIT-1003
3.7	Corrosion coupon	Metal Samples	Metal Samples	
3.8	Pig detector	Kidd	Kidd	ZIS-1701, 1702, 1703, ZS-1704
3.9	Pressure transmitter	Rosemount	Rosemount	PT-1702, 1703
3.10	Pressure gauge	Ash croft		PI-1701, 1702, 1703, 1704

Project Data Book

Table of Contents – Volume 2

3. Instrumentation (cont.)

	Description	Manufacturer	Vendor	Tag No.
3.11	Tank level transmitter	Pulsar	Davis & Davis	LT-1701, 1702
3.12	PLC	Allen-Bradley		
3.13	Radio			
3.14	DP transmitter	Rosemount	Rosemount	DPIT-1701
3.15	DP indicator			
3.16	Gas detector			
3.17	Room temperature transmitter		Minco	TT-1701,1702
3.18	Door alarm			ZA-1701,1702

4. Buildings

	Description	Manufacturer	Vendor	Tag No.
4.1	Trap building	Nucor	Alco	
4.2	Separator building	Nucor	Alco	
4.3	Control building	T. Ingram	T. Ingram	
4.4	TEG	Global Thermoelectric	Wyoming Service & Supply	
4.5	Catalytic heater	Bruest	Bruest	

5. Pipeline

	Description	Manufacturer	Vendor	Tag No.
5.1	End closure	TDW	TDW	
5.2	Rectifier			
5.3	CP test station			
5.4	Strain gage	Geo Durham	Geo Durham	
5.5	Inclinometer	Geo Durham	Geo Durham	
5.6	Anchor flange			
5.7	Anchor sleeve			

6. Miscellaneous

	Description	Manufacturer	Vendor	Tag No.
6.1	Valve shelter	T. Ingram	T. Ingram	
6.2	Pipe supports			
6.3	Hose connectors			
6.4	Tank containment	Brock		
6.5	Guardrail			
6.6	Paint			

21.9 STARTUP SEQUENTIAL PROCESS

The operating manual will include a step-by-step sequence of how to start up a complete system or one facility on a system from a status of nothing operating to full operation. The operation sequence will be accompanied by a schematic of the system or facility, which will graphically show all of the components of the facility and their relationship to one another.

In many cases, all or part of the startup sequence may be automated by a control system. For instance, when starting a pipeline mainline centrifugal pump, the following example summarizes the startup:

- If not already open, open the station mainline block valve(s).
- Close the pump discharge valves.
- Ensure that the pump suction valve is open.
- Push the pump start button.
- As the pump motor is coming up to speed, start opening the pump discharge valve.

If this were an automated process, after ensuring that the mainline block valve(s) were open and issuing the command to start the pump, the remainder of the sequence would be automated.

See Fig. 21.1 for the corresponding schematic.

21.10 SHUTDOWN SEQUENTIAL PROCESS

A procedure for shutting down the entire pipeline or a facility is also required. It is basically the reverse of starting the system.

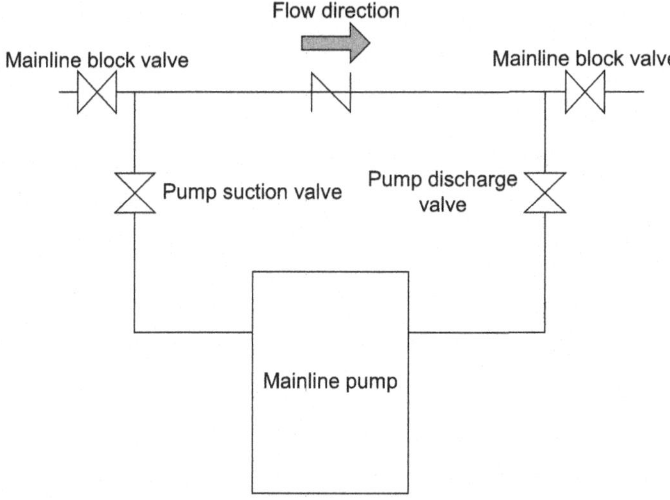

FIGURE 21.1 Schematic.

BIBLIOGRAPHY

ASME B31.4-2006, Pipeline Transportation Systems for Liquid Hydrocarbons and Other Liquids, American Society of Mechanical Engineers, New York, NY, 2006.

ASME B31.8-2007, Gas Transportation and Distribution Piping Systems, American Society of Mechanical Engineers, New York, NY, 2007.

CFR 49, Part 195, Transportation of Hazardous Liquids by Pipeline: Minimum Federal Safety Standards, U.S. Government Printing Office, Washington, DC.

CFR 49, Part 192, Transportation of Natural or Other Gas by Pipeline: Minimum Federal Safety Standards, U.S. Government Printing Office, Washington, DC.

Appendix 1

Chapter 1

OUTLINE OF DESIGN BASIS MANUAL (DBM)

1.0 Introduction
- Definitions
- Reference Documents
- Preliminary Engineering
- Regulatory Compliance
- Health and Safety
- Environmental Permitting
- Right-of-Way
- Quality Management Plan

2.0 Project Design Background
- Design Data
- Pipeline Hydraulics

3.0 Project Scope
- General Description
- Pipeline
- Pump Stations/Compressor Stations
- Mainline Valve Stations
- Pigging Facilities
- Heater Stations
- Storage Tanks
- Metering Facilities
- Supervisory Control and Data Acquisition (SCADA)
- Communications
- Power Requirement

4.0 Project Schedule

5.0 Codes and Standards

6.0 Operating and Maintenance Philosophy

7.0 Design Basis
- General
- Pipeline
- Pump Stations/Compressor Stations

- Mainline Valve Stations
- Pigging Facilities
- Heater Stations
- Cathodic Protection
- SCADA and Communications Systems
- Other Facilities

8.0 Appendices

- Pipeline Routing Maps
- Flow Diagrams
- Schematic Diagrams
- Plot Plans
- Pipeline Hydraulics

UNITS AND CONVERSIONS

Item	USCS Units	SI Units	USCS to SI Conversion	SI to USCS Conversion
Mass	slug (slug)	kilogram (kg)	1 slug = 14.594 kg	1 kg = 0.0685 slug
	1 US ton = 2000 lb	metric ton (t) = 1000 kg	1 US ton = 0.9072 t	1 t = 1.1023 US ton
	1 long ton = 2240 lb		1 long ton = 1.016 t	1 t = 0.9842 long ton
Weight	pound (lb)	Newton (N)	1 lb = 4.4482 N	1 kg = 2.205 lb
Length	inch (in)	millimeter (mm)	1 in = 25.4 mm	1 mm = 0.0394 in
	1 foot (ft) = 12 in	1 meter (m) = 1000 mm	1 ft = 0.3048 m	1 m = 3.2808 ft
	1 mile (mi) = 5280 ft	1 kilometer (km) = 1000 m	1 mi = 1.609 km	1 km = 0.6214 mi
Area	square foot (ft^2)	square meter (m^2)	1 ft^2 = 0.0929 m^2	1 m^2 = 10.764 ft^2
	1 acre = 43,560 ft^2	1 hectare (ha) = 10,000 m^2	1 acre = 0.4047 ha	1 ha = 2.4711 acre
Volume	cubic inch (in^3)	cubic millimeter (mm^3)	1 in^3 = 16387.0 mm^3	1 mm^3 = 6.1×10^{-5} in^3
	cubic foot (ft^3)	1 liter (L) = 1000 cm^3 (cc)	1 ft^3 = 0.02832 m^3	1 m^3 = 35.3134 ft^3
	1 US gallon (gal) = 231 in^3	1 cubic meter (m^3) = 1000 L	1 gal = 3.785 L	1 L = 0.2642 gal
	1 barrel (bbl) = 42 gal		1 bbl = 158.97 L = 0.15897 m^3	1 m^3 = 6.2905 bbl
	1 ft^3 = 7.4805 gal			
	1 bbl = 5.6146 ft^3			
Density	slug per cubic foot (slug/ft^3)	kilogram/cubic meter (kg/m^3)	1 slug/ft^3 = 515.38 kg/m^3	1 kg/m^3 = 0.0019 slug/ft^3
Specific weight	pound per cubic foot (lb/ft^3)	Newton per cubic meter (N/m^3)	1 lb/ft^3 = 157.09 N/m^3	1 N/m^3 = 0.0064 lb/ft^3
Viscosity (dynamic)	slug/ft·s	1 poise (P) = 0.1 Pa·s		1 cP = 6.7197×10^{-4} lb/ft·s
	lb·s/ft^2	1 centipoise (cP) = 0.01 P	1 lb·s/ft^2 = 47.88 N·s/m^2	1 N·s/m^2 = 0.0209 lb·s/ft^2
		1 poise = 1 dyne·s/cm^2	1 lb·s/ft^2 = 478.8 Poise	1 Poise = 0.00209 lb·s/ft^2
		1 poise = 0.1 N·s/m^2		
Viscosity (kinematic)	ft^2/s	m^2/s	1 ft^2/s = 0.092903 m^2/s	1 m^2/s = 10.7639 ft^2/s
	SSU*, SSF*	stoke (S), centistoke (cSt)		1 cSt = 1.076×10^{-5} ft^2/s
Flow rate	cubic foot/second (ft^3/s)	liter/minute (L/min)	1 gal/min = 3.785 L/min	1 L/min = 0.2642 gal/min
	gallon/minute (gal/min)	cubic meter/hour (m^3/h)	1 bbl/h = 0.159 m^3/h	1 m^3/h = 6.2905 bbl/h
	barrel/hour (bbl/h), bbl/day			
Force	pound (lb)	Newton (N) = kg·m/s^2	1 lb = 4.4482 N	1 N = 0.2248 lb

(Continued)

Item	USCS Units	SI Units	USCS to SI Conversion	SI to USCS Conversion
Pressure	pound/square inch, lb/in^2 (psi)	pascal (Pa) = N/m^2	1 psi = 6.895 kPa	1 kPa = 0.145 psi
	1 lb/ft^2 = 144 psi	1 kilopascal (kPa) = 1000 Pa		
		1 megapascal (MPa) = 1000 kPa		
		1 bar = 100 kPa	1 psi = 0.069 bar	1 Bar = 14.5 psi
		kilogram/sq. centimeter (kg/cm^2)	1 psi = 0.0703 kg/cm^2	1 kg/cm^2 = 14.22 psi
Velocity	foot/second (ft/s)	meter/second (m/s)	1 ft/s = 0.3048 m/s	1 m/s = 3.281 ft/s
	mile/hour (mi/h) = 1.4667 ft/s			
Work and energy	foot-pound (ft·lb)	joule (J) = N·m	1 Btu = 1055.0 J	1 kJ = 0.9478 Btu
	British Thermal Unit (Btu)			
	1 Btu = 778 ft·lb			
Power	ft·lb/min	joule/second (J/s)	1 Btu/h = 0.2931 W	1 W = 3.4121 Btu/h
	Btu/h	watt (W) = J/s		
	Horsepower (HP)	1 kilowatt (kW) = 1000 W	1 HP = 0.746 kW	1 kW = 1.3405 HP
	1 HP = 33,000 ft·lb/min			
Temperature	degree Fahrenheit (°F)	degree Celsius (°C)	1°F = 9/5°C + 32	1°C = (°F − 32)/1.8
	1 degree Rankin (°R) = °F + 460	1 Kelvin (K) = °C + 273	1°R = 1.8 K	1 K = °R/1.8
Thermal conductivity	Btu/h/ft/°F	W/m/°C	1 Btu/h/ft/°F = 1.7307 W/m/°C	1 W/m/°C = 0.5778 Btu/h/ft/F
Heat transfer coefficient	Btu/h/ft^2/°F	W/m^2/C	1 Btu/h/ft^2/F = 5.6781 W/m^2/C	1 W/m^2/C = 0.1761 Btu/h/ft^2/F
Specific heat	Btu/lb/°F	kJ/kg/°C	1 Btu/lb/°F = 4.1869 kJ/kg/°C	1 kJ/kg/°C = 0.2388 Btu/lb/F

Kinematic viscosity in SSU and SSF are converted to viscosity in cSt using the following formulas:

$$\text{Centistokes} = 0.226 \times \text{SSU} - 195/\text{SSU} \text{ for } 32 \leq \text{SSU} \leq 100 \quad \text{(A 1.1)}$$

$$\text{Centistokes} = 0.220 \times \text{SSU} - 135/\text{SSU} \text{ for } \text{SSU} > 100 \quad \text{(A 1.2)}$$

$$\text{Centistokes} = 2.24 \times \text{SSF} - 184/\text{SSF} \text{ for } 25 \leq \text{SSF} \leq 40 \quad \text{(A 1.3)}$$

$$\text{Centistokes} = 2.16 \times \text{SSF} - 60/\text{SSF} \text{ for } \text{SSF} > 40 \quad \text{(A 1.4)}$$

PHYSICAL PROPERTIES OF LIQUIDS AND GASES

Properties of Petroleum Liquids

Product	Viscosity (cSt) @ 60°F	°API Gravity	Specific Gravity @ 60°F
Regular gasoline			
Summer grade	0.70	62.0	0.7313
Interseasonal grade	0.70	63.0	0.7275
Winter grade	0.70	65.0	0.7201
Premium gasoline			
Summer grade	0.70	57.0	0.7467
Interseasonal grade	0.70	58.0	0.7165
Winter grade	0.70	66.0	0.7711
No.1 fuel oil	2.57	42.0	0.8155
No.2 fuel oil	3.90	37.0	0.8392
Kerosene	2.17	50.0	0.7796
Jet fuel JP-4	1.40	52.0	0.7711
Jet fuel JP-5	2.17	44.5	0.8040

Specific Gravity and API Gravity

Liquid	Specific Gravity (60°F)	API Gravity (60°F)
Propane	0.5118	N/A
Butane	0.5908	N/A
Gasoline	0.7272	63.0
Kerosene	0.7796	50.0
Diesel	0.8398	37.0
Light crude	0.8348	38.0
Heavy crude	0.8927	27.0
Very heavy crude	0.9465	18.0
Water	1.0000	10.0

Properties of Hydrocarbon Gases

			Vapor	Critical Constants			Ideal Gas, 14.696 psia, 60°F		
Gas	Formula	Molecular Weight	Pressure at 100°F (psia)	Pressure (psia)	Temperature (°F)	Volume (ft^3/lb)	Gravity (air = 1.00)	Volume (ft^3/lb)	Specific Heat (Btu/lb/°F)
Methane	CH_4	16.0430	5000	666.0	−116.66	0.0988	0.5539	23.654	0.52676
Ethane	C_2H_6	30.0700	800	707.0	90.07	0.0783	1.0382	12.620	0.40789
Propane	C_3H_8	44.0970	188.65	617.0	205.93	0.0727	1.5226	8.6059	0.38847
Isobutane	C_4H_{10}	58.1230	72.581	527.9	274.4	0.0714	2.0068	6.5291	0.38669
n-Butane	C_4H_{10}	58.1230	51.706	548.8	305.52	0.0703	2.0068	6.5291	0.39500
Isopentane	C_5H_{12}	72.1500	20.443	490.4	368.96	0.0684	2.4912	5.2596	0.38448
n-pentane	C_5H_{12}	72.1500	15.575	488.1	385.7	0.0695	2.4912	5.2596	0.38831
Neopentane	C_5H_{12}	72.1500	36.72	464.0	321.01	0.0673	2.4912	5.2596	0.39038
n-Hexane	C_6H_{14}	86.1770	4.9596	436.9	453.8	0.0688	2.9755	4.4035	0.38631
2-Methylpentane	C_6H_{14}	86.1770	6.769	436.6	435.76	0.0682	2.9755	4.4035	0.38526
3-Methylpentane	C_6H_{14}	86.1770	6.103	452.5	448.2	0.0682	2.9755	4.4035	0.37902
Neohexane	C_6H_{14}	86.1770	9.859	446.7	419.92	0.0667	2.9755	4.4035	0.38231
2,3-Dimethylbutane	C_6H_{14}	86.1770	7.406	454.0	440.08	0.0665	2.9755	4.4035	0.37762
n-Heptane	C_7H_{16}	100.2040	1.621	396.8	512.8	0.0682	3.4598	3.7872	0.38449
2-Methylhexane	C_7H_{16}	100.2040	2.273	396.0	494.44	0.0673	3.4598	3.7872	0.38170
3-Methylhexane	C_7H_{16}	100.2040	2.13	407.6	503.62	0.0646	3.4598	3.7872	0.37882
3-Ethylpentane	C_7H_{16}	100.2040	2.012	419.2	513.16	0.0665	3.4598	3.7872	0.38646
2,2-Dimethylpentane	C_7H_{16}	100.2040	3.494	401.8	476.98	0.0665	3.4598	3.7872	0.38651
2,4-Dimethylpentane	C_7H_{16}	100.2040	3.294	397.4	475.72	0.0667	3.4598	3.7872	0.39627
3,3-Dimethylpentane	C_7H_{16}	100.2040	2.775	427.9	505.6	0.0662	3.4598	3.7872	0.38306
Triptane	C_7H_{16}	100.2040	3.376	427.9	496.24	0.0636	3.4598	3.7872	0.37724
n-Octane	C_8H_{18}	114.2310	0.5371	360.7	564.15	0.0673	3.9441	3.322	0.38334
Di-isobutyl	C_8H_{18}	114.2310	1.1020	361.1	530.26	0.0676	3.9441	3.322	0.37571
Isooctane	C_8H_{18}	114.2310	1.7090	372.7	519.28	0.0657	3.9441	3.322	0.38222
n-Nonane	C_9H_{20}	128.2580	0.17155	330.7	610.72	0.0693	4.4284	2.9588	0.38248

n-Decane	$C_{10}H_{22}$	142.2850	0.06088	304.6	652.1	0.0702	4.9127	2.6671	0.38181
Cyclopentane	C_5H_{10}	70.1340	9.917	653.8	461.1	0.0594	2.4215	5.411	0.27122
Methylcyclopentane	C_6H_{12}	84.1610	4.491	548.8	499.28	0.0607	2.9059	4.509	0.30027
Cyclohexane	C_6H_{12}	84.1610	3.267	590.7	536.6	0.0586	2.9059	4.509	0.29012
Methylcyclohexane	C_7H_{14}	98.1880	1.609	503.4	570.2	0.0600	3.3902	3.8649	0.31902
Ethylene	C_2H_4	28.0540	1400	731.0	48.54	0.0746	0.9686	13.527	0.35789
Propylene	C_3H_6	42.0810	232.8	676.6	198.31	0.0717	1.4529	9.0179	0.35683
Butylene	C_4H_8	56.1080	62.55	586.4	296.18	0.0683	1.9373	6.7636	0.35535
Cis-2-butene	C_4H_8	56.1080	45.97	615.4	324.31	0.0667	1.9373	6.7636	0.33275
Trans-2-butene	C_4H_8	56.1080	49.88	574.9	311.8	0.0679	1.9373	6.7636	0.35574
Isobutene	C_4H_8	56.1080	64.95	580.2	292.49	0.0681	1.9373	6.7636	0.36636
1-Pentene	C_5H_{10}	70.1340	19.12	509.5	376.86	0.0674	2.4215	5.411	0.35944
1,2-Butadene	C_4H_6	54.0920	36.53	656.0	354	0.0700	1.8677	7.0156	0.34347
1,3-Butadene	C_4H_6	54.0920	59.46	620.3	306	0.0653	1.8677	7.0156	0.34223
Isoprene	C_5H_8	68.1190	16.68	582.0	403	0.0660	2.3520	5.571	0.35072
Acetylene	C_2H_2	26.0380		890.4	95.29	0.0693	0.8990	14.574	0.39754
Benzene	C_6H_6	78.1140	3.225	710.4	552.15	0.0531	2.6971	4.8581	0.24295
Toluene	C_7H_8	92.1410	1.033	595.5	605.5	0.0549	3.1814	4.1184	0.26005
Ethylbenzene	C_8H_{10}	106.1670	0.3716	523	651.22	0.0564	3.6657	3.5744	0.27768
o-Xylene	C_8H_{10}	106.1670	0.2643	541.6	674.85	0.0557	3.6657	3.5744	0.28964
m-Xylene	C_8H_{10}	106.1670	0.3265	512.9	650.95	0.0567	3.6657	3.5744	0.27427
p-Xylene	C_8H_{10}	106.1670	0.3424	509.2	649.47	0.0572	3.6657	3.5744	0.27470
Styrene	C_8H_8	104.1520	0.2582	587.8	703	0.0534	3.5961	3.6435	0.26682

(Continued)

Properties of Hydrocarbon Gases—cont'd

			Vapor	Critical Constants			Ideal Gas, 14.696 psia, 60°F		
Gas	Formula	Molecular Weight	Pressure at 100°F (psia)	Pressure (psia)	Temperature (°F)	Volume (ft^3/lb)	Gravity (air = 1.00)	Volume (ft^3/lb)	Specific Heat (Btu/lb/°F)
Isopropylbenzene	C_9H_{12}	120.1940	0.188	465.4	676.2	0.0569	4.1500	3.1573	0.30704
Methyl alcohol	CH_4O	32.0420	4.631	1174	463.01	0.0590	1.1063	11.843	0.32429
Ethyl alcohol	C_2H_6O	46.0690	2.313	891.7	465.31	0.0581	1.5906	8.2372	0.33074
Carbon monoxide	CO	28.0100		506.8	−220.51	0.0527	0.9671	13.548	0.24847
Carbon dioxide	CO_2	44.0100		1071	87.73	0.0342	1.5196	8.6229	0.19909
Hydrogen sulfide	H_2S	34.0820	394.59	1306	212.4	0.0461	1.1768	11.134	0.23838
Sulfur dioxide	SO_2	64.0650	85.46	1143	315.7	0.0305	2.2120	5.9235	0.14802
Ammonia	NH_3	17.0305	211.9	1647	270.2	0.0681	0.5880	22.283	0.49678
Air	N_2+O_2	28.9625		546.9	−221.29	0.0517	1.0000	13.103	0.2398
Hydrogen	H_2	2.0159		187.5*	−400.3	0.5101	0.06960	188.25	3.4066
Oxygen	O_2	31.9988		731.4	−181.4	0.0367	1.1048	11.859	0.21897
Nitrogen	N_2	28.0134		493	−232.48	0.0510	0.9672	13.546	0.24833
Chlorine	Cl_2	70.9054	157.3	1157	290.69	0.0280	2.4482	5.3519	0.11375
Water	H_2O	18.0153	0.95	3200.1	705.1	0.04975	0.62202	21.065	0.44469
Helium	He	4.0026		32.99	−450.31	0.2300	0.1382	94.814	1.24040
Hydrogen chloride	HCl	36.4606	906.71	1205	124.75	0.0356	1.2589	10.408	0.19086

ASTM METHOD FOR VISCOSITY VERSUS TEMPERATURE OF LIQUIDS

The ASTM method of calculating the viscosity variation with temperature for a petroleum liquid is explained below. This does not require the use of the special logarithmic graph paper.

$$\log\log(Z) = A - B\log(T) \quad \text{(A 1.5)}$$

where

log – Logarithm to base 10
Z – Variable that depends on viscosity of the liquid, v
v – Viscosity of liquid, cSt
T – Absolute temperature, °R or K
A and B – constants that depend on the specific liquid

The variable Z is defined as follows:

$$Z = (v + 0.7 + C - D) \quad \text{(A 1.6)}$$

where C and D are as follows:

$$C = \exp[-1.14883 - 2.65868(v)] \quad \text{(A 1.7)}$$

$$D = \exp[-0.0038138 - 12.5645(v)] \quad \text{(A 1.8)}$$

C, D, and Z are all functions of the kinematic viscosity.

Given two sets of temperature viscosity values (T_1, v_1) and (T_2, v_2), we can calculate the corresponding values of C, D, and Z from Eqs (A 1.6) to (A 1.8).

We can then come up with two equations using the pairs of (T_1, Z_1) and (T_2, Z_2) values by substituting these values into Eq. (A 1.5) as shown below:

$$\log\log(Z_1) = A - B\log(T_1) \quad \text{(A 1.9)}$$

$$\log\log(Z_2) = A - B\log(T_2) \quad \text{(A 1.10)}$$

From the above equations, the two unknown constants A and B can be easily calculated, since T_1, Z_1, and T_2, Z_2 values are known.

VISCOSITY OF A MIXTURE OF LIQUIDS USING BLENDING INDEX

In this method, a Blending Index is calculated for each liquid based on its viscosity. Next, the Blending Index of the mixture is calculated from the individual blending indices by using the weighted average of the composition of the mixture. Finally, the viscosity of the blended mixture is calculated using the Blending Index of the mixture. The equations used are as follows:

$$H = 40.073 - 46.414 \log_{10} \log_{10}(V + B) \quad \text{(A 1.11)}$$

$$B = 0.931\,(1.72)^V \quad \text{for} \quad 0.2 < V < 1.5 \quad \text{(A 1.12)}$$

$$B = 0.6 \quad \text{for} \quad V >= 1.5 \quad \text{(A 1.13)}$$

$$H_m = [H_1(pct_1) + H_2(pct_2) + \cdots]/100 \quad \text{(A 1.14)}$$

where

H, H_1, H_2, ... – Blending Index of liquids
H_m – Blending Index of mixture
B – Constant in Blending Index equation
V – Viscosity in cSt
pct_1, pct_2, ... – Percentage of liquids 1, 2, ... in blended mixture

VISCOSITIES OF COMMON HYDROCARBON GASES

Gas	Viscosity (cP)
Methane	0.0107
Ethane	0.0089
Propane	0.0075
Isobutane	0.0071
n-Butane	0.0073
Isopentane	0.0066
n-Pentane	0.0066
Hexane	0.0063
Heptane	0.0059
Octane	0.0050
Nonane	0.0048
Decane	0.0045
Ethylene	0.0098
Carbon monoxide	0.0184
Carbon dioxide	0.0147
Hydrogen sulphide	0.0122
Air	0.0178
Nitrogen	0.0173
Helium	0.0193

Note: Properties are based on 50°F temperature and atmospheric pressure.

GAS COMPRESSIBILITY FACTOR CALCULATION METHODS

Dranchuk, Purvis, and Robinson Method

In this method of calculating the compressibility factor, the coefficients A_1, A_2, etc., are used in a polynomial function of the reduced density ρ_r as follows:

$$Z = 1 + \left(A_1 + \frac{A_2}{T_{pr}} + \frac{A_3}{T_{pr}^3}\right)\rho_r + \left(A_4 + \frac{A_5}{T_{pr}}\right)\rho_r^2 + \frac{A_5 A_6 \rho_r^5}{T_{pr}} + \frac{A_7 \rho_r^3}{T_{pr}^3 (1 + A_8 \rho_r^2) e^{(-A_8 \rho_r^2)}} \quad \text{(A 1.15)}$$

where

$$\rho_r = \frac{0.27 P_{pr}}{Z T_{pr}} \quad \text{(A 1.16)}$$

and

$A_1 = 0.31506237$
$A_2 = -1.04670990$
$A_3 = -0.57832729$
$A_4 = 0.53530771$
$A_5 = -0.61232032$
$A_6 = -0.10488813$
$A_7 = 0.68157001$
$A_8 = 0.68446549$
P_{pr} = Pseudo-reduced pressure
T_{pr} = Pseudo-reduced temperature

Other symbols have been defined before.

American Gas Association (AGA) Method

The AGA Method for the compressibility factor uses a complicated mathematical algorithm and therefore does not lend itself easily to manual calculations. Generally, a computer program is used to calculate the compressibility factor. Mathematically, the AGA method is represented by the following function:

$$Z = \text{Function (gas properties, pressure, temperature)} \quad \text{(A 1.17)}$$

where gas properties include the critical temperature, critical pressure, and gas gravity.

The AGA-IGT, Report No. 10 describes in detail this method of calculating Z. This approach is valid for gas temperatures in the range of 30–120°F and for pressures not exceeding 1380 psig. The compressibility factor calculated using this method is quite accurate and generally within 0.03% of those calculated using the Standing-Katz chart in these range of temperatures and pressures. When temperatures and pressures are higher than these values, the compressibility factor calculated using this method is within 0.07% of the value obtained from Standing-Katz chart.

The reader may also refer to the AGA publication Report No. 8, Second Edition, November 1992 for more information on compressibility factor calculation methods.

California Natural Gas Association (CNGA) Method

This is a fairly simple equation for quickly calculating the compressibility factor, when the gas gravity, temperature and, pressure are known, as described in the following equation:

$$Z = \frac{1}{\left[1 + \left(\frac{P_{avg} 344400 (10)^{1.785G}}{T_f^{3.825}}\right)\right]} \tag{A 1.18}$$

where

P_{avg} – Average gas pressure, psig
T_f – Average gas temperature, °R
G – Gas gravity (air = 1.00)

Note that pressure used in this equation is the gauge pressure.

This formula for the compressibility factor is valid when the average gas pressure P_{avg} is more than 100 psig. For pressures less than 100 psig, Z is approximately equal to 1.00.

Chapter 3

UNDERGROUND UTILITY PERMIT

State of <State>

County of <County>

DATE ______________

PERMITTEE'S NAME ____

ADDRESS ______________________________ PHONE # ______________

Your request for permission to install a ______________________________
line along and/or across ____________________________ at ______________
is granted subject to the following terms and conditions:

IT IS UNDERSTOOD that the permittee will cause the installation at no expense whatsoever to <County> County and that the permittee will own and maintain the same after installation.

Where the installation crosses the roadway, it shall be encased in pipe of larger diameter and the crossing shall be as perpendicular to the roadway as physically possible. However, open cut shall be allowed up to the edge of the surface portion of the highway. All installations shall be installed by the method of boring or jacking through beneath the road surface, unless prior arrangements have been made and approved by the Road Supervisor. No water shall be used in the boring and no tunneling shall be permitted.

The ____ line shall be installed beneath the surface of the right-of-way at a minimum depth of 48 inches; unless the right-of-way ditch is full of dirt, permittee will need an additional depth of 48 inches; and the disturbed portion of the roadway shall be restored to its original condition. The back filling shall be made in 6" lifts and mechanically tamped and packed, and the last 12" of back fill shall be of stable granular material such as crushed rock or gravel.

Where installation crosses any ditches, canals or water carrying structures, where possible, it shall be pushed through and beneath in a pipe of larger diameter thereby eliminating the necessity of trenching. In no case shall the flow of water ever be impaired or interrupted.

This work must be accomplished in accordance with accepted good practices and conform to the recommendations of the national electric safety code and to such <State> statute as are applicable.

SPECIAL PROVISIONS: ____

____

____

The Permittee shall maintain the installation at all times and agrees to hold <County> County, the agencies thereof, and their officers and employees harmless from any and all loss and damage which may arise out of or be connected with the installation, maintenance, alteration, removal, or presence of the installation herein referred to or any work or facility connected therewith within the area covered by this permit.

This work shall be completed within ____________________________ from the above date. No work shall be allowed on Saturdays or Sundays. No open trench shall be permitted in traveled roadway after dark, unless otherwise specified in Special Provisions.

Permittee shall be required to shut off lines and remove all combustible materials from the highway right-of-way when requested to do so by the County of <County> because of necessary highway construction or maintenance operations.

If the County so requires, Permittee shall mark this installation with markers acceptable to the County of <County> at the locations designated by the County of <County>.

Permits involving encroachment on the national System of Interstate Defense Highways require concurrence by the U.S. Bureau of Public Road prior to the issuance of permit by the County of <County>.

The traveling public must be protected during this installation with proper warning signs or signals both day and night. Warning signs and signals shall be installed by and at the expense of the Permittee and in accordance with directions given by the Supervisor or his subordinate.

In the event any changes are made to this highway in the future that would necessitate removal for relocation of this installation, Permittee will do so promptly at his own expense upon written request from the County of <County>, State of <State>. The County will not be responsible for any damage that may result in the maintenance of the highway to installation placed inside <County> County right-of-way limits.

County of <County>

State of <State>

BY: ______________________ BY: ______________________
Road Supervisor <County> County Board Chairman

DATE: ____________________ DATE: ____________________

In accepting this Permit, the undersigned representing the Permittee, verifies that he has read and understood all of the foregoing provisions; that he has authority to sign for and bind the Permittee; and that by virtue of his signature, the Permittee is bound by all the conditions set forth herein.

BY: ______________________
Permittee

DATE: __________ PHONE # ______________

FEES ARE TO BE SUBMITTED WITH PERMIT APPLICATION.

UNDERGROUND PERMIT FEES

$ <Amount> – FOR BORING UNDER DIRT OR GRAVEL ROADS.
$ <Amount> – FOR DITCHING OR DIGGING IN THE RIGHT-OF-WAY.
PERMIT FOR LAYING LINES IN BARROW DITCH @ $<Amount>/FOOT.
PLUS <Amount> FOR DITCHING OR DIGGING IN RIGHT-OF-WAY.
PLUS ROAD CROSSING FEE.
PLUS ANY DAMAGES.

LICENSE AGREEMENT

THIS LICENSE AGREEMENT, effective the <Date> is executed by and between <Entity>, having its principal place of business at <Address>, (hereinafter referred to as "Licensor") and <Company>, a <State> corporation, having its principal place of business at <Address>, (hereinafter referred to as "Licensee").

Licensor and Licensee agree as set forth hereunder.

Subject to the terms, conditions and covenants hereinafter set forth, Licensor licenses and permits Licensee to use portions of Licensor's properties located in the County of <County>, State of <State> more particularly described as follows:

<Description>

Subject to the terms, conditions and covenants hereinafter set forth, Licensor licenses and permits Licensee to use and enjoy the License Areas for the construction, installation, maintenance, operations, alteration, repair, replacement and removal of a pipeline with valves, tie-overs and other appurtenant facilities for the transmission of natural gas (the "Permitted Activity").

In consideration of the foregoing license and permission, Licensor and Licensee agree to the following terms, conditions and covenants:

1. This license and permission shall be revocable only to the extent Licensee fails to observe or honor any term, condition or covenant hereinafter set forth.
2. The period of the license and permission granted herein shall be for <Number> years beginning on the date of execution of the License Agreement and continuing through until the same date of the <Number> year following the commencement date.
3. The license fee for the permission granted herein shall be <Amount> Dollars for the entire <Term> term payable at the time a fully-executed License Agreement is delivered to Licensee.
4. After the construction and installation of the pipeline, Licensee shall restore the disturbed License Areas in a manner as near as practicable to the original condition of the premises.
5. Licensee shall indemnify and save Licensor harmless from all claims, liability, loss damage cost or expense arising out of the negligent exercise of the license and permission granted herein to Licensee and its employees or contractors, excepting therefrom any such damage or loss that arise from the negligent act or omission of Licensor, its employees, representatives or contractors.
6. Licensee shall operate and maintain the pipeline in an orderly manner and in compliance with all applicable laws, rules and regulations.
7. Licensor agrees that its activities shall not interfere with Licensee's use of the License Areas and shall keep clear said areas for the free and full right of ingress and egress to, over and across the License Areas for Licensee's intended purpose.
8. All modifications and amendments to the License Agreement must be in writing and executed by Licensor and Licensee in order to be effective.

IN WITNESS WHEREOF, the parties hereto have hereunto caused this agreement to be executed by their respective authorized officials, effective as of the date first written above.

____________________	____________________
____________________	____________________
____________________	____________________
____________________	____________________

PERMIT INVESTIGATION REPORT

Page ________ of ______ Page(s), including attachments

PROJECT NAME: ____

PROJECT NUMBER: ____

TYPE PERMIT REQUIRED: _____________________________________

(Crossing, Longitudinal, Zoning, License/Franchise, Building, etc.)
(Attach Application)

DATE CONTACT MADE: ____

AGENCY NAME: ____

(City, County, State [Zoning, Planning Commissions, Franchise Depts., etc.], BLM, BIA, DOT, etc.)

PERSON CONTACTED: ____

(Name & Title)

Address: ____

Mailing (if different) ______________________________________

Telephone # ______________ Fax # __________ E-Mail ____________

TYPE PERMIT NEEDED: _______________________________________

(Temporary/Special Use, Crossing, Longitudinal Encroachment, Franchise/License, Drainage/Flood Plain Permits, Land Development Permits, Building Permit, etc.)

SETBACK REQUIREMENTS: _____________________________________

____

____

ZONING REQUIREMENTS: ______________________________________

____

____

Will setback variance(s) be required? Explain: ______________________

____

____

Will subdivision process be required? Explain: ______________________

____

____

List any visual, noise, or other requirements: ______________________

____

____

If approval required, who gives approval? __________________________

(City/County/Franchise Board of Commissioners, etc.)

If approval required, how soon can application get on agenda for next meeting? ______________________

____

When is the next meeting that will consider application? ______________________

Is a public hearing required? ☐ Yes ☐ No

How much notice to the public is required? ______________________

Approximate time to go through approval process: ______________________

Additional comments: ______________________

____

LIST ANY OTHER AGENCIES WHO ARE REQUIRED TO REVIEW AND/OR APPROVE PERMIT(S): ______________________

____

____

____

ADDITIONAL REQUIREMENTS: **(Attach Guidelines, Sample drawings, including typicals and Detailed plans)**

Drawing type(s): ______________________

[Alignment sheet(s), Cross-section, Profile, As-built, Mylars, etc.]

____

Narrative requirements: ______________________

Number of drawings required: ______________________

Must drawings be signed and sealed by registered engineer of state? ☐ Yes ☐ No

Are bond and insurance required? ☐ Yes ☐ No

Requirements of bond and insurance: ______________________

____

____

CONSTRUCTION REQUIRMENTS: ______________________

____

(Open cut, Bore, Distance of bore pits from edge of Pavement/Right-of-way, Traffic control, etc.)

Distance from construction site notifications required: ______________________

Number to call for location of facilities of other utilities: ______________________

(DigAlert, OneCall, etc.)

COMMENTS: ______________________

____

____

ATTACHMENTS: ______________________

(Applications, Rules and regulations, Sample drawings, Maps, Drawings reflecting utilities in area, Drainage plans, etc.)

____

THIS FORM COMPLETED BY: ______________ DATE: ______________

Chapter 4

RIGHT-OF-WAY BUDGET

Budget Index No.	Description	Date	Date	Date	Date	Date	Date	Date	Date	Total
	Land and R/W costs									
	Right-of-way	$0	$0	$0	$0	$0	$0	$0	$0	$0
	Damages	$0	$0	$0	$0	$0	$0	$0	$0	$0
	Site costs	$0	$0	$0	$0	$0	$0	$0	$0	$0
	Other land costs	$0	$0	$0	$0	$0	$0	$0	$0	$0
	Personnel and overhead									
	Management	$0	$0	$0	$0	$0	$0	$0	$0	$0
	Supervision	$0	$0	$0	$0	$0	$0	$0	$0	$0
	Agents	$0	$0	$0	$0	$0	$0	$0	$0	$0
	Support staff	$0	$0	$0	$0	$0	$0	$0	$0	$0
	Expenses									
	Mgmt/Supv travel	$0	$0	$0	$0	$0	$0	$0	$0	$0
	Agent expenses	$0	$0	$0	$0	$0	$0	$0	$0	$0
	Auto expenses	$0	$0	$0	$0	$0	$0	$0	$0	$0
	Office	$0	$0	$0	$0	$0	$0	$0	$0	$0
	Telephone	$0	$0	$0	$0	$0	$0	$0	$0	$0
	Office machines	$0	$0	$0	$0	$0	$0	$0	$0	$0
	Supplies	$0	$0	$0	$0	$0	$0	$0	$0	$0
	Miscellaneous	$0	$0	$0	$0	$0	$0	$0	$0	$0
	Recording	$0	$0	$0	$0	$0	$0	$0	$0	$0
	Outside costs									
	Legal	$0	$0	$0	$0	$0	$0	$0	$0	$0
	Appraisal	$0	$0	$0	$0	$0	$0	$0	$0	$0
	Agricultural consultant	$0	$0	$0	$0	$0	$0	$0	$0	$0
	Other consultant	$0	$0	$0	$0	$0	$0	$0	$0	$0
	Title research	$0	$0	$0	$0	$0	$0	$0	$0	$0
Total	**Total**	$0	$0	$0	$0	$0	$0	$0	$0	$0

TITLE RESEARCH CHECK LIST

For each tract run, check off the items listed below as they are completed. Get a copy of all effective documents on the subject property (i.e., current deed, outstanding mortgages and liens that have not been released, easements, leases, etc.)

Deed Records:

- ☐ Run reverse indices from present for prescribed period of time (no less than 30 years is advisable).
- ☐ Run direct indices to check for outsells (portions out of subject tract).
- ☐ Check for surface mining leases (coal, uranium, etc.)
- ☐ Attach copy of last deed in chain.
- ☐ Check for recorded tenant agreements (leases) and attach a copy
- ☐ Is plat attached to any instrument? If so, obtain a copy.
- ☐ Show all prior easements on property in chain of title and obtain copy of each.

Mortgage Records:

- ☐ Run present owner direct in mortgage records from acquisition date.
- ☐ Run prior owners direct in event last deed conveyed is subject to prior mortgage(s).

☐ Yes ☐ No Does existing Mortgage(s) prevent conveyance of any kind?

☐ Yes ☐ No Does Mortgage state that all or portion of conveyance monies payable toward mortgage?

Ad Valorem Taxes on Subject Real Property:

- ☐ Check tax assessment records for current taxpayer. (This can be used for a starting point when other ownership sources are not available.)
- ☐ Obtain tax certificate or other proof of the status of taxes.

☐ Check tax exempt records for cemeteries, churches, etc., located within tract.

☐ Get copy of tax map reflecting APN (Assessor Parcel Number) or other tax identification number.

Liens and Judgments:

☐ Check all judgment records.

☐ Check for mechanic's and materialmen's liens.

☐ Check for federal and state tax liens.

☐ Check lis pendens records.

Probate Records:

☐ Check probate records for current owner of record.

☐ Check probate records when skips or irregularities appear in the chain of title.

Present Ownership:

☐ Yes ☐ No Is record owner the same as the party paying taxes? If not, determine why.

☐ Yes ☐ No Is record owner a corporation? Obtain Articles of Incorporation.

☐ Yes ☐ No Is the ownership in trust? Does Trustee(s) have right to convey?

☐ Yes ☐ No Is any interest in the present ownership a minor?

☐ Yes ☐ No Was subject property acquired through Tax Sale?

☐ Yes ☐ No Is property under Contract for Deed? Who has right to convey an interest in contracted land, Contract Seller or Contract Purchaser? Get both to execute conveyance document or have the one with right to convey execute conveyance document and the other ratify the conveyance.

TITLE REPORT

DATE: ____________ TRACT R/W #: ____________
COUNTY: ____________ PROJECT NAME: ____________
STATE: ____________ PROJECT NUMBER: ____________
APN/Tax I.D. No. ____________ Long/Lat. of Site ____________
Title researched from ____________ to ________ for a total of ________ years
LEGAL DESCRIPTION: (may include here or attach document containing description) ____________

A.	CURRENT OWNERSHIP (add additional page, if necessary)

#1 NAME ____________
ALSO KNOWN AS OR
FORMERLY KNOWN AS ____________
PERCENT INTEREST ____________ SS/TAX ID # ____________
ADDRESS ____________
PHONE # ____________

#2 NAME ____________
ALSO KNOWN AS OR
FORMERLY KNOWN AS ____________
PERCENT INTEREST ____________ SS/TAX ID # ____________
ADDRESS ____________
PHONE # ____________

#3 NAME ____________
ALSO KNOWN AS OR
FORMERLY KNOWN AS ____________
PERCENT INTEREST ____________ SS/TAX ID # ____________
ADDRESS ____________
PHONE # ____________

#4 NAME ____________
ALSO KNOWN AS OR
FORMERLY KNOWN AS ____________
PERCENT INTEREST ____________ SS/TAX ID # ____________
ADDRESS ____________
PHONE # ____________

B. DOCUMENTS VESTING FEE INTEREST

[Add additional page, if necessary. Attach copy of each acquisition document and last conveyance document containing legal description if not contained in latest acquisition document(s)]

OWNER NUMBER(S) FROM "A" ABOVE: ____________

Title vested in ______________________________

____

(Name of Current Owner(s) as reflected in document including marital status or type of ownership (i.e., a single man, a single woman, a widower, a widow, a married man/woman dealing in his/her sole and separate property, husband and wife, tenants in common, joint tenants with right of survivorship, as community property with right of survivorship, etc.)

by virtue of ______________________________

[Title or type of vesting instrument (i.e., patent/land grant, warranty/general/special/quitclaim/sheriff's deed, testate succession, intestate succession, other legal action, etc.)]

dated __________ filed for record __________ in the __________ Records of __________ County, State of __________

from ______________________________

(Name of Grantor, etc.)

OWNER NUMBER(S) FROM "A" ABOVE: ____________

Title vested in ______________________________

____

(Name of Current Owner(s) as reflected in document including marital status or type of ownership (i.e., a single man, a single woman, a widower, a widow, a married man/woman dealing in his/her sole and separate property, husband and wife, tenants in common, joint tenants with right of survivorship, as community property with right of survivorship, etc.)

by virtue of ______________________________

[Title or type of vesting instrument (i.e., patent/land grant, warranty/general/ special/ quitclaim/sheriff's deed, testate succession, intestate succession, other legal action, etc.)]

dated __________ filed for record __________ in the __________ Records of __________ County, State of __________

from ______________________________

(Name of Grantor, etc.)

NOTES: ______________________________

C. DOCUMENTS VESTING LEASEHOLD INTEREST

Including Farming, Grazing, Mineral &/or Mining and other Leases with Surface Rights

(Add additional page, if necessary. Attach copy of each document)

#1 ______________________________

(Title of document)

dated __________ filed for record __________ in the __________ Records of __________ County, State of __________

from ______________________________

to ______________________________

#2	

(Title of document)

dated ________________ filed for record ________________ in the

________ Records of ________ County, State of ________

from ________________

to

NOTES: ________________

TO THE BEST OF MY KNOWLEDGE, THERE ARE NO UNRELEASED OR UNSATISFIED MORTGAGES, LIENS, JUDGMENTS OR OTHER ENCUMBRANCES AGAINST OR ON SUBJECT PROPERTY EXCEPT AS FOLLOWS FOR THE PERIOD SEARCHED:

D. MORTGAGES, LIENS, JUDGMENTS, TAX LIENS, EASEMENTS, LEASES (OTHER THAN LISTED IN "C" ABOVE) AND ANY OTHER ENCUMBRANCE UPON OR AGAINST THE PROPERTY

(Add additional page, if necessary. Attach copy of each document which is in effect and has not been released or otherwise has not been determined to unencumber subject property)

#1	

[Title of document (i.e., Deed of Trust, Mortgage, Mechanic's or Materialmen's Lien, Judgment, Lis Pendens, Right-of-Way Easement, Utility Easement, Powerline Easement, Fiber Easement, Communications Easement, etc.)]

dated ________________ filed for record ________________ in the

________ Records of ________ County, State of ________

from ________________

to

#2	

[Title of document (i.e., Deed of Trust, Mortgage, Mechanic's or Materialmen's Lien, Judgment, Lis Pendens, Right-of-Way Easement, Utility Easement, Powerline Easement, Fiber Easement, Communications Easement, etc.)]

dated ________________ filed for record ________________ in the

________ Records of ________ County, State of ________

from ________________

to

#3	

[Title of document (i.e., Deed of Trust, Mortgage, Mechanic's or Materialmen's Lien, Judgment, Lis Pendens, Right-of-Way Easement, Utility Easement, Powerline Easement, Fiber Easement, Communications Easement, etc.)]

dated ________________ filed for record ________________ in the

________ Records of ________ County, State of ________

from ________________

to

#4	
[Title of document (i.e., Deed of Trust, Mortgage, Mechanic's or Materialmen's Lien, Judgement, Lis Pendens, Right-of-Way Easement, Utility Easement, Powerline Easement, Fiber Easement, Communications Easement, etc.)]	
dated ____________ filed for record ____________ in the	
____________ Records of ____________ County, State of ____________	
from ____________	
to	

☐YES ☐NO	OUTSALES OR CONVEYANCES WHICH AFFECT THE DESCRIPTION OF THE PROPERTY ARE ATTACHED HERETO.
☐YES ☐NO	UNREDEEMED TAX SALES AFFECTING SAID PROPERTY.
☐YES ☐NO	TAXES ARE CURRENT.
☐YES ☐NO	TAXES ARE RENDERED IN OWNERSHIP AS REFLECTED IN "A." ABOVE.
	IF NO, SHOW NAME AND ADDRESS OF TAXPAYER: NAME: ____________ ADDRESS:
☐YES ☐NO	OWNERSHIP IS ESTATE, LIFE ESTATE, TRUST, COMPANY, CORPORATION, PARTNERSHIP, L.L.C., OR VIA CONTRACT FOR DEED, ETC. NOTE: GET PROOF OF CONVEYANCE RIGHTS.

THE ABOVE INFORMATION IS A RESULT OF THE TITLE SEARCH PERFORMED ON THE SUBJECT PROPERTY FOR THE PERIOD REQUESTED OR FROM THE DATE OF THE CURRENT OWNERSHIP'S VESTING DOCUMENT(S), WHICHEVER IS EARLIER. ALTHOUGH A THOROUGH EXAMINATION AND SEARCH WAS MADE FOR THE PERIOD COVERED, NO GUARANTEE IS MADE AS TO THE CORRECTNESS OF THE ABOVE INFORMATION UNLESS OTHERWISE AGREED TO IN A SEPARATE AGREEMENT BETWEEN ABSTRACTOR AND THE PARTY FOR WHICH THE ABSTRACT WAS PREPARED.

PREPARED AND SUBMITTED BY:

ABSTRACTOR: ____________________

PRINTED NAME: ____________________

REPRESENTING: ____________________

SURVEY PERMIT

State of <State> ____________
County <County> ____________
Parcel No. <Number> ____________

RECORD OWNER (Name, Address, Phone #)	TENANT (Name, Address, Phone #)

Brief Tract Description:

____
____
____
____
____

I/we hereby give my/our permission to <Company> to survey for a pipeline location across the above property in <County> County, State of <State>.

____
☐ Owner ☐ Tenant

____
☐ Owner ☐ Tenant

Date Signed:____________________________

____
☐ Owner ☐ Tenant

Date Signed:____________________________

____
R/W Agent

PIPELINE RIGHT OF WAY EASEMENT

State of <State>

County of <County>

Parcel No. <Number>

STATE OF <STATE>)
)
COUNTY OF <County>)

KNOW ALL MEN BY THESE PRESENTS, that for and in consideration of <Amount> Dollars ($<Amount>) in hand paid, the receipt of which is hereby acknowledged, <Name>, hereinafter called GRANTOR (whether one or more), hereby grants and conveys to <Company>, a corporation organized under the laws of the State of <State>, with its principal office at <City, State>, its successors and assigns, hereinafter called GRANTEE, the rights of way, easements and privileges to install, repair, maintain, operate and remove pipelines, valves, valve guards, drips, fittings, meters, and similar appurtenances on the following described property:

<Description>

with ingress and egress to and from the same.

TO HAVE AND TO HOLD unto said GRANTEE, its successors and assigns, so long as the same shall be used for the purposes aforesaid, and GRANTEE hereby agrees to pay any damages which may arise to crops, timber, or fences from the use of said premises for such purposes.

GRANTOR covenants and agrees that he will not impound water or construct buildings or structures of any type whatsoever on the above described facilities. This shall be a covenant running with the land and shall be binding on GRANTOR, his heirs and assigns.

Dated this <Date>.

<Signature Block>

<Acknowledgement>

VALVE EASEMENT

State of <State>

County <County>

Parcel No. <Number>

FOR AND IN CONSIDERATION OF THE SUM OF <Amount> Dollars ($<Amount>), the receipt of which is hereby acknowledged, <Name>, hereinafter called Grantors (whether one or more), hereby grant unto <Company>, a <State> corporation, hereinafter called Grantee, the right to construct, install, maintain, inspect, protect, operate, replace, change, and remove a gate valve, apparatus and equipment, including, if Grantee desires to do so, fences and structures to enclose the same, or any part thereof, on the following described land, of which Grantors warrant they are the owner in fee simple, situated in <County> County, State of <State>.

A tract of land described as follows:

<Description>

To be used in connection with and as long as any pipeline or pipelines now or hereafter constructed is or are maintained and operated by said Grantee on, over, and through said described land, together with the right of ingress and egress to and from said land for any and all purposes necessary and incident to the exercise by said Grantee of the rights granted by this contract.

The rights herein granted may be assigned in whole or in part.

The terms, conditions, and provisions of this contract shall extend to and be binding upon the heirs, executors, administrators, personal representatives, successors and assigns of the parties hereto.

Dated the <Date>.

<Signature Block>

<Acknowledgement>

SURFACE EASEMENT

STATE OF <STATE>)
)
COUNTY OF <County>)

KNOW ALL MEN BY THESE PRESENTS:

That the undersigned, <Name>, (hereinafter called GRANTORS, whether one or more) for and in consideration of the sum of <Amount> Dollars ($<Amount>) in hand paid, the receipt and sufficiency of which is hereby acknowledged, do hereby grant, bargain, sell and convey unto <Company>, a corporation, its successors and assigns, (hereinafter called GRANTEE) an easement in and to the tract of land hereinafter described for the exclusive use and occupancy thereof for the purpose of erecting and constructing thereon, and thereafter maintaining, operating, altering, repairing and removing a meter station and all appurtenances and equipment used, useful, or convenient in connection therewith, said tract of land being situated in <County> County, <State> and more particularly described as follows:

<Description>

together with the rights of ingress and egress thereto across adjacent lands at convenient points for the enjoyment of the uses, rights and privileges aforesaid; hereby releasing and waiving all rights under and by virtue of any applicable homestead exemption laws.

Grantee shall have the right to fence and enclose the above described tract of land within a single fence or to fence any of the facilities installed thereon in separate enclosures, and all other rights and benefits necessary or convenient for the full enjoyment and use thereof, including, but not by way of limitation, the right to erect a building or buildings and radio facilities thereon, the right to install pipelines, valves, conduits and meter runs above and below the surface thereof, Grantee having the right to select the route or location thereof, and the right, from time to time, to cut and remove all trees, undergrowth, and other obstructions, whether on the above described tract of land or adjacent thereto, which may injure, endanger or interfere with the rights herein granted, but Grantee shall pay to Grantors the reasonable value of any trees, undergrowth, or other obstructions cut or removed from any area adjacent to the above described tract.

TO HAVE AND TO HOLD said easement unto Grantee, its successors and assigns, until a meter station or other facility of Grantee shall be constructed thereon, and so long thereafter as a meter station or other facility of Grantee may be maintained thereon, and Grantors bind themselves, their successors, heirs and assigns to warrant and forever defend all and singular the said premises unto Grantee, its successors and assigns, against every person whomsoever lawfully claiming, or to claim the same, or any part thereof, by, through and under Grantors.

It is mutually understood and agreed that the person securing this grant is without authority from Grantee to make any agreement in respect of the subject matter hereof not herein expressed, and that the rights and interests of Grantee hereunder may be assigned in whole or in part.

Dated this <Date>.

<Signature Block>

<Acknowledgement>

GENERAL WARRANTY DEED

KNOW ALL MEN BY THESE PRESENTS:

THAT, for and in consideration of the sum of <Amount> Dollars ($<Amount>) and other good and valuable consideration, receipt of which is hereby acknowledged, <Name> does hereby GRANT, SELL AND CONVEY unto <Names>, as joint tenants, and not as tenants in common, whereby on the death of one, the survivor, the heirs and assigns of the survivor, to take the entire fee simple title of the following described real estate situate in <County> County, State of <State>, to-wit:

<Description>

TO HAVE AND TO HOLD the same as joint tenants, and not as tenants in common, with the fee simple title in the survivor, the heirs and assigns of the survivor, together with all and singular the tenements, hereditaments and appurtenances belonging or in any wise appertaining thereto;

AND Grantor does hereby WARRANT AND FOREVER DEFEND title to the above described premises, together with all and singular the rights and appurtenances thereto in any wise belonging, unto the said Grantee, their heirs and assigns forever, against every person whomsoever lawfully claiming or to claim the same or any part thereof and further warrants that said premises are free, clear and discharged and unencumbered of and from all former and other grants, titles, charges, judgements, estates, taxes, assessments and encumbrances of whatsoever nature and kind, EXCEPT easements, building restrictions of record and special assessments not yet due;

Dated this <Day> day of <Month>, <Year>.

<Signature Block>

<Acknowledgment>

PIPE AND STORAGE YARD LEASE

This Agreement dated this <Date>, between <Name>, having its principal offices <or residing> at <Address>, hereinafter called Lessor, and <Company>, a <State> corporation, having its principal offices at <Address>, hereinafter called Lessee.

WITNESSETH:

1. Lessor hereby leases to Lessee a parcel of land containing a total of <Number> acres of flat, vacant land suitable for Lessee's use, for the purposes of storing, welding & testing of pipe and valve assembly operations, and for storing of

other pipeline related materials including construction equipment. Lessee and its contractor will also be allowed to set up and maintain portable office trailers during the term(s) of this Lease to help facilitate Lessee's operations. The lease area is located at <Address>, and is more particularly shown as <Name>, on a map entitled <Name>, Drawing labeled <Number> attached hereto and made a part hereof (the "Lease Area").

2. Lessor guarantees that Lessee, its agents and assigns, will have unrestricted access to the Lease Area during the term(s) of this Lease, as extended.

3. The term of this Lease shall run from the <Date>, to <Date>, and for such term Lessee shall pay, in advance, the total sum of <Amount> Dollars. Said sum to be paid on or before the <Date>. The Lessee shall have the option to extend this Lease on a month to month basis for a period of <Period> months commencing on the <Date>, through the <Date>, to be exercised upon payment of <Amount> Dollars per month in advance.

4. Lessee shall have the right to remove shrubbery, trees and brush and to install and maintain a fence upon the Lease Area during the term(s) of this Lease. Any fence installed by Lessee will remain the property of Lessee and will be removed by Lessee prior to the termination of this Lease.

5. At the conclusion of Lessee's operations, Lessee will clean up the Lease Area in a workmanlike manner and remove all materials placed on the Lease Area by Lessee and restore the Lease Area as nearly as practicable to the same condition as prior to commencement of its operations, including leveling all ruts, seeding and removing all debris.

6. Lessee agrees to hold harmless and defend Lessor from the claims and demands of all persons arising out of its negligent operations in the Lease Area, during the term(s) of this Lease.

7. Lessee shall have the right to terminate this Lease if Lessor breaches any provisions herein and does not cure any such breach within <Number (#)> days written notice of such breach.

8. Lessor warrants that it is the owner of the premises of which the Lease Area is a part and has lawful authority to enter into this Lease and will permit Lessee to enjoy the Lease Area without interference.

9. Lessee shall not cause or permit the presence, use, disposal, storage, or release of any Hazardous Substances on or in the Lease Area other than fuel and fluids for vehicles, radioactive material necessary to operate X-ray equipment and painting and pipeline coating materials, oils and fluids for equipment and containerized gases (i.e. acetylene and oxygen cylinders), necessary for welding activities. The painting and pipeline coating materials and hydraulic oil and transmission fluids for equipment shall be stored in a totally enclosed structure with an impervious secondary containment. No equipment maintenance will be performed at the Lease Area. Lessee shall not violate any Environmental Law. Lessee shall promptly give Lessor written notice of any investigations, claim or other action by any government or regulatory agency or private party involving the Lease Area and any Hazardous Substance or Environmental Law of which Lessee has actual knowledge. If, whether during or subsequent to the term and any extended term of this Lease, any governmental or regulatory authority determines that any removal or other remediation of any Hazardous Substance affecting the Lease Area is necessary solely as a result of the activities of the Lessee, its agents or invitees, Lessee shall promptly take all necessary remedial actions in accordance with Environmental Law, and shall indemnify and hold Lessor harmless from all costs, expenses, fines and damages assessed in connection therewith, including Lessor's reasonable attorneys fees. This provision shall survive termination of this Lease.

As used in this Lease, "Hazardous Substances" are those substances defined as toxic or hazardous substances by Environmental Law and the following substances: gasoline, kerosene, other flammable or toxic petroleum products, toxic pesticides and herbicides, volatile solvents, materials containing asbestos or formaldehyde, and radioactive materials other than those previously mentioned. As used in this Lease, "Environmental Law" means federal laws and laws of the State in which the Property is located that relate to health, safety or environmental protection.

10. Lessor warrants that, to the best of Lessor's knowledge, (a) the Lease Area is free of Hazardous Substances and (b) there have been no decrees, injunctions, judgments, orders or writs of an environmental nature relating to the Lease Area, Lessor's adjacent property or their uses, and there are no lawsuits, claims, proceedings or investigations of an environmental nature relating to the Lease Area, Lessor's adjacent property or their uses. If the Lease Area or any portion thereof is rendered untenantable either as a result of the presence of the Hazardous Substances or removal thereof, the rent shall be abated in proportion to the area which has been rendered untenantable during such period of untenantability. In addition, if, whether during or subsequent to the term and any extended term of this Lease, any governmental or regulatory authority determines that any removal or other remediation of any Hazardous Substance affecting the Lease Area is necessary as a result of the activities of the Lessor, its agents or invitees, Lessor shall promptly take all necessary remedial actions in accordance with Environmental Law, and Lessor shall indemnify and hold Lessee harmless from and against all costs, expenses, fines and damages assessed including Lessee's reasonable attorney's fees, in connection with the presence of any Hazardous Substances existing on the Lease Area prior to commencement of this Lease, during the lease term if related to Lessor's activities or the activities of Lessor's agents or invitees, and after the lease termination, the latter subject to Lessee's obligations, if any, herein. This provision shall survive termination of this lease.

11. This Lease shall be binding upon the respective heirs, successors and assigns of the parties hereto.

Dated the date first above written.

<Signature Block>

<Acknowledgement>

DAMAGE RELEASE

State of	<State>
County	<County>
Parcel No.	<Tract No>

The undersigned, hereby acknowledge receipt from <Company> of the sum of <Amount> Dollars ($<Amount>) in full settlement of all claims for damages sustained by the undersigned by reason of the construction and testing of a pipeline across the lands referred to in that certain Right of Way Easement executed by the undersigned which granted a right of way for construction, maintenance and operation of a pipeline and other appurtenances in connection therewith across the real property of the undersigned. In consideration of such payment the undersigned hereby discharges and releases <Company> and its respective agents, servants, employees and contractors from all claims, demands, actions and causes of action which the undersigned has for off right of way and on right of way damage arising out of construction and testing of said pipeline.

It is understood that this release does not in any way cover damage which may be sustained from time to time by reason of the operation, maintenance, repair, altering, moving or removing of the said pipeline after the date hereof.

Dated this <Day> day of <Month>, <Year>.

____
Owner

____
Owner

____
Other

____
Date

ADVANCE DAMAGE RELEASE

State of <State>

County <County>

Parcel No. <Number>

For and in consideration of the sum of <Amount> ($<Amount>) Dollars paid by <Company> (herein designated "<Company>") to the undersigned, the receipt and sufficiency of which is hereby acknowledged, the undersigned do hereby release <Company> its successors and assigns, associates, associated and affiliated companies, contractors, agents and employees of and from any and all claims for damages which may have arisen heretofore or which may arise on account of the construction of a pipeline to be laid across the following described tract of land situated in <County> County, <State>:

<Land Description>

Dated this <Day> day of <Month>, <Year>.

____
Owner

____
Owner

____
Tenant

____
Other

____
Date

Appendix 4

Chapter 7

PIPE SEAM JOINT FACTORS

Specification	Pipe Class	Seam Joint Factor (E)
ASTM A53	Seamless	1
	Electric resistance welded	1
	Furnace lap welded	0.8
	Furnace butt welded	0.6
ASTM A106	Seamless	1
ASTM A134	Electric fusion arc welded	0.8
ASTM A135	Electric resistance welded	1
ASTM A139	Electric fusion welded	0.8
ASTM A211	Spiral welded pipe	0.8
ASTM A333	Seamless	1
ASTM A333	Welded	1
ASTM A381	Double submerged	1
	Arc welded	1
ASTM A671	Electric fusion welded	1
ASTM A672	Electric fusion welded	1
ASTM A691	Electric fusion welded	1
API 5L	Seamless	1
	Electric resistance welded	1
	Electric flash welded	1
	Submerged arc welded	1
	Furnace lap welded	0.8
	Furnace butt welded	0.6
API 5LX	Seamless	1
	Electric resistance welded	1
	Electric flash welded	1
	Submerged arc welded	1
API 5LS	Electric resistance welded	1
	Submerged arc welded	1

PIPELINE INTERNAL DESIGN PRESSURES AND TEST PRESSURES – USCS UNITS

Pipeline Internal Design Pressures and Test Pressures (USCS Units) of Pipe Material API 5LX-70 with SMYS of 70,000 psig

Diameter (in.)	Wall Thickness (in.)	Weight (lb/ft)	Internal Design Pressure (psig) Class 1	Class 2	Class 3	Class 4	Hydrostatic Test Pressure (psig) 90% SMYS	95% SMYS	100% SMYS
4.5	0.237	10.79	5309	4424	3687	2949	6636	7005	7373
	0.337	14.98	7549	6291	5242	4194	9436	9960	10484
	0.437	18.96	9789	8157	6798	5438	12236	12916	13596
	0.531	22.51	11894	9912	8260	6608	14868	15694	16520
6.625	0.250	17.02	3804	3170	2642	2113	4755	5019	5283
	0.280	18.97	4260	3550	2958	2367	5325	5621	5917
	0.432	28.57	6573	5477	4565	3652	8216	8673	9129
	0.562	36.39	8551	7126	5938	4750	10689	11282	11876
8.625	0.250	22.36	2922	2435	2029	1623	3652	3855	4058
	0.277	24.70	3237	2698	2248	1798	4047	4271	4496
	0.322	28.55	3763	3136	2613	2091	4704	4965	5227
	0.406	35.64	4745	3954	3295	2636	5931	6261	6590
10.75	0.250	28.04	2344	1953	1628	1302	2930	3093	3256
	0.307	34.24	2879	2399	1999	1599	3598	3798	3998
	0.365	40.48	3423	2852	2377	1901	4278	4516	4753
	0.500	54.74	4688	3907	3256	2605	5860	6186	6512
12.75	0.250	33.38	1976	1647	1373	1098	2471	2608	2745
	0.330	43.77	2609	2174	1812	1449	3261	3442	3624
	0.375	49.56	2965	2471	2059	1647	3706	3912	4118
	0.406	53.52	3210	2675	2229	1783	4012	4235	4458
	0.500	65.42	3953	3294	2745	2196	4941	5216	5490
14.00	0.250	36.71	1800	1500	1250	1000	2250	2375	2500
	0.312	45.61	2246	1872	1560	1248	2808	2964	3120
	0.375	54.57	2700	2250	1875	1500	3375	3563	3750
	0.437	63.30	3146	2622	2185	1748	3933	4152	4370
	0.500	72.09	3600	3000	2500	2000	4500	4750	5000
16.00	0.250	42.05	1575	1313	1094	875	1969	2078	2188
	0.312	52.27	1966	1638	1365	1092	2457	2594	2730
	0.375	62.58	2363	1969	1641	1313	2953	3117	3281
	0.437	72.64	2753	2294	1912	1530	3441	3633	3824
	0.500	82.77	3150	2625	2188	1750	3938	4156	4375
18.00	0.250	47.39	1400	1167	972	778	1750	1847	1944
	0.312	58.94	1747	1456	1213	971	2184	2305	2427
	0.375	70.59	2100	1750	1458	1167	2625	2771	2917
	0.437	81.97	2447	2039	1699	1360	3059	3229	3399
	0.500	93.45	2800	2333	1944	1556	3500	3694	3889
20.00	0.312	65.60	1572	1310	1092	874	1966	2075	2184
	0.375	78.60	1890	1575	1313	1050	2363	2494	2625
	0.437	91.30	2202	1835	1530	1224	2753	2906	3059
	0.500	104.13	2520	2100	1750	1400	3150	3325	3500
	0.562	116.67	2832	2360	1967	1574	3541	3737	3934
22.00	0.375	86.61	1718	1432	1193	955	2148	2267	2386
	0.500	114.81	2291	1909	1591	1273	2864	3023	3182
	0.625	142.68	2864	2386	1989	1591	3580	3778	3977
	0.750	170.21	3436	2864	2386	1909	4295	4534	4773
24.00	0.375	94.62	1575	1313	1094	875	1969	2078	2188
	0.437	109.97	1835	1530	1275	1020	2294	2422	2549
	0.500	125.49	2100	1750	1458	1167	2625	2771	2917
	0.562	140.68	2360	1967	1639	1311	2951	3114	3278
	0.625	156.03	2625	2188	1823	1458	3281	3464	3646
	0.750	186.23	3150	2625	2188	1750	3938	4156	4375
26.00	0.375	102.63	1454	1212	1010	808	1817	1918	2019
	0.500	136.17	1938	1615	1346	1077	2423	2558	2692
	0.625	169.38	2423	2019	1683	1346	3029	3197	3365
	0.750	202.25	2908	2423	2019	1615	3635	3837	4038

Pipeline Internal Design Pressures and Test Pressures (USCS Units) of Pipe Material API 5LX-70 with SMYS of 70,000 psig—cont'd

Diameter (in.)	Wall Thickness (in.)	Weight (lb/ft)	Internal Design Pressure (psig) Class 1	Class 2	Class 3	Class 4	Hydrostatic Test Pressure (psig) 90% SMYS	95% SMYS	100% SMYS
28.00	0.375	110.64	1350	1125	938	750	1688	1781	1875
	0.500	146.85	1800	1500	1250	1000	2250	2375	2500
	0.625	182.73	2250	1875	1563	1250	2813	2969	3125
	0.750	218.27	2700	2250	1875	1500	3375	3563	3750
30.00	0.375	118.65	1260	1050	875	700	1575	1663	1750
	0.500	157.53	1680	1400	1167	933	2100	2217	2333
	0.625	196.08	2100	1750	1458	1167	2625	2771	2917
	0.750	234.29	2520	2100	1750	1400	3150	3325	3500
32.00	0.375	126.66	1181	984	820	656	1477	1559	1641
	0.500	168.21	1575	1313	1094	875	1969	2078	2188
	0.625	209.43	1969	1641	1367	1094	2461	2598	2734
	0.750	250.31	2363	1969	1641	1313	2953	3117	3281
34.00	0.375	134.67	1112	926	772	618	1390	1467	1544
	0.500	178.89	1482	1235	1029	824	1853	1956	2059
	0.625	222.78	1853	1544	1287	1029	2316	2445	2574
	0.750	266.33	2224	1853	1544	1235	2779	2934	3088
36.00	0.375	142.68	1050	875	729	583	1313	1385	1458
	0.500	189.57	1400	1167	972	778	1750	1847	1944
	0.625	236.13	1750	1458	1215	972	2188	2309	2431
	0.750	282.35	2100	1750	1458	1167	2625	2771	2917
42.00	0.375	166.71	900	750	625	500	1125	1188	1250
	0.500	221.61	1200	1000	833	667	1500	1583	1667
	0.625	276.18	1500	1250	1042	833	1875	1979	2083
	0.750	330.41	1800	1500	1250	1000	2250	2375	2500
	1.000	437.88	2400	2000	1667	1333	3000	3167	3333

For other values of SMYS, multiply table values of pressures by the factor (SMYS/70,000).
For other wall thickness t_1*, multiply table values of pressures by the factor* (t_1/t)*, where* t *is the table wall thickness.*

Example:

For X-80 pipe, NPS 30 pipe, 0.400 in. wall thickness,
Class 1 design pressure = 1680 × (80,000/70,000) × (0.400/0.500) = 1536 psig.

PIPELINE INTERNAL DESIGN PRESSURES AND TEST PRESSURES – SI UNITS

Pipeline Internal Design Pressures and Test Pressures (SI Units) of Pipe Material API 5LX-70 with SMYS of 483 MPa

Diameter (DN, mm)	Wall Thickness (mm)	Weight (kg/m)	Internal Design Pressure (kPa) Class 1	Class 2	Class 3	Class 4	Hydrostatic Test Pressure (kPa) 90% SMYS	95% SMYS	100% SMYS
100	6.0198	16.03	36,631	30,526	25,438	20,350	45,788	48,332	50,876
	8.5598	22.27	52,087	43,406	36,171	28,937	65,108	68,726	72,343
	11.0998	28.18	67,543	56,286	46,905	37,524	84,428	89,119	93,809
	13.4874	33.45	82,071	68,393	56,994	45,595	102,589	108,289	113,988
150	6.35	25.29	26,246	21,872	18,226	14,581	32,808	34,630	36,453
	7.112	28.20	29,396	24,496	20,414	16,331	36,744	38,786	40,827
	10.9728	42.46	45,353	37,794	31,495	25,196	56,691	59,841	62,990
	14.2748	54.08	59,001	49,168	40,973	32,778	73,751	77,849	81,946
200	6.35	33.23	20,160	16,800	14,000	11,200	25,200	26,600	28,000
	7.0358	36.70	22,337	18,614	15,512	12,410	27,922	29,473	31,024
	8.1788	42.43	25,966	21,638	18,032	14,426	32,458	34,261	36,064
	10.3124	52.96	32,740	27,283	22,736	18,189	40,925	43,198	45,472

(Continued)

Pipeline Internal Design Pressures and Test Pressures (SI Units) of Pipe Material API 5LX-70 with SMYS of 483 MPa—cont'd

Diameter (DN, mm)	Wall Thickness (mm)	Weight (kg/m)	Internal Design Pressure (kPa) Class 1	Class 2	Class 3	Class 4	Hydrostatic Test Pressure (kPa) 90% SMYS	95% SMYS	100% SMYS
250	6.35	41.66	16,175	13,479	11,233	8,986	20,219	21,342	22,465
	7.7978	50.88	19,863	16,552	13,794	11,035	24,828	26,208	27,587
	9.271	60.16	23,615	19,679	16,400	13,120	29,519	31,159	32,799
	12.7	81.34	32,350	26,958	22,465	17,972	40,437	42,684	44,930
300	6.35	49.60	13,638	11,365	9,471	7,576	17,047	17,994	18,941
	8.382	65.05	18,002	15,001	12,501	10,001	22,502	23,752	25,002
	9.525	73.65	20,456	17,047	14,206	11,365	25,571	26,991	28,412
	10.3124	79.54	22,148	18,456	15,380	12,304	27,684	29,222	30,760
	12.7	97.21	27,275	22,729	18,941	15,153	34,094	35,988	37,882
350	6.35	54.56	12,420	10,350	8,625	6,900	15,525	16,388	17,250
	7.9248	67.78	15,500	12,917	10,764	8,611	19,375	20,452	21,528
	9.525	81.09	18,630	15,525	12,938	10,350	23,288	24,581	25,875
	11.0998	94.07	21,710	18,092	15,077	12,061	27,138	28,645	30,153
	12.7	107.13	24,840	20,700	17,250	13,800	31,050	32,775	34,500
400	6.35	62.49	10,868	9,056	7,547	6,038	13,584	14,339	15,094
	7.9248	77.68	13,563	11,302	9,419	7,535	16,953	17,895	18,837
	9.525	92.99	16,301	13,584	11,320	9,056	20,377	21,509	22,641
	11.0998	107.94	18,996	15,830	13,192	10,554	23,745	25,065	26,384
	12.7	123.00	21,735	18,113	15,094	12,075	27,169	28,678	30,188
450	6.35	70.43	9,660	8,050	6,708	5,367	12,075	12,746	13,417
	7.9248	87.59	12,056	10,046	8,372	6,698	15,070	15,907	16,744
	9.525	104.90	14,490	12,075	10,063	8,050	18,113	19,119	20,125
	11.0998	121.81	16,886	14,071	11,726	9,381	21,107	22,280	23,452
	12.7	138.87	19,320	16,100	13,417	10,733	24,150	25,492	26,833
500	7.9248	97.49	10,850	9,042	7,535	6,028	13,563	14,316	15,070
	9.525	116.80	13,041	10,868	9,056	7,245	16,301	17,207	18,113
	11.0998	135.68	15,197	12,664	10,554	8,443	18,996	20,052	21,107
	12.7	154.74	17,388	14,490	12,075	9,660	21,735	22,943	24,150
	14.2748	173.38	19,544	16,287	13,572	10,858	24,430	25,787	27,145
550	9.525	128.70	11,855	9,880	8,233	6,586	14,819	15,643	16,466
	12.7	170.61	15,807	13,173	10,977	8,782	19,759	20,857	21,955
	15.875	212.03	19,759	16,466	13,722	10,977	24,699	26,071	27,443
	19.05	252.94	23,711	19,759	16,466	13,173	29,639	31,285	32,932
600	9.525	140.61	10,868	9,056	7,547	6,038	13,584	14,339	15,094
	11.0998	163.42	12,664	10,554	8,795	7,036	15,830	16,710	17,589
	12.7	186.48	14,490	12,075	10,063	8,050	18,113	19,119	20,125
	14.2748	209.05	16,287	13,572	11,310	9,048	20,358	21,489	22,621
	15.875	231.86	18,113	15,094	12,578	10,063	22,641	23,898	25,156
	19.05	276.75	21,735	18,113	15,094	12,075	27,169	28,678	30,188
650	9.525	152.51	10,032	8,360	6,966	5,573	12,539	13,236	13,933
	12.7	202.35	13,375	11,146	9,288	7,431	16,719	17,648	18,577
	15.875	251.70	16,719	13,933	11,611	9,288	20,899	22,060	23,221
	19.05	300.56	20,063	16,719	13,933	11,146	25,079	26,472	27,865
700	9.525	164.41	9,315	7,763	6,469	5,175	11,644	12,291	12,938
	12.7	218.23	12,420	10,350	8,625	6,900	15,525	16,388	17,250
	15.875	271.54	15,525	12,938	10,781	8,625	19,406	20,484	21,563
	19.05	324.36	18,630	15,525	12,938	10,350	23,288	24,581	25,875
750	9.525	176.32	8,694	7,245	6,038	4,830	10,868	11,471	12,075
	12.7	234.10	11,592	9,660	8,050	6,440	14,490	15,295	16,100
	15.875	291.38	14,490	12,075	10,063	8,050	18,113	19,119	20,125
	19.05	348.17	17,388	14,490	12,075	9,660	21,735	22,943	24,150
800	9.525	188.22	8,151	6,792	5,660	4,528	10,188	10,754	11,320
	12.7	249.97	10,868	9,056	7,547	6,038	13,584	14,339	15,094
	15.875	311.22	13,584	11,320	9,434	7,547	16,980	17,924	18,867
	19.05	371.98	16,301	13,584	11,320	9,056	20,377	21,509	22,641
850	9.525	200.12	7,671	6,393	5,327	4,262	9,589	10,122	10,654
	12.7	265.84	10,228	8,524	7,103	5,682	12,785	13,496	14,206
	15.875	331.06	12,785	10,654	8,879	7,103	15,982	16,869	17,757
	19.05	395.78	15,342	12,785	10,654	8,524	19,178	20,243	21,309

Pipeline Internal Design Pressures and Test Pressures (SI Units) of Pipe Material API 5LX-70 with SMYS of 483 MPa—cont'd

Diameter (DN, mm)	Wall Thickness (mm)	Weight (kg/m)	Internal Design Pressure (kPa) Class 1	Class 2	Class 3	Class 4	Hydrostatic Test Pressure (kPa) 90% SMYS	95% SMYS	100% SMYS
900	9.525	212.03	7,245	6,038	5,031	4,025	9,056	9,559	10,063
	12.7	281.71	9,660	8,050	6,708	5,367	12,075	12,746	13,417
	15.875	350.90	12,075	10,063	8,385	6,708	15,094	15,932	16,771
	19.05	419.59	14,490	12,075	10,063	8,050	18,113	19,119	20,125
1050	9.525	247.74	6,210	5,175	4,313	3,450	7,763	8,194	8,625
	12.7	329.32	8,280	6,900	5,750	4,600	10,350	10,925	11,500
	15.875	410.41	10,350	8,625	7,188	5,750	12,938	13,656	14,375
	19.05	491.01	12,420	10,350	8,625	6,900	15,525	16,388	17,250
	25.4	650.71	16,560	13,800	11,500	9,200	20,700	21,850	23,000

For other values of SMYS, multiply table values of pressures by the factor (SMYS/483).
For other wall thickness t_1*, multiply table values of pressures by the factor* (t_1/t)*, where* t *is the table wall thickness.*

Example:

(a) For X-80 pipe, SMYS = 552 MPa, DN 400 pipe, 6.35 mm wall thickness,
Class 1 design pressure = 10,868 × (552/483) = 12,420 kPa.

(b) For X-70 pipe, DN 600 pipe, 10 mm wall thickness,
Class 1 design pressure = 14,490 × (10/12.7) = 11,409 kPa.

Appendix 5

Chapter 8

EXPLICIT FRICTION FACTOR EQUATIONS

Churchill Equation

This equation was first reported in *Chemical Engineering* magazine in November 1977. Unlike the Colebrook–White equation that requires trial and error solution, this equation is explicit in f as follows:

$$f = [(8/\mathrm{R})^{12} + 1/(\mathrm{A}+\mathrm{B})^{3/2}]^{1/12} \qquad \text{(A 5.1)}$$

where

$\mathrm{A} = [2.457 \log_e(1/((7/\mathrm{R})^{0.9} + (0.27e/D))]^{16}$

$\mathrm{B} = (37{,}530/\mathrm{R})^{16}$

The Churchill equation for friction factor appears to correlate well with the Colebrook–White equation.

Swamee–Jain Equation

This equation appeared in the 1976 *Journal of the Hydraulics Division* of the ASCE. It is found to be the best and easiest of all explicit equations for calculating the friction factor, and it correlates well with the Colebrook–White equation.

$$f = 0.25/[\log_{10}(e/3.7D + 5.74/\mathrm{R}^{0.9})]^2 \qquad \text{(A 5.2)}$$

HAZEN–WILLIAMS C-FACTOR

Pipe Material	C-Factor
Smooth pipes (all metals)	130–140
Smooth wood	120
Smooth masonry	120

(Continued)

Pipe Material	C-Factor
Vitrified clay	110
Cast iron (old)	100
Iron (worn/pitted)	60–80
Polyvinyl chloride (PVC)	150
Brick	100

GAS FLOW PRESSURE DROP EQUATIONS

American Gas Association (AGA) Equation

The AGA equation requires the transmission factor F to be calculated using two different equations. First, F is calculated for the rough pipe law (referred to as the fully turbulent zone). Next, F is calculated based on the smooth pipe law (referred to as the partially turbulent zone). Finally, the smaller of the two values of the transmission factor is used in the general flow Eq. (8.57) through Eq. (8.60) for pressure drop calculation.

For the fully turbulent zone, the following formula for F is based only on the relative roughness e/D.

$$F = 4\log_{10}\left(\frac{3.7D}{e}\right) \tag{A 5.3}$$

For the partially turbulent zone, F is calculated as follows, using the Reynolds number, a parameter D_f known as the pipe drag factor, and the Von Karman smooth pipe transmission factor F_t.

$$F = 4\,D_f\log_{10}\left(\frac{R}{1.4125F_t}\right) \tag{A 5.4}$$

where

$$F_t = 4\log_{10}\left(\frac{R}{F_t}\right) - 0.6 \tag{A 5.5}$$

where

F_t – Von Karman smooth pipe transmission factor
D_f – Pipe drag factor that depends on the Bend index (BI) of the pipe

The pipe drag factor D_f is a parameter that takes into account the number of bends, fittings, etc. Its value ranges from 0.90 to 0.99. The Bend index is the sum of all the angles and bends in the pipe segment, divided by the total length of the pipe section under consideration, as follows:

$$BI = \frac{\text{total degrees of all bends in pipe section}}{\text{total length of pipe section}} \tag{A 5.6}$$

The value of the pipe drag factor D_f is generally chosen from Table A 5.1.

TABLE A 5.1 Bend Index and Drag Factor

	Bend Index		
	Extremely Low (5° to 10°)	Average (60° to 80°)	Extremely High (200° to 300°)
Bare steel	0.975–0.973	0.960–0.956	0.930–0.900
Plastic lined	0.979–0.976	0.964–0.960	0.936–0.910
Pig burnished	0.982–0.980	0.968–0.965	0.944–0.920
Sand-blasted	0.985–0.983	0.976–0.970	0.951–0.930

Note: The drag factors above are based on 40-ft joints of pipelines and mainline valves at 10-mi spacing.

Weymouth Equation

The Weymouth equation is used for high pressure, high flow rate, and large-diameter gas gathering systems. This formula directly calculates the flow rate through a pipeline for given values of gas gravity, compressibility, inlet and outlet pressures, pipe diameter and length. In USCS units, the Weymouth equation is stated as follows:

$$Q = 433.5E\left(\frac{T_b}{P_b}\right)\left(\frac{P_1^2 - e^s P_2^2}{GT_f L_e Z}\right)^{0.5} D^{2.667}(\text{USCS}) \quad (A\,5.7)$$

where

Q – Volume flow rate, standard ft^3/day (SCFD)
E – Pipeline efficiency, a decimal value less than or equal to 1.0
P_b – Base pressure, psia
T_b – Base temperature, °R (460 + °F)
P_1 – Upstream pressure, psia
P_2 – Downstream pressure, psia
G – Gas gravity (air = 1.00)
T_f – Average gas flow temperature, °R (460 + °F)
L_e – Equivalent length of pipe segment, mi
Z – Gas compressibility factor, dimensionless
D – Pipe inside diameter, in.

where the equivalent length L_e and parameter s were defined earlier in Chapter 8, Eqs (8.61) and (8.62).

The Weymouth transmission factor in USCS units is

$$F = 11.18(D)^{1/6}(\text{USCS}) \quad (A\,5.8)$$

In SI units, the Weymouth equation is as follows:

$$Q = 3.7435 \times 10^{-3} E\left(\frac{T_b}{P_b}\right)\left(\frac{P_1^2 - e^s P_2^2}{G T_f L_e Z}\right)^{0.5} D^{2.667}\,(\text{SI}) \tag{A 5.9}$$

where

Q – Gas flow rate, standard condition m^3/day
T_b – Base temperature, K (273 + °C)
P_b – Base pressure, kPa
T_f – Average gas flow temperature, K (273 + °C)
P_1 – Upstream pressure, kPa
P_2 – Downstream pressure, kPa
L_e – Equivalent length of pipe segment, km

Other symbols are as defined previously.
The Weymouth transmission factor in SI units is

$$F = 6.521(D)^{1/6}\,(\text{SI}) \tag{A 5.10}$$

Panhandle A Equation

The Panhandle A Equation was developed for use in natural gas pipelines for Reynolds numbers in the range of 5–11 million. This equation does not use a pipe roughness, but includes an efficiency factor. The general form of the Panhandle A equation in USCS units is as follows:

$$Q = 435.87E\left(\frac{T_b}{P_b}\right)^{1.0788}\left(\frac{P_1^2 - e^s P_2^2}{G^{0.8539} T_f L_e Z}\right)^{0.5394} D^{2.6182}\,(\text{USCS}) \tag{A 5.11}$$

where

Q – Volume flow rate, standard ft^3/day (SCFD)
E – Pipeline efficiency, a decimal value less than 1.0
P_b – Base pressure, psia
T_b – Base temperature, °R (460 + °F)
P_1 – Upstream pressure, psia
P_2 – Downstream pressure, psia
G – Gas gravity (air = 1.00)
T_f – Average gas flow temperature, °R (460 + °F)
L_e – Equivalent length of pipe segment, mi
Z – Gas compressibility factor, dimensionless
D – Pipe inside diameter, in.

Other symbols are as defined previously.

In SI units, the Panhandle A equation is

$$Q = 4.5965 \times 10^{-3} E \left(\frac{T_b}{P_b} \right)^{1.0788} \left(\frac{P_1^2 - e^s P_2^2}{G^{0.8539} T_f L_e Z} \right)^{0.5394} D^{2.6182} \text{ (SI)} \qquad \text{(A 5.12)}$$

where

Q – Gas flow rate, standard condition m^3/day
E – Pipeline efficiency, a decimal value less than 1.0
T_b – Base temperature, K (273 + °C)
P_b – Base pressure, kPa
T_f – Average gas flow temperature, K (273 + °C)
P_1 – Upstream pressure, kPa (absolute)
P_2 – Downstream pressure, kPa (absolute)
L_e – Equivalent length of pipe segment, km

Other symbols are as defined previously.

An equivalent transmission factor for Panhandle A equation in USCS units is as follows:

$$F = 7.2111 E \left(\frac{QG}{D} \right)^{0.07305} \text{ (USCS)} \qquad \text{(A 5.13)}$$

And in SI units, it is

$$F = 11.85 E \left(\frac{QG}{D} \right)^{0.07305} \text{ (SI)} \qquad \text{(A 5.14)}$$

Panhandle B Equation

The Panhandle B Equation, also known as the revised Panhandle equation is used in large diameter, high pressure transmission lines. In fully turbulent flow, it is found to be accurate for values of Reynolds number in the range of 4–40 million. This equation in USCS units is as follows:

$$Q = 737 E \left(\frac{T_b}{P_b} \right)^{1.02} \left(\frac{P_1^2 - e^s P_2^2}{G^{0.961} T_f L_e Z} \right)^{0.51} D^{2.53} \text{ (USCS)} \qquad \text{(A 5.15)}$$

where

Q – Volume flow rate, standard ft^3/day (SCFD)
E – Pipeline efficiency, a decimal value less than 1.0
P_b – Base pressure, psia
T_b – Base temperature, °R (460 + °F)
P_1 – Upstream pressure, psia
P_2 – Downstream pressure, psia
G – Gas gravity (air = 1.00)

T_f – Average gas flow temperature, °R (460 + °F)
L_e – Equivalent length of pipe segment, mi
Z – Gas compressibility factor, dimensionless
D – Pipe inside diameter, in.

Other symbols are as defined previously.

In SI units, the Panhandle B equation is

$$Q = 1.002 \times 10^{-2} E \left(\frac{T_b}{P_b}\right)^{1.02} \left(\frac{P_1^2 - e^s P_2^2}{G^{0.961} T_f L_e Z}\right)^{0.51} D^{2.53} \text{ (SI)} \qquad \text{(A 5.16)}$$

where

Q – Gas flow rate, standard condition m^3/day
E – Pipeline efficiency, a decimal value less than 1.0
T_b – Base temperature, K (273 + °C)
P_b – Base pressure, kPa
T_f – Average gas flow temperature, K (273 + °C)
P_1 – Upstream pressure, kPa (absolute)
P_2 – Downstream pressure, kPa (absolute)
L_e – Equivalent length of pipe segment, km
Z – Gas compressibility factor at the flowing temperature, dimensionless

Other symbols are as defined previously.

The equivalent transmission factor for the Panhandle B equation in USCS units is

$$F = 16.7E\left(\frac{QG}{D}\right)^{0.01961} \text{ (USCS)} \qquad \text{(A 5.17)}$$

In SI units, it is as follows:

$$F = 19.08E\left(\frac{QG}{D}\right)^{0.01961} \text{ (SI)} \qquad \text{(A 5.18)}$$

Institute of Gas Technology (IGT) Equation

The IGT equation proposed by the Institute of Gas Technology is also known as the IGT distribution equation and is stated as follows for USCS units:

$$Q = 136.9E\left(\frac{T_b}{P_b}\right)\left(\frac{P_1^2 - e^s P_2^2}{G^{0.8} T_f L_e \mu^{0.2}}\right)^{0.555} D^{2.667} \text{ (USCS)} \qquad \text{(A 5.19)}$$

where

Q – Volume flow rate, standard ft^3/day (SCFD)
E – Pipeline efficiency, a decimal value less than 1.0

P_b – Base pressure, psia
T_b – Base temperature, °R (460 + °F)
P_1 – Upstream pressure, psia
P_2 – Downstream pressure, psia
G – Gas gravity (air = 1.00)
T_f – Average gas flow temperature, °R (460 + °F)
L_e – Equivalent length of pipe segment, mi
Z – Gas compressibility factor, dimensionless
D – Pipe inside diameter, in.
μ – Gas viscosity, lb/ft · s

Other symbols are as defined previously.
In SI units the IGT equation is expressed as follows

$$Q = 1.2822 \times 10^{-3} E \left(\frac{T_b}{P_b}\right) \left(\frac{P_1^2 - e^s P_2^2}{G^{0.8} T_f L_e \mu^{0.2}}\right)^{0.555} D^{2.667} \text{(SI)} \qquad \text{(A 5.20)}$$

where

Q – Gas flow rate, standard condition m³/day
E – Pipeline efficiency, a decimal value less than 1.0
T_b – Base temperature, K (273 + °C)
P_b – Base pressure, kPa
T_f – Average gas flow temperature, K (273 + °C)
P_1 – Upstream pressure, kPa (absolute)
P_2 – Downstream pressure, kPa (absolute)
L_e – Equivalent length of pipe segment, km
μ – Gas viscosity, Poise

Other symbols are as defined previously.

FRICTIONAL PRESSURE DROP IN LIQUIDS – USCS UNITS

Pressure Drop in psi/mi for Water (Sg = 1.0), Friction Factor $f = 0.01$

Flow Rate	Inside Diameter (in.)											
(bbl/h)	4	6	8	10	12	13.5	15.5	17.5	19	23	29	35
100	3.40	0.45	0.11	0.03	0.01	0.01	0.00	0.00	0.00	0.00	0.00	0.00
200	13.61	1.79	0.43	0.14	0.06	0.03	0.02	0.01	0.01	0.00	0.00	0.00
500	85.08	11.20	2.66	0.87	0.35	0.19	0.10	0.05	0.04	0.01	0.00	0.00
1000	340.31	44.81	10.63	3.48	1.40	0.78	0.39	0.21	0.14	0.05	0.02	0.01
2000	1361.25	179.26	42.54	13.94	5.60	3.11	1.56	0.85	0.56	0.22	0.07	0.03
3000	3062.81	403.33	95.71	31.36	12.60	6.99	3.51	1.91	1.27	0.49	0.15	0.06
4000	5445.00	717.04	170.16	55.76	22.41	12.43	6.23	3.40	2.25	0.87	0.27	0.11
5000	8507.81	1120.37	265.87	87.12	35.01	19.43	9.74	5.31	3.52	1.35	0.42	0.17
6000	12251.25	1613.33	382.85	125.45	50.42	27.98	14.02	7.64	5.07	1.95	0.61	0.24

For a liquid other than water, multiply table values of pressure drop by the specific gravity. For different pipe inside diameter *D*, multiply table values of pressure drop by $(D_t/D)^5$, where D_t is the diameter in the table. For different friction factor *f*, multiply table values of pressure drop by (f/0.01). Extreme values of pressure drop are shown shaded.

Pressure Drop in psi/mi for Water (Sg = 1.0), Friction Factor $f = 0.02$

Flow Rate	Inside Diameter (in.)											
(bbl/h)	4	6	8	10	12	13.5	15.5	17.5	19	23	29	35
100	6.81	0.90	0.21	0.07	0.03	0.02	0.01	0.00	0.00	0.00	0.00	0.00
200	27.23	3.59	0.85	0.28	0.11	0.06	0.03	0.02	0.01	0.00	0.00	0.00
500	170.16	22.41	5.32	1.74	0.70	0.39	0.19	0.11	0.07	0.03	0.01	0.00
1000	680.63	89.63	21.27	6.97	2.80	1.55	0.78	0.42	0.28	0.11	0.03	0.01
2000	2722.50	358.52	85.08	27.88	11.20	6.22	3.12	1.70	1.13	0.43	0.14	0.05
3000	6125.63	806.67	191.43	62.73	25.21	13.99	7.01	3.82	2.53	0.97	0.31	0.12
4000	10890.00	1434.07	340.31	111.51	44.81	24.87	12.46	6.79	4.50	1.73	0.54	0.21
5000	17015.63	2240.74	531.74	174.24	70.02	38.86	19.48	10.62	7.04	2.71	0.85	0.33
6000	24502.50	3226.67	765.70	250.91	100.83	55.96	28.04	15.29	10.13	3.90	1.22	0.48

For a liquid other than water, multiply table values of pressure drop by the specific gravity. For different pipe inside diameter *D*, multiply table values of pressure drop by $(D_t/D)^5$, where D_t is the diameter in the table. For different friction factor *f*, multiply table values of pressure drop by (f/0.02). Extreme values of pressure drop are shown shaded.

FRICTIONAL PRESSURE DROP IN LIQUIDS – SI UNITS

Pressure Drop in kPa/km for Water (Sg = 1.0), Friction Factor $f = 0.01$

Flow Rate	Inside Diameter (mm)											
m³/h	100	150	200	250	300	338	394	445	483	584	737	889
10	6.25	0.82	0.20	0.06	0.03	0.01	0.01	0.00	0.00	0.00	0.00	0.00
20	24.99	3.29	0.78	0.26	0.10	0.06	0.03	0.01	0.01	0.00	0.00	0.00
50	156.19	20.57	4.88	1.60	0.64	0.35	0.16	0.09	0.06	0.02	0.01	0.00
60	224.91	29.62	7.03	2.30	0.93	0.51	0.24	0.13	0.09	0.03	0.01	0.00
80	399.84	52.65	12.50	4.09	1.65	0.91	0.42	0.23	0.15	0.06	0.02	0.01
100	624.75	82.27	19.52	6.40	2.57	1.42	0.66	0.36	0.24	0.09	0.03	0.01
200	2499.00	329.09	78.09	25.59	10.28	5.66	2.63	1.43	0.95	0.37	0.11	0.05
500	15618.75	2056.79	488.09	159.94	64.27	35.40	16.45	8.95	5.94	2.30	0.72	0.28
600	22491.00	2961.78	702.84	230.31	92.56	50.98	23.69	12.89	8.56	3.31	1.03	0.41
800	39984.00	5265.38	1249.50	409.44	164.54	90.64	42.11	22.91	15.21	5.89	1.84	0.72
1000	62475.00	8227.16	1952.34	639.74	257.10	141.62	65.80	35.80	23.77	9.20	2.87	1.13

For a liquid other than water, multiply table values of pressure drop by the specific gravity. For different pipe inside diameter D, multiply table values of pressure drop by $(D_t/D)^5$, where D_t is the diameter in the table. For different friction factor f, multiply table values of pressure drop by $(f/0.01)$. Extreme values of pressure drop are shown shaded.

Pressure Drop in kPa/km for Water (Sg = 1.0), Friction Factor $f = 0.02$

Flow Rate	Inside Diameter (mm)											
m³/h	100	150	200	250	300	338	394	445	483	584	737	889
10	12.50	1.65	0.39	0.13	0.05	0.03	0.01	0.01	0.00	0.00	0.00	0.00
20	49.98	6.58	1.56	0.51	0.21	0.11	0.05	0.03	0.02	0.01	0.00	0.00
50	312.38	41.14	9.76	3.20	1.29	0.71	0.33	0.18	0.12	0.05	0.01	0.01
60	449.82	59.24	14.06	4.61	1.85	1.02	0.47	0.26	0.17	0.07	0.02	0.01
80	799.68	105.31	24.99	8.19	3.29	1.81	0.84	0.46	0.30	0.12	0.04	0.01
100	1249.50	164.54	39.05	12.79	5.14	2.83	1.32	0.72	0.48	0.18	0.06	0.02
200	4998.00	658.17	156.19	51.18	20.57	11.33	5.26	2.86	1.90	0.74	0.23	0.09
500	31237.50	4113.58	976.17	319.87	128.55	70.81	32.90	17.90	11.88	4.60	1.44	0.56
600	44982.00	5923.56	1405.69	460.62	185.11	101.97	47.38	25.78	17.11	6.62	2.07	0.81
800	79968.00	10530.77	2499.00	818.87	329.09	181.27	84.22	45.83	30.42	11.77	3.68	1.44
1000	124950.00	16454.32	3904.69	1279.49	514.20	283.24	131.60	71.60	47.53	18.39	5.75	2.25

For a liquid other than water, multiply table values of pressure drop by the specific gravity. For different pipe inside diameter D, multiply table values of pressure drop by $(D_t/D)^5$, where D_t is the diameter in the table. For different friction factor f, multiply table values of pressure drop by $(f/0.02)$. Extreme values of pressure drop are shown shaded.

Chapter 9

EQUIVALENT LENGTH OF SERIES PIPES

USCS Units

Inside Diameter	Base Inside Diameter (in.)											
(in.)	4	6	8	10	12	13.5	15.5	17.5	19	23	29	35
4.0	1.0000	7.5938	32.0000	97.6563	243.0000	437.89	873.69	1602.84	2418.07	6285.49	20030.42	51290.89
6.0	0.1317	1.0000	4.2140	12.8601	32.0000	57.67	115.05	211.07	318.43	827.72	2637.75	6754.36
8.0	0.0313	0.2373	1.0000	3.0518	7.5938	13.68	27.30	50.09	75.56	196.42	625.95	1602.84
10.0	0.0102	0.0778	0.3277	1.0000	2.4883	4.48	8.95	16.41	24.76	64.36	205.11	525.22
12.0	0.0041	0.0313	0.1317	0.4019	1.0000	1.80	3.60	6.60	9.95	25.87	82.43	211.07
13.5	0.0023	0.0173	0.0731	0.2230	0.5549	1.00	2.00	3.66	5.52	14.35	45.74	117.13
15.5	0.0011	0.0087	0.0366	0.1118	0.2781	0.50	1.00	1.83	2.77	7.19	22.93	58.71
17.5	0.0006	0.0047	0.0200	0.0609	0.1516	0.27	0.55	1.00	1.51	3.92	12.50	32.00
19.0	0.0004	0.0031	0.0132	0.0404	0.1005	0.18	0.36	0.66	1.00	2.60	8.28	21.21
23.0	0.0002	0.0012	0.0051	0.0155	0.0387	0.07	0.14	0.26	0.38	1.00	3.19	8.16
29.0		0.0004	0.0016	0.0049	0.0121	0.02	0.04	0.08	0.12	0.31	1.00	2.56
35.0		0.0001	0.0006	0.0019	0.0047	0.01	0.02	0.03	0.05	0.12	0.39	1.00

Example: 1 ft of 10-in. pipe is equivalent to 0.0778 ft of 6-in. pipe.

SI Units

Inside Diameter	Base Inside Diameter (mm)											
(mm)	100	150	200	250	300	343	394	445	483	584	737	889
100	1.0000	7.5938	32.0000	97.6563	243.0000	474.76	949.47	1745.02	2628.67	6793.04	21743.90	55527.59
150	0.1317	1.0000	4.2140	12.8601	32.0000	62.52	125.03	229.80	346.16	894.56	2863.39	7312.27
200	0.0313	0.2373	1.0000	3.0518	7.5938	14.84	29.67	54.53	82.15	212.28	679.50	1735.24
250	0.0102	0.0778	0.3277	1.0000	2.4883	4.86	9.72	17.87	26.92	69.56	222.66	568.60
300	0.0041	0.0313	0.1317	0.4019	1.0000	1.95	3.91	7.18	10.82	27.95	89.48	228.51
343	0.0021	0.0160	0.0674	0.2057	0.5118	1.00	2.00	3.68	5.54	14.31	45.80	116.96
394	0.0011	0.0080	0.0337	0.1029	0.2559	0.50	1.00	1.84	2.77	7.15	22.90	58.48
445	0.0006	0.0044	0.0183	0.0560	0.1393	0.27	0.54	1.00	1.51	3.89	12.46	31.82
483	0.0004	0.0029	0.0122	0.0372	0.0924	0.18	0.36	0.66	1.00	2.58	8.27	21.12
584	0.0001	0.0011	0.0047	0.0144	0.0358	0.07	0.14	0.26	0.39	1.00	3.20	8.17
737		0.0003	0.0015	0.0045	0.0112	0.02	0.04	0.08	0.12	0.31	1.00	2.55
889		0.0001	0.0006	0.0018	0.0044	0.01	0.02	0.03	0.05	0.12	0.39	1.00

Example: 1 m of 250-mm pipe is equivalent to 2.4883 m of 300-mm diameter pipe.

EQUIVALENT DIAMETER OF PARALLEL PIPES

USCS Units

Inside	Inside Diameter (in.)											
Diameter (in.)	4	6	8	10	12	13.5	15.5	17.5	19	23	29	35
4.0	5.28	6.79	8.54	10.39	12.30	13.75	15.71	17.67	19.15	23.12	29.08	35.06
6.0	6.79	7.92	9.38	11.03	12.81	14.18	16.06	17.97	19.42	23.32	29.22	35.17
8.0	8.54	9.38	10.56	11.98	13.58	14.86	16.62	18.45	19.85	23.64	29.46	35.35
10.0	10.39	11.03	11.98	13.20	14.60	15.76	17.40	19.11	20.44	24.11	29.79	35.60
12.0	12.30	12.81	13.58	14.60	15.83	16.87	18.36	19.96	21.21	24.71	30.24	35.94
13.5	13.75	14.18	14.86	15.76	16.87	17.81	19.20	20.71	21.90	25.26	30.64	36.26
15.5	15.71	16.06	16.62	17.40	18.36	19.20	20.45	21.83	22.94	26.11	31.29	36.76
17.5	17.67	17.97	18.45	19.11	19.96	20.71	21.83	23.09	24.11	27.09	32.04	37.35
19.0	19.15	19.42	19.85	20.44	21.21	21.90	22.94	24.11	25.07	27.90	32.67	37.86
23.0	23.12	23.32	23.64	24.11	24.71	25.26	26.11	27.09	27.90	30.35	34.65	39.46
29.0	29.08	29.22	29.46	29.79	30.24	30.64	31.29	32.04	32.67	34.65	38.27	42.50
35.0	35.06	35.17	35.35	35.60	35.94	36.26	36.76	37.35	37.86	39.46	42.50	46.18

Example: Equivalent diameter of equal lengths of 12-in. diameter pipe and 10-in. diameter pipe in parallel is 14.60 in.

SI Units

Inside	Inside Diameter (mm)											
Diameter (mm)	100	150	200	250	300	343	394	445	483	584	737	889
100	131.95	169.78	213.46	259.83	307.55	349.21	399.07	449.23	486.75	586.82	739.00	890.51
150	169.78	197.93	234.41	275.85	320.18	359.73	407.73	456.52	493.22	591.73	742.48	893.14
200	213.46	234.41	263.90	299.62	339.55	376.18	421.48	468.19	503.65	599.71	748.18	897.48
250	259.83	275.85	299.62	329.88	365.10	398.35	440.37	484.45	518.28	611.06	756.37	903.73
300	307.55	320.18	339.55	365.10	395.85	425.64	464.10	505.18	537.11	625.90	767.23	912.07
343	349.21	359.73	376.18	398.35	425.64	452.59	487.98	526.36	556.51	641.45	778.77	921.01
394	399.07	407.73	421.48	440.37	464.10	487.98	519.89	555.06	583.05	663.12	795.12	933.79
445	449.23	456.52	468.19	484.45	505.18	526.36	555.06	587.18	613.02	688.08	814.32	948.97
483	486.75	493.22	503.65	518.28	537.11	556.51	583.05	613.02	637.32	708.66	830.43	961.83
584	586.82	591.73	599.71	611.06	625.90	641.45	663.12	688.08	708.66	770.59	880.24	1002.32
737	739.00	742.48	748.18	756.37	767.23	778.77	795.12	814.32	830.43	880.24	972.48	1079.76
889	890.51	893.14	897.48	903.73	912.07	921.01	933.79	948.97	961.83	1002.32	1079.76	1173.04

Example: Equivalent diameter of equal lengths of 250-mm diameter pipe and 300-mm diameter pipe in parallel is 365.10 mm.

BRAKE POWER REQUIRED FOR WATER

USCS Units

Flow Rate	Differential Pressure (psig)											
(bbl/h)	100	200	300	400	500	600	700	800	900	1000	1100	1200
100	5.10	10.21	15.31	20.42	25.52	30.62	35.73	40.83	45.94	51.04	56.15	61.25
200	10.21	20.42	30.62	40.83	51.04	61.25	71.46	81.67	91.87	102.08	112.29	122.50
500	25.52	51.04	76.56	102.08	127.60	153.12	178.64	204.16	229.69	255.21	280.73	306.25
1000	51.04	102.08	153.12	204.16	255.21	306.25	357.29	408.33	459.37	510.41	561.45	612.49
2000	102.08	204.16	306.25	408.33	510.41	612.49	714.58	816.66	918.74	1,020.82	1,122.91	1,224.99
3000	153.12	306.25	459.37	612.49	765.62	918.74	1,071.87	1,224.99	1,378.11	1,531.24	1,684.36	1,837.48
4000	204.16	408.33	612.49	816.66	1,020.82	1,224.99	1,429.15	1,633.32	1,837.48	2,041.65	2,245.81	2,449.98
5000	255.21	510.41	765.62	1,020.82	1,276.03	1,531.24	1,786.44	2,041.65	2,296.86	2,552.06	2,807.27	3,062.47
6000	306.25	612.49	918.74	1,224.99	1,531.24	1,837.48	2,143.73	2,449.98	2,756.23	3,062.47	3,368.72	3,674.97

Brake horsepower required for water (Sg = 1.00) efficiency = 80%.
For other values of efficiency *E* percent, multiply table values by (80/*E*).

SI Units

Flow Rate	Differential Pressure (kPa)											
(m^3/h)	500	1000	2000	3000	4000	5000	6000	7000	8000	9000	10000	12000
100	17.36	34.72	69.44	104.17	138.89	173.61	208.33	243.06	277.78	312.50	347.22	416.67
200	34.72	69.44	138.89	208.33	277.78	347.22	416.67	486.11	555.56	625.00	694.44	833.33
500	86.81	173.61	347.22	520.83	694.44	868.06	1,041.67	1,215.28	1,388.89	1,562.50	1,736.11	2,083.33
1000	173.61	347.22	694.44	1,041.67	1,388.89	1,736.11	2,083.33	2,430.56	2,777.78	3,125.00	3,472.22	4,166.67
2000	347.22	694.44	1,388.89	2,083.33	2,777.78	3,472.22	4,166.67	4,861.11	5,555.56	6,250.00	6,944.44	8,333.33
3000	520.83	1,041.67	2,083.33	3,125.00	4,166.67	5,208.33	6,250.00	7,291.67	8,333.33	9,375.00	10,416.67	12,500.00
4000	694.44	1,388.89	2,777.78	4,166.67	5,555.56	6,944.44	8,333.33	9,722.22	11,111.11	12,500.00	13,888.89	16,666.67
5000	868.06	1,736.11	3,472.22	5,208.33	6,944.44	8,680.56	10,416.67	12,152.78	13,888.89	15,625.00	17,361.11	20,833.33
6000	1,041.67	2,083.33	4,166.67	6,250.00	8,333.33	10,416.67	12,500.00	14,583.33	16,666.67	18,750.00	20,833.33	25,000.00

Brake power (kW) required for water (Sg = 1.00) efficiency = 80%.
For other values of efficiency *E* percent, multiply table values by (80/*E*).

Chapter 11

LINE FILL VOLUME PER MILE OF PIPE – USCS UNITS

Inside Diameter (in.)	Line Fill (bbl)
4.0	82.06
6.0	184.64
8.0	328.26
10.0	512.90
12.0	738.58
13.5	934.76
15.5	1232.24
17.5	1570.76
19.0	1851.57
23.0	2713.24
29.0	4313.49
35.0	6283.03

LINE FILL VOLUME PER KILOMETER OF PIPE – SI UNITS

Inside Diameter (mm)	Line Fill (m^3)
100.0	7.86
150.0	17.67
200.0	31.42
250.0	49.09
300.0	70.70
343.0	92.41
394.0	121.94
445.0	155.55
483.0	183.25
584.0	267.90
737.0	426.66
889.0	620.80

SPECIFIC SPEED OF CENTRIFUGAL PUMPS

Capacity	Head per Stage (ft)								
(gal/min)	50	100	150	200	250	300	350	400	500
100	1893.3	1125.8	830.6	669.4	566.2	493.9	439.9	398.0	336.7
200	2677.5	1592.1	1174.6	946.7	800.8	698.4	622.2	562.9	476.1
400	3786.6	2251.5	1661.2	1338.8	1132.5	987.7	879.9	796.0	673.4
500	4233.6	2517.3	1857.2	1496.8	1266.1	1104.3	983.7	890.0	752.8
600	4637.7	2757.6	2034.5	1639.7	1387.0	1209.7	1077.6	974.9	824.7
800	5355.1	3184.2	2349.2	1893.3	1601.5	1396.9	1244.4	1125.8	952.3
1000	5987.2	3560.0	2626.5	2116.8	1790.6	1561.7	1391.2	1258.7	1064.7
1200	6558.6	3899.8	2877.2	2318.8	1961.5	1710.8	1524.0	1378.8	1166.3
1400	7084.1	4212.2	3107.7	2504.6	2118.6	1847.9	1646.1	1489.3	1259.8
1500	7332.8	4360.1	3216.8	2592.5	2193.0	1912.7	1703.9	1541.5	1304.0
1800	8032.6	4776.2	3523.9	2840.0	2402.3	2095.3	1866.5	1688.7	1428.4
2000	8467.2	5034.6	3714.5	2993.6	2532.3	2208.6	1967.5	1780.0	1505.7

Pump speed $N = 3560$ RPM.
For other pump speed N_1, multiply table values by $(N_1/3560)$.

SUCTION SPECIFIC SPEED OF CENTRIFUGAL PUMPS

Capacity	NPSH Required (ft)								
(gal/min)	10	20	30	40	50	60	70	80	100
100	6330.7	3764.2	2777.2	2238.2	1893.3	1651.3	1471.0	1330.9	1125.8
200	8952.9	5323.4	3927.6	3165.3	2677.5	2335.3	2080.4	1882.1	1592.1
400	12661.3	7528.5	5554.4	4476.5	3786.6	3302.7	2942.1	2661.7	2251.5
500	14155.8	8417.1	6210.0	5004.8	4233.6	3692.5	3289.4	2975.9	2517.3
600	15506.9	9220.5	6802.8	5482.5	4637.7	4044.9	3603.3	3259.9	2757.6
800	17905.9	10646.9	7855.1	6330.7	5355.1	4670.7	4160.7	3764.2	3184.2
1000	20019.4	11903.6	8782.3	7077.9	5987.2	5222.0	4651.9	4208.6	3560.0
1200	21930.1	13039.7	9620.5	7753.5	6558.6	5720.4	5095.9	4610.2	3899.8
1400	23687.2	14084.5	10391.4	8374.7	7084.1	6178.7	5504.2	4979.6	4212.2
1500	24518.6	14578.8	10756.1	8668.6	7332.8	6395.6	5697.3	5154.4	4360.1
1800	26858.8	15970.3	11782.7	9496.0	8032.6	7006.0	6241.1	5646.4	4776.2
2000	28311.6	16834.2	12420.1	10009.7	8467.2	7385.0	6578.7	5951.8	5034.6

Pump speed $N = 3560$ RPM; single suction pump.
For other pump speed N_1, multiply table values by $(N_1/3560)$.

Appendix 8

Chapter 12

COMPRESSOR POWER (USCS UNITS)

Compressor Power in HP/MMSCFD for 80% Adiabatic Efficiency and $Z = 0.85$ with Gas Specific Heat Ratio of 1.2

Pressure	Suction Temperature (°F)									
Ratio (P_2/P_1)	50	60	70	75	80	85	90	95	100	110
1.2	8.60	8.77	8.93	9.02	9.10	9.19	9.27	9.36	9.44	9.61
1.4	16.07	16.39	16.70	16.86	17.02	17.17	17.33	17.49	17.65	17.96
1.6	22.70	23.15	23.59	23.82	24.04	24.26	24.48	24.71	24.93	25.38
1.8	28.68	29.24	29.80	30.08	30.36	30.65	30.93	31.21	31.49	32.05
2.0	34.12	34.79	35.46	35.79	36.13	36.46	36.80	37.13	37.47	38.14
2.2	39.13	39.90	40.66	41.05	41.43	41.82	42.20	42.58	42.97	43.73
2.4	43.77	44.63	45.49	45.92	46.35	46.78	47.20	47.63	48.06	48.92
2.6	48.10	49.04	49.99	50.46	50.93	51.40	51.87	52.35	52.82	53.76
2.8	52.16	53.18	54.21	54.72	55.23	55.74	56.25	56.76	57.28	58.30
3.0	55.99	57.09	58.18	58.73	59.28	59.83	60.38	60.93	61.48	62.57

For different adiabatic efficiency Eff (%), multiply table values of power by (80/Eff).
For different compressibility factor Z, multiply table values of power by (Z/0.85).

Compressor Power in HP/MMSCFD for 80% Adiabatic Efficiency and $Z = 0.85$ with Gas Specific Heat Ratio of 1.3

Pressure	Suction Temperature (°F)									
Ratio (P_2/P_1)	50	60	70	75	80	85	90	95	100	110
1.2	8.65	8.82	8.99	9.07	9.16	9.24	9.33	9.41	9.50	9.66
1.4	16.25	16.57	16.89	17.04	17.20	17.36	17.52	17.68	17.84	18.16
1.6	23.05	23.51	23.96	24.18	24.41	24.64	24.86	25.09	25.31	25.77
1.8	29.23	29.81	30.38	30.67	30.95	31.24	31.53	31.81	32.10	32.67
2.0	34.91	35.59	36.28	36.62	36.96	37.30	37.64	37.99	38.33	39.01
2.2	40.16	40.94	41.73	42.13	42.52	42.91	43.31	43.70	44.09	44.88
2.4	45.05	45.94	46.82	47.26	47.70	48.15	48.59	49.03	49.47	50.35
2.6	49.65	50.62	51.59	52.08	52.57	53.05	53.54	54.03	54.51	55.49
2.8	53.97	55.03	56.09	56.62	57.15	57.68	58.21	58.73	59.26	60.32
3.0	58.07	59.21	60.35	60.91	61.48	62.05	62.62	63.19	63.76	64.90

For different adiabatic efficiency Eff (%), multiply table values of power by (80/Eff).
For different compressibility factor Z, multiply table values of power by (Z/0.85).

The compressor power in HP/MMSCFD is based on Eq. (12.14) for an adiabatic efficiency of 80% and average compressibility factor of 0.85.

COMPRESSOR POWER (SI UNITS)

Compressor Power in kW/Mm³/day for 80% Adiabatic Efficiency and $Z = 0.85$ with Gas Specific Heat Ratio of 1.2

Pressure	Suction Temperature (°C)							
Ratio (P_2/P_1)	5	15	20	25	30	35	40	45
1.2	222.21	230.21	234.20	238.20	242.20	246.19	250.19	254.19
1.4	415.43	430.38	437.85	445.32	452.79	460.26	467.73	475.21
1.6	586.87	607.98	618.53	629.09	639.64	650.20	660.75	671.31
1.8	741.28	767.95	781.28	794.61	807.94	821.28	834.61	847.94
2.0	882.00	913.73	929.59	945.46	961.32	977.18	993.05	1008.91
2.2	1011.45	1047.83	1066.02	1084.21	1102.40	1120.59	1138.79	1156.98
2.4	1131.43	1172.13	1192.48	1212.83	1233.17	1253.52	1273.87	1294.22
2.6	1243.35	1288.07	1310.43	1332.80	1355.16	1377.52	1399.88	1422.25
2.8	1348.31	1396.81	1421.06	1445.31	1469.56	1493.81	1518.06	1542.31
3.0	1447.20	1499.26	1525.28	1551.31	1577.34	1603.37	1629.40	1655.43

For different adiabatic efficiency Eff (%), multiply table values of power by (80/Eff).
For different compressibility factor Z, multiply table values of power by (Z/0.85).

Compressor Power in kW/Mm³/day for 80% Adiabatic Efficiency and $Z = 0.85$ with Gas Specific Heat Ratio of 1.3

Pressure	Suction Temperature (°C)							
Ratio (P_2/P_1)	5	15	20	25	30	35	40	45
1.2	223.52	231.56	235.58	239.60	243.62	247.64	251.66	255.69
1.4	419.99	435.09	442.65	450.20	457.76	465.31	472.86	480.42
1.6	595.91	617.35	628.07	638.78	649.50	660.22	670.94	681.66
1.8	755.66	782.84	796.43	810.02	823.61	837.20	850.79	864.38
2.0	902.28	934.73	950.96	967.19	983.42	999.64	1015.87	1032.10
2.2	1038.02	1075.36	1094.02	1112.69	1131.36	1150.03	1168.70	1187.37
2.4	1164.57	1206.46	1227.41	1248.35	1269.30	1290.25	1311.19	1332.14
2.6	1283.26	1329.42	1352.50	1375.58	1398.66	1421.74	1444.82	1467.90
2.8	1395.11	1445.30	1470.39	1495.48	1520.57	1545.67	1570.76	1595.85
3.0	1500.98	1554.98	1581.97	1608.97	1635.96	1662.96	1689.96	1716.95

For different adiabatic efficiency Eff (%), multiply table values of power by (80/Eff).
For different compressibility factor Z, multiply table values of power by (Z/0.85).

The compressor power in kW/Mm³/day is based on Eq. (12.15) for an adiabatic efficiency of 80% and average compressibility factor of 0.85.

BRAKE POWER (HP/MMSCFD) FOR RECIPROCATING COMPRESSORS – USCS UNITS

Compression Ratio/	Number of Stages		
Stage	1	2	3
1.1	24.20	52.27	79.86
1.2	26.40	57.02	87.12
1.3	28.60	61.78	94.38
1.4	30.80	66.53	101.64
1.5	33.00	71.28	108.90
1.6	35.20	76.03	116.16
1.7	37.40	80.78	123.42
1.8	39.60	85.54	130.68
2.8	61.60	133.06	203.28
2.9	63.80	137.81	210.54
3.0	66.00	142.56	217.80

BRAKE POWER (KW/Mm^3/DAY) FOR RECIPROCATING COMPRESSORS – SI UNITS

Compression Ratio/	Number of Stages		
Stage	1	2	3
1.1	637.64	1377.30	2104.20
1.2	695.60	1502.50	2295.49
1.3	753.57	1627.71	2486.78
1.4	811.54	1752.92	2678.08
1.5	869.51	1878.13	2869.37
1.6	927.47	2003.34	3060.66
1.7	985.44	2128.55	3251.95
1.8	1043.41	2253.76	3443.24
2.8	1623.08	3505.84	5356.15
2.9	1681.04	3631.05	5547.44
3.0	1739.01	3756.26	5738.73

The brake power above is based on Eq. (12.26) for large, slow-speed reciprocating compressors and gas specific gravity of 0.65.

Appendix 9

Chapters 17 and 18

PIPE HOOP STRESS AND PIPE LONGITUDINAL STRESS FORMULA

We will derive the pipe stress formulas resulting from the pipeline's internal pressure (P). These pipe stress formulas are important to help show that pipe pressure testing alone is not sufficient to ensure girth weld integrity and also to demonstrate the importance and use of pipe hoop stress in pipeline design.

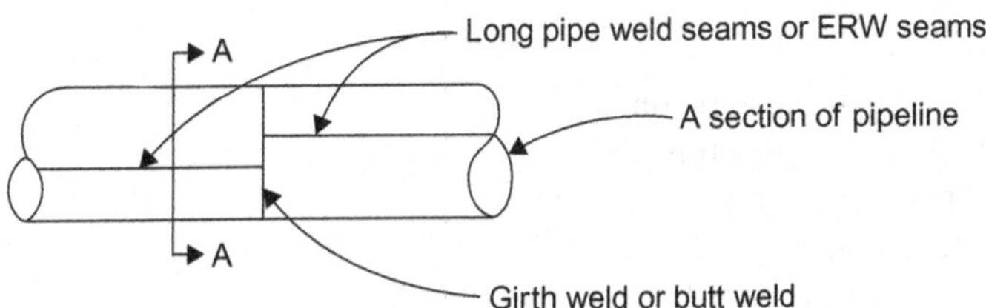

FIGURE A 9.1 Section of pipeline showing longitudinal and girth welds.

Assume the pipeline is statically in equilibrium, supported by the pipe trench and compacted fill, around and over the pipe. In this case, the main forces on the pipe are from internal pressure, inside the pipe. There is an additional force on the pipe from the weight of pipe cover material, but part of that weight is supported by the bridging effect at the trench sidewalls. So with only 3 or 4 ft of cover material, the external pressure on the pipe can be neglected.

The pipe stress (called hoop stress) caused by the pipe's internal pressure can be determined using Statics. We will treat the pipe as a thin-walled vessel. First, we take a cut through the pipe and call it section A-A, and then look inside at the forces due to internal pressure (P). We then take a vertical cut through section A-A. Then, we take half of the section away, and draw a free body diagram with the summation of forces equal to zero (statically in equilibrium).

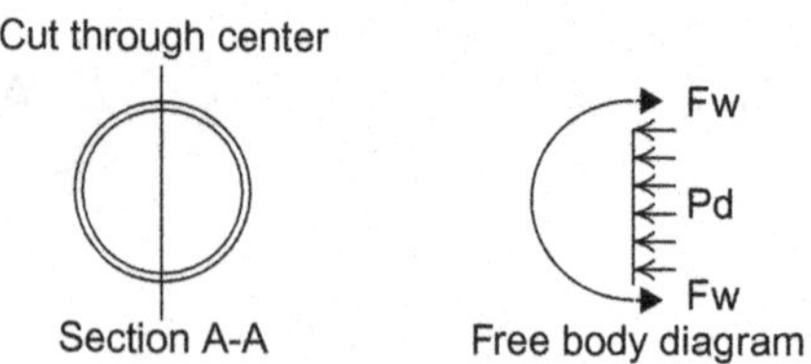

FIGURE A 9.2 Free body diagram at Section A-A.

where P_d is pressure distribution.

Therefore,

$$\text{The summation of forces} = 0$$

$$\sum \text{Forces} = \text{pressure force} - 2 \text{ pipe wall forces} = 0$$

$$\text{Pressure force} = 2 \text{ pipe wall forces}$$

$$P \times A_p = 2 \times A_w \times \sigma_H$$

$$P \times \text{ID} \times L = 2 \times t \times L \times \sigma_H$$

$$P = 2t\,\sigma_H/\text{ID} \qquad \text{(A 9.1)}$$

where

σ_H is the pipe stress (hoop stress)
A_w is pipe wall area (one side).
A_p is pipe area, the pressure acts on
L is pipe length.

Because the pipe is thin walled, we can substitute ID with the outside diameter (D), which slightly reduces the allowable pressure P, making P more conservative in design.

$$\sigma_H = PD/(2t) \qquad \text{(A 9.2)}$$

This is known as the hoop stress formula.

PIPELINE DESIGN

To develop criteria for determining pipeline design pressure (P_d), a design factor $F = 0.72$ is used for onshore pipeline per code, but for high-risk locations, such as offshore, water crossings, and in urban locations, $F = 0.60$ and sometimes F can be even smaller. Also some pipe must be de-rated as a result of its longitudinal weld-seam strength. This is done by using a weld joint factor E. With pre-1970 seam-welded pipe, E can range from 0.60 to 1.00, so careful analysis is necessary regarding the use of the seam factor E. For seamless pipe and post-1970 ERW pipe $E = 1.00$.

PIPELINE DESIGN PRESSURE (P_d)

$P_d = 2 \times t \times \sigma_H \times E \times F/D$ (now including the design factor and weld seam factor). The hoop stress (σ_H) can be replaced with SMYS, which is the specified minimum yield strength of the pipe to be used, determined at the factory. After making that substitution,

$$P_d = (2 \times t \times \text{SMYS}/D) \times E \times F \tag{A 9.3}$$

LONGITUDINAL STRESS

Now, let us look at the effect of internal pressure on the pipe in the longitudinal direction. We will derive the equation for longitudinal stress.

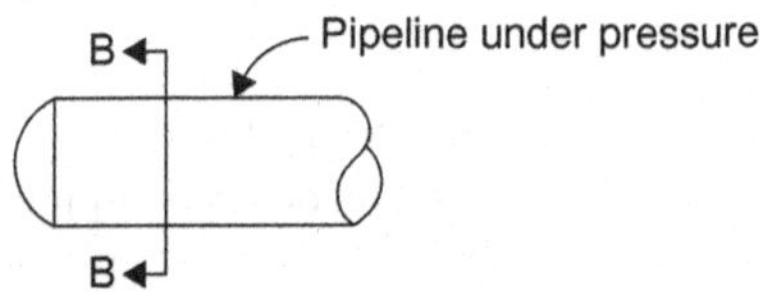

FIGURE A 9.3 Showing pipeline under pressure at Section B-B.

Think of the pipe as having an end cap, and looking inside of the pipe toward the end cap. We will cut through the pipe at Section B-B.

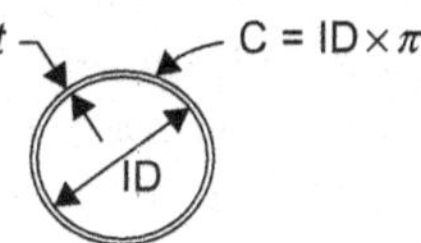

FIGURE A 9.4 Showing Section B-B of pipeline.

The pressure is distributed across the inside area of the pipe producing the pressure force and that force is opposed by the pipe wall.

$$\text{Summation of forces} = 0 \text{ (statically in equilibrium)}$$
$$\text{Pressure force} = \text{pipe wall force}$$
$$P \times \text{area inside of pipe} = \text{area of pipe wall} \times \sigma_L \text{ (the longitudinal stress)}$$
$$P \times \cancel{\pi} \times \cancel{\text{ID}}^2/4 = t \times \text{C} \times \sigma_L = t \times \cancel{\pi} \times \cancel{\text{ID}} \times \sigma_L$$
$$P \times \text{ID}/4 = t \times \sigma_L$$

Let ID $= D$ for thin wall pipe:

$$\sigma_L = PD/4t \tag{A 9.4}$$

Now, equate the two pipe stresses:

$$P = 4 \times t \times \sigma_L / D = 2 \times t \times \sigma_H / D \qquad \text{(A 9.5)}$$

As you can see from Eq. (A 9.5) above, for a given pipeline pressure (P) and pipe geometry (t and D), σ_L, the longitudinal stress, must be one half of σ_H, the hoop stress.

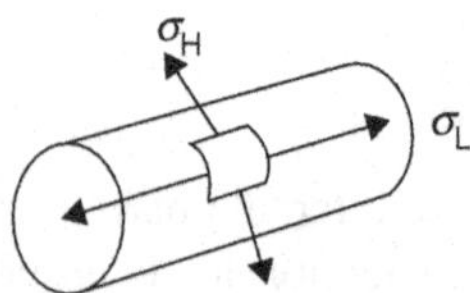

FIGURE A 9.5 Showing hoop stress and longitudinal stress at the pipe wall.

For the same internal pressure, the hoop stress is twice as large as the longitudinal stress. And because hoop stress is dominate, pipe defects that line up in the pipe's longitudinal direction, such as corrosion pits and poor quality longitudinal seam welds, are critical to pipeline integrity because the hoop stress acts directly on them.

Also because longitudinal stress produces the force that pulls girth (or butt) welds apart, and because it is one half the strength of hoop stress, defects in the girth weld are not as critical from a static pipeline point of view. So the chances are, during hydrostatic test, the test pressure will not cause failure in a girth weld even though severe defects exists, but for example, a bad longitudinal seam weld might be pulled apart by hoop stress and split the pipe wide open.

Ground movement is another matter. Defects in the girth weld or the heat-affected zone of the weld caused by hydrogen cracking, or corrosion pitting will weaken the pipeline and allow failure to occur by bending and shear loads, during ground movement. This ground movement is frequently caused by earthquakes and landslides, but even vibration from railroad crossings or nearby heavy truck traffic can load up the pipe and cause failure.

So you can see that pipeline girth welds must be as strong as the pipe itself in order to withstand bending and shear loads, as mentioned earlier. That is part of the reason why API 1104, the pipeline welding standard, has been so focused on weld NDT and defect criteria.

Index

Page numbers in *italics* indicate figures, tables and foot notes